Climate Change
and Biodiversity

Climate

Change

EDITED BY
THOMAS E. LOVEJOY &
LEE HANNAH

and

Biodiversity

YALE UNIVERSITY PRESS
NEW HAVEN & LONDON

Set in Joanna type by The Composing Room of Michigan,
Inc., Grand Rapids, Michigan.
Printed in the United States of America by Sheridan
Books, Ann Arbor, Michigan.

The Library of Congress has cataloged the hardcover
edition as follows:

Climate change and biodiversity / edited by Thomas E.
Lovejoy and Lee Hannah.
    p.   cm.
    Includes bibliographical references and index.
    ISBN 0-300-10425-1 (cloth)
    1. Biological diversity. 2. Climatic changes—
Environmental aspects. I. Lovejoy, Thomas E.
II. Hannah, Lee Jay.
QH541.15.B56C62 2004
577.2′2—dc22

                                2004043536

A catalogue record for this book is available from
the British Library.

The paper in this book meets the guidelines for
permanence and durability of the Committee on
Production Guidelines for Book Longevity of the
Council on Library Resources.

ISBN-13: 978-0-300-11980-0 (pbk. : alk. paper)
ISBN-10: 0-300-11980-1 (pbk. : alk. paper)

10 9 8 7 6 5 4

Contents

Preface

The idea for a new book on climate change and biological diversity grew out of discussions we had during a hike in the Succulent Karoo of South Africa with a group of fellow climate change biologists at the dawn of the twenty-first century. We realized that so much had been learned and published since the publication of _Global Warming and Biological Diversity_ (R. L. Peters and T. E. Lovejoy, Yale University Press, 1992) that a whole new book was required. So the two of us took on this challenging task with a sense of great urgency, because the evidence of biodiversity responding to current climate change is rapidly accumulating.

Much of what was prophecy in the earlier book has now come to pass, while conventional wisdom at that time has been overturned in other cases. Species range shifts due to climate change were widely projected in the first book, and dozens have now been documented. It is now even more firmly established that species respond to climate change individually, not as coherent communities. Habitat loss and fragmentation in synergy with climate change was seen to be a paramount issue in the early 1990's and that view is equally valid today. In other areas, views have changed over the intervening years. Coral bleaching, which ravaged reefs in the 1990s, was recognized as a threat in 1992, but the eminence of the crisis was not foreseen. Most of us involved in the first book thought that human-induced climate change would be faster than anything experienced in tens of millions of years. Thanks to Greenland ice core records interpreted in the early 1990s, we now know that in some areas very rapid, decadal and millennial climate changes have taken place frequently in the past.

The sum of these new and established

2005

views reinforces the importance of climate change to conservation efforts. It is now clear that climate change is the major new threat that will confront biodiversity this century, and that if greenhouse gas emissions run unchecked until 2050 or beyond, the long-term consequences for biodiversity will be disastrous. At the time of the first book, the relationship between biodiversity and a limit to greenhouse gas concentrations in the atmosphere had not been given much thought, while it is now clear that biological systems are important barometers of what constitutes "dangerous" climate change. With these and other developments in mind, we set out to create a new landmark reference in the field.

Our goal was to develop an overview of what we now know in climate change biology, past, present and future, as well as to explore the conservation and policy implications of that knowledge. To kick off the project, we held a small workshop of leaders in the field to advise us on content and structure of this new volume. Our advisory meeting took place in Annapolis, Maryland, in early 2001, barely six months after the birth of the idea for the new book. A rough outline of the book emerged from this meeting, as well as a list of possible chapter authors. As we later narrowed this roster of contributors, we endeavored to include most of the major experts on the topic, although it was impossible to include everyone. The response from potential authors was overwhelmingly positive, and collaboration with dozens of our friends and colleagues began. Throughout, it was a pleasure to work with contributors who conduct some of the most interesting and innovative research today, and to revisit seminal papers for the case studies. We organized the book around the following questions: What have we learned about the present? What insights have we gained from studies of past climate change and biotic response? And, most important, what does the future hold? Sections on conservation and policy

response emerged from discussions and an impromptu workshop with conservation leaders, all of whom agreed with our premise that it is not too soon to be making conservation decisions with the effects of climate change in mind.

The manuscript swelled, but we pressed on, working rapidly, and submitted it to Yale University Press late in 2002. Three helpful reviews guided us toward a more focused and tractable volume. We labored for several months to meet the spirit of the reviewers' comments. We certainly tried the patience of some authors in the process, but it was clearly worth the effort. Some chapters were put into a companion publication that appeared just in time for the World Parks Congress in September 2003, and limited copies of that volume are still available to purchasers of this book, by request to Conservation International (*Climate Change and Biodiversity: Synergistic Impacts*, Advances in Applied Biodiversity Science 4, Washington, D.C.: Conservation International). The contents of that companion volume are summarized here as Chapter 18. In the end, a tight, interwoven story of climate change effects on biodiversity emerged in the final manuscript. The evidence in the following chapters represents a synthesis by some of the best minds in the field, alternately summarizing the state of knowledge and suggesting emerging areas of concern.

The process of creating this book has brought home to us the immense importance of the intersection of two global issues often dealt with in isolation. Existing trajectories of habitat loss combined with climate change are clearly "a no-brainer for an extinction spasm," to borrow Steve Schneider's characterization of the problem. The most recent global analyses indicate that the majority of the earth's surface is now dominated by human activities, and that habitats are dangerously fragmented. These trends are having an impact not only on biodiversity, but on an entire range of ecosystem services, such as clean water, that have tremendous implications for the quality of human life, particularly of the poor. As the following chapters demonstrate, climate change will act in synergy with these forces in many cases, further damaging biodiversity—a great loss in its own right and one accompanied by cascading effects on human and eco-nomic values. If ever there is to be sustain-able development, minimizing the negative impacts of the climate change–biodiversity interaction will lie at the center of it. We dedicate this volume to our children (in Lee's case) and grandchildren (in Tom's case) in the hope that they will live in a world where that will have happened.

Acknowledgments

We extend special thanks to our friends who have so ably advised us in the Annapolis meeting and thereafter—Steve Schneider, Terry Root, Hank Shugart, Diana Wall, Bert Drake and Brian Huntley. For their continuing insights and good company, as well as helping germinate the idea of this book, we also thank the Karoo group—Guy Midgley, Jon Lovett, Dan Scott, William Bond, and Ian Woodward. Jean Black went the extra mile for us as editor, while Morgan Hutchison, Melissa Thomas and Tenley Wurglitz have our gratitude for their tireless efforts on some of the most difficult aspects of production. We often thought of Rob Peters as we worked on this project; we are happy to count him as a friend and a source of inspiration. Funding from the Center for Applied Biodiversity Science at Conservation International and a grant from the Henry Luce Foundation has supported many aspects of the preparation of this book.

PART I

Introduction

CHAPTER ONE

Biodiversity and Climate Change in Context

LEE HANNAH, THOMAS E. LOVEJOY,
AND STEPHEN H. SCHNEIDER

Biodiversity is continually transformed by a changing climate. Conditions change across the face of the planet, sometimes rapidly, sometimes slowly, sometimes in large increments, sometimes in small increments, resulting in the rearrangement of biological associations. But now a new type of climate change, brought about by human activities, is being added to this natural variability, threatening to accelerate the loss of biodiversity already under way due to other human stressors.

Biodiversity is the sum of the species, ecosystems, and genetic diversity of Earth. It often is also considered to include biological processes, some of which may operate at scales larger than single ecosystems. Biodiversity is not evenly distributed on the planet. Relatively small areas, particularly tropical uplands, contain a disproportionate share of restricted range endemic species (species unique to a limited area). As a result, these areas have high numbers of species with small ranges, as well as a complement of widespread species, resulting in concentrations of species richness and endemism. Many high biodiversity areas are also under high human threat, and these have been designated "biodiversity hotspots" (Myers et al. 2000). They are distributed throughout the tropics, and in higher latitude mountains that have escaped glaciation (Plate 1).

Human development has transformed and fragmented the natural landscape on which biodiversity depends, creating altered conditions and "islands" of isolated habitats in the hotspots and many other areas (Earn et al. 2000). At the same time, exotic species have been introduced beyond their natural biogeographic boundaries, and a host of chemicals for which many species have no evolutionary experience have been released (Mooney and Hobbs 2000). The combination of these stressors gives rise to the well-known problems of attempting to practice conservation in the face of human disturbance (Vitousek et al. 1997). Now there is yet another element of stress that must be accounted for—human-induced climate change.

Climate is now warming rapidly—so rapidly that the effects are perceptible within a single human lifetime or within the history of a people (IPCC 2001). With this new change come alterations in biodiversity already facing multiple threats. This synergy—between "normal" stresses like habitat fragmentation and altered climate—poses a new challenge to conservation (Peters and Darling 1985; Peters and Lovejoy 1992; Hannah et al. 2002a). With species being increasingly isolated in fragments, a rapidly changing climate will force migration; but unlike past migrations, in the future species will find factories, farms, freeways, and urban settlements in their path.

The synergy between climate change and habitat fragmentation is the most threatening aspect of climate change for biodiversity, and is a central challenge facing conservation. Current observations provide a clear signal that change is already under way, but over a very short time span—the few tens of years in which human-induced climate change has been measurable (Part II). Fuller understanding of climate change and biodiversity dynamics can come from the study of past changes (Part III). Models of future climate and biological systems provide additional insights, and allow quantitative and geographically explicit impacts and solutions to be explored (Part IV). These paths of understanding—from paleoecology to present-day changes, from modeling to multidisciplinary synthesis—and the conservation strategies that evolve from that understanding (Parts V and VI) are the topics of this volume.

THE CHANGE

Climate change and its biological consequences are under way. Changes in the physiology, phenology, and distributions of species are evidence of changes to biodiversity that have occurred within the past few decades, directly attributable to recent temperature trends (Hughes 2000; Root et al. 2003; Parmesan and Yohe 2003). Future changes are very likely to be greater, as both habitat fragmentation and climatic change intensify. Our ability to conserve the living resources of the planet will be increasingly tied to our ability to manage climate change, and to manage the biotic changes associated with it.

Knowledge of physiology and biogeography of various species leads to expectations of how they will respond to climate change—particularly temperature trends. Extensive analyses of the impact of recent climate change on biodiversity have revealed exactly the sorts of changes predicted. Species ranges shifting poleward and upslope has been observed in birds, marine communities, butterflies (Fig. 1.1), and other insects.

Not only must species find suitable habitats in the altered climatic future, but different taxa will experience differential rates of movement. Mobile species like birds can change their ranges quickly, whereas species like reptiles and plants are likely to move more slowly. Thus, in addition to the stresses that individual species may encounter in responding to climate, differential rates of response imply that current communities of species will be disaggregated. Any ecosystem functions derived from particular communities of species will also be altered, in proportion to the speed and scale of climate change. Current data on changes in individual species show clear responses, but the consequences of the tearing apart of communities of species as now constituted are largely unknown.

An example of these changes comes

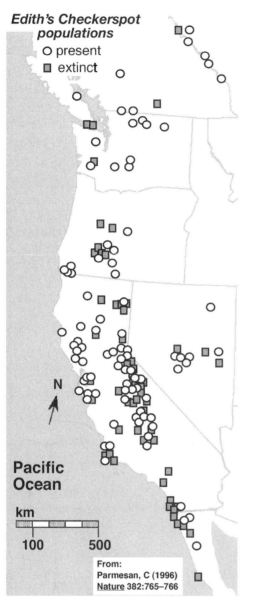

Figure 1.1. Range shift of Edith's checkerspot butterfly (*Euphydryas editha*). Population extinctions were high in the southern and lowland parts of this species' range, as predicted by climate change theory. (For full discussion, see Chapter 4 case study.) *Source:* Courtesy of Camille Parmesan, Climate and species' range, 1996, *Nature*, 382, 765–766.

from coral reefs, spectacularly diverse systems that turn out to be particularly sensitive to rapid increases in temperature. Corals expel their symbiotic algae when sudden warming occurs, often resulting in the death of the coral. The very extreme El Niño events of the past few decades, coupled with overall increasing temperature trends, have devastated coral reefs across the tropics. Moreover, this damage combines with impacts of other human stresses on reef systems, including overfishing, pollutant runoff, poisoning, and dynamite-based fishing techniques. These disturbances of coral reef communities and increasing temperatures and sea levels are exemplary of synergy between reinforcing threats.

LEARNING FROM THE PAST

Biologists have sought to understand the potential impacts of various climate futures through the use of analogies to the past and insights from theory. Perhaps the most important insight from past change is that species will respond to climate independently of one another. During a significant portion of the ice age and present interglacial transition, the distribution and combinations of pollen types provided no analog associations to today's vegetation communities (Overpeck et al. 1991). That is, when species moved, they moved at different rates and directions, not as groups. Consequently, the groupings of species during the transition period were often dissimilar to those present today. This suggests that in the future, independent, individualistic species range shifts and novel species associations may be expected as well.

The past also offers insights on rates of change, both in climates and in biotic response. Climate has changed rapidly in the past (Figs. 1.2 and 1.3), and biodiversity has kept pace with relatively few extinctions (Huntley, this volume; Bush, this vol-

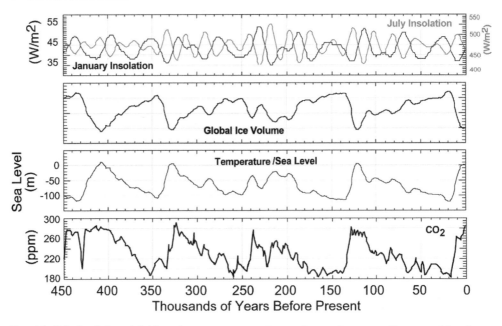

Figure 1.2. Solar insolation, global ice volume, temperature, sea level, and CO_2 levels for the northern hemisphere over the past 450,000 years. Note that here temperature is approximated by sea level and changes are therefore highly smoothed—there have been many large, rapid climate changes, which are not represented (see Fig. 1.3 for examples of rapid changes).

ume; Roy and Pandolfi, this volume). This indicates that there are inherent capacities for dealing with dynamic change that can be incorporated into conservation strategies. Habitat loss now greatly constrains natural movements in response to climate change, however, at the levels of individual organisms, populations, and entire species. In addition, climate change will be moving beyond the range defined by envelopes of past natural variability (Bush 2004; Overpeck et al., this volume).

Rapid regional temperature changes in the past have included transitions of up to 10°C or more in a matter of years or decades. Rapid global changes have generally been of lesser magnitude, but were often large. These are frequently associated with changes in global climate patterns or deep ocean circulation. The past 11,000 years have been unusually warm and stable, which means that there are few precedents from the past for possible future change (Fig 1.3).

Past vegetation responses to climatic change at a sustained average rate of 1°C per millennium indicate that credible predictions of future vegetation changes cannot neglect transient (i.e., time-evolving) dynamics of the ecological system. Furthermore, because the forecasted global average rate of temperature increase over the next century or two exceeds the typical sustained average rates experienced during the last 120,000 years, it is unlikely that paleoclimatic conditions reconstructed from millennial time scale conditions would be near analogs for a rapidly changing anthropogenically warmed world. Future climates may not only be quite different from recent climates, they may also be quite different from those inferred from paleoclimatic data and from those to which some existing species are evolutionarily adapted. Therefore, past changes provide only a backdrop or context to gauge future changes—not primarily as a precise analog, but rather as a means to verify the behavior of models of climate or ecosystem dynamics that are then used to project the future conditions

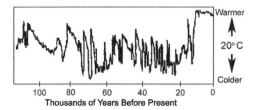

Figure 1.3. Temperature change over the past 150,000 years as interpreted from a Greenland ice core oxygen isotope proxy. Note that the resolution of this record is much greater than that shown in Figure 1.2, and reveals the many large, abrupt changes that have characterized past climate. *Source:* Courtesy of Wallace Broecker.

given the rapid time frame of human-induced climate change.

VIEWS OF THE FUTURE

Understanding the future course of these and other changes requires projections of future human disturbances, including climate. To predict our climatic future, we first need to project scenarios of human emissions and land use changes, and then to use these changes as inputs to computer models of the Earth's climate, the most comprehensive of which are General Circulation Models (GCMs) and Coupled Atmosphere Ocean General Circulation Models (AOGCMs).

GCM projections of the magnitude of global temperature change vary between models (Grassl 2000). There is also inter-model variability in projections of the regional distribution of changes. This inter-model variability is not great for estimates of global mean temperature change in the 2030–2050 range. Yet there is considerable regional variability between models, even in these "near-term" projections. In other words, even when models agree quite closely about mean global temperature change, their predictions for a particular region may vary considerably. Plate 2 illustrates mean global temperature changes projected for 2030–2050 by leading

GCMs and an ensemble of GCMs under a "business-as-usual" emissions scenario.

Model differences become greater farther into the future. Plate 3 illustrates the projections of temperature change for 2080–2100 using the same models and the same emissions scenario. These variations in model predictions introduce uncertainty into analyses of possible biological effects of future climate change.

While global temperature changes remain the political yardstick for the overall magnitude of climate change policy, global mean temperature is not the most useful index for biodiversity (Hannah et al. 2002b). Global temperature means hide patterns in climate change that are important for biodiversity. For instance, there is broad agreement among GCMs that nighttime temperatures will rise more than daytime temperatures, with important implications for biological impacts. Since low temperatures occur at night and many plants are limited by minimum temperature, frost or freezing, warming of nighttime temperatures has significant implications for determining changes in plant distributions.

Global temperature means also mask regional variations and trends important to species and ecosystems. Regional variability is a source of uncertainty, but it too is very relevant for the understanding of biodiversity impacts. For instance, temperature change is expected to be greater over land than over the ocean. Since they cover two-thirds of the Earth's surface, the thermally massive oceans exert a strong restraining influence on the rate of global mean temperature increase. Global mean temperature also is not uniformly distributed across land and oceans, particularly during the transition of the climate to a new equilibrium. Except over massive ice sheets, the land usually warms faster than the oceans. Therefore, global mean temperature rise understates the change likely to be experienced by most terrestrial ecosystems. This may be illustrated by chart-

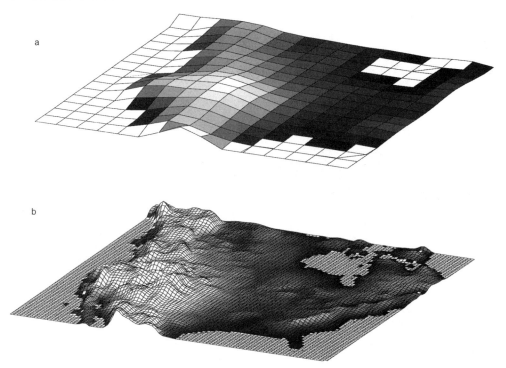

Figure 1.4. The difference in resolution between a GCM and an RCM. (a) terrain over North America on a GCM model grid of 400-km horizontal resolution; (b) the same terrain at an RCM model grid of 50-km horizontal resolution. *Source:* Courtesy of Prudence N. Foster.

ing projected temperature changes over land and oceans by latitude (as in Plate 1, which also indicates the latitudinal location of biodiversity hotspots).

Further, the impact of climate change on biodiversity depends on much more than temperature change. Precipitation, extreme events, and other climate factors may be more important than temperature in some settings (Easterling et al. 2000). Regional variation is high for all of these variables. Here uncertainty is even greater between regions and among models. Plate 4 illustrates precipitation changes predicted for 2030–2050 by two leading GCMs and an ensemble of GCM runs under the "business-as-usual" emissions scenario.

To understand regional variability at a scale appropriate for impact assessment,

projections at a scale finer than that which GCMs can provide is essential. The scale of GCM projections is too coarse for many regional impact applications. For instance, all the mountains of western North America from the Sierra Nevada to the Rockies are represented as a single large "hump" in most GCMs (Fig. 1.4). Finer-scale GCMs improve this representation but are still much coarser than large ecological divisions such as watersheds. Unfortunately, computing time required increases rapidly as model scale becomes smaller, so another solution to obtaining fine-scale results for regional analysis is needed.

One approach is to use regional-scale models driven by GCMs (Raper and Giorgi, this volume). This approach uses a regional climate model (RCM) very similar in principle to a GCM. The RCM covers only a limited region, but at a higher resolution (Fig. 1.4). It translates GCM values (typically averaged over a few hundred kilometers in scale) into a finer-scale mesh (10- to 50-kilometer grid cells) that is much closer to the dimensions of most ecological applications (Root and Schneider, 1993). The RCM grids are still bigger than individual clouds or trees, but such methods do bring climate model scales and ecological response scales much closer. RCMs are also valuable in reproducing the effects of land cover change (for instance, forest clearing to pasture) on rainfall and other regional climate variables.

Changing Biodiversity at a Global Scale

Models of global vegetation change may be based on GCM projections and known limits of biomes or plant physiological response. Such models have progressed from simple correlative models to more complex dynamic models. The more sophisticated, dynamic models are referred to as Dynamic Global Vegetation Models (DGVMs). They use mathematical models of photosynthesis, plant carbon balance, and other factors to simulate the form of vegetation at a particular location (Betts and Shugart, this volume). Plate 5 shows how well six of these models simulate current global vegetation patterns.

DGVMs are useful for modeling the future effects of both increased atmospheric CO_2 and climate change on global vegetation (Cramer et al. 2000). Plate 6 illustrates future vegetation changes simulated by a composite of the six DGVMs in Plate 5 in response to changes in both climate and CO_2. Often these models show northern vegetation types expanding toward the pole, and some transition of tropical moist forests to savannas.

Still more sophisticated models represent the effects of these changes in vegetation on atmospheric CO_2 and climate (Woodward et al. 1998). These "carbon-cycle feedbacks" tend to exacerbate the effects of climate change. For example, loss of forest to savannas can release CO_2 to the atmosphere, accelerating warming and driving further forest loss. Plate 7 shows the results of one carbon-cycle feedback model. It indicates a major accentuation of

forest loss in the Amazon—a serious potential impact that is the subject of current research. The importance of these model results is underscored by observations of recent changes in Amazon forest composition that may well be the result of elevated CO_2 levels (Phillips et al. 2003; Laurance et al. 2004).

Focusing on Species Range Shifts

Global modeling has been useful in obtaining an overview of possible gross changes in growth form and vegetation distributions, yet conservation planning requires information about species range shifts at specific sites and within specific regions. Accordingly, more detailed, regional modeling of biotic changes is of fundamental importance in understanding the impacts of climate change on biodiversity (Hannah et al. 2002b).

Regional climate model projections or downscaled GCM projections may be used to drive vegetation models similar to those used at the global scale. Figure 1.5 illustrates the result of a simple correlative biome modeling exercise for South Africa (Rutherford et al. 1999). The results show striking changes that suggest priorities for research (such as areas in which investigation of impacts on individual species may be warranted), illustrating the utility of even relatively simple fine-scale analysis.

Species range shift modeling can reveal impacts on biodiversity in much greater detail (Peterson et al. this volume). Correlative or "climate envelope" methods may be applied to species modeling, or statistical techniques such as general additive modeling (GAM) or general linear modeling (GLM) may be used to simulate the distribution of suitable climate space for a species in the future. Plate 8 illustrates the results of such an effort for a plant from the Protea family in South Africa. These changes may be substantial, and the pattern is one of individualistic range shifts. Detailed models such as these help shed

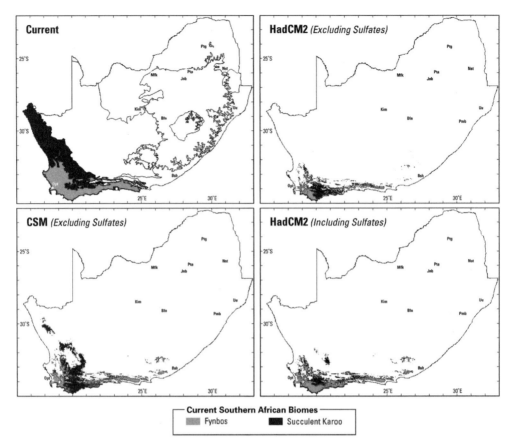

Figure 1.5. Biome change in the Fynbos and Succulent Karoo of South Africa under alternative future climate scenarios. This analysis was conducted as part of the South Africa Country Study on Climate Change. It used correlative biome "envelope" modeling and down-scaled climate data. The white areas that replace the two biomes in future vegetation projections are climatic conditions that have no biome correlate at present in South Africa. These may correspond to more arid vegetation types now typical of Botswana or Namibia.

light on possible future changes in biodiversity and provide spatially explicit projections for use in conservation planning. Modeling of multiple species and multiple regions has suggested that climate change may approach or surpass habitat loss as the leading cause of species extinctions sometime this century (Thomas et al. 2004).

CONSERVATION IN A DYNAMIC CLIMATE

Challenges confronting conservation increase with climate change. Habitat loss, the greatest threat to biodiversity, will interact with climate change to make range shifts very difficult in many areas. Other nonclimate stressors to biodiversity, such as alien invasive species, pollution, and overhunting, may also have synergistic effects with climate change. Systems that are under multiple stresses often behave in unpredictable ways, such that the sum of the stresses may push them past thresholds and into new states. Since climate change is expected to lead to individualistic species range shifts resulting in the development of new species associations and communities, the threat of unstable transitions into new states is all the more likely.

Further, climate change will have an impact on human systems, resulting in changes, often increases, in impacts on biodiversity. In the tropical areas highest in biodiversity, farmers are less likely to have the resources or information needed to adapt effectively to changing conditions, and are more likely to rely on natural resources as a fallback source of income. In these regions, crop failures may precipitate increased forest or wildlife exploitation. Even temporary shortfalls in food or income may result in permanent loss of forest cover or biodiversity due to the long-term impacts of land clearing and overharvesting.

Conservationists are entering a new era of conservation, one in which last-ditch stands to save species where they currently exist may not be enough (Lovejoy, this volume). In this era, letting natural processes operate unfettered may result in species or whole associations running into the ecological equivalent of a brick wall, as natural processes take the system toward range shifts and reassociation, while habitat loss prevents migration or replacement of species.

Coordination among and between natural areas will take on increasing importance. Management in one area for transition to a new vegetation type or retention of current vegetation will need to be coordinated with management in other natural areas to ensure that change in a region is managed harmoniously.

Difficult theoretical questions will evolve—such as, "What is natural vegetation?" When climate conditions have no historic analog and species can no longer move freely into and out of areas, natural precedents may provide no ready answer to this question. Managing to maintain conditions of a historical reference point (e.g., pre-European contact) would be trying to force vegetation to match a climate that no longer exists. Yet estimating what species would find new climates suitable and how they would interact in competition with one another is well beyond our current abilities to predict. Where there may be strong recreational values associated with particular biodiversity values (e.g., tree cover), there may be strong social pressures to retain these elements, even when natural processes are driving transitions toward new states (e.g., grassland).

Conservation responses to these problems must have two components—one, adapting conservation strategies to deal with dynamic biodiversity; and two, constraining greenhouse gas levels within bounds that keep biological changes manageable. When biodiversity behaves in new ways in response to new climates, it may be difficult to be sure what is manageable. Nonetheless, many authors in this volume believe that if climate goes very far outside the bounds within which it has stayed over the past few interglacial periods, widespread extinctions may result—clearly an unmanageable and untenable situation for biodiversity.

GLOBAL RESPONSES TO A GLOBAL PROBLEM

Conservation strategies that respond to climate dynamics have been termed climate change–integrated conservation strategies (CCS) (Hannah et al. 2002a). CCS respond to fundamental challenges to conservation posed by climate change, including rapid climate change, the magnitude of such change, and range shifts across landscapes that have been heavily fragmented by habitat loss. These strategies include addition of protected areas to ensure that species are afforded fundamental protection in both present and future ranges; creating connectivity in the matrix of land uses outside of protected areas; and coordinating management action and monitoring at the regional scales at which climate change operates (Hannah and Hansen, this volume). Planning for these activities is supported by detailed analysis of past, present, and modeled future responses of biodiversity to climate change.

Constraining greenhouse gas concentrations is a larger and more daunting task. Limiting climate change requires stabilizing greenhouse gas concentrations. This requires major changes beyond signing on to the Kyoto Accord—which is but the first step in international cooperation to conserve the atmospheric commons. It implies the evolution of a society that becomes carbon-neutral on a global scale in this century. This would mean phasing out all fossil fuel-burning vehicles, aircraft, and electricity generating facilities or implementing permanent carbon sequestration—probably underground—on a massive scale (Watson, this volume). While these may sound like science fiction scenarios to biologists who are not energy experts, if the alternative is the inexorable and unmanageable loss of biodiversity, biologists have a strong reason to advocate exactly these changes. Understanding what that means and the alternatives is an important part of responding to the challenge that climate change poses for biodiversity.

One part of the solution is not to delay further. The longer action is delayed, the shorter the time will be available to implement alternative energy sources or sequestration activities. Because long-term economic growth is essentially exponential, even seemingly large costs incurred early only delay slightly the time at which defined levels of economic well-being are attained (Azar and Schneider 2002). Other parts of the solution are properly worked out in public policy debate, a debate in which biologists have been largely silent, but which holds enormous implications for the future of biodiversity.

STRUCTURE OF THIS VOLUME

The chapters that follow are organized into six parts. Chapter 2 examines the question, "What is climate change?" It describes the basic components of the climate system, their drivers, and how human activities are influencing climate. Thereafter, each part opens with a short chapter introducing relevant principles from climate change science, to help the reader place the corresponding biological changes in context. The parts are organized with the most recent changes first, followed by known past changes, projected future changes, and chapters dealing with conservation and policy response.

Part II examines biotic responses to climate change already under way. This establishes the immediacy of the problem climate change presents to biodiversity. Species are shifting ranges (Chapter 4), changing timing of life cycles (Chapter 5), and even evolving (Chapter 6) in response to this mounting threat.

Placing these present changes in context, and understanding possible future change, requires an understanding of past natural responses to climate change—the topic of Part III. These chapters (7–12) outline known changes in past climate and their effects on biodiversity, thus providing analogies to possible future change. The authors tackle Northern Hemisphere, tropical, Southern Hemisphere, and marine systems in turn. Their contributions make it clear that our conservation systems should be taking greater consideration of climate change, even if human activities were not causing its acceleration.

Part IV looks at projected future impacts of climate change on biodiversity. An overview of climate models precedes discussion of single species models, correlative biome models (Chapter 14), and models of dynamic response (Chapter 15). Possible impacts on marine and freshwater systems are treated in turn (Chapters 16 and 17), followed by examination of synergistic effects (Chapter 18).

The final two parts examine conservation responses, including both modifications of field conservation efforts and strategies (Part V) and the need to limit human-induced climate change by stabilizing levels of greenhouse gases in the atmosphere (Part VI).

Some chapters in each part are complemented by case studies. These studies are intended to illuminate methods of research that are mentioned in the chapters, as well as to present landmark research results from the literature on climate change and biodiversity. For example, Parmesan (Part II) presents a synopsis of her classic study of range shifts in Edith's checkerspot butterfly, as well as presenting insight into the methods she used in that research. Each case study is intended to provide the reader with an interlude between, and greater insight into, the surrounding chapters.

It has been more than 10 years since the publication of *Global Warming and Biological Diversity* (Peters and Lovejoy 1992). The field has grown immensely in this time, representing the work of thousands of researchers. Much of what the authors of that first volume imagined has now been confirmed. Much has been learned that is new, and yet our understanding is still dwarfed by the vast and complex changes to come. In the chapters and case studies that follow, over two dozen authors attempt to synthesize what they and their colleagues have learned, and to provide some preview of what is left to be discovered. Should they inspire another generation of researchers and conservationists, the most important purpose of this book will have been achieved.

REFERENCES

Azar, C., and S. H. Schneider. 2002. Are the economic costs of stabilising the atmosphere prohibitive? *Ecological Economics* 42:73–80.

Bush, M. V., M. R. Silman, and D. H. Urrego. 2004. 48,000 years of climate and forest change in a biodiversity hot spot. *Science* 303:827–829.

Cramer, W., A. Bondeau, F. I. Woodward, I. C. Prentice, R. A. Betts, V. Brovkin, P. M. Cox, V. Fischer, J. A. Foley, A. D. Friend, C. Kucharik, M. R. Lomas, N. Ramankutty, S. Sitch, B. Smith, A. White, and C. Young-Molling. 2000. Global response of terrestrial ecosystem structure and function to CO_2 and climate change: results from six dynamic global vegetation models. *Global Change Biology* 7:357–373.

Earn, D. J., S. A. Levin, and P. Rohani. 2000. Coherence and Conservation. *Science* 290:1360–1364.

Easterling, D. R., G. A. Meehl, C. Parmesan, S. A. Changnon, T. R. Karl, and L. O. Mearns. 2000. Atmospheric science: Climate extremes: Observations, modeling, and impacts. *Science Washington D C* 289:2068–2074.

Grassl, H. 2000. Status and improvements of coupled general circulation models. *Science* 288:1991–1997.

Hannah, L., G. F. Midgley, T. E. Lovejoy, W. J. Bond, M. Bush, J. C. Lovett, D. Scott, and F. I. Woodward. 2002a. Conservation of biodiversity in a changing climate. *Conservation Biology* 16:264–268.

Hannah, L., G. F. Midgley, and D. Millar. 2002b. Climate change-integrated conservation strategies. *Global Ecology & Biogeography* 11:485–495.

Hughes, L. 2000. Biological consequences of global warming: Is the signal already apparent? *Trends in Ecology and Evolution* 15:56–61.

IPCC. Climate Change 2001: The Scientific Basis; contribution of Working Group I to the Third Assessment Report of the Intergovernmental Panel on Climate Change. 2001. Port Chester, N.Y.: Cambridge University Press. Ref Type: Report

Laurance, W. F., A. A. Oliveira, S. G. Laurance, R. Condit, H. E. M. Nascimento, A. Sanchez-Thorin, T. E. Lovejoy, A. Andrade, S. D'Angelo, J. E. Ribeiro, and C. W. Dick. 2004. Pervasive alteration of tree communities in undisturbed Amazonian forests. *Nature* 428:171–174.

Mooney, H. A., and R. J. Hobbs. 2000. Invasive species in a changing world. Washington, D.C.: Island Press,

Myers, N., R. A. Mittermeier, C. G. Mittermeier, G. A. B. Da Fonseca, and J. Kent. 2000. Biodiversity hotspots for conservation priorities. *Nature* 403:853–858.

Overpeck, J. T., P. J. Bartlein, and T. Webb, III. 1991. Potential magnitude of future vegetation change in eastern North America: comparisons with the past. *Science* 254:692–695.

Parmesan, C., and G. Yohe. 2003. A globally coherent fingerprint of climate change impacts across natural systems. *Nature* 421:37–42.

Peters, R. L., and J. D. S. Darling. 1985. The greenhouse effect and nature reserves. *BioScience* 35:707–717.

Peters, R. L., and T. E. Lovejoy. 1992. Global warming and biological diversity. London: Yale University Press.

Peterson, A. T., M. A. Ortega-Huerta, J. Bartley, V. Sanchez-Cordero, J. Soberon, R. H. Buddemeier, and D. R. Stockwell. 2002. Future projections for Mexican faunas under global climate change scenarios. *Nature* 416:626–629.

Phillips, O., R. Vasquez Martinez, L. Arroyo, T. R. Baker, T. Killeen, S. Lewis, Y. Malhi, A. Mendoza, D. Neill, P. Vargas, M. Alexiades, C. Ceron, A. Di Fiore, T. Erwin, A. Jardim, W. A. Palacios, M. Saldias, and B. Vinceti. 2004. Increasing dominance of large lianas in Amazonian forests. *Nature* 418:770–774.

Root, T. L., and S. H. Schneider. 1993. Can large-scale climatic models be linked with multi-scale ecological studies? *Conservation Biology* 7:256–270.

Root, T., J. T. Price, K. R. Hall, S. H. Schneider, C. Rosenzweig, and J. A. Pounds. 2003. Fingerprints of global warming on wild animals and plants. *Nature* 421:57–60.

Rutherford, M. C., L. W. Powrie, and R. E. Schulze. 1999. Climate change in conservation areas of South Africa and its potential impact on floristic composition: A first assessment. *Diversity and Distributions* 5:253–262.

Thomas, C. D., A. Cameron, R. E. Green, M. Bakkenes, L. Beaumont, A. Grainger, Y. Collingham, B. F. Erasmus, M. Ferreira de Siqueira, L. Hannah, L. Hughes, B. Huntley, A. T. Peterson, A. van Jaarsveld, G. F. Midgely, L. Miles, M. Ortega-Huerta, O. Phillips, and S. Williams. 2004. Extinction risk from climate change. *Nature* 427:145–148.

Vitousek, P. M., H. A. Mooney, J. Lubchenco, and J. M. Melillo. 1997. Human domination of earth's ecosystems. *Science* 277:494–499.

Woodward, F. I., M. R. Lomas, and R. A. Betts. 1998. Vegetation-climate feedbacks in a greenhouse world. *Philosophical Transactions of the Royal Society of London, Series B* 353:29–39.

CHAPTER TWO

What Is Climate Change?

THOMAS R. KARL
AND KEVIN E. TRENBERTH

There is clear evidence for substantial variations in the Earth's climate, both globally and regionally, that range from years to many millennia in duration. Numerous types of evidence have been analyzed, including instrumental records, historical records, and paleoclimate data (indicators used to infer past climates). Recently, human activities have been identified as likely contributors to global as well as regional climate change (IPCC 2001). This means that to understand climate variations and change, it is essential to assess the climate's sensitivity to a variety of factors, both human and natural (Karl and Trenberth, 2003). This gives rise to some fundamental questions that we address in this chapter concerning the components of the climate system, their variability, climate change and its causes, the ability to predict variations and change, and how those will be manifested and experienced.

CLIMATE SYSTEM COMPONENTS

Climate involves variations in which the atmosphere is influenced by and interacts with other parts of the climate system, and "external" forcings (Fig. 2.1). The internal interactive components in the climate system include the atmosphere, the oceans, sea ice, the land and its features (including the vegetation, albedo, biomass, and ecosystems), snow cover, land ice, and hydrology (including rivers, lakes, and surface and subsurface water). The components normally regarded as external to the system include the sun and its output; the Earth's rotation; sun-Earth geometry and the slowly changing orbit; and the physical components of the Earth system, such as the distri-

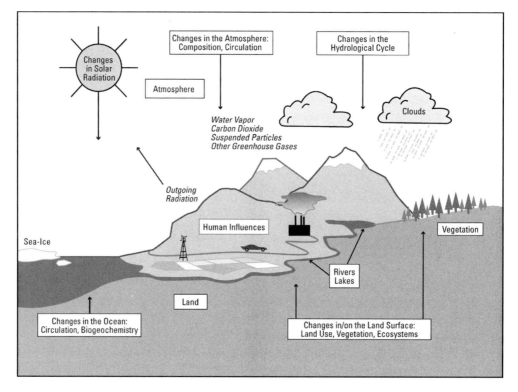

Figure 2.1. The components of the climate system are schematically illustrated along with some of their main interactions and origins of change. *Source:* Adapted from Trenberth, Houghton, and Filho 1996.

bution of land and ocean, the geographic features on the land, the ocean bottom topography and basin configurations, and the mass and basic composition of the atmosphere and ocean. These components determine the mean climate, which may vary due to natural causes. Perturbations in incident solar radiation or the emergent infrared radiation lead to what is known as radiative forcing of the system. Changes in the energy from the sun or changes in atmospheric composition due to natural events like volcanoes are possible examples. Human activities release gases into the atmosphere that affect radiative forcing, and are regarded as external to the climate system.

The source of energy that drives the climate is the radiation from the sun. The average amount of energy incident on a level surface outside the atmosphere is 342 W m^{-2}. This unit is Watts per square meter, which refers to the energy rate per unit area. Integrated over the Earth, this amounts to about 175 PetaWatts (Peta is 10^{15}, or 1 followed by 15 zeros). The largest power stations have a capacity of about 1000 MegaWatts, and so this amount of energy is equal to 175 million of those power stations! About 31 percent is scattered or reflected back to space by molecules, tiny airborne particles (known as aerosols) and clouds in the atmosphere, or by the Earth's surface, which leaves about 235 W m^{-2} on average to warm the Earth's surface and atmosphere (Fig. 2.2). The incoming energy is balanced on average by the Earth radiating the same amount of energy back to space. It does this by emitting thermal "long-wave" radiation in the infrared part of the spectrum. The bulk of the Earth's radiation to space is intercepted by the atmosphere and reemitted both up and down, making the atmosphere considerably warmer than it otherwise would be. This blanketing is

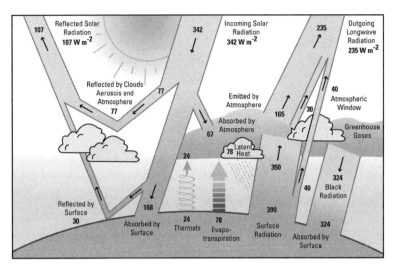

Figure 2.2. The Earth's radiation balance. The net incoming solar radiation of 342 W m^{-2} is partially reflected by clouds and the atmosphere, or at the surface, but 45 percent is absorbed by the surface. *Source:* Adapted from Kiehl and Trenberth 1997.

known as the natural greenhouse effect. The most important greenhouse gases are water vapor, which gives rise to about 60 percent of the current greenhouse effect, and carbon dioxide (CO_2), which accounts for about 26 percent (Kiehl and Trenberth 1997).

Clouds absorb and emit thermal radiation and have a blanketing effect similar to that of the greenhouse gases. But clouds are also bright reflectors of solar radiation and thus also act to cool the surface. While on average there is strong cancellation between the two opposing effects, the net global effect of clouds in our current climate, as determined by space-based measurements, is a small cooling of the surface. The energy received at the surface varies both latitudinally and longitudinally, owing to sun–Earth geometry, the atmospheric effects discussed above, and the distribution of land (including geographic features), ocean, and sea ice, which differentially reflect and absorb incoming solar radiation. This gives rise to numerous atmospheric and oceanic circulation features.

CLIMATE SYSTEM TELECONNECTIONS

Because of the Earth's rotation, the atmospheric circulation of the planet is intricately linked through large-scale waves. Consequently, it is natural for features of one sign (such as cold northerlies in the Northern Hemisphere) to occur simultaneously with features of the opposite sign (warm southerlies) downstream. On seasonal and longer time scales, these linkages are often referred to as teleconnections. An analogy can be made to a seesaw: as one side goes up, the other goes down. In the atmosphere and ocean, the seesaw may be linked across the globe in complex ways because of the Earth's shape, rotation, and land–ocean differences.

Perhaps the best known of the teleconnections is the El Niño phenomenon. A major redistribution of planetary heat and moisture occurs in the atmosphere and ocean about every three to seven years through a phenomenon centered in the Equatorial Pacific known as El Niño, and, in its opposite phase, as La Niña (Fig. 2.3). El Niño in the ocean is accompanied by a change in atmospheric pressure and winds throughout the tropics and subtropics, called the Southern Oscillation. Combined, they give rise to the dominant natural coupled mode of the climate system

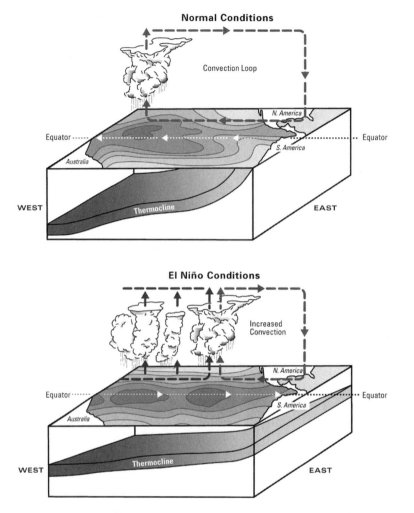

Figure 2.3. Schematic of the change in the tropical Pacific from normal to El Niño conditions. Shown are the patterns of sea surface temperatures (SSTs), with warmest values of 29°C in the west (darkest shades). The thermocline, above which the ocean is well mixed, is indicated. *Source:* Copyright University Corporation for Atmospheric Research, reprinted by permission.

called El Niño–Southern Oscillation (ENSO). ENSO has impacts throughout much of the world. As sea surface temperatures (SSTs) rise in the central and eastern Pacific, the ocean thermocline deepens. The main center of atmospheric convection and rainfall follows the warm waters into the central Pacific and changes atmospheric heating patterns, which, in turn, change the atmospheric circulation around

the world. El Niño is a dominant cause of droughts and floods in various parts of the world. So is La Niña, except in places different from those of El Niño.

El Niño also plays a prominent role in modulating carbon dioxide exchanges with the atmosphere. The normal upwelling of cold nutrient-rich and CO_2-rich waters in the tropical Pacific is suppressed during El Niño. The presence of nutrients and sunlight fosters development of phytoplankton, zooplankton, and thus primary productivity, to the benefit of many fish species. Thus, El Niño has a significant effect on many species. Multi-decadal variations of ENSO are linked to pronounced variations of temperature, pressure, and circulation in the North Pa-

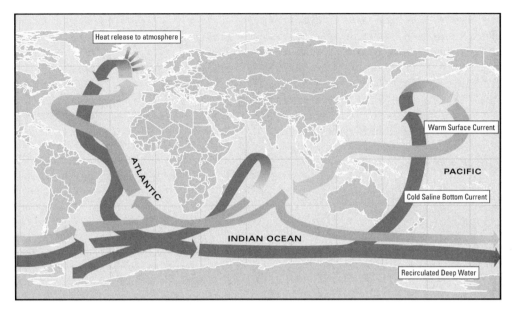

Figure 2.4. Schematic depiction of the global ocean circulation and the embedded Atlantic thermohaline circulation. Lighter shades indicate near-surface currents. *Source:* Courtesy E. Maier-Reimer. Reprinted by permission.

cific, referred to as the Pacific Decadal Oscillation (PDO).

Other important teleconnections include the North Atlantic Oscillation (NAO) and the Pacific–North American (PNA). The NAO is related to the strength of the surface westerly winds over the North Atlantic and outbreaks of cold arctic air over Europe. The NAO is closely related to a "northern annular mode," sometimes called the Arctic Oscillation, which involves the westerlies and incursions of arctic air around the whole hemisphere (hence, "annular"–related to a ring). The PNA teleconnection pattern is associated with a deeper than normal Aleutian low and a region of lower than normal pressure over the Southeast United States, while pressure is higher than normal over western North America. The PNA is associated with large-scale temperature and precipitation anomalies in the North Pacific and North American regions.

The Atlantic thermohaline circulation is an oceanic teleconnection that involves global-scale overturning in seawater driven by differences in density that arise from temperature (thermal) and salinity (haline) effects. Cold waters at high latitude are supplemented by salt from brine rejection as sea ice forms to produce very dense waters that sink in the North Atlantic. A schematic view of the global ocean circulation and the embedded thermohaline circulation is given in Figure 2.4.

IDENTIFYING CLIMATE CHANGE

Forms of Climate Change

It is convenient to define the form and shape of climate changes and variations from a statistical time series perspective. Climate variations and changes can be manifested in various ways, as depicted in Figure 2.5. One example of a regular periodic climate variation is the seasonal cycle, which certainly represents a large change in climatic variables during the course of the year, with enormous impacts on the biota (Fig. 2.5a). Normally, however, an inter-annual time frame is the minimum time used to describe climate variability and change. It is rare for the climate to manifest a periodicity like the seasonal cycle, and most often, on inter-annual and greater time scales, variations are at best quasi-periodic. Such is the case with the El Niño / La Niña cycle. An El Niño event has occurred about every three to seven years over the past few hundred years. These events would be defined as a climate variation. We distinguish between climate variations and changes from the perspective of semi-permanence. For example, even though there were enormous consequences for the biota, the decadal time-scale droughts in the 1930s across much of the United States and the Sahel drought that peaked in the 1970s and 1980s are regarded as climate variations, not changes. During these droughts, the climate was significantly different from that of previous decades, but when viewed over century time scales, they are seen as variations, not semi-permanent changes from the mean climate state. If, on the other hand, the climate is shown to be nonstationary within the time domain of interest, then discontinuities, jumps, and trends (Fig. 2.5b and c) can all be manifestations of climate change. A new baseline climate emerges with such a climate change. Sometimes, however, when the time domain expands, what was once viewed as a discontinuity or trend may reveal itself to be part of a cycle, like the advance and retreat of glaciers. It is important to consider that changes are not limited to changes in the average, but are the full description of a statistical distribution. This means, for example, that climate change and variations can also be defined by changes in variability about the mean (Fig. 2.5d).

Even after the time domain of interest has been defined for a climate change or variation, it is important to recognize and classify its spatial extent. For example, it is possible to have different responses from region to region, reflecting boundaries of the climate change or variation. It is even possible to have changes of opposite sign,

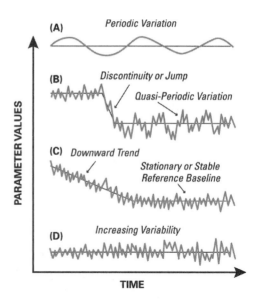

Figure 2.5. Types of climate variations.

such as a northward shift in a storm track that creates drier, less stormy conditions to the south, but wetter, more stormy conditions to the north. Similarly, a specific climate variable, like temperature, may show marked change, while another variable, like wind speed, may show little change over the same time domain, yet we would classify the climate as having changed if any one of the parameters required to define climate changes.

To ascertain climate change, we must consider a number of issues: the time scale of the change or variation; the space scale over which it is occurring; the statistical form of the change; the ensemble of climate variables that constitutes the climate at any given time and region, such as temperature, precipitation, winds, moisture, solar radiation; and all the attributes given to these elements, including tornadoes, thunderstorms, hurricanes, blizzards, and so on. In such a framework, weather is viewed as one snapshot of an ensemble of events that, when integrated over space and time, defines the climate.

Understanding and Predicting Climate Variations and Change

Our understanding of the causes of climate variations and change is limited in part because there are many possible causes, typically disguised by complex interactions. They include changes external to the climate system as well as internal drivers, which force the climate to behave in specific ways. Changes in the external climate forcings can affect variables internal to the climate system in complex ways that often lend themselves to climate feedback effects. This occurs when changes in one part of the climate system act to amplify (positive feedback) or mitigate (negative feedback) the initial direction of change. Perhaps the clearest example of a climate feedback relates to snow and ice cover. As global temperatures cool, snow and ice increase, reflecting more solar radiation back to space, leading to a positive feedback by enhancing the initial cooling. The inherent coupling of the atmosphere, ocean, cryosphere (the portion of the earth where water or land is frozen), and land surface leads to complex interactions and feedbacks.

A key challenge is to isolate (if possible) the contribution that specific mechanisms make to climate change and to identify cause and effect. This requires knowledge of the changes and variations of the external forcings controlling the climate, and a comprehensive understanding of the feedbacks and natural internal variability. Many of the laws governing climate change and the processes involved can be quantified and linked by mathematical equations. Understanding the effect of forcings, feedbacks, and internal variability requires a hierarchy of mathematical climate models of increasing complexity and reality, including coupled models that are capable of addressing the interactions among the various components of the climate system. Coupled climate models can include physics, chemistry, and biogeochemistry. Because of the long time integrations and

large range of space scales (from summertime convective clouds to large-scale winter storms) required to understand climate, solutions to these equations often require enormous computer resources.

Model simulations of climate over specified periods can be verified and validated against the observational record. Models that can describe the nature of climate change, variations, and steady-state climate conditions serve not only as a measure of our understanding, but as a tool to increase this understanding. Once evaluated, they can then be used for predictive purposes. Given specific forcing scenarios, models can provide viable climate-response scenarios. They are the primary means we have to predict climate, although ultimately prediction is likely to be achieved through a variety of means, including the observed rate of global climate change.

CLIMATE FORCINGS, FEEDBACKS, AND CLIMATE RESPONSE

The Swedish chemist Svante Arrhenius (1896) was the first to express concern about possible climate effects of increased concentrations of carbon dioxide in the atmosphere. He recognized that greenhouse gases, such as carbon dioxide, trap outgoing radiation to space and by reradiating it, cause a warming at the surface. His concerns appear justified, as in recent decades changes in atmospheric composition have been the largest forcing factors affecting global climate. The amount of carbon dioxide in the atmosphere has increased by about 31 percent since the beginning of the industrial revolution, mainly because of the combustion of fossil fuels and the removal of forests. In the absence of unforeseen technological development or controls, future projections indicate that atmospheric CO_2 will double from preindustrial values within the next 50 to 100 years. Several other greenhouse gases are also increasing in concentration in the atmosphere because of human activities (especially biomass burning, agriculture, animal husbandry, and fossil fuel use and industry, and through creation of landfills and rice paddies). These gases, which include methane, nitrous oxide, chlorofluorocarbons (CFCs), and tropospheric ozone, tend to reinforce the changes caused by increased carbon dioxide. However, the observed decreases in lower stratospheric ozone since the 1970s, caused principally by human-introduced CFCs and halons, contribute to a small cooling.

Human activities also affect the amount of aerosol in the atmosphere, which influences climate in other ways (Ramanathan et al. 2001). A direct effect of some aerosols is the scattering of some solar radiation back to space, which tends to cool the Earth's surface. Other aerosols directly absorb solar radiation, leading to local heating of the atmosphere, and some absorb and emit thermal radiation. A further influence of aerosols is that many act as nuclei on which cloud droplets condense, affecting the number and size of droplets in a cloud, and, hence, alter the reflection and the absorption of solar radiation by the cloud and also the lifetime of a cloud and its ability to precipitate.

Aerosols occur in the atmosphere because of natural phenomena; for instance, they are blown off the surface of deserts or dry regions. The eruption of Mt. Pinatubo in the Philippines in June 1991 added considerable amounts of aerosol into the stratosphere that, for several years, scattered solar radiation, leading to a loss of radiation at the surface and a cooling there. Human activities contribute to aerosol particle formation mainly through injection of sulfur dioxide into the atmosphere (which contributes to acid rain), particularly from power stations and through biomass burning. Because anthropogenic aerosols are mostly introduced near the Earth's surface where they can be washed out of the atmosphere by rain, they typically remain in the atmosphere for only a few days, and they tend to be concentrated near their sources, such as industrial regions. The radiative forcing therefore possesses a very strong regional pattern, and the presence of aerosols can help mask temporarily any global warming arising from increased greenhouse gases.

The increases in greenhouse gases in the atmosphere and changes in aerosol content produce a change in the radiative forcing. The determination of the climatic response to this change in forcing is complicated by internal feedbacks. If, for instance, the amount of carbon dioxide in the atmosphere were suddenly doubled, but other things remained the same, some outgoing long-wave radiation (about 4 W m^{-2}) would be trapped in the atmosphere. To restore the radiative balance, the atmosphere must warm up, and, in the absence of other changes, the warming at the surface and throughout the troposphere would be about 1.2°C. In reality, many other factors will change, and various feedbacks come into play, so that the central estimate of the average global warming for doubled carbon dioxide is 2.5 to 3.0°C (IPCC 1996; NRC 2001). This key value is the *climate sensitivity* to a doubling of carbon dioxide, but uncertainties are considerable and place it within the range from 1.5 to 4.5°C as determined by coupled climate system models. In other words, the net effect of the feedbacks is positive and amplifies the direct response by roughly a factor of 2.5.

On time scales of thousands of years, the amount of radiation reaching the Earth is affected by variations in the Earth's orbit, known as "Milankovitch cycles" (Fig. 2.6). Presently, the Earth orbits the sun with its axis tilted by 23.5° and an elliptical orbit, but it wobbles like a spinning top in space (Fig 2.6a). The Earth comes closest to the sun around January 3 and is farthest away on July 5 (Northern Hemi-

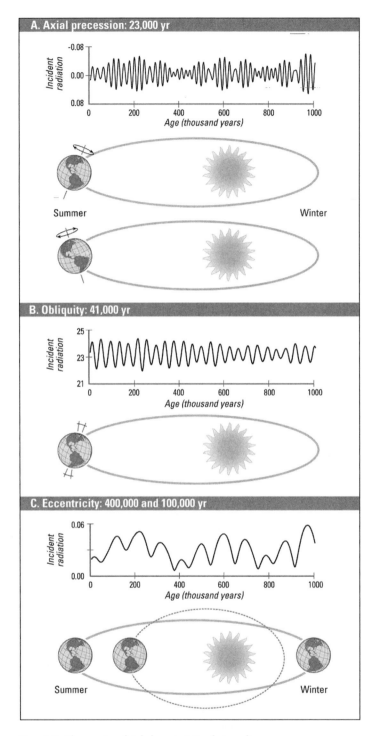

Figure 2.6. Changes in orbital characteristics that result in variation in incoming solar radiation (total solar irradiance) (W m^{-2}) reaching the top of the atmosphere. *Source*: Courtesy of James Zachos, reprinted with permission from *Science*.

sphere summer), but because of the wobble, this precesses with time, as the North Pole traces a circle in space with a cycle of about 22,000 years. Additionally, the Earth's tilt varies from 21.1° to 24.5° over the course of 41,000 years (Fig. 2.6b), and the eccentricity of its orbit (Fig. 2.6c) changes from nearly circular (eccentricity = 0.00) to more elliptical (eccentricity = 0.06), with cycles of about 100,000 and 400,000 years. With a nearly circular orbit, it matters little when the Earth is closest to the sun, but as the eccentricity increases, precession becomes more important. The combined effect of these factors changes the amount of solar radiation reaching the top of the atmosphere. Most significant are the differences in solar radiation in one hemisphere compared to the other and the seasonal changes.

Figure 2.6 reveals an interesting relationship between the volume of ice and the amount of solar radiation received in the Northern Hemisphere summer. About 20,000 years ago, the last glacial maximum began to retreat as summer insolation increased. Similar connections occurred during previous glacial retreats, for example, 135,000 years ago and 200,000 years ago. This observation has led to the theories that major glaciation is dominated by Northern Hemisphere insolation because more land is available on which snow and ice can accumulate. It is noteworthy that changes in insolation for mid and high latitudes are an order of magnitude greater than changes for the globe. Hence, the importance of the Milankovitch cycles has to come about through nonlinear amplifications in the climate system. These include changes in surface albedo from changes in vegetation, snow, and ice (e.g., Lahr and Foley 1994). Once a major continental ice sheet builds up, then changes in atmospheric and oceanic circulation also occur for a variety of reasons (e.g., differing ocean–land temperature contrast affecting the monsoons, changes

in elevation of the land surface due to ice thickness, etc.).

Global climate models have had some successes in trying to simulate the advance and retreat of the glacial cycles (e.g., Ganopolski and Rahmstorf 2001), but they have not been able to fully simulate the onset or exit from a major glacial cycle. Part of this problem may stem from the inability to run complete climate models over this long time span. Another factor may be feedback from changes in atmospheric composition of important greenhouse gases like CO_2 and CH_4 (and ultimately water vapor) through soil respiration, which could be the catalyst needed for glacial advance or retreat. Other types of models—for instance, relatively simple multiple-state threshold models—have been able to reproduce onset and exit from glacial cycles with greater success. However, these models show that the current conditions of solar forcing have few precedents, making it difficult to predict when the current interglacial (warm) period might end, even had there been no human interference with atmospheric processes (Paillard 2001).

Sometimes changes in the amount of solar radiation reaching the Earth's atmosphere occur because of changes in the energy output from the sun itself. For example, it has been shown from direct measurements that total solar output varies over a period of 11 years on the order of 0.1 percent, and indirect measurements also suggest that solar irradiance can vary over longer periods of time. During the course of the last millennium, solar irradiance may have varied by as much as a few tenths of a percent (IPCC 2001), contributing to climate variations and changes as it interacts with the complex array of other forcings and feedbacks.

THE MANIFESTATION OF CLIMATE CHANGE AND VARIATIONS

Effects on the Hydrological Cycle

How climate change manifests itself is critically important to understanding its impact on biodiversity. For example, increased heating leads naturally to expectations for increases in global mean temperatures, but other changes in weather are also important. Increases in greenhouse gases in the atmosphere produce global warming through an increase in downward infrared radiation, and thus not only raise surface temperatures, but also enhance the hydrological cycle as much of the heating at the surface goes into evaporating surface moisture. Global temperature increases signify increases in the water-holding capacity of the atmosphere, and, together with enhanced evaporation, this means that the actual atmospheric moisture should increase. It follows that naturally occurring droughts are likely to be exacerbated by enhanced drying. Thus, droughts, such as those triggered by El Niño, are likely to set in more quickly, plants will wilt sooner, and the droughts may become more extensive and last longer with global warming. Once the land is dry, then all the solar radiation goes into raising temperature, bringing on sweltering heat waves and increasing potential for wildfires. However, aerosols of all kinds are apt to decrease surface heating and, thus, regionally cut down on evapotranspiration.

Globally, there must be an increase in precipitation to balance the enhanced evaporation. Most (perhaps 75 percent) of the moisture in an extratropical storm (storms outside the tropics) comes from moisture stored in the atmosphere when it began. This increased moisture provides fuel for storms and enhances rainfall and snowfall intensity, increasing risk of flooding (Trenberth 1998; IPCC 2001; National Assessment Synthesis Team 2001). In many parts of the world, heavy rainfall events and total precipitation steadily increased throughout the twentieth century (Karl et al. 1995; Karl and Knight 1998; Groisman et al. 1999; IPCC 2001). More generally, as the mean climate changes, there is likely to be an amplified change in extremes (Katz 1999).

Effects on Atmospheric Circulation and Teleconnections

Hemispheric average temperatures are dominated by changes over land, which are larger than those over the oceans, and thus are greatest when the changes over Eurasia and North America are in phase. This "cold ocean–warm land" pattern has been shown to be linked to changes in the atmospheric circulation and, in particular, to the tendency for the NAO to be in its positive phase.

Similarly, the PNA teleconnection pattern in a positive phase is often associated with a negative Southern Oscillation index, or, equivalently, the warm El Niño phase of the El Niño–Southern Oscillation (ENSO). As El Niño is involved with movement of heat in the tropical Pacific Ocean, increased heating from the buildup of greenhouse gases can interfere (see Fig. 2.3). Climate models certainly show changes in El Niño with global warming, but none simulates it with sufficient fidelity for scientists to have much confidence in the results. It is likely that changes in ENSO will occur, but their nature, how large and rapid they will be, and their implications for regional climate around the world are quite uncertain and vary from model to model.

Perhaps the main way that global warming and El Niño may reinforce one another and greatly affect the environment and society is through their common influences on the carbon cycle. The decrease in outgassing of CO_2 from the oceans during El Niño (Feely et al. 1999) is enough to decrease the buildup of CO_2 in the atmosphere from 1.5 to 0.7 ppm yr^{-1}. El Niño also influences the incidence of fires, which result in more CO_2 emissions, while it changes rainfall and temperatures over land, through the teleconnections, where CO_2 uptake by the terrestrial biosphere is enhanced (Braswell et al. 1997).

The patterns of temperature change associated with these atmospheric circulation changes are similar to what might be expected directly from increases in greenhouse gases, as the land should warm first. Consequently, such changes complicate the interpretation of cause and effect. Hence, the detection of the anthropogenic signal is masked by the nature of the observed circulation changes, at least in the northern winter season. But then a vital question is whether the observed changes in PNA and the NAO (and other modes of the atmosphere) and ENSO (a natural coupled atmosphere–ocean mode) are a consequence of global warming itself.

There is evidence that changes in the NAO may be, at least in part, linked to changes in tropical sea surface temperatures (Hoerling et al. 2001), which may in turn be increased by global warming. In the Southern Hemisphere, similar kinds of changes take place in which the westerlies become stronger, and this is referred to as the "Southern Hemisphere annular mode" (SAM) (Thompson and Solomon 2002). Changes in stratospheric ozone may also be a factor in atmospheric circulation and changes in the polar vortex in both hemispheres.

On time scales of centuries, the continuing increase of greenhouse gases in the atmosphere may cause the climate system to respond through changes in the ocean circulation and, in particular, a slowdown in the Atlantic thermohaline circulation. An increase in rainfall is expected in mid to high latitudes as the climate warms because warmer air can hold more moisture, and this may result in a freshening of the ocean at the same time that it warms, reducing the thermohaline circulation. Because this circulation transports heat poleward, in part through warm waters in the Gulf Stream, any changes can have profound regional effects in Europe and surrounding areas. There remain considerable uncertainties about how important this is, although it seems likely to be more important with rapid and large increases in greenhouse gases.

Climate change is expected to alter the preferred circulation features. This can be manifested through a decrease or an increase in the range of some types of extremes. For example, we have already observed during the twentieth century in North America (and other areas) that the diurnal temperature range decreased as cloud amounts and temperatures increased. This decreased the likelihood of frosts, which can be very important for killing insects and fungal diseases that can profoundly affect fauna and flora. The wide range of natural variability associated with all the elements that constitute day-to-day weather (e.g., temperature, precipitation, solar radiation, wind speed, humidity) means that climate changes and variations are likely to be complex.

SUMMARY

Change is a natural part of the history of the Earth's climate. Climate involves variations in which the atmosphere is influenced by and interacts with various parts of the climate system, and "external" forcings. To adequately characterize climate change, the time and space scales over which the change occurs must be known as well as the statistical form of the change—for example, monotonic, nonlinear, oscillatory, and so on. This includes the ensemble of variables that constitute climate, such as temperature, precipitation, winds, moisture and solar radiation. A key challenge is to isolate the contributions of specific external forcings and internal feedbacks. A large number of analyses now indicate that human-induced changes to atmospheric composition are currently changing the climate and are expected to continue to do so into the foreseeable future. As the changes progress, it is likely to be manifested in naturally occurring circulation features, such as ENSO, the NAO, the PDO, ocean circulation, and the typical sequences of day-to-day weather. These changes are expected to have major impacts. For example, the hydrologic cycle will be affected through changes in temperature, evapotranspiration, precipitation (amounts, frequencies, and rates), the severity of droughts, snow cover duration and extent, and other factors. The changes in climate will be complex, and the reactions of the biota will vary considerably in space and time.

REFERENCES

Arrhenius, S. 1896. On the influence of carbonic acid in the air upon temperature of the ground. *Philosoph. Mag.* 41:237–275.

Berger, A. L. 1978. Long-term variations of daily insolation and Quaternary Climatic changes. *J. Atmos. Sciences* 35:2362–2367.

Braswell, B. H., D. S. Schimel, E. Linder, and B. Moore III, 1997. The response of global terrestrial ecosystems to interannual temperature variability. *Science* 278:870–872.

Feely, R. A., R. Wanninkof, T. Takahashi, and P. Tans. 1999. Influence of El Niño on the equatorial Pacific contribution to atmospheric CO_2 accumulation. *Nature* 398:597–601.

Ganopolski, A., and S. Rahmstorf. 2001. Rapid changes of glacial climate simulated in a coupled climate model. *Nature* 409:153–158.

Groisman, P. Ya, T. R. Karl, D. R. Easterling, R. W. Knight, P. B. Hamason, K. J. Hennessy, R. Suppiah, C. M. Page, J. Wibig, K. Fortuniak, V. N. Razuvaev, A. Douglas, E. Førland, and P. M. Zhai. 1999. Change in the probability of heavy precipitation: Important indicators of climatic change. *Clim. Change* 42:243–283.

Hansen, J., M. Sato, A. Lacis, and R. Ruedy. 1997. The missing climate forcing. *Phil. Trans. R. Soc. Lond.* 352:231–240.

Hoerling, M. P., J. W. Hurrell, and T. Xu. 2001. Tropical origins for recent North Atlantic climate change. *Science* 292:90–92.

Imbrie, J., Hays, J. D., Martinson, D. G., McIntyre, A., Mix, A. C., Morley, J. J., Pisias, N. G., and Prell, W. L. 1984. The orbital theory of Pleistocene climate: Support from a revised chronology of the marine $\mu^{18}O$ record. In *Milankovitch and Climate*, A. L. Berger, ed., pp. 269–305. Boston: D. Reidel.

IPCC (Intergovernmental Panel of Climate Change). 1996. *Climate Change 1995: The Science of Climate Change.* Eds. J. T. Houghton, et al. Cambridge, U.K.: Cambridge University Press.

IPCC. 2001. *Climate Change 2001. The Scientific Basis.* Eds. J. T. Houghton, et al. Cambridge, U.K.: Cambridge University Press.

Karl, T. R., and K. E. Trenberth. 2003. Modern global climate change. *Science* 302:1719–1723.

Karl, T. R., and R. W. Knight. 1998. Secular trends of precipitation amount, frequency and intensity in the USA. *Bull. Am. Meteorol. Soc.* 79:231–242.

Karl, T. R., R. W. Knight, and N. Plummer. 1995. Trends in high frequency climate variability in the twentieth century. *Nature* 377:217–220.

Katz, R. W. 1999. Extreme value theory for precipitation: Sensitivity analysis for climate change. *Adv. in Water Res.* 23:122–139.

Kiehl, J. T., and K. E. Trenberth. 1997. Earth's annual global mean energy budget. *Bull. Am. Meteorol. Soc.* 78:197–208.

Lahr, M. M., and R. A. Foley. 1994. Multiple dispersals and modern human origins. *Evolutionary Anthropology* 3:48–60.

National Assessment Synthesis Team. 2001. *Climate Change Impacts on the United States: The Potential Consequences of Climate Variability and Change.* Report for the US Global Change Research Program, New York: Cambridge University Press.

NRC (National Research Council). 2001. *Climate Change Science: An Analysis of Some Key Questions.* Washington, D.C.: National Academy Press.

Paillard D. 2001. Glacial cycles: Toward a new paradigm. *Review of Geophysics* 39:325–346.

Paillard D., L. Labeyrie, and P. Yiou. 1996. Macintosh Program performs time-series analysis. *Eos Transactions,* American Geophysical Union, 77:379.

Ramanathan, V., P. J. Crutzen, J. T. Kiehl, and D. Rosenfeld. 2001. Aerosols, climate and the hydrological cycle. *Science* 294:2119–2124.

Thompson D. W. J., and S. Solomon. 2002. Interpretation of recent Southern Hemisphere climate change. *Science* 296:895–899.

Trenberth, K. E. 1998. Atmospheric moisture residence times and cycling: Implications for rainfall rates with climate change. *Climatic Change* 39:667–694.

Trenberth, K. E., G. W. Branstator, D. Karoly, A. Kumar, N-C. Lau, and C. Ropelewski. 1998. Progress during TOGA in understanding and modeling global teleconnections associated with tropical sea surface temperatures. *J. Geophys. Res.* 103:14291–14324.

Trenberth, K. E., J. T. Houghton, and L. G. Meira Filho. 1996. The climate system: An overview. In *Climate Change 1995: The Science of Climate Change*, J. T. Hought on, et al., eds., pp. 51–64. Cambridge, U.K.: Cambridge University Press.

PART II

Present Changes

One of the most important landmarks in the literature of climate change biology is a review produced by Lesley Hughes (2000) for *Trends in Ecology and Evolution* titled "Biological Consequences of Global Warming: Is the Signal Already Apparent?" This piece opened the new century (or closed the old, for those who keep close count) with a clear statement that multiple lines of evidence point to biological response to climate change. This conclusion was subsequently affirmed by more detailed analyses of published records by Terry Root (Root et al. 2003) and Camille Parmesan that showed undeniable "fingerprints" of climate change in observed range shifts and life-cycle changes (phenology). The work of these and other scientists has confirmed that all types of biological change expected with a warming climate are occurring, including range shifts, changes in phenology, evolutionary change, and extinctions.

This recent and compelling evidence opens the body of this book as a means of underscoring the immediacy of the problem facing biodiversity. Climate change impacts are no longer theoretical or far in the future. They are here and now, steadily mounting. The urgency of response sends biologists to the past, searching for relevant insights, and inspires the construction of models of biological change in the future. Conservationists must also take notice, because these results indicate that the systems they seek to preserve may already be changing.

This section opens with a review of the evidence of recent climate change, then proceeds to chapters examining the biological consequences of that change. Subsequent sections will follow the same format—an opening chapter on climate change science (past reconstructions in Part III and future projections in Part IV), followed by chapters exploring the bio-

29

logical responses that have been observed or are expected. In this way, readers will be introduced to the fundamentals of climatology necessary to understand the associated biological changes described in the later chapters. Parallels between the two disciplines emerge in this format—for instance, issues of detection and attribution, as described by Camille Parmesan in Chapter 4.

Throughout the book, case studies are interspersed with chapters to illustrate the challenges climate change biologists face and the methods they use. The cases briefly examine a single study or a suite of research in greater detail than is possible in the chapters. The studies have been selected to represent many of the classic papers in the field, providing glimpses into landmark research. In this part, case studies feature Camille Parmesan's pioneering range shift demonstration in Edith's checkerspot butterfly and Alan Pounds's insightful documentation of the climate link to amphibian declines and avian changes at the Monteverde cloud forest in Costa Rica.

The changes described in the chapters and case studies in this part reside on a continuum of biotic responses to climate change that extend into the deep past and will continue as long as there is life on earth. Understanding this change in context requires understanding of past climate shifts and their biotic impacts, as well as projecting possible future climate alterations and their impacts on biodiversity. But it begins in the present, right now, with changes under way that will alter the course of biodiversity and its conservation forever.

REFERENCES

Hughes, L. 2000. Biological consequences of global warming: Is the signal already apparent? *Trends in Ecology and Evolution* 15:56–61.

Parmesan, C., and G. Yohe. 2003. A globally coherent fingerprint of climate change impacts across natural systems. *Nature* 421:37–42.

Root, T., J. T. Price, K. R. Hall, S. H. Schneider, C. Rosenzweig, and J. A. Pounds. 2003. Fingerprints of global warming on wild animals and plants. *Nature* 421:57–60.

CHAPTER THREE

Recent Climate Trends

MIKE HULME

Knowing the climatic history of our planet is a necessary condition for understanding why climates change over time, and how they may change in the future. In order to test theories of climate change, and to understand the relationship between climate and life, we need reliable records of how climate has changed in the past.

Most formal measurements of the elements of planetary and regional climates—local weather—are made at specific locations, although space-borne instruments are adding a new dimension to our measuring ability. Furthermore, until very recently, directional changes in climates over a single human life span (30 to 60 years) have been small and generally not recognized by individuals. And, of course, no individual humans have experienced the substantial changes in climates that have occurred over centuries and millennia.

Direct and accurate measurements of weather developed hand in hand with the invention of new measuring instruments during the 1600s and 1700s. These measurements became routine and standardized for some parts of the world during the 1800s, yet only became truly global during the 1900s. During the last decades of the twentieth century, it became possible using digital technology to collate and analyze many millions of these individual meteorological readings, allowing us to build up an unprecedented picture of global climate and how this climate changed during the previous 150 years (Jones et al. 1999).

Over this same period—the 1970s onward—it has also been possible to use new satellite-borne measurements of the Earth's atmosphere and oceans to obtain alternative estimates of some aspects of planetary climate. These satellite measurements are much shorter in duration than many surface instrumental records, yet they provide in many cases a more panoramic and comprehensive picture of how some climatic elements have changed over the past 25 years. Quantifying changes in regional or global climate that took place before about 1800 has been much more problematic and has required a range of specialized methods and techniques. In all cases, these have entailed inferring some aspect(s) of climate from other biogeophysical measurements—proxies—which are climate-sensitive to some degree. The most frequently used proxies are tree ring widths and densities, borehole temperatures, ice cores, peat bog sediments, and ocean floor sediments (Bradley 1999).

This chapter summarizes what we have observed about Earth's climate over recent years and decades, focusing specifically on the changes in climate over time rather than on the differences in climate from place to place. The majority of evidence is drawn from the past 150 years (Overpeck, Cole, and Bartlein, this volume, examine the longer climatic history of the planet), and most of the sources used are conventional instrumental measurements. The chapter is divided into four sections focusing, respectively, on changes in temperature, the water cycle, climatic variability, and weather extremes.

The changes described in this chapter have occurred over no more than about four generations—they have been the changes in climate experienced by our parents, our grandparents, and our great grandparents. Other chapters (in Part IV) explore whether or not the changes in climate to be experienced in the future by our children, grandchildren, and great grandchildren are likely to be greater than these and what implications they will have for the remaining time of life on Earth.

IS THE PLANET WARMING?

Primary Evidence

The annual average surface (1.5 meters) air temperature of the planet is probably the single most widely used indicator of the state of global climate. This quantity—the near-surface air temperature averaged over all land, ocean, and ice surfaces—is estimated to have averaged 14.0°C during the reference period 1961–1990 (Jones et al. 1999). More important, the compilation of instrumental temperature measurements since 1860 tells us that this quantity has increased by between 0.4° and 0.8°C over the last 140 years, and that the planetary temperature during the decade of the 1990s averaged about 14.5°C (Fig. 3.1). This warming of the surface of the planet has by no means been constant over time, nor has it been geographically uniform. The most rapid periods of warming occurred between 1910 and 1940 and again between 1970 and the present, both periods warming globally at the rate of about 0.1°C per decade. This warming has been more pronounced over land areas than over the oceans and, during the most recent warming episode, has been especially pronounced over the large continental interiors of the American and Eurasian landmasses.

It is also noteworthy that the warming trend has not been equally pronounced in nighttime and daytime temperatures. Especially since the 1970s, nighttime warming has been about twice as rapid (0.2°C per decade) than daytime warming, a phenomenon that for many mid-latitude regions has resulted in a reduction in the frequency of frosts and a lengthening of the frost-free season. For example, over much of northwestern Europe, the frequency of frosts at the end of the twentieth century was between 25 and 50 percent fewer than at the end of the nineteenth century. This has also meant that although daytime maximum temperatures have increased over many parts of the planet's surface, the

global temperature record

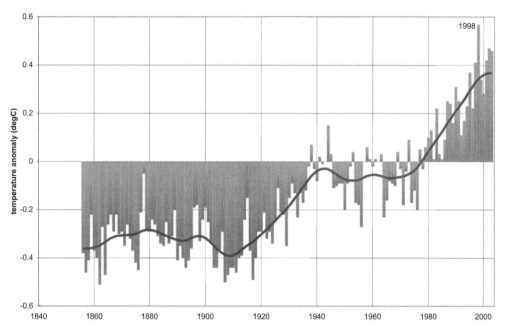

Figure 3.1. The Earth's surface air temperature for the period 1860–2001, expressed as deviations from the 1961–1990 average temperature of 14.0°C. The smooth curve emphasizes 30-year trends. *Source:* Hadley Centre and Climatic Research Unit, School of Environmental Sciences, UEA.

proportionate increase in frequency of extreme daily maximum temperatures has not been as large as the proportionate decrease in the frequency of extreme daily minimum temperatures.

Supplementary Evidence

A warming of planetary climate might be expected to induce changes in a number of other biogeophysical phenomena that should be measurable: for example, sea level, glacier mass balances, sea-ice thickness and/or extent (Johannessen et al. 2004), snow cover and/or duration, the behavior of biological systems, and so on. Many, although not all, of these phenomena that have been monitored appear to be changing in ways that are consistent with a warming climate (IPCC 2001).

Changes in worldwide sea level can be driven by changes in ocean temperature, by changes in land ice, or by changes in surface or subsurface freshwater storage. Locally, sea level can be affected by isostatic land movements, tectonic activities, and groundwater extraction. Based on tide-gauge records from around the world, and after correcting for land movements, the index of global-average sea level showed an annual average rise of between 1 and 2 mm during the twentieth century. Although this rise was greater than during the nineteenth century, there was no measurable acceleration in the rate of rise as the twentieth century progressed. This increase in global sea level is consistent with the observed increase in subsurface ocean temperatures—measured accurately since the 1960s—which, in turn, is consistent with a warming surface air climate (Cabanes et al. 2001).

Snow cover over the Northern Hemisphere has decreased in extent by about 10 percent since the late 1960s, and there was a widespread retreat of mountain glaciers in nonpolar regions during the twentieth

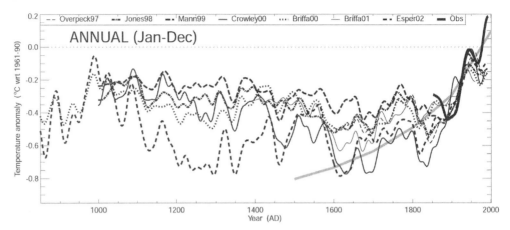

Figure 3.2. Northern Hemisphere annual surface air temperature for the period AD 1000–2000, expressed as smoothed (50-year filter) deviations from the 1961–1990 average temperature. Several different reconstructions are shown; the smooth gray line is estimated from borehole profiles. *Source:* Based on Briffa and Osborn (2002), *International Journal of Climatology,* reprinted with permission from the Royal Meteorological Society.

century. Many changes were also observed in regional biological systems—species distributions, population sizes, and phenology—changes that, more often than not, are consistent with the global and regional warming of climate (Parmesan and Yohe 2003).

Warming in the Context of the Past Millennium

Although this chapter is primarily concerned with climatic trends over the past 200 years, it is worth noting how unusual the behavior of planetary temperature has been over this period of human history in the context of the past millennium. This millennial context is used as a comparator, since we now have available to us (on the basis of proxy records of climate) reasonably robust reconstructions of temperature, at least for the Northern Hemisphere, for this 1000-year period (Mann and Jones 2003). Annually, average hemispheric temperature varied by about

±0.3°C during the past millennium; when averaged over decades, these variations were about ±0.1°C. Nowhere in this thousand-year reconstructed temperature series has a directional change in average hemispheric temperature of magnitude 0.4° to 0.8°C—such as has occurred since 1860—been observed. On this basis, we can state with some confidence that, for the Northern Hemisphere, the year 1998 and the decade of the 1990s were the warmest of the past millennium and that, at least for this hemisphere, the sustained warming that occurred during the 1910–1940 and post-1970 periods has not occurred during the previous 900 years (Fig. 3.2). We do not yet possess robust estimates of annual hemispheric temperatures prior to AD 1000, nor do we yet have sufficient proxy indicators from the Southern Hemisphere to construct a truly global series for the past millennium or longer.

The Upper Atmosphere

Warming of the atmosphere not only has occurred at the surface, but has been detected up to a level of about 8 km. Over the period for which reliable weather balloon measurements are available since the 1950s, the rate of warming through this layer of the atmosphere—the troposphere—has been similar to that observed at the surface. For the past 20 years additional measurements of air temperature through the depth of the atmosphere have become available from satellites (Christy et al. 2001), and while these also show warming in the troposphere when averaged globally, there are some differences in the regional magnitudes of warming compared to those measured by weather balloons, especially in the tropics. The reasons for these differences remain to be identified. Between about 8 and 11 km no significant trends in temperature have been observed. Higher up in the atmosphere, however, above about 11 km in the lower stratosphere, a substantial cooling (0.5°C per decade or more) of temperatures over recent decades has been measured by both balloons and satellites. This switch in trend from warming in the troposphere to cooling in the stratosphere is fully consistent with what is expected from the observed reductions in stratospheric ozone and increases in tropospheric greenhouse gases. Indeed, these contrasting observed trends confirm that our understanding of the radiative properties of a changing atmosphere is well-founded.

IS THE HYDROLOGICAL CYCLE INTENSIFYING?

Changes in the heat balance of the planet—as implied by a warming Earth—are likely to have implications for the hydrological cycle. Not only will evaporation from the ocean surface and the water-holding capacity of a warmer atmosphere increase—leading to an overall potential increase in precipitation—but changes in atmospheric circulation in response to warming might lead to a redistribution of precipitation so that some regions become drier and some wetter. An overall intensification of the hydrological cycle might also be expected to lead to an increase in the frequency of the more intense precipitation events.

Terrestrial Precipitation

The quality of instrumental measurements related to precipitation is inferior to those for temperature, even though they are more numerous, and we have virtually no direct surface-based historical measurements of precipitation over ocean areas or over much of the two polar ice caps. It is nevertheless possible to compile for the major continental landmasses large-area indexes of measured precipitation that are representative of the last 100 years or so (New et al. 2001). Averaged across all extra-polar land areas, precipitation showed little trend during the twentieth century, at most a modest increase of perhaps 2 percent. When analyzed by latitude band, however, stronger trends emerged. Annual precipitation increased in most mid to high-latitude regions by between 5 and 10 percent over the past 100 years, and, at least in some of these regions (for example, northern Europe), this was accompanied by a change in the seasonality toward greater winter half-year precipitation and reduced summer half-year precipitation. Conversely, in subtropical latitudes in both hemispheres—although especially the northern—there were decreases in precipitation, most notably in the African Sahel, where precipitation declined by up to 30 percent during the 1970s–1990s.

Although broadly coherent across these large-area latitude bands, these observed changes in precipitation were less spatially uniform at small scales than were the observed warming trends in temperature, and the much greater variability of precipitation from year to year and from decade to decade means that interpreting the significance of these precipitation changes is more difficult than it is for temperature.

Measurements of daily or hourly precipitation amounts have not yet been compiled or systematically analyzed for all regions, especially in the tropics, although for the majority of countries where they have been scrutinized, these data support the notion of an increase in the frequency of heavy rainfall events. This is most clearly the case in mid to high latitudes of the Northern Hemisphere, where the heaviest of precipitation events would appear to be up to 5 percent more frequent at the end of the twentieth century than at the beginning (Easterling et al. 2000). Changes in drought frequency have been less easy to detect and would appear to be less systematic than the increases in heavy rainfall. Consistent with reductions in overall precipitation, however, regions such as the African Sahel, parts of southeast India, and southeast Africa had more frequent dry spells in the latter half of the twentieth century compared to the first half.

Ocean Precipitation

The lack of extensive long-term observations over the oceans makes the identification of oceanic trends in precipitation difficult. Long surface measurements for some tropical islands suggest increasing precipitation in the eastern Pacific and decreasing precipitation in the western Pacific. These trends are likely to be related to changes in the El Niño–Southern Oscillation. Satellite measurements can be used to estimate precipitation over the ocean, but only since the late 1970s. These measurements suggest that, when averaged across the tropical oceans, precipitation has increased over the past 20 years. Satellite estimates of precipitation over higher-latitude ocean regions are less robust, and results are less conclusive.

IS CLIMATE BECOMING MORE VARIABLE?

Climates may change without becoming more variable. Conversely, climates may become more variable without mean climatic conditions altering. Climate varies on all time scales—from the daily variability associated with individual weather systems, through to the regular intra-annual variability induced by the seasonal cycle, the irregular multi-year variability induced by coupled atmosphere–ocean oscillations such as the El Niño–Southern Oscillation (ENSO) and the North Atlantic Oscillation, to multi-decadal and multi-century variability induced by low-frequency changes in climate system dynamics. This spectrum of behavior is quite natural for a complex nonlinear system, driven by an external energy source in the sun. Although it is relatively straightforward to define the spectral characteristics of these different modes of variability, it is much more difficult to detect whether these characteristics are changing over time. At least for the lowest-frequency variations, very long—multi-century—time series of reliable observed data would be needed to detect such changes. Nevertheless, since we are observing changes in many of the first-order statistics of the planetary climate regime, it is worth asking whether we are also seeing systematic changes in second- and third-order statistics associated with these different modes of variability.

El Niño–Southern Oscillation

One of the easiest modes of variability in the climate to analyze are multi-year quasi-periodic oscillations of the coupled ocean–atmosphere system. These oscillations can be defined relatively unambiguously using indexes of large-scale atmosphere behavior, typically based on surface (or sometimes upper atmosphere) pressure fields. ENSO has been the longest studied of these oscillations, and ENSO events—both warm phase El Niño and cold phase La Niña—are the primary cause of inter-annual climate variability over most of the tropics, and possibly some extra-tropical regions also. Warm episodes of ENSO have been more frequent, persistent, and intense since the mid-1970s compared with those of the

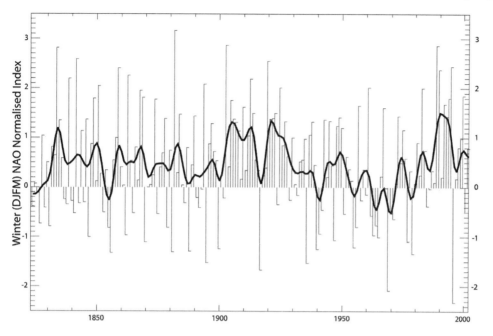

Figure 3.3. An index of the North Atlantic Oscillation, averaged over December–March, reconstructed back to 1826/27. The smooth curve emphasizes 10-year trends. *Source:* Climatic Research Unit, School of Environmental Sciences, UEA.

previous 100 years. These episodes, most notably those in 1972/73, 1982/83, 1991–94, and 1997/98, were associated with significant regional climate anomalies in many parts of the tropics—in particular, droughts in southern Africa, Indonesia, and eastern Australia, and flooding along the Pacific coast of northwestern South America, East Africa, and California (Glantz 2000). Warm excursions of sea surface temperatures associated with these ENSO episodes, exacerbated by rising background ocean temperatures, particularly during the 1990s, induced coral bleaching episodes in many tropical ocean reefs (Hoegh-Guldberg, this volume).

North Atlantic Oscillation

Another natural mode of variability in the climate system, the North Atlantic Oscillation (NAO), has been well studied and monitored in recent years. A high phase of

the NAO is associated with very mild, wet, and windy winters in northwestern Europe, a cold phase with dry and cold winter conditions. For this region of the world, the NAO is the major cause of interannual variability in winter climate. As with ENSO, the behavior of the NAO over recent decades appears to have been unusual in that the index that describes this phenomenon displays a multi-decade trend toward more positive values, starting in the early 1970s (Fig. 3.3). Only very few of the past 30 years have experienced a negative value of the NAO. This has certainly had major repercussions for the winter climate of northwestern Europe, which has correspondingly become milder and wetter.

Thermohaline Circulation

One final aspect of climatic variability worth mentioning is the behavior of the overturning circulation of the North Atlantic Ocean. This behavior is part of a larger inter-hemispheric oceanic feature called the thermohaline circulation (THC), which is responsible for delivering mild, maritime climates to latitudes 50° to

60°N in northwestern Europe that otherwise would be several degrees colder than they are. In popular terminology, this phenomenon is called the "Gulf Stream." It is known that the THC has been unstable in the past and, again, has varied on time scales from decades to centuries and millennia. Measuring the strength of this circulation has been a serious challenge for oceanographers, and even now we have no single robust measure of its global, or even regional, behavior. One index that *has* recently been developed describes the return flow of subsurface water through the Faroe Bank Channel between Greenland and Scotland in the northern Atlantic Ocean (Hansen et al. 2001). This index suggests a weakening of this return flow on the order of 20 percent over the past 50 years. If continued, and if this is indeed indicative of the wider behavior of the THC, this trend might have major implications for the stability of the northwestern European climate. The connections between the THC and the North Atlantic Oscillation are not well understood, one reason being that their modes of variability operate on quite different time scales. This remains one of the many facets of the climate system that continue to intrigue scientists.

IS WEATHER BECOMING MORE EXTREME?

This question is slightly different from, although related to, the previous one about changes in climate variability. It is possible for climates to change little in their variability from year to year or from decade to decade, and yet the very occasional extreme weather event might become more or less frequent or more or less severe. On the other hand, it is possible for there to occur changes in the frequency or severity of extreme weather events and yet for the overall annual or decadal variability of climate not to alter much. The two questions *are* related, however, to the extent that if

any extreme weather event is defined in absolute terms (e.g., a daily temperature higher than 35°C) rather than in relative terms (e.g., a daily temperature greater than 10°C above the average), then a simple change in the mean climate, and/or in its variability, will almost certainly alter the frequency of the extremes (Meehl et al. 2000).

Cyclones and Storms

Tropical cyclones and severe mid-latitude storms are two important extremes of weather, and we need to know whether their frequency and/or severity is changing. We do not have long enough records of these phenomena, however, to be able to detect confidently any long-term changes. Changes in both tropical and extra-tropical storm intensity and frequency are dominated by inter-decadal to multi-decadal variations, and no significant trends over the twentieth century have been evident (IPCC 2001). For hurricanes making landfall in the United States, for example, while the years 1996, 1997, and 1999 all had more than twice the long-term annual average (2.3 hurricanes per year), the 1950s also experienced several years with well-above-average hurricane frequency, and the long-term trends are inconclusive. For mid-latitude storms, while there appears to be some evidence for increased activity in the latter part of the twentieth century in the Northern Hemisphere, there seems to have been decreased activity in the Southern Hemisphere. And the changes, again, are not so large as to be detectable from what might simply be quite natural decadal variations.

Daily Extremes

As mentioned earlier, there have been changes in the frequency of daily temperature and precipitation extremes—a reduction in very cold winter days has been nearly universal, and increases in very hot summer days have been found in a majority of regions where the data have been analyzed (Frich et al. 2002). When high temperatures are combined with high humidities, additional stresses may be created for humans and for other living organisms. Where such combined extreme weather events have been analyzed, their frequency has also increased in recent decades, for example, during summer over most of the United States. No systematic or widespread changes have been found in the frequency of other types of small-scale severe weather phenomena, such as tornadoes, thunder days, or hail events, at least on the basis of the relatively small number of studies conducted.

Overall, we are probably not in a position to generalize that the variability of global climate has increased as it has warmed, nor to state that we have seen consistent and widespread increases in the frequency and severity of extreme weather. For some regions and for some dimensions of climate variability and extreme weather, we undoubtedly have seen statistically significant changes over the past 50 years; yet for other regions and for other types of events, the evidence remains unconvincing. For some regions and events, the data simply have not been analyzed.

SUMMARY

This chapter has shown, using both direct and indirect evidence, that the surface climate of the planet has warmed discernibly over the past 100 years, and very markedly over the past 30 years. This large-scale planetary warming has been unprecedented during at least the previous 1000 years. There is also evidence that the global hydrological cycle has been intensifying, although this has led to different regional trends in precipitation—some areas drying and some wetting. Evidence for changes in the variability of climate is more ambiguous, although several large-scale phenomena, such as the El Niño–Southern Oscillation and the North Atlantic Oscillation, have changed their behavior over the past few decades.

Some weather extremes in some regions have become more frequent or more severe, for example, heavy precipitation in northern mid-latitudes. However, detecting statistically significant changes in extreme weather behavior for some variables and some regions is difficult, and much historical data remain unanalyzed in any systematic way.

The observations of climate over the past 200 years have emphasized the complexity of the planetary climate system and the difficulty of defining what a "normal" climatic regime is. It is quite clear, however, that on time scales from years to millennia and longer, climates around the world are fundamentally variable. This inconstancy of climate occurs for a variety of entirely natural reasons, as it has always done, but also now as a potential consequence of human interventions in the climate system. Previous conventions (such as engineering standards for flood control) that have implied that statistics of climate defined over a period of 30 years or so provide human society with a sound and robust basis for future planning, design, and management are likely to prove fundamentally unsound. Conservation strategies that assume a stable climate may be equally unwise.

There remains an urgent need to maintain and improve our climate-observing system if we are to provide the needed information about our changing climate for climate analysts, for climate modelers, and for our planners, managers, and engineers. More archival weather data need digitizing and analyzing, especially for countries in the tropics and especially with regard to extreme weather events, and we need to maintain and extend the range of climate-related variables that we routinely monitor using both surface-based and satellite-based measuring instruments. As with the human body, we need to carefully and routinely measure and monitor the behavior of the climate system to allow early detection of the symptoms associated with possible dangerous interference by humans with the natural functioning of our planet.

REFERENCES

Bradley, R. S. 1999. *Paleoclimatology: Reconstructing climates of the Quaternary.* San Diego: Academic Press.

Briffa, K. R., and T. J. Osborn. 2002. Blowing hot and cold. *Science* 295:2227–2228.

Cabanes, C., A. Cazenave, and C. Le Provost. 2001. Sea level rise during past 40 years determined from satellite and in situ observations. *Science* 294:840–842.

Christy, J. R., D. E. Parker, S. J. Brown, I. Macadam, M. Stendel, and W. B. Norris. 2001. Differential trends in tropical sea surface and atmospheric temperatures since 1979. *Geophys. Res. Lett.* 28:183–186.

Easterling, D. R., G. A. Meehl, C. Parmesan, S. Changnon, T. R. Karl, and L. O. Mearns. 2000. Climate extremes: Observations, modeling and impacts. *Science* 289:2068–2074.

Frich, P., L. V. Alexander, P. Della-Marta, B. Gleason, M. Haylock, A. Klein-Tank, and T. Peterson. 2002. Observed coherent changes in climatic extremes during the second half of the twentieth century. *Climate Research* 19:193–212.

Glantz, M. 2000. Currents of change: Impacts of El Niño and La Niña on society (2nd edition). Cambridge, UK: Cambridge University Press.

Hansen, B., W. R. Turrell, and S. Osterhuis. 2001. Decreasing overflow from the Nordic seas into the Atlantic Ocean through the Faroe Bank channel since 1950. *Nature* 411:927–930.

IPCC. 2001. Climate change 2001: The scientific basis. Summary for Policymakers and Technical Summary. Cambridge, U.K.: Cambridge University Press.

Johannessen, O. M., L. Bengtsson, M. W. Miles, S. I. Kuzmina, V. A. Semenov, G. V. Alekseev, A. P. Nagurnyi, V. F. Zakharov, L. P. Bobylev, L. H. Pettersson, K. Hasselmann, and H. P. Cattle. 2004. Arctic climate change—observed and modelled temperature and sea ice variability. *Tellus* 56A:1–18.

Jones, P. D., M. New, D. E. Parker, S. Martin, and I. G. Rigor. 1999. Surface air temperature and its changes over the past 150 years. *Revs. Geophysics* 37:173–199.

Mann, M. E., and P. D. Jones. 2003. Global surface temperatures over the past two millennia. *Geophys. Res. Letts.* 30(15), 1820. doi: 10.1029/2003GLO17814.

Meehl, G. A., T. Karl, D. W. Easterling, S. Changnon, R. Jr., Pielke, D. Changnon, J. Evans, P. Ya. Groisman, T. R. Knutson, K. E. Knukel, L. O. Mearns, C. Parmesan, R. Pulwarty, T. Root, R. T. Sylves, P. Whetton, and F. Zwiers. 2000. An introduction to trends in extreme weather and climate events: Observations, socio-economic impacts, terrestrial ecological impacts and model projections. *Bull. Amer. Meteor. Soc.* 81:413–416.

New, M., M. Todd, M. Hulme, and P. D. Jones. 2001. Precipitation measurements and trends in the twentieth century *Int. J. Climatol.* 21:1899–1922.

Parmesan, C., and G. Yohe. 2003. A global coherent fingerprint of climate change impacts across natural systems. *Nature* 421:37–42.

CHAPTER FOUR

Biotic Response: Range and Abundance Changes

CAMILLE PARMESAN

Humans have been altering species' distributions and influencing local population abundances for hundreds (perhaps thousands) of years. Hunting pressures and habitat alteration have frequently caused species to disappear locally, often as a prelude to global extinction. Given the magnitude of these and other human activities, is it possible to detect an influence of climate change on species' distributions? Even given that a response is detected, what is its relative impact with respect to other anthropogenic stressors? If the impacts of climate change are relatively weak and often buried within a general framework of human-mediated habitat deterioration, is climate change, in itself, important? Addressing these questions is essential for determining the conservation implications of climate change.

The first question has been addressed in several recent reviews, which show that twentieth-century climate change has had a wide range of consequences and has had an impact on many diverse taxa in disparate geographic regions (Hughes 2000; Easterling et al. 2000b; Peñuelas and Filella 2001; IPCC 2001b; Walther et al. 2002; Root et al. 2003; Parmesan and Yohe 2003). However, this knowledge has been gained only by overcoming the enormous difficulties that biologists have had in tackling the question of climate change impacts. A basic problem is that studies used to assess climate change impacts in natural systems are, of necessity, correlational rather than experimental. Therefore, interpretation of biological changes requires less direct, more inferential methods of scientific inquiry. Further, detecting significant trends in long-term data sets is particularly difficult when data are often patchy in quantity and quality, and when natural yearly fluctuations are typically noisy. Once change has been detected, attribution requires consideration of multiple nonclimatic factors that may confound the effects of climate.

It is no surprise, then, that detection and attribution of a climate "signal" in natural systems has been a challenge for climate change biologists. A strong picture is emerging, but the evidence still only touches a fraction of all species. This chapter begins with a review of the impacts that recent climate change has had on species' distributions and local abundances, with an emphasis on linkages to underlying mechanisms. It concludes with a general discussion of detection and attribution from a biological perspective, as this issue is rarely directly tackled in the literature.

OBSERVED RANGE CHANGES AND MECHANISTIC LINKS

Studies showing the impacts of climate change on biodiversity number in the hundreds and are increasing every year. The scales of study have varied from local to regional to continental, and from 20 years to more than 100 years. For studies of distributional and abundance changes, the observed range shifts have been correlated with climatic trends in the study region for a given time period. Assigning climate change as the cause of the observed biotic changes has often had a deeper basis, such as a known mechanistic link between climatic variables and biology of the study species (Parmesan et al. 2000; Easterling et al. 2000b; Inouye 2000; Ottersen et al. 2001).

Expected distributional shifts in warming regions would be poleward and upward range shifts. However, very few studies actually document species' range shifts (but see Parmesan 1996; Parmesan et al. 1999). Most studies fall into two types: those that infer large-scale range shifts

from small-scale range boundary observations, and those that infer range shifts from changes in species' composition (abundances) in a local community. Studies that have looked for such shifts have typically been conducted for a part of the northern range boundary for northern hemisphere species (with the total study area often determined by a political boundary such as state, province, or country lines).

Community-level studies are used to infer range shifts when they are located at an ecotone involving species having fundamentally different geographic ranges: higher versus lower latitudes, or upper versus lower altitudes. The logic of looking for community changes at ecotones rather than at random geographical points is that dynamics at range boundaries are expected to be more influenced by climate than are dynamics within the interior of a species range (MacArthur 1972). Thus, community changes at major ecotones are more likely to stem from climate change impacts than are randomly picked communities. If a site at such an ecotone is warming, the expectation is that species' representation within the local community would shift in a nonrandom manner: species typically inhabiting lower latitudes and altitudes would increase in abundance, while those typically inhabiting higher latitudes and altitudes would decline (Peters 1992; Schneider 1993). Individual studies are too numerous to review exhaustively; therefore, the following studies were chosen to exemplify such mechanistic links. (For more examples of distribution and abundance changes, see Hoegh-Guldberg, this volume, and Pounds, this volume).

Observed Range Shifts and Trends in Local Abundance

On a continental scale, movements of entire species' ranges have been found in butterflies in both North America and Europe, where two-thirds of the 58 species studied have shifted their ranges northward (Parmesan 1996; Parmesan et al. 1999). Seventy years of published studies document the limiting effects of temperature on butterfly population dynamics, particularly at northern range edges (Pollard 1988; Dennis 1993; Parmesan, 2003). The northern boundaries of many European butterflies are correlated with summertime temperature isotherms (Thomas 1993). Populations toward the northern boundary become increasingly confined to the warmest micro-climates (e.g., short turf and south-facing hills) (Thomas 1993; Warren et al. 2001). Transplants beyond the northern boundary have failed to sustain breeding colonies, even when the habitat appeared suitable (Ford 1945).

On a regional scale, a study of the 59 breeding bird species in Great Britain showed both expansions and contractions of the different northern range boundaries, but northward movements were of a greater magnitude than were southward movements. Thus, the average boundary change (over all 59 species) was a mean northward shift of 18.9 km over a 20-year period (Thomas and Lennon 1999). For a few well-documented species, it has been shown that in the northern United Kingdom, boundaries have tracked winter temperatures for more than 130 years (Williamson 1975). Further, higher reproductive success has been linked to warmer springtime temperatures (Visser et al. 1998). Physiological studies have indicated that the northern boundaries of North American songbirds may generally be limited by winter nighttime temperatures (Root 1988; Burger 1998).

On a larger scale, across northern Canada the red fox has expanded northward over the past 70 years, while the Arctic fox has contracted toward the Arctic Ocean (Hersteinsson and MacDonald 1992). The timings of the boundary changes have tracked warming phases. In the past, occasional accidental transplants

of the Arctic fox southward from its range limit had succeeded, provided that the red fox was locally absent. However, prior to recent climatic warming, multiple accidental transplants of red fox north of its range limit had failed. The red fox has physical attributes which make it less well-adapted to cold conditions than the Arctic fox (e.g., longer ears and limbs). In manipulated encounters, the red fox is competitively dominant over the Arctic fox. From this, it has been inferred that the expansion of the red fox is due to warming trends, causing the competitively inferior arctic fox to retreat northward (Hersteinsson and MacDonald 1992).

Montane studies have generally been scarcer and less well-documented, but (as expected) these show a general movement of species upward in elevation. Lowland birds have begun breeding on mountain slopes in Costa Rica (Pounds et al. 1999; Pounds, this volume), alpine flora have expanded toward the summits in Switzerland (Grabherr et al. 1994), and Edith's Checkerspot butterfly has shifted upward by 105 m in the Sierra Nevada mountains of California (Parmesan 1996) (Fig. 4.1). Range contraction and population declines have been observed in situations with physical limitations on range movement. Most cloud-forest-dependent amphibians have declined or gone extinct on a mountain in Costa Rica (Pounds et al. 1999; Pounds, this volume). Low-elevation pine forests in Florida have contracted to hilltops as sea-level rise led to toxic levels of salination (Ross et al. 1994).

Local community-level studies also suggest climate-mediated species' shifts. Changes in the structure of single communities indicate that the more warm-adapted species present at a given study site are flourishing, while more cold-adapted species are declining. In Alaska, a short-term heat-treatment experiment resulted in a decline in tundra species (Chapin et al. 1995), and shrubs have been increasing at the expense of tundra at several sites (Chapin et al. 1995,

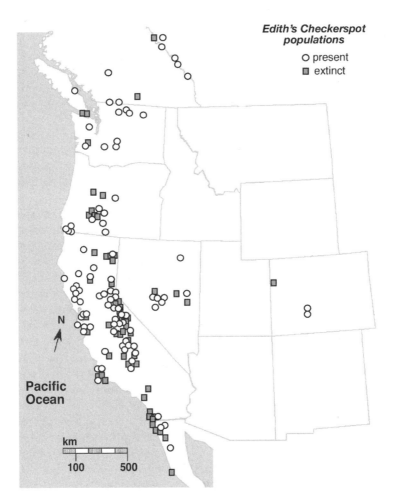

Edith's Checkerspot populations

○ present
▣ extinct

Pacific
Ocean

N

km
100   500

Sturm et al. 2001). Off the California coast, zooplankton have declined and intertidal invertebrate and fish communities have become increasingly dominated by "southerly" species (Roemmich and McGowan 1995; Holbrook et al. 1997; Sagarin et al. 1999).

Complex Responses

Changes in treeline have been more complex. Multiple studies at high latitudes in the northern hemisphere have shown twentieth-century poleward shifts (often inferred from growth indices of individuals at the range boundary). Further, several species have shown similar patterns of increased growth at treeline during the early warming in the 1930s and 1940s and during the recent warming of the past 20 years (Shiyatov 1983; Meshinev et al. 2000; Kullman 2001).

However, interpretation of these trends has not always been straightforward, as many species responded to early-twentieth-century warming, but have exhibited less pronounced (or absent) response in recent (warm) decades (Hamburg and Cogbill 1988; Kullman 1986; Kullman 1990; Innes 1991; Lescop-Sinclair and

Payette 1995; Briffa et al. 1998). The simplest resolution of this apparent contradiction is that some localities mimicked regional warming in the 1930s and 1940s but have gone counter to regional climate trends and shown cooling since the 1970s. This has been reflected in stable growth of individuals at those treelines (Kullman 1991; Kullman 1993).

For other studies, the explanation may lie in differences in rainfall during the two warm periods. At sites in Alaska, more recent decades have been relatively dry, which is believed to have prevented trees from responding to current warming as they did before (Briffa et al. 1998; Barber et al. 2000). This is in contrast to treelines in the arid southwest that have shown unprecedented tree ring growth at high elevations, and which had not only warming temperatures but also unprecedented rainfall (Swetnam and Betancourt 1998).

Precipitation changes are implicated in other vegetation shifts as well. Major cover changes in a protected part of the Sonoran desert in the southwestern United States were attributed to an extended drought period from the 1940s to 1970s (Turner 1990). This drought not only had an impact on very long-lived species (such as Creosote bush, *Larrea tridentata*), but at another southwestern U.S. site, rapidly and permanently shifted the ecotone between lower elevation juniper woodland and higher elevation ponderosa pine by more than 2 km (Allen and Breshears 1998). More recent trends in the same region toward increased precipitation seem to be driving vegetation composition at a second protected site, which, in turn, appears to be altering the relative abundances of species within rodent, reptile, and ant communities (Brown et al. 1997).

MULTI-SPECIES PERSPECTIVE

Studies that examine long-term trends across a suite of species offer invaluable insight into the strength and breadth of climate change impacts. The key advantage is that these studies document species that have not responded to climate change, along with those exhibiting responses. This allows an unbiased estimate of overall impact of recent climate change to be made. Parmesan and Yohe (2003) estimated overall impacts by conducting meta-analyses on a set of more than 30 studies covering more than 1700 species and a wide variety of taxa (insects, vertebrates, and plants). Choice of studies was conservative, with only those meeting stringent criteria included (more than 20 years of data and most covering multiple species and multiple sites). Still, the studies were not uniform, operating at many different scales (local to continental) and suffering, to various degrees, from confounding influences. Yet, the overall picture is of similarity in recent observed changes, in spite of very different levels of detail, design, and scale.

This analysis showed that more than half of the species exhibited significant changes in their phenologies and/or their distributions over the past 20 to 140 years. These changes are not random, but are systematically in the direction expected from regional changes in the climate. Responses were documented in every category, across diverse ecosystems (from temperate terrestrial grasslands to marine intertidal to tropical cloudforest), and in many types of organisms (e.g., birds, butterflies, sea urchins, trees, and mountain flowers) (Table 4.1).

A quantitative analysis of range boundary changes found boundaries to be moving, on average, 6.1 km northward per decade ($P < 0.02$, Parmesan and Yohe 2003). This climate-consistent shift emerged despite there being many nonresponders (stable range boundaries) and some southward movements. A rough estimate from this study is that 41 percent of all wild species have been affected by recent climate change (Table 4.1). This is an

Table 4.1. A Global Summary of Observed Changes from More than 30 Studies

Climate Change Prediction	Changed as Predicted	Changed Opposite to Prediction	P-value
Extensions of poleward or upper species' range boundaries; retractions of equatorial or lower-range boundaries (177 changed/279 total)	81% of changes 51% of total	19% of changes 12% of total	$<0.1 \times 10^{-12}$
Community (abundance) predictions: cold-adapted species declining and warm-adapted species increasing (283 changed/641 total)	82% of changes 36% of total	18% of changes 8% of total	$<0.1 \times 10^{-12}$

Note: Of 920 species, 373 (41%) showed change in the predicted direction. Data given in literature for a minimum of 1598 species or functional groups showed that 944 (59%) had indicated detectable phenological or distributional change, either in the direction predicted by regional climate changes or in an opposite direction. Phenological data not shown. Source: Adapted from Parmesan and Yohe 2003.

unexpectedly strong response, suggesting that current warming trends are swamping other, potentially counteractive, global change forces. Thus, the more visible aspects of human impacts on species' distributions, such as habitat loss, are not completely masking the impacts of climate change.

Differentiating Diagnostic Patterns

Important diagnostic patterns, specific to climate change impacts, have helped identify a fingerprint of global warming as the driver of the observed changes in natural systems (Parmesan and Yohe 2003). These include differential responses of cold-adapted and warm-adapted species at the same location, and the tracking of decadal temperature swings. For the latter, very long time series are essential (more than 70 years). For example, a typical pattern observed in Britain, Sweden, and Finland was northward shift of the northern range boundaries of birds and butterflies during two twentieth-century warming periods (1930–1945; 1975–1999), and southward shifts during the intervening cooling period (1950–1970). There were no instances of the opposite pattern (i.e.,

boundaries shifting southward in warm decades and northward in cool decades). In total, such diagnostic "sign-switching" responses were observed in 294 species, spread across the globe and ranging from oceanic fish to tropical birds to European butterflies (Parmesan and Yohe 2003). This specific pattern of long-term biological trends is uniquely predicted by climate trends—no other known driver could be responsible.

RATES OF CHANGE: DISPERSAL ABILITY AND RANGE MOVEMENT

Theory indicates that a highly dispersive species would be better able to exploit small, isolated patches of habitat, even if its habitat requirements were fairly restrictive (Thomas 2000). We already have evidence that the rate of range movement is very idiosyncratic both within and among broad taxonomic groups of different mobility. For instance, looking at the average movement over all butterflies, the mean change in geographic location of the range is of the same order as the regional mean change in temperature isotherms—about 100 km for both over 100 years (Parmesan 1996;

Parmesan et al. 1999). A 70-year time series of marine species abundances and sea temperatures from Britain allowed a more detailed analysis, as many sites had monthly census data (Southward et al. 1995). As with the butterfly data sets, responses were found to be rapid. Warm-water species increased in abundance and cold-water species declined during short-term periods of ocean warming (1920–1960; 1981–1995), while the opposite occurred during cooling periods (1960–1981).

A study of alpine plants in Switzerland, however, found that the rate of colonization toward higher elevations was about half that expected from the local rise in temperature (Grabherr et al. 1994). Variation within European butterflies parallel this trend. There was a trend for species with naturally low dispersal to be less responsive to regional warming than species with naturally high dispersal (Parmesan et al. 1999). Out of 10 species of Lycaenidae (Blues and Coppers—typically weak flyers), half showed stable distributions and half had moved northward in range. In contrast, among 15 species of Nymphalidae (typically strong flyers), 80 percent had shifted their ranges northward. Though not significant (N = 25; log-likelihood ratio test, G = 3.6, P = 0.17), this trend suggests that dispersal abilities among butterflies may be an important component of the ability to respond.

Trees in Sweden exhibited a particularly delayed response to local climate change. In this case, new seedling establishment beyond the current treeline had a lag-time of 30 years (Kullman 2001). This, however, is not easily explained by dispersal abilities, as the same species showed rapid response to early-twentieth-century warming.

ROLES OF EXTREME WEATHER AND CLIMATE EVENTS

Wild species have been studied with respect to climate tolerances for more than a hundred years. Many of these studies have been conducted under rigorous experimental conditions, such as laboratory manipulations of temperature and water conditions, and transplants of organisms into unoccupied habitat or outside of the species' current range. Collectively, these studies have shown that upper and lower temperature and precipitation thresholds are extremely important in determining the present distributions of wild species (Precht et al. 1973; Weiser 1973; Woodward 1987; Hoffman and Parsons 1997). Understanding organisms' responses to climate *change*, however, requires an understanding not only of an organism's "static" response to climate conditions, but also of an organism's dynamic response to year-to-year fluctuations as well as to long-term trends.

Empirical evidence abounds that it is climate variability—extreme climate years (e.g., droughts and ENSO events) and runs of extreme weather (e.g., numbers of days below freezing)—which influences species dynamics, behaviors, and distributions far more than yearly mean climate (reviewed by Easterling et al. 2000b; Inouye 2000; Parmesan et al. 2000; Otterson et al. 2001). Since extreme weather and climate events appear to be increasing in magnitude and frequency (reviewed by Easterling et al. 2000a,b; Hulme, this volume), understanding their impacts on natural systems becomes crucial for understanding observed long-term trends in wild species. Dynamic responses to these climate regime shifts could include population resiliency to crashes brought about by extreme weather, plastic adaptation of individuals, as well as true evolution of a population to a new climate.

COMPLEX INTERPLAY: HABITAT CHANGE, MICRO-CLIMATE, AND RANGE MOVEMENT

In northern Europe, some species that specialize in hot micro-climates may actually be experiencing cooling trends due to changes in human land management (Thomas 1993; Warren 1995). For example, the Lycaenid butterfly *Lysandra bellargus* requires extremely short turf where the soil and host plants are especially warm. In northern Europe, this requirement has long restricted the insects to chalk downs. However, vegetation height has recently increased in these habitats as sheep and rabbit grazing has declined, resulting in local micro-climate cooling and thereby necessitating active management of the few remaining butterfly populations. Northward range expansion of this species in Great Britain is impeded by patchiness of the very restricted habitats it requires—short, south-facing turf (C. D. Thomas and J. A. Thomas, 1998, personal communication). Thus, for *L. bellargus*, the regional warming trend in Britain would have to be much greater than has yet been observed, perhaps as much as a further rise of 3°C, before it would compensate for the microclimatic cooling caused by land use changes.

Similarly, the Nymphalid butterfly *Argynnis paphia* requires open woodland so that sun may penetrate to the forest floor where its larvae feed on violets. Closing of the forest canopy due to changes in woodland management has shaded the host plants and cooled the micro-climate in Great Britain (Thomas 1993). *A. paphia* is only now showing signs of northward expansion, a response lag to the warming trend of more than 20 years (C. D. Thomas, 2000, personal communication).

Alternatively, warming trends may change "suitable" habitat, as it did for the silver-spotted skipper butterfly (*Hesperia comma*) in Britain. Development of offspring in this species was historically restricted to the hottest micro-climates (south-facing chalk slopes). Recent northern range expansion, however, has coincided with colonization of nonsouthern slopes (Thomas et al. 2001). Simulation models of response to observed climate change based solely on previously mea-

sured thermal tolerances (i.e., with no adaptation) closely matched the observed expansion of 16.4 km (model prediction 14.4 km) (Warren et al. 2001). This butterfly has apparently experienced an increase in available habitat due to an increase in thermally suitable micro-sites, allowing it to increase in abundance and expand northward in spite of being a poor disperser.

ISSUES OF DETECTION AND ATTRIBUTION

Many of the studies mentioned above highlight the issues of detection (identification of a trend) and attribution (identifying its causal factors) in climate change biology. Issues of detection and attribution in the biological realm parallel those in the climate realm (see Overpeck et al., this volume; Hulme, this volume). Further, because climate change biology has immediate policy relevance, it is essential that the detection and attribution processes be transparent.

In both the biological and climate change spheres, detection of a climate change signal relies on statistical analyses for long-term trends. Interpretation of such analyses must take into account problems of data quality and quantity, which, again, are similar across climate and biological databases. Detailed biological records go back to the 1700s, when the first researchers began to systematically record the timing of biological events and the locations of species. Variance in historical records is due partly to genuine variance of the biological trait as well as to errors or biases due to changes in the recorder, in methods of recording, local urban expansion, and other landscape changes. There are large differences in the length of the records (some observational data going back 300 years). Interest in making such records has gradually increased through time, with substantial increases at the beginning of the twentieth century.

Once observed, a trend must be attributed to a causal factor. Here again, a strong parallel exists between climate change biology and the physical sciences focused on climate change. For both disciplines, the challenge lies in deciphering the causes of twentieth-century trends from correlational data. For both, conclusions come from scientific inference rather than direct experimental manipulation. Biologists rely upon three main lines of evidence to ascribe biotic changes to local or regional climate changes:

- A large body of theory which links known regional climate changes to oberved biotic changes (*paleological studies, biogeographic theory, biogeographic models*). (See Part III, "Learning from the Past," this volume)

- Known fundamental mechanistic links between thermal/precipitation tolerances and the study species

- Direct observations of climate effects

For biological data as well as climate data, finding a global climate change signal necessitates large-scale syntheses which assess the effects over many hundreds of species and regions.

Issues of Scale

Few distributional studies meet the ideal of having long-term data (more than 50 years) taken at regular intervals (yearly) over large geographic areas (across the species' natural range). Because of these data limitations, many studies extrapolate range shifts from small-scale range boundary studies. There are potential problems with this, as changes along one part of a species' range may not reflect whole-range trends (Fig. 4.2).

To examine whether such issues of scale might alter a study's conclusions, Parmesan and colleagues (1999) may be used as an example. In this study of 57 species of European butterflies, data were available from both the northern and southern range boundaries of 35 species (Parmesan et al. 1999). Trends at the northern boundary were mimicked at the southern boundary for only half (17 species), causing possible problems of interpretation. However, for most species the discrepancies between range edges consisted of one boundary being stable while the other shifted as expected from warming trends. Because all the shifts were northward, having data from only one range edge would have resulted in less data but would not have biased the conclusion.

One exception was the map butterfly (*Araschnia levana*). Expansion at its northern boundary into Fenno-scandia appeared to indicate a response to regional warming, but equally strong expansion southward into Spain is contra-indicative of a climate change response (Fig. 4.3). Though this example brings a cautionary tale, there is such strong consistency among the 34 remaining species that the odd behavior of *A. levana*, in itself, does not alter the study's conclusions of general northward range shifts (Parmesan et al. 1999).

Issues of Confounding Factors

Many wild species have shifted their ranges in concert with regional climate shifts. However, the leap from correlation to causation is a particularly difficult phenomenon to document, because land use change has clearly affected the distributions of many wild species over the twentieth century. It is important, then, to consider the complicating influences of urbanization, conversion of land to agriculture, contaminants, naturally occurring pathogens, overgrazing, and invasion by exotics from other continents. The influence of these other factors may never be completely ruled out; however, they can

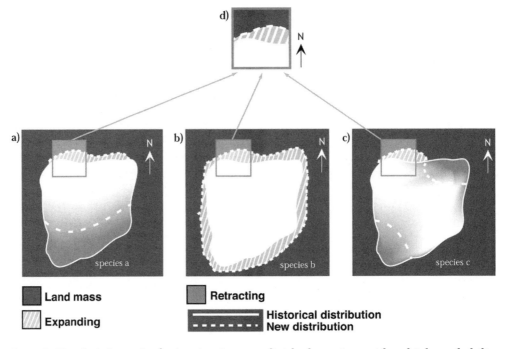

Land mass

Expanding

Retracting

Historical distribution
New distribution

Figure 4.2. Hypothetical scenarios for three imaginary species. The black squares represent generic continental landmasses, each containing a species whose range spans the bulk of the continent (e.g., a species that covers most of western Europe, from Spain to Great Britain). (a) Species A shown disappearing from its southern range edge and expanding along its northern range edge. (b) Species B shown expanding along all range edges, in all directions. (c) Species C shown disappearing from parts of its southern edge as well as parts of its northern edge, and expanding along only one portion of its northern edge. (d) A blowup of the northwestern corner of the range for all three species, showing only the expansion along that part of the northern range edge.

be minimized to the extent that any signal due to climatic factors can be discerned.

Positive Publishing Bias

As with any phenomena, there is less incentive to research or to publish a "nonresult" (i.e., documentation of no response to climate change) than it is to publish an observed response. This is perhaps most evident for evolutionary responses, with only a few studies having been published. These have (inevitably) been studies of in-

dividual species, with a high probability that only positive results will have been written up in the context of climate change (de Jong and Brakefield 1998; Rodríguez-Trelles and Rodríguez 1998; Thomas et al. 2001; Thomas, this volume).

For other types of response, however, studies are increasingly of multi-species assembleges (e.g., birds of Monteverde Preserve, Pounds et al. 1999; Pounds, this volume) or even of more complex, multicommunity ecosystems (e.g., coral reefs, Hoegh-Guldberg 1999). Such studies, by incorporating data across species, demonstrate which species/systems are nonresponsive as well as which are exhibiting change. Syntheses of such studies allowed an estimate to be made that 41 percent of wild species have responded to recent climate change (Parmesan and Yohe 2003).

Determining "Cause" of Nonresponsiveness

Findings of no detectable change in longterm studies are not easy to interpret. Species or systems that appear nonresponsive do not represent a single "result," as

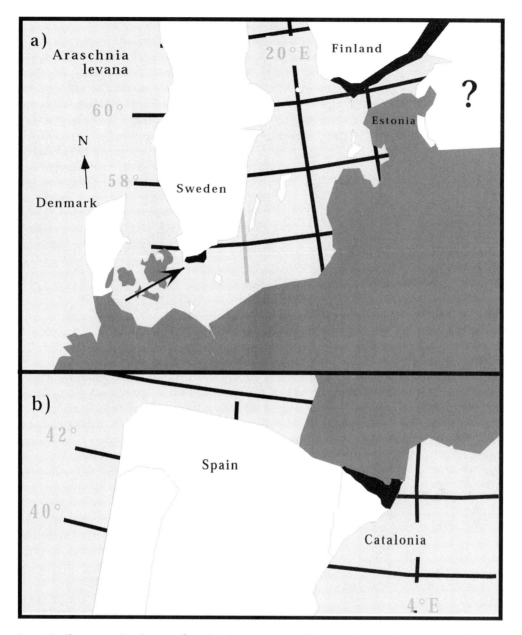

Figure 4.3. Changes in distribution of *Araschnia levana* over the twentieth century. Black = areas of extension since 1970; dark gray = distribution from 1900– 1969. (a) Northern range boundary: first seen in Sweden in 1970s, disappeared in cold summer of 1987, returned in 1992 with steady expansion ever since; first seen in Finland in 1973, established breeding by 1983; arrow indicates route of expansion from SE Denmark across islands and into Sweden. (b) Southern range boundary: first seen in Pyrenees in 1962, steady expansion into Spain ever since.

apparent stability could arise from a diversity of situations: (1) The species' phenology, abundance, or distribution is not driven by climatic factors; (2) The species is actually changing, but poor data resolution could not detect small changes; or (3) The species' phenology, abundance, or distribution is driven by climatic factors, but is failing to respond to current climate change.

Failure to respond could stem from a lag in response time, natural or anthropogenic barriers to dispersal, or genetic or physiological constraints. Lags are expected when limited dispersal capabilities retard poleward/upward colonization (Warren et al. 2001) or when a necessary resource has slower response time than that of the focal species (Parmesan et al. 1999). For example, ranges of many butterfly species in northern Europe are limited by the distributions of their host plants, and range limits of plants appear to change more slowly than those of butterflies (Grabherr et al. 1994; Parmesan et al. 1999). Habitat fragmentation and large bodies of water are among the major barriers to dispersal (Malcolm et al., this volume). Inability to respond may result even when local climate change is stressing the population. This could be important in populations that have been recently reduced in size due to human activities. These artificially small populations could suffer reduced genetic variation, thereby restricting capacity for an evolutionary response, and are at risk of extinction.

Collaborative Research

Having established that a substantial proportion of wild species are responding to current climate trends, the future challenge lies in predicting which species are most or least susceptible to projected climate change. Such understanding is necessary to evaluate options to help mitigate potential adverse effects on the more sensitive species or ecosystems. Recent policy emphasis on global climate change has brought together climate scientists and biologists at both the national and international levels. This is perhaps best evidenced by the hundreds of scientists who worked together for the Third Assessment Report of the Intergovernmental Panel on Climate Change (2001a,b). Improved cross-discipline communication and active collaborations have led to more complex, biologically relevant climate analyses conducted at ecologically important scales (e.g., Pounds et al. 1999 coupled with Still et al. 1999). This is an important shift; complemented by prior physiological and ecological experiments, it allows for a better understanding of the mechanistic bases for the observed responses of wild life to ongoing climatic change. Better mechanistic understanding is essential to reduce uncertainty in predicting species range shifts (see Peterson et al., this volume, and Midgley and Millar, this volume). Successful mitigation of future climate change impacts will depend heavily on further collaborative research into the biological impacts of climate change.

SUMMARY

Global coherency in patterns of biological change, a substantial literature linking climate variables with ecological and physiological processes, and tell-tale diagnostic fingerprints allow a causal link to be made between twentieth-century climate change and biological impacts. The patterns of change alone are evidence of a climate change "signal." Poleward range shifts have been observed in many species, with very few documentations of shifts toward the equator. Species compositions within communities have altered in concert with local temperature rise. At single sites, species from lower latitudes have tended to increase in abundance as those residing predominantly at higher latitudes suffer local declines. Consistency across studies indicates that local and regional results reflect global trends.

Though these trends may appear small compared to the massive changes in species distributions caused by habitat loss and land use modifications, that does not mean that they are unimportant. Weak but persistent forces will have a major impact on long-term trajectories, perhaps eclipsing the (apparent) importance of strong but short-lived forces (Parmesan and Yohe 2003). Continuation of the types of distributional changes which have been occurring is likely to alter species interactions, destabilize communities, and drive major biome shifts.

REFERENCES

Allen, C. D., and D. D. Breshears. 1998. Drought-induced shift of a forest-woodland ecotone: Rapid landscape response to climate variation. *Proc. Natl. Acad. Sci. USA* 95:14839–14842.

Barber, V. A., G. P. Juday, and B. P. Finney. 2000. Reduced growth of Alaskan white spruce in the twentieth century from temperature-induced drought stress. *Nature* 405:668–673.

Briffa, K. R., F. H. Schweingruber, P. D. Jones, T. J. Osborn, S. G. Shiyatov, and E. A. Vaganov. 1998. Reduced sensitivity of recent tree-growth to temperature at high northern latitudes. *Nature* 391:678–682.

Brown, J. H., T. J. Valone, and C. G. Curtin. 1997. Reorganization of an arid ecosystem in response to recent climate change. *Proc. Natl. Acad. Sci. U.S.A.* 94:9729–9733.

Burger, M. 1998. Physiological mechanisms limiting the northern boundary of the winter range of the northern cardinal (*Cardinalis cardinalis*). Ph.D. Dissertation, University of Michigan.

Chapin, F. S. III, G. R. Shaver, A. E. Giblin, K. G. Nadelhoffer, and J. A. Laundre. 1995. Response of arctic tundra to experimental and observed changes in climate. *Ecology* 76:694–711.

de Jong, P. W., and P. M. Brakefield. 1998. Climate and change in clines for melanism in the two-spot ladybird, *Adalia bipunctata* (Coleoptera: Coccinellidae). *Proceedings of the Royal Society London* 265:39–43.

Dennis, R. L. H. 1993. *Butterflies and Climate Change.* Manchester: Manchester University Press.

Easterling, D. R., J. L. Evans, P. Y. Groisman, T. R. Karl, K. E. Kunkel, and P. Ambenje. 2000a. Observed variability and trends in extreme climate events: A

brief review. *Bull. Amer. Meteorological Soc.* 80:417–425.

Easterling, D. R., G. A. Meehl, C. Parmesan, S. A. Chagnon, T. R. Karl, and L. O. Mearns. 2000b. Climate extremes: Observations, modeling, and impacts. *Science* 289:2068–2074.

Ford, E. B. 1945. *Butterflies.* London: Collins.

Grabherr G., M. Gottfried, and H. Pauli. 1994. Climate effects on mountain plants. *Nature* 369:448.

Hamburg, S. P., and C. V. Cogbill. 1988. Historical decline of red spruce population and climatic warming. *Nature* 331:428–431.

Hersteinsson, P., and D. W. MacDonald. 1992. Interspecific competition and the geographical distribution of red and arctic foxes *Vulpes vulpes* and *Alopex lagopus. Oikos* 64:505–515.

Hoegh-Guldberg, O. 1999. Climate change, coral bleaching and the future of the world's coral reefs. *Mar. Freshwater Res.* 50:839–866.

Hoffman, A. A., and P. A. Parsons. 1997. *Extreme Environmental Change and Evolution.* Cambridge, U.K.: Cambridge University Press.

Holbrook, S. J., R. J. Schmitt, and J. S. Stephens Jr. 1997. Changes in an assemblage of temperate reef fishes associated with a climatic shift. *Ecological Applications* 7:1299–1310.

Hughes, L. 2000. Biological consequences of global warming: Is the signal already apparent? *Trends Ecol. Evol.* 15:56–61.

Innes, J. L. 1991. High-altitude and high-latitude tree growth in relation to past, present and future global climate change. *The Holocene* 1,2:168–173.

Inouye, D. W. 2000. The ecological and evolutionary significance of frost in the context of climate change. *Ecology Letters* 3:457–463.

Intergovernmental Panel on Climate Change Third Assessment Report. 2001a. *Climate Change 2001: Impacts, Adaptation, and Vulnerability.* J. J. McCarthy, O. F. Canziani, N. A. Leary, D. J. Dokken, K. S. White, eds. Cambridge: Cambridge University Press.

Intergovernmental Panel on Climate Change Third Assessment Report. 2001b. *Climate Change 2001: The Science of Climate Change.* J. T. Houghton, Y. Ding, D. J. Griggs, M. Noguer, P. J. van der Linden, X. Dai, K. Maskell, C. A. Johnson, eds. Cambridge: Cambridge University Press.

Kullman, L. 1986. Recent tree-limit history of *Picea abies* in the southern Swedish Scandes. *Canadian Journal of Forest Research* 16:761–771.

Kullman, L. 1990. Dynamics of altitudinal tree-limits in Sweden: A review. *Norsk Geografisk Tidsskrift* 44:103–116.

Kullman, L. 1991. Pattern and process of present tree-limits in the Tärna regiona, southern Swedish Lapland. *Fennia* 169:25–38.

Kullman, L. 1993. Tree limit dynamics of *Betula pubescens* spp. *Tortuosa* in relation to climate variability: Evidence from central Sweden. *Journal of Vegetation Science* 4:765–772.

Kullman L. 2001. 20th century climate warming and tree-limit rise in the southern Scandes of Sweden. *AMBIO* 30(2):72–80.

Lescop-Sinclair, K., and S. Payette. 1995. Recent advance of the arctic treeline along the eastern coast of Hudson Bay. *J. Ecology* 83:929–936.

MacArthur, R. H. 1972. *Geographical Ecology*. New York: Harper and Row.

Meshinev T., I. Apostolova, and E. Koleva. 2000. Influence of warming on timberline rising: A case study on *Pinus peuce* Griseb. in Bulgaria. *Phytocoenologia* 30(3–4):431–438.

Otterson, G., B. Planque, A. Belgrano, E. Post, P. C. Reid, and N. C. Stenseth. 2001. Ecological effects of the North Atlantic Oscillation. *Oecologia* 128:1–14.

Parmesan, C. 1996. Climate and species' range. *Nature* 382:765–766.

Parmesan, C. 2003. Butterflies as bio-indicators of climate change impacts. In *Evolution and Ecology Taking Flight: Butterflies as Model Systems*, C. L. Boggs, W. B. Watt, and P. R. Ehrlich, eds. Chicago, Ill.: University of Chicago Press.

Parmesan, C., T. L. Root, and M. Willig. 2000. Impacts of extreme weather and climate on terrestrial biota. *Bull. American Meteorological Soc.* 81:443–450.

Parmesan, C., N. Ryrholm, C. Stefanescu, J. K. Hill, C. D. Thomas, H. Descimon, B. Huntley, L. Kaila, J. Kullberg, T. Tammaru, W. J. Tennent, J. A. Thomas, and M. Warren. 1999. Poleward shifts in geographical ranges of butterfly species associated with regional warming. *Nature* 399:579–583.

Parmesan, C., and G. Yohe. 2003. A globally coherent fingerprint of climate change impacts across natural systems. *Nature* 421:37–42.

Peñuelas, J., and I. Filella. 2001. Responses to a warming world. *Science* 294:793–795.

Peters, R. L. 1992. Conservation of biological diversity in the face of climate change. In *Global Warming and Biological Diversity*, R. L. Peters and T. E. Lovejoy, eds. New Haven: Yale University Press.

Pollard, E. 1988. Temperature, rainfall and butterfly numbers. *J. Applied Ecology* 25:819–828.

Pounds, J. A., M. P. L. Fogden, and J. H. Campbell. 1999. Biological responses to climate change on a tropical mountain. *Nature* 398:611–615.

Precht, H., J. Christophersen, H. Hensel, and W. Larcher. 1973. *Temperature and Life*. New York: Springer-Verlag.

Rodríguez-Trelles, F., and M. A. Rodríguez. 1998. Rapid micro-evolution and loss of chromosomal diversity in *Drosophila* in response to climate warming. *Evolutionary Ecology* 12:829–838.

Roemmich, D., and J. McGowan. 1995. Climatic warming and the decline of zooplankton in the California Current. *Science* 267:1324–1326.

Root, T. 1988. Energy constraints on avian distributions and abundances. *Ecology* 69:330–339.

Root, T. L., J. T. Price, K. R. Hall, S. H. Schneider, C. Rosenzweig, and J. A. Pounds. 2003. Fingerprints of global warming on wild animals and plants. *Nature* 421:57–60.

Ross, M. S., J. J. O'Brien, L. Da Silveira, and L. Sternberg. 1994. Sea-level rise and the reduction in pine forests in the Florida keys. *Ecological Applications* 4(1):144–156.

Sagarin, R. D., J. P. Barry, S. E. Gilman, and C. H. Baxter. 1999. Climate-related change in an intertidal community over short and long time scales. *Ecological Monographs* 69:465–490.

Schneider, S. H. 1993. Scenarios of global warming. In *Biotic Interactions and Global Change*, P. M. Kareiva, J. G. Kingsolver, and R. B. Huey, eds. Sunderland, Mass.: Sinauer Associates.

Shiyatov, S. G. 1983. Experience of use of old photographs for the study of changes of forest vegetation on the high-altitude range of its distribution. In *Floristic and Geobotanical Researches on the Urals* Sverdlovsk, pp. 76–109.

Southward, A. J., S. J. Hawkins, and M. T. Burrows. 1995. Seventy years' observations of changes in distribution and abundance of zooplankton and intertidal organisms in the western English Channel in relation to rising sea temperature. *J. Thermal Biol.* 20:127–155.

Still, C. J., P. N. Foster, and S. H. Schneider. 1999. Simulating the effects of climate change on tropical montane cloud forests. *Nature* 398:608–610.

Sturm, M., C. Racine, and K. Tape. 2001. Increasing shrub abundance in the Arctic. *Nature* 411:546–547.

Swetnam, T. W., and J. L. Betancourt. 1998. Mesoscale disturbance and ecological response to decadal climatic variability in the American Southwest. *J. of Climate* 11:3128–3147.

Thomas, C. D. 2000. Dispersal and extinction in fragmented landscapes. *Proc. Royal Society of London B* 267:139–145.

Thomas, C. D., E. J. Bodsworth, R. J. Wilson, A. D. Simmons, Z. G. Davies, M. Musche, and L. Conradt. 2001. Ecological processes at expanding range margins. *Nature* 411:577–581.

Thomas, C. D., and J. J. Lennon. 1999. Birds extend their ranges northwards. *Nature* 399:213.

Thomas, J. A. 1993. Holocene climate changes and warm man-made refugia may explain why a sixth of British butterflies possess unnatural early-successional habitats. *Ecography* 16:278–284.

Turner, R. M. 1990. Long-term vegetation change at a fully protected Sonoran desert site. *Ecology* 71(2):464–477.

Visser, M. E., A. J. van Noordwijk, J. M. Tinbergen, and C. M. Lessels. 1998. *Proc. R. Soc. Lond. B* 265:1867–1870.

Walther, G.-R., E. Post, A. Menzel, P. Convey, C. Parmesan, F. Bairlen, T. Beebee, J. M. Fromont, and O. Hoegh-Guldberg. 2002. Ecological responses to recent climate change. *Nature* 416:389–395.

Warren, M.S. 1995. Managing local microclimates for the high brown fritillary, *Argynnis adippe*. In *Ecology and Conservation of Butterflies*, A. S. Pullin, ed., pp. 198–210. London: Chapman and Hall.

Warren, M. S., J. K. Hill, J. A. Thomas, J. Asher, R. Fox, B. Huntley, D. B. Roy, M. G. Telfer, S. Jeffcoate, P. Harding, G. Jeffcoate, S. G. Willis, J. N. Greatorex-Davies, D. Moss, and C. D. Thomas. 2001. Rapid responses of British butterflies to opposing forces of climate and habitat change. *Nature* 414:65–69.

Weiser, W., ed. 1973. *Effects of Temperature on Ectothermic Organisms*. New York: Springer-Verlag.

Williamson, K. 1975. Birds and climatic change. *Bird Study* 3:143–164.

Woodward, F. I. 1987. *Climate and Plant Distribution*. Cambridge, U.K.: Cambridge University Press.

Detection at Multiple Levels: *Euphydryas editha* and Climate Change

Camille Parmesan

Edith's checkerspot butterfly (*Euphydryas editha*) presents a model example of how detailed ecological knowledge of individuals and populations can be integrated to both predict and understand species-level responses to climate change. This species has been studied for 40 years by dozens of researchers. In particular, a great deal is known about the effects of weather events and yearly climate variability on behavior, physiology, individual fitness, and population dynamics. This rich knowledge base has allowed a mechanistic link to be made between large-scale patterns of distribution change in E. *editha* and long-term (twentieth-century) climate trends.

Empirical studies on E. *editha* populations have elucidated subtle and complex interactions between climatic variability and the survival of both individuals (Singer 1972, 1983, 1994; Weiss et al. 1988; Moore 1989) and populations (Singer 1971; Singer and Ehrlich 1979; Ehrlich et al. 1980; Dobkin et al. 1987; Weiss et al. 1987, 1988; Singer and Thomas 1996; Thomas et al. 1996; Boughton 1999; McLaughlin et al. 2002). A recurring theme from observational and manipulative experiments is that the relationship between climate and survival of E. *editha* is typically mediated not by direct effects of temperature or precipitation on the insect, but by their indirect effects on timing of the butterfly's life-cycle relative to that of their host and nectar plants. At low to moderate elevation sites it is routine that the majority of larvae starve because they are not large enough to enter their summer resting stage (diapause) when their annual host plants senesce in late springtime (Singer 1971, 1972). The degree of mortality at this stage depends on the phase relationship between egg hatching and host senescence. Because this phase relationship is strongly affected by temperature and precipitation, both year-to-year variation and systematic trends in climate will affect this synchrony (Singer 1971, 1972, 1983; Singer and Ehrlich 1979; Mackay 1985a,b; Weiss et al. 1987, 1988; Moore 1989; Boughton 1999).

Local topographic diversity (north and south facing hillsides and "mima mound" terrain) and variation among habitats in vegetation cover can give as much variation of temperature as found across years. Temperature differences of 2°–4°C have been measured across a single landscape, resulting in differences in larval body temperature of more than 3°C, with hotter temperatures shortening larval duration by several days over a 10–14 day period (Weiss et al. 1988; Boughton 1999). Thus, the ultimate population response to systematic climatic trends depends on the interplay between (1) host-plant distribution across the micro- and macro-topographic landscape, (2) larval and adult dispersal, and (3) female choice of oviposition sites (Singer 1971, 1972; Mackay 1985a; Weiss et al. 1988; Boughton 1999).

The extreme sensitivity of E. *editha* to the synchrony between their life cycle and that of their hosts has rendered local extinctions a natural part of E. *editha* biology. Direct observations of population extinctions over the years indicate that many of them were caused by extreme climate years and unusual weather events (Singer and Ehrlich 1979; Ehrlich et al. 1980; Singer and Thomas 1996; Thomas et al. 1996; McLaughlin et al. 2002). In a large metapopulation in the Sierra Nevada mountains (California), all local populations using the host-plant *Collinsia torreyi* went extinct in response to three distinct extreme climate events (Singer and Thomas 1996; Thomas et al. 1996; Peterson et al., this volume). The 1975–1977 severe drought over California caused the extinction of 5 out of 21 surveyed populations (Singer and Ehrlich 1979; Ehrlich et al. 1980). Wet years have had opposite effects in different habitat types, causing population crashes in the San Francisco Bay area (*E. e. bayensis*; Dobkin et al. 1987) and population booms in Mexico (*E. e. quino*; Murphy and White 1984). Modeling of population dynamics at one locality has shown that one of these extinction events was hastened by an increased *variance* in rainfall (McLaughlin et al. 2002).

If extinctions are part of *E. editha* biology, then so too must be colonizations. The wet-year boom of Mexican populations was associated with rare long-distance dispersal (Murphy and White 1984). But insect–host-plant synchrony plays a strong role in whether such dispersal is followed by successful colonization. The recolonization of empty *Collinsia* habitat in the Sierra Nevada metapopulation was impeded because potential immigrants arrived too late in the season, when the annual *Collinsia* hosts were already close to senescence (Boughton 1999). It was a question not of host-plant suitability, but of timing. Not surprisingly, recolonizations are indeed rare, estimated at only 14 percent over a 30-year period (Parmesan 1996).

A general link between these apparently disparate population extinctions and human-mediated climate change can be sought by scaling up from the population level and looking for patterns across multiple extinction events and over a large geographic area across which regional climate change has been documented. By this means an attempt can be made to detect predicted poleward/upward distribution shifts of the entire species range. In the case of *E. editha*, this attempt was facilitated by the extensive prior knowledge of population distributions gathered by lepidopterists who were intrigued by the pronounced phenotypic variability of the adult size and wing pattern across the species' range (Singer et al. 1995). The diversity of habitats used and the sedentary nature of the insects had led to the evolution of phenotypically distinct "ecotypes," each adapted to a different habitat type, over distances as short as 10 km (Singer et al. 1995). Early lepidopterists labeled each of these "ecotypes" as a separate species and collected as many of these "species" as possible. The result is an abundance of historical records spread fairly evenly across the western half of the known range of *E. editha*.

Modern censuses of these historical sites reveal an asymmetrical pattern of population extinctions on a continental scale (Parmesan 1996). While climate change could have caused subtle responses, or responses idiosyncratic for each habitat type, the actual pattern of extinctions in *E. editha* was quite simple: population extinctions were four times as high along the southern range boundary (in Baja, Mexico) than along the northern range boundary (in Canada), and nearly three times as high at lower elevations (below 8,000 feet) than at higher elevations (from 8,000 to 12,500 feet) (Fig. 4.4; Parmesan 1996). In concert with global warming predictions, this extinction process had effectively shifted the range of *E. editha* both northward and upward in elevation since the late 1800s (Parmesan 1996).

The magnitude of shift (in mean location of populations) was 92 km toward the north and 124 m upward in elevation. This closely matches the observed warming trend over the same region, in which the mean yearly temperature isotherms had shifted 105 km northward and 105 m upward (Karl et al. 1996). Further, the altitudinal cline in frequency of population extinctions had a breakpoint at 2,400 m

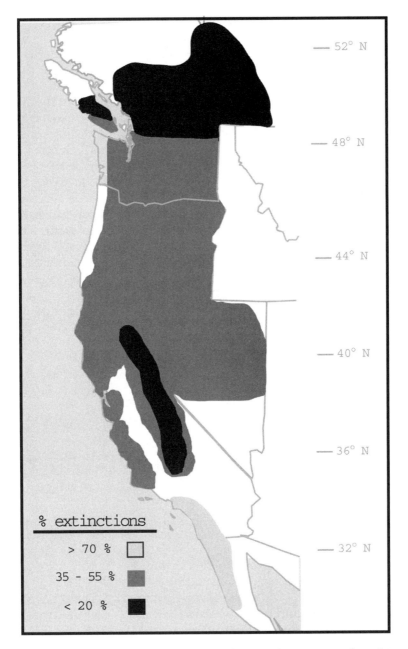

% extinctions

> 70 %

35 - 55 %

< 20 %

Figure 4.4. Patterns of population extinctions of Euphydryas editha from 1860 to 1996. Each shaded area represents multiple populations. Shades of gray represent the proportion of populations extinct in a given area during the period 1993–1996 that were previously recorded as present during the period 1860–1983.

(fewer extinctions at the higher elevations—Fig. 4.4). This breakpoint correlates with that for changes in snowpack depth and timing of snowmelt across the Sierra Nevada mountains. Below 2,440 m, snowpack had become 14 percent lighter and melt date had advanced by one week during the twentieth century (P < 0.05 for both). In contrast, snowpack had become 8 percent heavier and melt date had not changed above 2,440 m (Johnson 1998).

The substantial background knowledge of this species made it possible to control for habitat degradation. Habitat suitability could be assessed based on previously determined criteria of host- and nectar-plant quality and density. If a historic record was at a site which was no longer suitable habitat for this species, it was eliminated from the study. Separate analyses showed that other anthropogenic factors (such as proximity to large urban areas) were not associated with the observed extinction patterns (Parmesan, unpublished). Regional climate warming is, by default, the most likely cause of the observed distributional shift.

These prior empirical studies suggest a mechanistic explanation for the observed continental-scale distributional shift. In the mountains, the trend toward lighter winter snowpack in mid-elevation populations has likely caused an increase in detrimental "false spring" events. Conversely, the trend toward heavier snowpack at the highest elevations delays flight season until the most climatically suitable (warm and sunny) months of July and August, assisting rapid, unimpeded growth of offspring up to diapause (C. Parmesan, unpublished). In the southernmost populations, a gradual warming / drying trend (Karl et al. 1996) has likely led to a steady shortening of the window of time in which the host is edible, causing increased larval mortality and, hence, increased population extinctions. Thus, systematic climatic trends are highly plausible drivers of the northward and upward range shift of E. editha.

This suite of studies makes it clear that, unlike many other insects, E. editha has evolved a life-history strategy that places them "at the limits of their ecological tolerance," making them particularly susceptible to extreme climate years (Singer 1971). In the short term, the high extinction rate of southern E. editha populations implies a tight adaptation to local conditions, with little evidence for flexibility in its behavior or life-history strategies (Singer 1994). However, E. editha has also demonstrated a remarkable ability for rapid evolution under novel conditions (Singer et al. 1993), which makes long-term response to climate change more difficult to predict.

REFERENCES

Anderson, A., A. E. Allen, M. Dodero, T. Longcore, D. D. Murphy, C. Parmesan, G. Pratt, and M. C. Singer. 2003. Recovery Plan for the Quino Checkerspot Butterfly (Euphydryas editha quino). Portland, Ore.: U.S. Fish and Wildlife Service.

Boughton, D. A. 1999. Empirical evidence for source-sink dynamics in a butterfly: Temporal barriers and alternative states. Ecology 80:2727–2739.

Dobkin, D. S., I. Olivieri, and P. R. Ehrlich. 1987. Rainfall and the interaction of microclimate with larval resources in the population dynamics of checkerspot butterflies (Euphydryas editha) inhabiting serpentine grassland. Oecologia 71:161–166.

Ehrlich, P. R., D. D. Murphy, M. C. Singer, C. B. Sherwood, R. R. White, and I. L. Brown. 1980. Extinction, reduction, stability and increase: The responses of checkerspot butterfly populations to the California drought. Oecologia 46:101–105.

Intergovernmental Panel on Climate Change Third Assessment Report. 2001. Climate Change 2001: The Science of Climate Change. J. T. Houghton, Y. Ding, D. J. Griggs, M. Noguer, P. J. van der Linden, X. Dai, K. Maskell, and C. A. Johnson, eds. Cambridge: Cambridge University Press.

Johnson, T. 1998. Snowpack accumulation trends in California. M. S. thesis, Bren School of Environmental Sciences, University of California at Santa Barbara.

Karl, T. R., R. W. Knight, D. R. Easterling, and R. G. Quayle. 1996. Indices of climate change for the

United States. *Bulletin American Meteorological Society* 77:279–292.

Mackay, D. A. 1985a. Conspecific host discrimination by ovipositing *Euphydryas editha* butterflies: Its nature and its consequences for offspring survivorship. *Res. Popul. Ecol.* 27:87–98.

Mackay, D. A. 1985b. Within-population variation in pre-alighting search behavior and host plant selection by ovipositing *Euphydryas editha* butterflies. *Ecology* 66:142–151.

McLaughlin, J. F., J. J. Hellmann, C. L. Boggs, and P. R. Ehrlich. 2002. Climate change hastens population extinctions. *Proc. National Acad. Sci. USA* 99(9):6070–6074.

Moore, S. D. 1989. Patterns of juvenile mortality within an oligophagous insect population. *Ecology* 70:1726–1731.

Murphy, D. D., and R. R. White. 1984. Rainfall, resources, and dispersal in southern populations of *Euphydryas editha* (Lepidoptera: Nymphalidae). *Pan-Pacific Entomologist* 60:350–354.

Parmesan, C. 1996. Climate and species' range. *Nature* 382:765–766.

Singer, M. C. 1971. Population dynamics and host relations of the butterfly *Euphydryas editha*. Ph.D. dissertation, Stanford University.

Singer, M. C. 1972. Complex components of habitat suitability within a butterfly colony. *Science* 173:75–77.

Singer, M. C. 1983. Determinants of multiple host use by a phytophagous insect population. *Evolution* 37:389–403.

Singer, M. C. 1994. Behavioral constraints on the evolutionary expansion of insect diet; a case history from checkerspot butterflies. In *Behavioral Mechanisms in Evolutionary Ecology*, L. Real, ed., pp. 279–296. Chicago: University of Chicago Press.

Singer, M. C., and P. R. Ehrlich. 1979. Population dynamics of the checkerspot butterfly *Euphydryas editha*. *Fortschritte der Zoologie* 25:53–60.

Singer, M. C., C. D. Thomas, and C. Parmesan. 1993. Rapid human-induced evolution of insect-host associations. *Nature* 366:681–683.

Singer, M. C., R. R. White, D. A. Vasco, C. D. Thomas, and D. A. Boughton. 1995. Multi-character ecotypic variation in Edith's checkerspot butterfly. In *Ecogeographic Races*, A. R. Kruckeberg, and R. B. Walker, eds., pp. 101–104. New York: American Association for the Advancement of Science.

Singer, M. C., and C. D. Thomas. 1996. Evolutionary responses of a butterfly metapopulation to human and climate-caused environmental variation. *American Naturalist* 148:S9–S39.

Thomas, C. D., M. C. Singer, and D. Boughton. 1996. Catastrophic extinction of population sources in a butterfly metapopulation. *American Naturalist* 148:957–975.

Weiss, S. B., D. D. Murphy, and R. R. White. 1988. Sun, slope and butterflies: Topographic determinants of habitat quality for *Euphydryas editha*. *Ecology* 69:1486–1496.

Weiss, S. B., R. R. White, D. D. Murphy, and P. R. Ehrlich. 1987. Growth and dispersal of larvae of the checkerspot butterfly *Euphydryas editha*. *Oikos* 50:161–166.

White, R. R., and M. P. Levin. 1980. Temporal variation in vagility: Implications for evolutionary studies. *American Midland Naturalist* 105:348–357.

CHAPTER FIVE

Present and Future Phenological Changes in Wild Plants and Animals

TERRY L. ROOT AND LESLEY HUGHES

Many seasonal biological phenomena such as plant growth, flowering, animal reproduction, and migration depend on accumulated temperature—organisms require the appropriate amount of heat at the required times to develop from one point to another in their life cycle (Peñuelas and Filella 2001). These phenomena are therefore expected to respond sensitively to climate warming. This chapter first explores how future climatic and atmospheric change may affect the phenology of both plants and animals, and in particular, the synchrony between interacting species. It then reviews recent evidence that the anomalous warming of the past few decades has already affected species' phenology.

PLANTS

Climate change will occur, and in many cases already has occurred, within the lifetime of long-lived plants. Phenological changes may be the primary short-term response. The most important environmental cues affecting plant life cycles are photoperiod and temperature, and less commonly, moisture availability. In temperate zones, the timing of spring-growth phases, such as budding, leafing, and flowering, is mainly a response to accumulated temperature, or total heat, above a threshold level (Peñuelas and Filella 2001). Experimental data for many plant species indicate that after release from winter dormancy and above the heat threshold, there is a linear dependence between the rate of development and temperature (Heikinheimo and Lappalainen 1997).

There is a general expectation that

warming will advance the timing of most seasonal events in plant life cycles. For example, the positive relationship between the onset of flowering and spring temperature suggests that flowering of many species in Britain may advance up to 25 days with a 2.5°C warming (Fitter et al. 1995; Sparks et al. 2000). Increasing temperature might also be expected to bring about earlier budburst dates, but this may not always be the case. The buds of most temperate tree species require chilling followed by warming to release winter dormancy. If climatic warming means that buds are inadequately chilled, then the buds will remain partially dormant in spring and will require a longer time period to experience enough change in temperature to reach budburst. With a warming climate, the date of budburst could, therefore, be unchanged or occur later in some tree species (Murray et al. 1989).

Phenological events in the autumn are also expected to change, with events such as leaf color and leaf fall occurring later. Indeed, this has already been seen and is expected to become more pronounced as the globe continues to warm. The combination of both earlier spring events and later autumn events creates a longer greening season, which has been recorded by several different large-scale studies (Schwartz 1998).

Plant phenology is expected to be most sensitive to warming at higher latitudes because temperature increases will be most pronounced in these regions and because phenological responses are most closely tied to temperature in more seasonal environments. In arctic and alpine zones, where the growing season is short, cool, and curtailed at both ends by subzero temperatures, increased growth and reproductive output as a result of experimental temperature elevation have been demonstrated for numerous species (Tøtland 1997). Direct effects of warming are likely to be particularly obvious at high elevations where primary productivity is strongly limited by the snow-free growing season, and where spring snowmelt serves as a discrete environmental cue initiating growth and flowering (e.g., Price and Waser 1998; Inouye et al. 2002).

In the tropics, phenology is most often related to rainfall (Reich 1995). Any substantial increase in the length of the dry season will have direct impact on the phenology of plants in this region. A dry period of inadequate length or intensity may fail to trigger or synchronize flowering, and may also fail to reduce herbivore populations during the main period of leaf expansion (Corlett and Lafrankie 1998).

Plant phenology will also be influenced by increases in atmospheric CO_2 concentration, although generalizations are difficult to make. The effects of elevated CO_2 on the timing of flowering, budburst, and senescence appear highly species-specific, with acceleration, delay, or no effect seen in various experimental studies (Murray and Ceulemans 1998; Ward and Strain 1999)

ANIMALS

Animal life cycles also depend on climate. Warming is expected to allow insects and other ectothermic animals to pass through their juvenile stages faster, thus becoming adults more quickly, which could result in smaller body size and possibly allow some species to undergo more generations per year (Ayres and Lombardero 2000; Peñuelas and Filella 2001; Bales et al. 2002). Higher ambient body temperatures will likely reduce the time for winged insects, such as butterflies, to reach flight threshold, allowing an increase in flight-dependent activities, such as mate location, dispersal, predator evasion, and egg-laying (Dennis 1993). Diapause, the equivalent of dormancy in plants, will probably also be affected (Tauber et al. 1986).

Advances in leafing, flowering, fruiting, and appearance of insects are likely to make food supplies available earlier for many animal species. Some migratory species may be particularly vulnerable to mismatches between resource availability and life history. In the temperate zone, migration for several species seems to be cued by day length (Coppack et al. 2001). These animals may, therefore, arrive at their breeding grounds at an inappropriate time to exploit the earlier emergence of leaves or insects due to warmer climates (Visser et al. 1998). These mismatches may mean that climate change becomes a particularly serious threat to some migratory species that winter in warmer locations, such as the tropics.

For endothermic animals (i.e., birds and mammals), ambient temperatures influence the energy expended to maintain homeostasis (constant body temperature). As the globe warms, animals will probably shift both their ranges and densities. Species will be able to move into regions that are warmed, and retreat from areas that become too warm. Species within a community will exhibit differential movement (Graham and Grimm 1990; Overpeck et al. 1992), owing to the fact that the metabolic energy expenditures of various species in response to changing temperatures will be species-specific. This differential shifting could easily cause a tearing apart of present-day communities, resulting in an uncoupling of predator–prey interactions and reequilibration of competitive interactions.

OBSERVED PHENOLOGICAL CHANGES

Evidence is accumulating that many phenological changes are already occurring in response to the relatively modest warming of the past few decades. While other climatic changes, such as precipitation, are important, the biological effects of temperature are better understood for most of the organisms examined. Consequently, this chapter focuses primarily on altera-tions associated with temperature changes.

Shifts in life cycle timing have been noted in locations from around the globe (e.g., Asia, Australia, Europe, North America, Russia) for all major groups of animals and plants (invertebrates, amphibians, reptiles, birds and mammals, and trees, shrubs, forbs, and grasses). For example, changes have been observed in the timing of events such as maximum zooplankton biomass in the North Pacific (Mackas et al. 1998), peak insect abundance (which reflects the time of emergence from dormant life stages) in Europe (Sparks and Yates 1997) and New Zealand (White and Sedcole 1991), calling by frogs (which reflects timing of breeding) in North America (Gibbs and Breisch 2001), migration arrival and departure of birds in Europe (Bezzel and Jetz 1995; Visser et al. 1998) and North America (Ball 1983; Bradley et al. 1999), breeding of birds in the United Kingdom (Thompson et al. 1986; Crick et al. 1997), Germany (Ludwichowski 1997) and North America (Brown et al. 1999; Dunn and Winkler 1999), and budburst and blooming by trees in North America (Beaubien and Freeland 2000) and Asia (Kai et al. 1996).

Chorusing behavior in frogs appears to be triggered by temperature (Busby and Brecheisen 1997). Between 1980 and 1998, the time of arrival of sexually mature common toads (*Bufo bufo*) at breeding ponds in the United Kingdom was highly correlated with the mean temperatures in the 40 days preceding their arrival (Reading 1998). Two frog species, at their northern range limit in the United Kingdom, spawned 2 to 3 weeks earlier in 1994 than in 1978 (Beebee 1995). Three newt species also showed highly significant trends toward earlier breeding, with the first individuals arriving 5 to 7 weeks earlier over the course of the same study period. This study also examined temperature data, finding strong correlations with average minimum temperature in March and April and maximum temperature in

March (positive) for the two frogs with significant trends, and a strong negative correlation between lateness of pond arrival and average maximum temperature in the month before arrival for the newts. Using less precise methods, a family of naturalists in England recorded the timing of first frog and toad croaks each year for the period 1736–1947 (Sparks and Carey 1995). The date of spring calling for these amphibians occurred earlier over time, and was positively correlated with spring temperature.

Many aspects of breeding in some birds seem to be associated with temperatures. In southern Germany, the number of Reed Warblers (*Acrocephalus scirpaceus*) fledging early in the season increased significantly between 1976 and 1997, probably due to long-term increases in spring temperatures (Bergmann 1999). The spring arrival of this warbler was earlier in warm years. Also in Germany, Winkel and Hudde (1996) documented significant advances in hatching dates of Nuthatches (*Sitta europea*) over the period 1970–1995. These advances correlated with a general warming trend. Some birds in both the United Kingdom (Crick et al. 1997) and the United States (Brown et al. 1999) are now breeding 10 days earlier. Migratory patterns of birds in Africa are changing (Gatter 1992).

Differential changes among species could easily be disruptive to communities, which in turn would most likely alter the structure and functioning of most, if not all, of the world's ecosystems. Competitive interactions could be disrupted, or predator–prey relationships could become uncoupled. For example, phenological changes may be causing mismatching in the timing between the breeding of Pied Flycatcher (*Ficedula hypoleuca*) in the United Kingdom and other species in their communities (Both and Visser 2001). These birds do not seem to be shifting their spring migration arrival time. The peak abundance of a caterpillar that is fed to their young, however, is occurring earlier due to earlier emergence dates. Consequently, the birds are having to nest more quickly once they arrive at their breeding grounds in order to have the necessary food source for their chicks. With continued warming, the date of emergence of caterpillars will likely continue to shift earlier—so much earlier that the flycatchers might not be able to build nests, lay eggs, and have the eggs hatch in time for their young to take advantage of the high abundance of caterpillars. Fewer caterpillars at this key time could negatively affect the population size of this species, and the decline of predation on the caterpillars could greatly increase their population size, which could be detrimental to the plants in the community.

DERIVATION OF THE AVERAGE CHANGE IN SPRING PHENOLOGIES

Analysis of multiple studies (meta-analysis) has been used to determine the overall magnitude of phenological change across 64 studies (Root et al. 2003). All of these studies investigated the timing of events after 1951, for a total of 694 species or groups of species. Analysis by the regression-slope model (Raju et al. 1986) indicates that a statistically significant change toward earlier timing of spring events has occurred for these species over the past 50 years. The estimated mean number of days changed per decade for all species, showing that the change in spring phenology is 5.3 (SE ± 0.09) days earlier (Fig. 5.1). Most species show a change between 2 to 7 days per decade earlier in the occurrence of a particular spring trait. Meta-analysis allows the inclusion of findings from the literature that are either statistically significant or that just show a trend.

Given the fact that the higher latitudes have warmed more than the lower latitudes in the past half-century (IPCC 2001), it is expected that the phenological

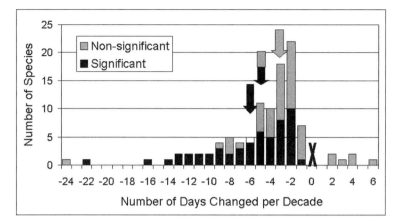

Figure 5.1. Frequency of species and groups of species (see text) exhibiting a temperature-related trait changing plotted against number of days changed in 10 years. The arrows indicate the means of the different groups (black for significant, gray for nonsignificant, and black and gray for the overall mean). The "X" indicates data were not included for those species showing no change (see Root et al. 2003).

responses would be larger nearer the poles and not as pronounced closer to the equator. Root and colleagues (2003) divided the data along the 50° latitude parallels. The high-latitude mean (-5.5 SE $\pm$ 0.11) was found to be larger than the low-latitude change (-4.4 SE $\pm$ 0.19), as expected. Both trends were statistically significant.

The studies reporting spring phenology data investigated species and populations from all major taxa. The sample size was large enough to examine the estimated mean of the shift due to climate change separately for invertebrates, amphibians, birds, trees, and other plants (Fig. 5.2). Three of the five means (invertebrates -5.41 ± 0.19; amphibians -4.83 ± 0.53; plants other than trees -5.12 ± 0.13) hover around -5.08 ± 0.09 days, which is the estimated mean for all taxa combined. Birds, however, show a statistically significantly earlier estimated mean than the others (-6.6 ± 0.2), while the estimated mean for trees is significantly later than the others (-2.6 ± 0.04). More

work is needed to help determine what these differences mean biologically.

FUTURE EXPECTATIONS

Projected future rapid climate change could soon become a major concern, especially because it will be occurring in concert with other already well-established stressors of species, particularly habitat destruction. Research and conservation attention thus needs to be focused not only on each of these stressors individually, but on the synergism of several pressures that together are likely to prove the greatest challenge to animal and plant conservation in the twenty-first century. Because anticipa-

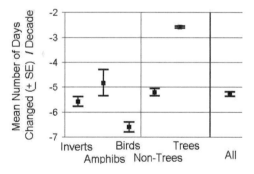

Figure 5.2. Means $\pm$ standard errors for the given groups of species. The "All" category includes only those species tallied in the groups of species (i.e., data for the one mammal, two fish, and zooplankton are not included).

tion of changes improves the capacity to manage—acting proactively rather than reactively—it behooves us to increase our understanding about the responses of plants and animals to a changing climate. This understanding, coupled with further documentation of change, may well indicate a need for actions to modify conservation efforts and future planning to account for climate change, and/or slow the projected rate of warming.

SUMMARY

Global warming has already had significant impacts on the timing of species' life cycles. In the future, the impacts are expected to be much more extreme. In many plant species, the timing of spring-growth phases—such as budding, flowering, and fruiting—is a response to accumulated temperature, or total heat, above a threshold level. In animals the timing of migrations, breeding, emergence, and meta-

Figure 5.3. Ecological consequences of climate warming for plant and animal phenology. *Source:* Redrawn from Peñuelas and Filella, 2001, with permission.

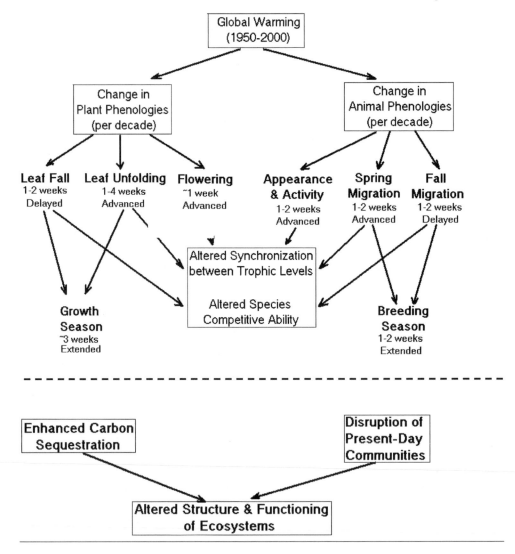

morphosis is already showing changes that are quite strong. These phenological events have been and will continue to be concomitant with delays in autumn events such as leaf color, leaf fall, and migrations. Earlier phenological changes in the spring and later in the autumn mean that there can be longer breeding periods for both plants and animals. Additionally, the largest phenological changes will be more extreme at higher latitudes and altitudes, because the warming of the globe is more intense in these regions than nearer the equator, and similarly for higher elevations. In the tropics, changes in the timing and intensity of precipitation may be more critical than temperature changes.

The anomalous warming of the last few decades has already had significant effects on the life cycles of many species (Fig. 5.3). Meta-analyses provide a way to combine results from various studies, whether significant or not, to quantify an underlying consistent shift, or "fingerprint," among species from different taxa examined at disparate locations. From 64 studies comprising data for 694 species or species groups, Root and colleagues (2003) determined the overall magnitude of phenological change for plants and animals, in terms of the number of days changed per decade, over the period 1951–2001. The number of days changed per decade ranged from 24 days per decade earlier to 6 days later, with the average shift per decade in the timing of events, such as breeding or blooming, of 5.3 ± 0.1 days earlier (N = 694). The balance of evidence, therefore, suggests that a significant impact from climatic warming is already discernible in the form of long-term, large-scale alteration of animal and plant populations. The observed consistent broad-scale patterns of change strongly suggest that recent temperature trends are the most likely explanation for these phenomena. Clearly, if such climatic and ecological changes are now being detected, when the globe has warmed by an esti-mated average of only 0.6°C, then many more far-reaching impacts on species and ecosystems are expected to occur in response to the future warming predicted by IPCC (2001), which is stated to be as high as 6±C by 2100.

ACKNOWLEDGMENTS

T. L. R. acknowledges partial support for this work from the U.S. Environmental Protection Agency, the Winslow Foundation, and the University of Michigan. We both thank Stephen H. Schneider, Lee Hannah, and Tom Lovejoy for helping us with earlier drafts of this chapter.

REFERENCES

Ayres, M. P., and M. J. Lombardero. 2000. Assessing the consequences of global change for forest disturbance from herbivores and pathogens. *The Science of the Total Environment* 262:263–286.

Ball, T. 1983. The migration of geese as an indicator of climate change in the southern Hudson Bay region between 1715 and 1851. *Climatic Change* 5:85–93.

Beaubien, E. G., and H. J. Freeland. 2000. Spring phenology trends in Alberta, Canada: Links to ocean temperature. *International Journal of Biometeorology* 44:53–59.

Beebee, T. J. C. 1995. Amphibian breeding and climate. *Nature* 374:219–220.

Bergmann, F. 1999. Long-term increase in numbers of early-fledged Reed Warblers (*Acrocephalus scirpaceus*) at Lake Constance (Southern Germany) [German]. *Journal Fuer Ornithologie* 140:81–86.

Bezzel, E., and W. Jetz. 1995. Delay of the autumn migratory period in the Blackcap (*Sylvia atricapilla*) 1966–1993: A reaction to global warming? *Journal Fuer Ornithologie* 136:83–87.

Both, C., and M. E. Visser. 2001. Adjustment to climate change is constrained by arrival date in a long-distance migrant bird. *Nature* 411:296–298.

Bradley, N. L., A. C. Leopold, J. Ross, and W. Huffaker. 1999. Phenological changes reflect climate change in Wisconsin. *Proceedings of the National Academy of Sciences of the United States of America* 96:9701–9704.

Brown, J. L., S.-H. Li, and N. Bhagabati. 1999. Long-term trend toward earlier breeding in an American bird: A response to global warming? *Proceedings of the National Academy of Sciences of the United States of America* 96:5565–5569.

Busby, W. H., and W. R. Brecheisen. 1997. Chorusing phenology and habitat associations of the crawfish frog, *Rana areolata* (Anura: Ranidae), in Kansas. *Southwestern Naturalist* 42:210–217.

Coppack, T., F. Pulido, and P. Bertold. 2001. Photoperiodic responses to early hatching in a migratory bird species. *Oecologia* 128:181–186.

Corlett, R. T., and J. V. Lafrankie. 1998. Potential impacts of climate change on tropical Asian forests through an influence on phenology. *Climatic Change* 39:439–453.

Crick, H. Q., C. Dudley, D. E. Glue, and D. L. Thomson. 1997. UK birds are laying eggs earlier. *Nature* 388:526.

Dennis, R. L. H. 1993. *Butterflies and Climate Change.* Manchester, N.Y.: Manchester University Press.

Dunn, P. O., and D. W. Winkler. 1999. Climate change has affected the breeding date of tree swallows throughout North America. *Proceedings of the Royal Society of London—Series B* 266:2487–2490.

Fitter, A. H., R. S. R. Fitter, I. T. B. Harris, and M. H. Williamson. 1995. Relationships between first flowering date and temperature in the flora of a locality in central England. *Functional Ecology* 9:55–60.

Gatter, W. 1992. Timing and patterns of visible autumn migration: Can effects of global warming be detected? *Journal Fuer Ornithologie* 133:427–436.

Gibbs, J. P., and A. R. Breisch. 2001. Climate warming and calling phenology of frogs near Ithaca, New York, 1900–1999. *Conservation Biology* 15:1175–1178.

Graham, R. W., and E. C. Grimm. 1990. Effects of global climate change on the patterns of terrestrial biological communities. *Trends in Ecology and Evolution* 5:289–292.

Heikinheimo, M., and H. Lappalainen. 1997. Dependence of the flower bud burst of some plant taxa in Finland on effective temperature sum: Implications for climate warming. *Annales Botanici Fennici* 34:229–243.

Inouye, D. W., M. A. Morales, and G. J. Dodge. 2002. Variation in timing and abundance of flowering by *Delphinium barbeyi* Huth (Ranunculaceae): The roles of snowpck, frost, and La Nina, in the context of climate change. *Oecologia* 130:543–550.

Intergovernmental Panel on Climate Change (IPCC). 2001. *Climate Change 2001: The Science of Climate Change.* New York: Cambridge University Press.

Kai, K., M. Kainuma, and N. Murakoshi. 1996. Effects of global warming on the phenological observation in Japan. In: *Climate Change and Plants in East Asia*, K. Omasa, K. Kai, H. Taoda, and Z. Uchijima, eds., pp 85–92. Tokyo: Springer.

Ludwichowski, I. 1997. Long-term changes of wing-length, body mass and breeding parameters in first-time breeding females of goldeneyes (*Bucephala clangula clangula*) in northern Germany. *Vogelwarte* 39:103–116.

Mackas, D. L., R. Goldblatt, and A. G. Lewis. 1998. Interdecadal variation in developmental timing of *Neocalanus plumchrus* populations at Ocean Station P in the subarctic North Pacfic. *Canadian Journal of Fisheries and Aquatic Science* 55:1878–1893.

McCleery, R. H., and C. M. Perrins. 1998. . . . temperature and egg-laying trends. *Nature* 391:30–31.

Murray, M. B., M. G. R. Cannell, and R. I. Smith. 1989. Date of bud burst of fifteen tree species in Britian following climatic warming. *Journal of Applied Ecology* 26:693–700.

Murray, M. B., and R. Ceulemans. 1998. Will tree foliage be larger and live longer? In: *European Forests and Global Change: The Likely Impacts of Rising CO_2 and temperature*, P. G. Jarvis, ed., pp. 94–125. Cambridge: Cambridge University Press.

Overpeck, J. T., R. S. Webb, and T. Webb III. 1992. Mapping eastern North American vegetation change over the past 18,000 years: No analogs and the future. *Geology* 20:1071–1074.

Peñuelas, J., and I. Filella. 2001. Responses to a warming world. *Science* 294:793–795.

Peñuelas, J., I. Filella, and P. Comas. 2002. Changed plant and animal life cycles from 1952 to 2000 in the Mediterranean region. *Global Change Biology* 8:531–544.

Price, M. V., and N. M. Waser. 1998. Effects of experimental warming on plant reproductive phenology in a subalpine meadow. *Ecology* 79:1261–1271.

Raju, N. S., R. Fralicx, and S. D. Steinhaus. 1986. Covariance and regression slope models for studying validity generalization. *Applied Psychological Measurement* 10:195–211.

Reading, C. J. 1998. The effect of winter temperatures on the timing of breeding activity in the common toad *Bufo bufo. Oecologia* 117:469–475.

Reich, P. B. 1995. Phenology of tropical forests: Patterns, causes and consequences. *Canadian Journal of Botany* 73:164–174.

Root, T. L., J. T. Price, K. R. Hall, S. H. Schneider, C. Rosenzweig, J. L. Pounds. 2003. Fingerprints of global warming on wild animals and plants. *Nature* 421:57–60.

Schwartz, M. D. 1998. Green-wave phenology. *Nature* 394:839–840.

Sparks, T. H., and P. D. Carey. 1995. The responses of species to climate over two centuries: An analysis of the Marsham phenological record. *Journal of Ecology* 83:321–329.

Sparks, T. H., E. P. Jeffree, and C. E. Jeffree. 2000. An examination of the relationship between flowering times and temperature at the national scale using long-term phenological records from the UK. *International Journal of Biometeorology* 44:82–87.

Sparks, T. H., and T. J. Yates. 1997. The effect of spring temperature on the appearance dates of British butterflies 1883–1993. *Ecography* 20:368–374.

Tauber, M. J., C. A. Tauber, and S. Masaki. 1986. *Seasonal Adaptations of Insecta.* Oxford: Oxford University Press.

Thompson, D. B. A., P. S. Thompson, and D. Nethersole-Thompson. 1986. Timing of breeding and breeding performance in a population of greenshank. *Journal of Animal Ecology* 55:181–199.

Tøtland, O. 1997. Effects of flowering time and temperature on growth and reproduction in *Leontodon autumnalis* var. *taraxaci*, a late-flowering alpine plant. *Arctic and Alpine Research* 29:285–290.

Visser, M. E., A. J. Vannoordwijk, J. M. Tinbergen, and C. M. Lessells. 1998. Warmer springs lead to mistimed reproduction in Great Tits (*Parus major*). *Proceedings of the Royal Society of London—Series B: Biological Sciences* 265:1867–1870.

Ward, J. K., and B. R. Strain. 1999. Elevated CO_2 studies: Past, present and future. *Tree Physiology* 19:211–220.

White, E. G., and J. R. Sedcole. 1991. A 20-year record of alpine grasshopper abundance, with interpretations for climate change. *New Zealand Journal of Ecology* 15:139–152.

Winkel, W., and H. Hudde. 1996. Long-term changes of breeding parameters of Nuthatches *Sitta europaea* in two study areas of northern Germany. *Journal Fuer Ornithologie* 137:193–202.

Responses of Natural Communities to Climate Change in a Highland Tropical Forest

J. Alan Pounds, Michael P. L. Fogden, and Karen L. Masters

Changes in species distribution and abundance in Costa Rica's Monteverde Cloud Forest suggest that global warming is already affecting natural communities in the highland tropics. The changes in local geographic range and population density suggestive of climate effects at Monteverde span a broad taxonomic range. Observations pertain to insects, salamanders, frogs, toads, lizards, snakes, birds, tree squirrels (Pounds and Crump 1994; Pounds et al. 1997; Pounds et al. 1999; Pounds 2000; M. Fogden, unpublished), and bats (Timm and Laval 2000). The data are most extensive and systematic for frogs, toads, anoline lizards, and breeding birds.

The principal trends for breeding birds reflect a tendency for species characteristic of the lower mountain slopes to increase in abundance at higher elevations (Pounds et al. 1999). In the earliest years of study (1979–1981), these cloud forest–intolerant "premontane" species nested in habitats below 1470 m. Many, however, have extended their local ranges upslope, colonizing a 40-ha cloud forest plot at 1540 m ("lower montane wet forest" of Holdridge 1967). The colonizing species include both forest interior specialists and birds that frequent clearings, and represent faunas of both the Caribbean and Pacific slopes as well as a variety of families and feeding guilds. Although a small influx occurred in the early 1980s, the major changes began in the late 1980s.

In contrast to these premontane birds, species that constituted the original breeding fauna at 1540 m show no general tendency to increase or decrease in abundance in the study area. Nevertheless, some species, including resplendent quetzales (Pharomachrus mocinno) and golden-bellied flycatchers (Myiodynastes hemichrysus), have declined. Several mountaintop species, including collared redstarts (Myioborus torquatus), ruddy treerunners (Margarornis rubiginosus), and hairy woodpeckers (Picoides villosus), have disappeared from this site as they have receded up the mountain slopes. Accordingly, at higher localities, sooty-capped bush-tanagers (Chlorospingus pileatus) and fiery-throated hummingbirds (Panterpe insignis) have retreated upslope since the early 1980s (G. Murray and W. Busby, personal communication), while yellow-thighed finches (Pselliophorus tibialis) have declined dramatically (M. Fogden, unpublished).

Highland specialists among the anoline lizards have likewise receded up the mountain slopes (Pounds et al. 1999; Pounds 2000). The change is evident in a 30-ha plot, overlapping that for birds, established for a community study in the early 1980s (Pounds 1988). At that time, the most abundant lizards in this plot were the cloud forest anole (Norops tropidolepis) and the montane anole (N. altae), which are endemic to the mountains of Costa Rica (Savage 2002). Both had declined by 1988 and had disappeared from this site by 1996; they currently persist in wetter areas farther upslope and to the east. Presently, the most common species in the 1540-m plot—the lichen anole (N. intermedius), the ground anole (N. humilis), and the blue-eyed anole (N. woodi)—are found in greater numbers in premontane areas downslope.

Massive declines of frogs and toads were apparent by 1990. A multispecies population crash in 1987 led to the disappearance of the endemic golden toad (*Bufo periglenes*) and many other species (Crump et al. 1992; Pounds and Crump 1994; Pounds et al. 1997; Pounds et al. 1999; Pounds 2000). Twenty of 50 were missing throughout surveys of a 30-km² area during 1990–1994, and there is still little sign of recovery 15 years after this crash. Tests of null models based on long-term studies of other amphibian assemblages suggest that the number of disappearances is improbable in the context of normal demographic variability. Moreover, surviving populations for which baseline data exist have fluctuated far below pre-crash levels, undergoing simultaneous downturns in 1994 and again in 1998.

In retrospect, the onset of the amphibian declines may have occurred in 1983. That year, a reduction in the numbers of Fleischmann's and Emerald glass frogs (*Hyalinobatrachium fleischmanni* and *Centrolenella prosoblepon*) active along the Río Guacimal was noted. Subsequent observations, reported by Hayes (1991), show that this reduction represented not only a hiatus in activity but also a decline in abundance. Glass frogs suffered additional declines in the late 1980s, and their numbers remain far below those observed by Jacobson (1985) and Hayes (1991).

To identify climate variables appropriate for testing whether the observed population changes are related to global warming requires an understanding of how this warming is manifested in tropical cloud forests. These forests receive much of their moisture in the form of clouds and mist. At Monteverde, the northeast trade winds carry moisture up the mountain slopes, where it condenses to form a large orographic cloud bank (Clark et al. 2000). Even during the dry season, which typically lasts from January to early May, moisture from this cloud bank ordinarily keeps the forest wet. However, if temperature-dependent relative humidity profiles change in response to warming trends, cloud formation heights may shift upward, reducing critical moisture inputs (Pounds et al. 1999; Still et al. 1999; Foster 2001).

A decline in mist frequency at the John H. Campbell Weather Station on Monteverde's upper Pacific slope (1540 m elevation) is consistent with this lifting-cloud-base hypothesis (Pounds et al. 1999). The weather station is adjacent to the study plots for anoline lizards and breeding birds and lies within the larger amphibian study area. Although the standard rain gauges at this station underestimate windblown moisture inputs, they record signatures of mist events. Because days with no measurable precipitation correspond to intervals with little or no mist, the number of these dry days provides a negative index of mist frequency. Dry days have increased in frequency since the early 1970s and have increasingly coalesced into dry periods (Fig. 5.4). Whereas mist-free periods in the 1970s rarely exceeded two days, they have recently lasted up to three weeks.

The climatic patterns imply an important effect of large-scale oceanic and atmospheric warming. Although lowland deforestation upwind of Monteverde may contribute to the observed drying trend (Lawton et al. 2001), mist frequency is negatively correlated with tropical sea surface temperatures, which have risen sharply since the mid-1970s (IPCC 2001). Warm episodes of the El Niño/Southern Oscillation contribute to the dry weather, yet a strong drying trend is evident after El Niño effects are taken into account. The most extreme conditions are a result of these effects acting in concert with the underlying trend.

To examine whether the observed biological changes are related to the climatic patterns, we tested for statistical associations between these changes and various climatic variables (Pounds et al. 1999). In a multiple regression analysis, the total number of dry days was by far the best predictor of upslope movements:

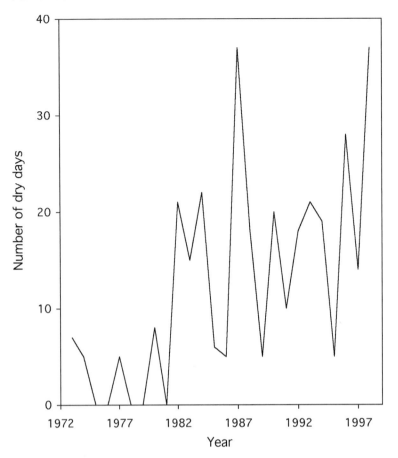

Figure 5.4. Trends and fluctuations in the number of dry days. Values are numbers of dry days (i.e., ones with no measurable precipitation) forming periods ≥ 5 days during the dry season. Regardless of the minimum period considered (≥ 1, ≥ 2, . . . ≥ 10 dry days), there is a significant trend toward more dry days. Much of the year-to-year variation superimposed on this drying trend is related to the El Niño/Southern Oscillation.

from 1982 to 1999, species composition changed in virtual phase-lock with climate change.

The diversity of these biological changes also implies that no single chain of events can account for them all. The patterns suggest instead that climate change may have orchestrated them through various mechanisms, probably involving indirect effects mediated through ecological interactions. In the case of breeding birds, for example, changes in the food availabil- ity might be important. For amphibians, evidence that epidemic disease is a factor in the declines in Central America (Berger et al. 1998; Lips 1998, 1999) suggests that climate change may contribute to declines by increasing the probability of pathogen outbreaks. This hypothesis is consistent with the growing evidence that climate in- fluences host-pathogen relationships in various ways (Dobson and Carper 1992; Epstein 1999; Kiesecker et al. 2001; Pounds 2001; Blaustein and Kiesecker 2002; Harvell et al. 2002).

The suggestion that climate change may be affecting natural populations through di- verse, often indirect, mechanisms carries important implications for our ability to predict future biological responses. We can test predictions concerning direct responses to warming, including changes in phenol- ogy and movements of populations poleward and up mountain slopes (e.g., Walther et al. 2002). When ecological interactions change, however, predicting the outcome is more difficult. The difficulty may be especially great in the tropics, where these interactions are complex and often poorly understood. The observed changes in distribution and abundance of the various animal species at Monteverde provide a clear warning that biodiversity is at risk if climate change continues unabated.

REFERENCES

Atwood, J. T. 2000. Orchids. In *Monteverde: Ecology and Conservation of a Tropical Cloud Forest*, N. M. Nadkarni, and N. T. Wheelwright, eds., p. 74. New York: Oxford University Press.

Benzing, D. H. 1998. Vulnerabilities of tropical forests to climate change: The significance of resident epiphytes. *Climatic Change* 39:519–540.

Berger, L., R. Speare, P. Daszak, D. E. Green, A. A. Cunningham, C. L. Goggin, R. Slocombe, M. A. Ragan, A. D. Hyatt, K. R. McDonald, H. B. Hines, K. R. Lips, G. Marantelli, and H. Parkes. 1998. Chytridiomycosis causes amphibian mortality associated with population declines in the rainforests of Australia and Central America. *Proceeding of the National Academy of Sciences USA* 95:9031–9036.

Blaustein, A. R., and J. M. Kiesecker. 2002. Complexity in conservation: Lessons from the global decline of amphibian populations. *Ecology Letters* 5:597–608.

Clark, K. L., R. O. Lawton, and P. R. Butler. 2000. The physical environment. In *Monteverde: Ecology and Conservation of a Tropical Cloud Forest*, N. M. Nadkarni, and N. T. Wheelwright, eds., pp. 15–38. New York: Oxford University Press.

Crump, M. L., F. R. Hensley, and K. L. Clark. 1992. Apparent decline of the golden toad: Underground or extinct? *Copeia* 1992:413–420.

Dobson, A., and R. Carper. 1992. Global warming and potential changes in host-parasite and disease-vector relationships. In *Global Warming and Biological Diversity*, R. L. Peters, and T. E. Lovejoy, eds., pp. 201–214. New Haven, CT: Yale University Press.

Dressler, R. L. 1993. *Phylogeny and Classification of the Orchid Family*. Portland, OR: Dioscorides Press.

Epstein, P. R. 1999. Climate and health. *Science* 285:347–348.

Foster, P. N. 2001. The potential negative impacts of global climate change on tropical montane cloud forests. *Earth-Science Reviews* 55:73–106.

Harvell, C. D., C. E. Mitchell, J. R. Ward, S. Altizer, A. P. Dobson, R. S. Ostfeld, and M. D. Samuel. 2002. Climate warming and disease risks for terrestrial and marine biota. *Science* 296:2158–2162.

Hayes, M. P. 1991. A study of clutch attendance in the neotropical frog Centrolenella fleischmanni (Anura: Centrolenidae). Ph.D. thesis, University of Miami, Coral Gables, Florida.

Holdridge, L. R. 1967. *Life Zone Ecology*. Tropical Science Center, San José, Costa Rica.

Intergovernmental Panel on Climate Change (IPCC). 2001. Climate Change 2001, The Scientific Basis. Third Assessment Report. Cambridge University Press, Cambridge.

Jacobson, S. K. 1985. Reproductive behavior and male mating success in two species of glass frogs (Centrolenidae). *Herpetologica* 41:396–404.

Kiesecker, J. M., A. R. Blaustein, and L. K. Belden. 2001. Complex causes of amphibian declines. *Nature* 410:681–684.

Lawton, R. O., U. S. Nair, R. A. Pielke Sr., and R. M. Welch. 2001. Climatic impact of tropical lowland deforestation on nearby montane cloud forests. *Science* 294:584–587.

Lips, K. R. 1998. Decline of a tropical montane amphibian fauna. *Conservation Biology* 12:106–117.

Lips, K. R. 1999. Mass mortality and population declines of anurans at an upland site in western Panama. *Conservation Biology* 13:117–125.

Loope, L. L., and T. W. Giambelluca. 1998. Vulnerability of island tropical montane cloud forests to climate change, with special reference to East Maui, Hawaii. *Climatic Change* 39:503–517.

Pounds, J. A. 1988. Ecomorphology, locomotion, and microhabitat structure: Patterns in a tropical mainland *Anolis* community. *Ecological Monographs* 58:299–320.

Pounds, J. A. 2000. Amphibians and reptiles. In *Monteverde: Ecology and Conservation of a Tropical Cloud Forest*, N. M. Nadkarni, and N. T. Wheelwright, eds., pp. 149–177. New York: Oxford University Press.

Pounds, J. A. 2001. Climate and amphibian declines. *Nature* 410:639–640.

Pounds, J. A., and M. L. Crump. 1994. Amphibian declines and climate disturbance: The case of the golden toad and the harlequin frog. *Conservation Biology* 8:72–85.

Pounds, J. A., M. P. L. Fogden, and J. H. Campbell. 1999. Biological response to climate change on a tropical mountain. *Nature* 398:611–615.

Pounds, J. A., M. P. L. Fogden, J. M. Savage, and G. C. Gorman. 1997. Tests of null models for amphibian declines on a tropical mountain. *Conservation Biology* 11:1307–1322.

Savage, J. M. 2002. *The Amphibians and Reptiles of Costa Rica: A Herpetofauna Between Two Continents, Between Two Seas*. Chicago: University of Chicago Press.

Still, C. J., P. N. Foster, and S. H. Schneider. 1999. Simulating the effects of climate change on tropical montane cloud forests. *Nature* 398:608–610.

Timm, R. M., and R. K. Laval. 2000. Mammals. In *Monteverde: Ecology and Conservation of a Tropical Cloud Forest*, N. M. Nadkarni, and N. T. Wheelwright, eds., pp. 223–244. New York: Oxford University Press.

Walther, G., E. Post, P. Convey, A. Menzel, C. Parmesan, T. J. C. Beebee, J. Fromentin, O. Hoegh-Guldberg, and F. Bairlein. 2002. Ecological response to recent climate change. *Nature* 416:389–395.

CHAPTER SIX

Recent Evolutionary Effects of Climate Change

CHRIS D. THOMAS

The impacts of climate change on the distributions and abundances of species are already widespread (Hughes 2000; Walther et al. 2002; Hewitt and Nichols, this volume), and even greater changes are predicted for the future (Peterson et al. 2002; Peterson et al., this volume; Betts and Shugart, this volume). Changes in population sizes and distributions imply variation in reproductive success and survival—in other words, changes in fitness. Because individuals and populations of a species will vary in their chances of survival and reproduction when the climate changes, evolutionary responses to climate change are likely to be widespread, if not ubiquitous.

The characteristics, or phenotypes, of all individual animals and plants are determined by the interaction between their genes and the environment. The physical environment experienced by each individual includes temperature, rainfall, day length, and geological substrate, while the biotic environment includes the presence of food, competitors, and natural enemies. Most long-standing populations are likely to have experienced some degree of local selection, so that traits which confer high fitness in a particular place (i.e., in a particular physical and biotic environment) are likely to have been favored. When the physical environment systematically changes, as with climate change, selection pressures will change, and we would expect evolutionary responses to take place. Even if a particular species is not directly limited by climate in a particular location, other plants and animals in the same area will be, altering the identities of the other species present and their relative abundances. Therefore, the biotic environment will be changed for virtually all species in all locations as a consequence of climate change, applying selection pressures to be able to coexist with new combinations of species and with new relative abundances of those that were already present. For example, the montane cloud forests at Monteverde in Costa Rica have become dryer and contain increasing numbers of formerly low-elevation birds (Pounds, this volume). The montane bird species potentially face a diversity of selection pressures, some of them associated with changes in the composition of the bird fauna, some of them associated with changes in the physical environment.

When the environment changes, the phenotypes of organisms may change in direct response, without the need for any genetic change. Most insects, for example, grow faster and have shorter generation times at higher environmental temperatures, until the temperature becomes so hot that excess temperature becomes directly deleterious (e.g., Taylor 1981; Bryant, Thomas, and Bale 1997, 2002). However, the existence of flexible responses (phenotypic plasticity) does not preclude the existence of genetic variation. Individuals, families, and populations may differ in their temperature-dependent development rates (e.g., Turnock and Boivin 1997), so that a change in temperature would lead to selection favoring some genotypes over others. The presence of phenotypic plasticity makes the situation complicated but does not alter the basic expectation that evolutionary responses to climate change are likely to be widespread.

This proposition makes for something of a conundrum. If climate change almost inevitably alters selection pressures, and hence results in evolutionary change, why are evolutionary responses to climate variation not more evident in the geological record? The paradigm of palaeobiologists is that over the past million years of glacial (colder) and inter-glacial (warmer, as now) climates, animals and plants have moved their distributions in response to climate rather than evolved new adaptations where they are (see Section III). This implies that at least some attributes of organisms that allow them to live in certain types of climate do not evolve readily, despite historical changes in climate every 10,000 to 100,000 years, generating repeated changes in selection pressures.

A possible solution to this conundrum is that there may be differences in the ability of populations to undertake evolutionary responses to climate change, depending on whether they are in the climatic core of their distribution or near the climatic edge of their distribution. In core areas, variation may be maintained by stabilizing or fluctuating selection, combined with gene flow from other populations that occur in either more (e.g., dryer) or less (e.g., wetter) extreme environments. Therefore, a change in selection may generate rapid evolutionary change, because variation is already present that would permit an evolutionary response in either direction. The situation at the "edge" is potentially quite different. Prior to a change in the environment, marginal populations are more likely to be under directional selection, but are failing to respond either because there is no useful variation to draw upon or because gene flow from core habitats is swamping out new, marginal adaptations (reviewed by Butlin et al. 2003).

The ability of different "edge" populations to undertake evolutionary responses is likely to vary between those that experience improving conditions and those that experience deteriorating conditions. When conditions improve, the environment becomes more like that experienced in the core, and the combination of changed selection and gene flow from core-like populations may permit a rapid evolutionary response. When conditions deteriorate, the environment becomes unlike that experienced by any other population. In this case, there is little or no potential for gene flow to introduce useful genetic variation; populations are more likely to become extinct than to adapt fully to the new conditions (although successful adaptation will *sometimes* be achieved).

This chapter describes some of the growing evidence that recent climatic change is already resulting in evolutionary responses.

EVOLUTION WITHIN EXISTING POPULATIONS

Over moderate lengths of time, we know that many climate-related traits can evolve. Latitudinal differences exist in animal responses to day length (Noda 1992; Blackenhorn and Fairbairn 1995) and temperature (McColl and McKechnie 1999), and in traits that affect the ability of species to adjust their body temperature, such as body size (Loeschcke, Bundgaard, and Barker 2000; Zwaan et al. 2000) and darkness (Dennis and Shreeve 1989; Dennis 1993). Many of these latitudinal differences exist in parts of Eurasia and North America that were covered in ice some 25,000 years ago, and must have arisen there within the last approximately 13,000 years, through a combination of responses to local selection and colonization from different population sources. Therefore, it is realistic that climate change could result in widespread changes in these traits, such as photoperiodism, over quite short time periods. For example, *Drosophila subobscura* evolved a temperature-related latitudinal gradient in adult size within two decades of its introduction to the New World (Huey et al. 2000).

Responses to Day Length

Under warmer conditions in the northern hemisphere, plants have flowered earlier, insects have emerged earlier in the season, amphibians have returned to their breeding ponds earlier, migrant birds have returned earlier in spring, and nonmigrant birds have nested earlier in spring (e.g., Hughes 2000; Fitter and Fitter 2002; Walther et al. 2002). In some cases (e.g., flowering of an individual long-lived plant), it is immediately clear that these changes in the timing of the life cycle represent purely plastic responses to the environment, with no genetic change involved.

However, the potential for evolutionary change in the timing of the life cycle is great. Most animals and plants use some combination of day length and temperature to act as cues to initiate particular stages of the life cycle, such as bud burst and flowering in plants, or diapause induction in insects (e.g., Takeda and Skopik 1997; Bale et al. 2002; Karlsson and Werner 2002; Root and Hughes, this volume). Particularly in autumn, shortening day length (photoperiod) is a reliable predictor of when an animal should prepare for winter (Saboureau et al. 1999; Pulido et al. 2001; Ansart, Vernon, and Daguzan 2001). Therefore, populations may come under strong selection to adjust the cues used to initiate overwintering behavior, and evolutionary responses could potentially be rapid. For example, in a breeding experiment, Pulido and colleagues (2001) showed that they could delay the onset of autumn migration behavior in blackcap warblers, *Sylvia atricapilla*, by more than one week, after only two generations of artificial selection.

The clearest example of recent evolution in the timing of the life cycle comes from a study of pitcher-plant mosquitoes, *Wyeomyia smithii*, in the eastern United States (Bradshaw and Holzapfel 2001). This mosquito carries out its entire preadult life cycle in the water-filled leaves of the purple-leaved pitcher plant, *Sarracenia purpurea*. Larvae of this mosquito show a latitudinal (and altitudinal) response to day length, such that they enter an overwintering diapause state at an earlier date in the north (and at high elevation) than in the south, enabling the mosquitoes to exploit the longer development period that is available in the warmer south. Responses to day length were heritable within populations, and crosses between populations produced offspring with intermediate responses to day length (Hard, Bradshaw, and Holzapfel 1993; Lair, Bradshaw, and Holzapfel 1997).

Mosquito larvae were reared in controlled environments in the laboratory in 1972, 1988, 1993, and 1996, so most of the differences between these time periods are likely to be genetic in origin. Remarkably, the populations evolved shorter day-length requirements to initiate diapause. For example, seven populations between 30 and 40°N that were sampled in both 1972 and 1996 all had shorter day-length requirements in 1996. These differences corresponded to initiating diapause about nine days later in 1996 than in 1972. In other words, much of the shift in timing of the life cycle in autumn can be attributed to an evolutionary response to the increased length of the growing season. Cases of life cycle changes being completely dependent on genetic changes are likely to be rare, but some element of genetic change is likely to be commonplace.

Choice of Habitat and Host Plants

Many animals and plants appear to select their habitats at least partly on the basis of microclimate. For example, at the cold edges of their distributions, many species occupy warm microhabitats, such as hot south-facing hillsides with short vegetation (in the northern hemisphere) (e.g., Pigott 1968; Thomas 1993; Thomas et al. 1999). Meanwhile, species found on cool, north-facing slopes may be more typically associated with habitats that are commoner much farther to the north (Billings 1952). When the climate changes, so should the choice of microclimate. Sometimes this can be achieved by instantaneous behavioral flexibility, or through habitat-specific population dynamics. However, choice of habitat may also involve behavioral responses to nonthermal aspects of habitat that will not immediately change when the climate changes. In this circumstance, a change in climate may lead to natural selection for changes in habitat choice.

This has recently been observed in the checkerspot butterfly, *Euphydryas editha*, a butterfly which shows geographic and genetic variation in its choice of host plants and habitats (e.g., White and Singer 1974; Singer, Ng and Thomas 1988; Singer, Moore, and Ng 1991; Singer and Parmesan 1993; Singer et al. 1994; Parmesan case study, this volume). A population in the Sierra Nevada in California occupied two types of habitat in the 1980s—rocky outcrops and forest clearings—but was eliminated from the clearings by three unusual climatic events that took place between 1989 and 1992 (Singer and Thomas 1996; Thomas, Singer, and Boughton 1996). Early springs caused massive mortality of adult butterflies in the clearing habitat because the butterflies emerged before adult nectar was available in 1989, and because the butterflies were buried for a week under snow in 1990. These two events reduced breeding densities in the clearings to approximately one-hundredth of the density that occurred in this habitat in the mid-1980s (Thomas, Singer, and Boughton 1996). To cap it all, a severe summer frost late in the season in 1992 killed all the host plants in the clearings (but not on the outcrops), whereupon the caterpillars starved to death. All the clearing populations went extinct. Microclimate and habitat differences between the clearings and outcrops meant that the later-emerging butterflies on the outcrops were largely unaffected by these events.

Different host plant species were used for egg-laying in the two habitats, and females differed in their choice of host plant. Females that chose to lay on *Collinsia torreyi* plants in the clearing habitats left no offspring. This resulted in very strong selection that favored females that chose *Pedicularis semibarbata* plants in the outcrop habitat. The population subsequently became restricted to laying eggs only on the host plants available on the outcrops; climatic variation had brought about an evolutionary change in habitat choice at the population level.

Similar scenarios are likely to occur in many other species because climatic events rarely have identical effects on populations in different habitats and microhabitats. Whenever there is a genetic contribution to habitat choice, climate variation is likely to result in evolutionary change.

Arrival of New Species

Climate change will bring with it massive changes in the composition of biological communities (Peterson et al. 2002; Parts III and IV, this volume). For populations that survive where they are, the arrival of new food items, competitors, predators, and diseases may result in evolutionary as well as ecological responses. There are so many possible consequences that it will be difficult to generalize. The following example simply illustrates that the arrival of new species can generate strong selection, resulting in evolutionary changes within existing populations.

A different population of E. *editha* recently changed its diet in response to plant invasion. In the 1980s, it undertook an evolutionary change in egg-laying preference (which was heritable), switching from the small, native annual *Collinsia parviflora* to the introduced European weed *Plantago lanceolata* (Singer, Ng, and Thomas 1988; Singer, Thomas, and Parmesan 1993). Larval survival was higher on *P. lanceolata* than on *C. parviflora* because the introduced perennial plant remained edible for caterpillars during summer droughts, so that selection favored females that laid their eggs on *P. lanceolata*. During the 1980s,

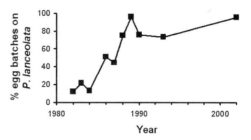

Figure 6.1. Switch in diet by the butterfly *Euphydryas editha* from the native plant *Collinsia parviflora* to the introduced plant *Plantago lanceolata*. Caterpillars died during summer droughts on the annual *C. parviflora* but survived better on the perennial *P. lanceolata*. *Data source:* Singer, Thomas, and Parmesan (1993); 2001 data point, M. C. Singer, unpublished.

the population evolved from one in which only about 5 percent of females preferred to lay on *P. lanceolata* to one in which over half of the females preferred this plant (Singer, Thomas, and Parmesan 1993). Between 1982–83 and 1989–93, the percentage of egg batches estimated to be laid on *P. lanceolata* increased from 15 percent to 80 percent (Singer, Thomas, and Parmesan 1993) and had risen to 96 percent by 2002 (M. C. Singer, personal communication) (Fig. 6.1). Over twenty years (generations), the butterfly more or less completed a switch in diet, now feeding on an introduced plant that allows the butterflies to survive better when faced with summer droughts. Although the arrival of *P. lanceolata* was not, in this case, due to climate change, the difference in the fitness of caterpillars feeding on them did depend on the responses of the two plants to climate.

EVOLUTION AS RANGES EXPAND

Just as climate change will eliminate species from parts of their existing distributions, it will provide opportunities for colonization of new habitats within the existing distribution (e.g., habitats that were previously too cool or wet) and of new areas beyond the existing distribution that become climatically suitable. Most colonization, especially long-distance colonization, is likely to be achieved by a small number of individuals. By chance, the average (genetically based) characteristics of the new populations may therefore differ from those of the species as a whole.

However, it will not all be left to chance. It is unlikely that all individuals within populations, or that all populations within a species, will have traits that are equally suited to colonization or the establishment of new populations. Almost any character that could increase the rate of range expansion in the direction where the climate has recently become more suitable could be favored, through selection favoring successive generations of successful colonists. Here, two traits are examined that are most obviously linked to colonization ability—flight capacity and habitat choice—but there are probably many other traits that are also under strong selection.

Flight Capacity at Expanding Range Margins

Many species have colonizing and "stay-at-home" forms, such as aphids with and without wings, and thistles with and without winged seeds: these are often associated with variation in habitat stability (e.g., Zera and Denno 1997). Mobile forms of species are likely to be favored when new suitable areas become available for colonization as a result of climate change. However, there are costs of dispersal. Dispersing individuals may fail to find any new suitable habitat and die, and even if they are successful they will use up energy that could otherwise be used for reproduction—there is usually a trade-off between dispersal and reproductive capacity (Zera and Denno 1997). Once dispersive individuals have founded a population, thereafter less dispersive forms may be at a selective advantage (Cody and Overton 1996). Because of this selection toward "stay-at-home" morphs after a

population has been established, evolutionary changes during range expansions may be transient but nonetheless critical to the ability of a species to shift its distribution.

Over the past 10,000 years, the lodgepole pine gradually colonized northward in North America. Trees in the south produce seeds that are relatively large, have relatively small winged protuberances, and are unlikely to disperse far. Trees close to the northern edge of the range produce seeds that are relatively small and have larger wings: both of these traits are likely to increase the distances over which the seeds are dispersed by wind (Cwynar and MacDonald 1987). Presumably, as the climate warmed and the distribution began to shift northward, the most dispersive seeds would have founded each new population to the north, establishing a latitudinal gradient in dispersal capacity.

Two species of bush crickets have recently started to expand their distributions extremely rapidly in Britain, spreading at a rate of approximately 10 km per year (generation) (Marshall and Haes 1988; Haes and Harding 1997; Widgery 2000). In the long-winged cone-head, *Conocephalus discolor*, there are two morphs of this species: "long-winged" and "extra-long-winged." The extra-long-winged form can fly for long periods, but the long-winged form appears to be capable of only very short flights (A. D. Simmons, personal communication). The much stronger flying extra-long-winged forms now occur at increased frequency in the recently colonized populations at the expanding range boundary (Fig. 6.2a). In the case of Roesel's bush cricket, *Metrioptera roeselii*, the expansion rate of about 10 km per year was particularly perplexing because virtually all individuals of this species that had ever been seen in Britain (and elsewhere in Europe) had wings that came only halfway down the abdomen. These short wings were entirely useless for flight, and it is difficult to imagine the insect walking across a field, let alone 10 km. However, flight-worthy long-winged forms are now regularly observed in the most recently founded populations (Fig. 6.2b), close to the expanding range boundary (Thomas et al. 2001). New populations must be founded by rare long-winged individuals (and by extra-long-winged individuals in the cone-head), applying selection that favors genotypes that produce high proportions of long-winged individuals. The proportion of long-winged individuals in the population then declines as the time since population establishment increases, presumably because the short-winged individuals breed at a higher rate. Work on the basis of differences in wing morph frequencies is still underway, but preliminary evidence suggests that there is at least some genetic basis for the differences observed among the populations (A. D. Simmons, personal communication).

Studies of two species of butterflies have produced comparable results. In butterflies, the thorax consists largely of flight muscles, and the abdomen consists mainly of reproductive structures. Therefore, differences in the relative size of the thorax and abdomen can be regarded as an approximate indicator of a butterfly's investment in flight versus reproduction. In the speckled wood *Pararge aegeria*, populations from areas of recent colonization are characterized by relatively large thoraxes, particularly in the females (Hill, Thomas, and Blakeley 1999). Females from expanding populations appear, morphologically, to be strong fliers. Because these insects were all reared in the same environment as one another in the laboratory, the differences between core and marginal populations are likely to represent genetic differences in morphology. In a similar study, Hill, Thomas, and Lewis (1999) reared silver-spotted skippers *Hesperia comma* in a common environment and examined the relative size of their thoraxes and abdomens. The butterflies that came from the area with the greatest rate of range expansion had the largest thoraxes (most

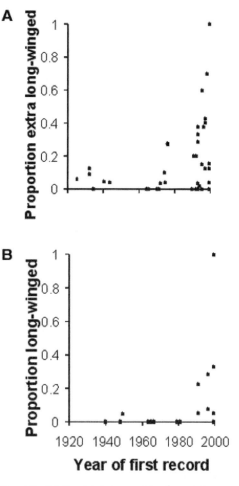

Figure 6.2. Relationship between date of population establishment and the proportion of (a) extra-long-winged (*Conocephalus discolor*) and (b) long-winged (b, *Metrioptera roeselii*) bush crickets. Recently colonized populations at the expanding range margins contain increased fractions of flying individuals. Source: Modified from Thomas et al. (2001), courtesy of A. D. Simmons.

flight-worthy), and those from the region where the distribution was most stable had the smallest thoraxes (Hill, Thomas, and Lewis 1999).

In each of these examples, more dispersive morphs are found at the expanding margins of the species' geographic distributions. The differences appear to stem, at least partly, from evolutionary changes that have arisen during range expansion.

Not only may dispersal capacity evolve, but the direction of travel may also come under selection. The best example comes from the blackcap, *Sylvia atricapilla*, a small European bird with a variable migration strategy and a genetically controlled direction of migration (Berthold et al. 1992; Berthold and Pulido 1994; Berthold 1998). Most of the birds breeding in Germany migrate in a southwesterly direction, toward Spain and Portugal. However, from the 1960s onward, increasing numbers of blackcaps have been observed overwintering in Britain. The percentage of banded birds from parts of Germany and Austria found wintering in Britain rose from 0 percent before 1960 to around 10 percent by 1990, with these individuals spending the winter some 1500 km farther north than most of the rest of the population. Birds caught overwintering in Britain were bred in captivity by Berthold and his colleagues, who found that the naïve offspring of these birds showed a west-northwesterly migratory direction. This direction would take them from southern Germany to Britain. They were genetically programmed to migrate in this new direction, and the shift in destination represented an evolutionary change. During this period, British winters have become milder and shorter, and climate change seems to be the most plausible explanation underlying the 1500-km northward shift in migratory destination.

Habitat Choice and Range Expansion

The ability of a species to expand into regions that have recently become climatically suitable is likely to depend jointly on the rate of dispersal and on the availability of habitat for the species to colonize. To a large extent, habitat availability will depend on human activities and land uses that are unrelated to climate. However, individuals and populations vary in their specificity of habitat and resource use. Any traits that allow a species to occupy a larger number of habitats or a habitat that is particularly common in the area of range expansion would be likely to increase the rate of spread (Thomas et al. 2001).

The brown argus butterfly, *Aricia agestis*, expanded northward at great speed between 1970–82 and 1995–99 (Asher et al. 2001). This was achieved by a switch in habitat in central England, away from traditional chalk grassland habitats where it lays its eggs on rock rose plants *Helianthemum nummularium*, to colonize field margins, waste ground, grassy woodland paths, and various other habitats (Thomas et al. 2001). These new habitats were colonized by a form of the butterfly that preferred to lay eggs on *Geranium molle*, a plant that was widely distributed in these new habitats (Thomas et al. 2001; E. J. Bodsworth, personal communication). Prior to the expansion, some populations of this butterfly in Britain did naturally lay eggs on *Geranium* plants, but the proportion of *Geranium*-feeding populations has increased dramatically over the past 20 years. This represents an evolutionary change in the frequency of populations with different host choices.

Responses of Different Species

It should be appreciated, however, that many species will not evolve fundamentally increased dispersal rates or habitat changes, and that dispersal and habitat availability are likely to remain major constraints limiting the ability of many or most species to colonize new areas of suitable climate space. Species that differ in their habitat and dispersal characteristics will differ in their abilities to respond to climate change. Evolutionary responses (defined broadly) to climate change include the differential survival of populations, subspecies, species, and higher taxonomic groupings, in addition to evolutionary change within individual populations.

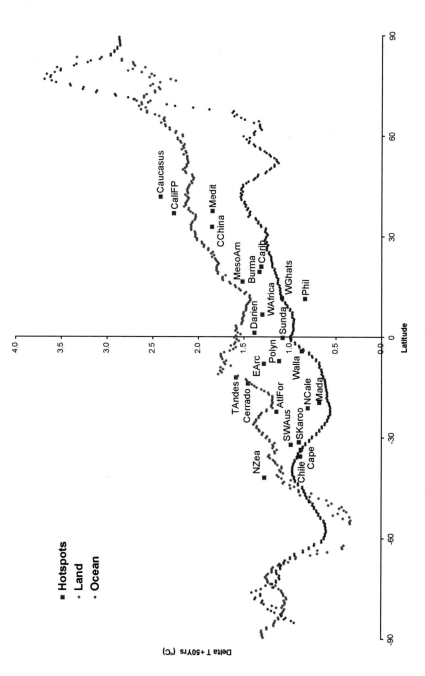

Plate 1. Latitudinal patterns of temperature change and biodiversity hotspots. Green dots indicate latitudinal mean temperature change over land as modeled by the HadCM3 AOGCM. Blue dots indicate latitudinal mean temperature change over ocean. Red squares indicate mean temperature change in biodiversity hotspots. The hotspots: AtlFor = Atlantic Forest of Brazil, Burma = Indo-Burma, CChina = Mountains of Central China, CaliFP = California Floristic Province, Cape = Cape Floristic Province, Carib = Caribbean, EArc = Eastern Arc Mountains of Tanzania, Mada = Madagascar, Medit = Mediterranean, NCale = New Caledonia, NZea = New Zealand, Phil = Philippines, Polyn = Polynesia, SWAus = Southwest Australia, SKaroo = Succulent Karoo, MesoAm = Mesoamerica, Sunda = Sundaland, TAndes = Tropical Andes, Walla = Wallacea, WAfrica = West Africa, WGhats = Western Ghats[o]

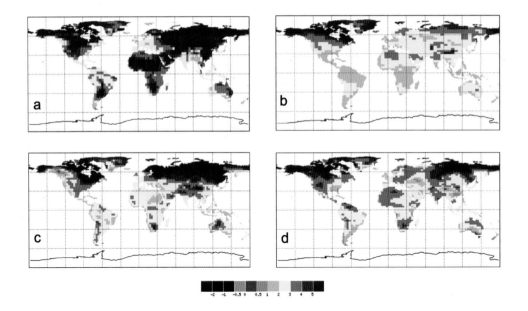

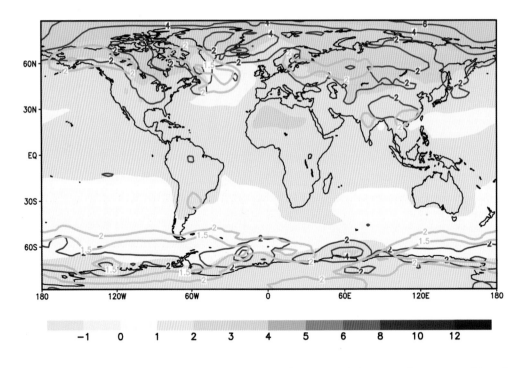

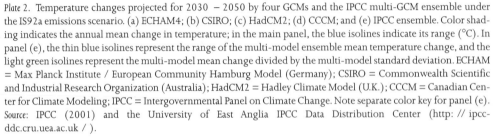

Plate 2. Temperature changes projected for 2030 − 2050 by four GCMs and the IPCC multi-GCM ensemble under the IS92a emissions scenario. (a) ECHAM4; (b) CSIRO; (c) HadCM2; (d) CCCM; and (e) IPCC ensemble. Color shading indicates the annual mean change in temperature; in the main panel, the blue isolines indicate its range (°C). In panel (e), the thin blue isolines represent the range of the multi-model ensemble mean temperature change, and the light green isolines represent the multi-model mean change divided by the multi-model standard deviation. ECHAM = Max Planck Institute / European Community Hamburg Model (Germany); CSIRO = Commonwealth Scientific and Industrial Research Organization (Australia); HadCM2 = Hadley Climate Model (U.K.); CCCM = Canadian Center for Climate Modeling; IPCC = Intergovernmental Panel on Climate Change. Note separate color key for panel (e). *Source:* IPCC (2001) and the University of East Anglia IPCC Data Distribution Center (http: // ipcc-ddc.cru.uea.ac.uk /).

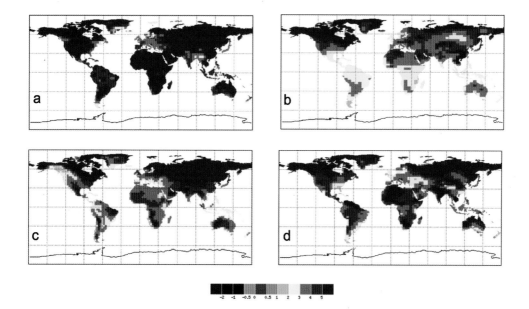

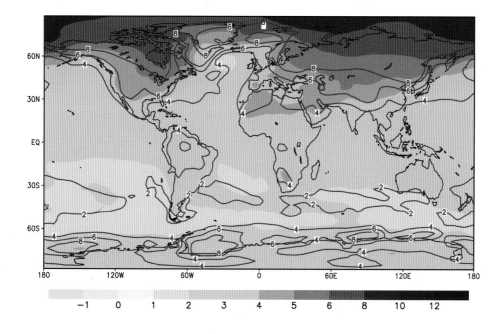

Plate 3. Temperature changes projected for 2080 − 2100 by four GCMs and the IPCC multi-GCM ensemble under IS92a or SRES A2 scenarios. (a) ECHAM4; (b) CSIRO; (c) HadCM2; (d) CCCM; and (e)IPCC ensemble. Color shading indicates the annual mean change in temperature; in the main panel, the blue isolines indicate its range (°C). The IS92a and SRES A2 scenarios project trends in social factors governing greenhouse gas production that are relatively similar to those of the present, and assume that there is no active effort to curtail greenhouse gas release. ECHAM = Max Planck Institute / European Community Hamburg Model (Germany); CSIRO = Commonwealth Scientific and Industrial Research Organization (Australia); HadCM2 = Hadley Climate Model (U.K.); CCCM = Canadian Center for Climate Modeling; IPCC = Intergovernmental Panel on Climate Change. Source: IPCC (2001) and the University of East Anglia IPCC Data Distribution Center (http: // ipcc-ddc.cru.uea.ac.uk).

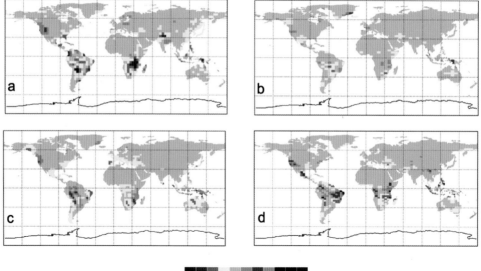

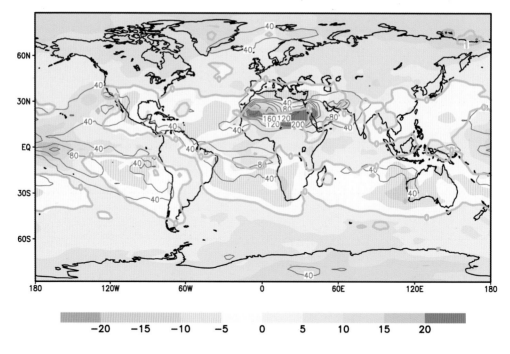

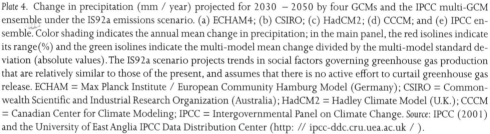

Plate 4. Change in precipitation (mm / year) projected for 2030 − 2050 by four GCMs and the IPCC multi-GCM ensemble under the IS92a emissions scenario. (a) ECHAM4; (b) CSIRO; (c) HadCM2; (d) CCCM; and (e) IPCC ensemble. Color shading indicates the annual mean change in precipitation; in the main panel, the red isolines indicate its range (%) and the green isolines indicate the multi-model mean change divided by the multi-model standard deviation (absolute values). The IS92a scenario projects trends in social factors governing greenhouse gas production that are relatively similar to those of the present, and assumes that there is no active effort to curtail greenhouse gas release. ECHAM = Max Planck Institute / European Community Hamburg Model (Germany); CSIRO = Commonwealth Scientific and Industrial Research Organization (Australia); HadCM2 = Hadley Climate Model (U.K.); CCCM = Canadian Center for Climate Modeling; IPCC = Intergovernmental Panel on Climate Change. *Source:* IPCC (2001) and the University of East Anglia IPCC Data Distribution Center (http: // ipcc-ddc.cru.uea.ac.uk /).

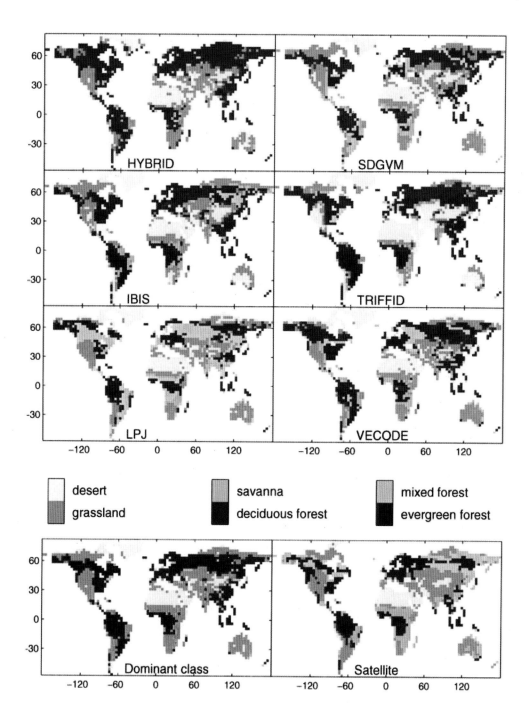

Plate 5. Comparison of vegetation type simulated by six dynamic global vegetation models (DGVMs), in comparison with the vegetation class on which most models agreed ("Dominant class," lower left) and vegetation derived from satellite image analysis ("Satellite," lower right). *Source:* Cramer et al. (2001). Copyright Blackwell Scientific Publishers. Reprinted with permission.

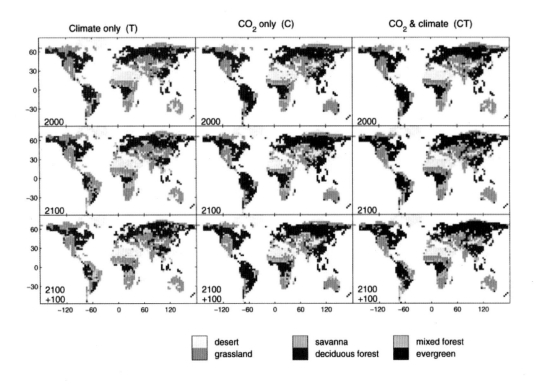

Plate 6. Ensemble DGVM comparison of the effects of climate change only, CO_2 change only, and climate change and CO_2 change combined. Vegetation types are the "Dominant class" as determined from an ensemble of projections (see Plate 5, lower left panel). *Source*: Cramer et al. (2001). Copyright Blackwell Scientific Publishers. Reprinted with permission.

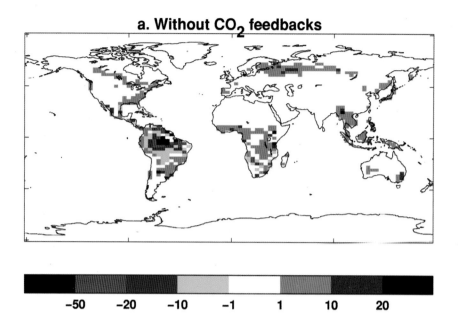

a. Without CO₂ feedbacks

−50	−20	−10	−1	1	10	20

b. With CO₂ feedbacks

−50	−20	−10	−1	1	10	20

Plate 7. Global vegetation change incorporating the effects of carbon-cycle feedbacks. (a) Change in broadleaf tree cover from 1860 − 2100 simulated by the HadCM3LC coupled climate − carbon-cycle model (Cox et al. 2000) without CO_2-climate feedbacks. (b) Further changes when CO_2-climate feedbacks are included. Note the severe changes in forest cover in the Amazon in panel (b). (See also Betts and Shugart, this volume.)

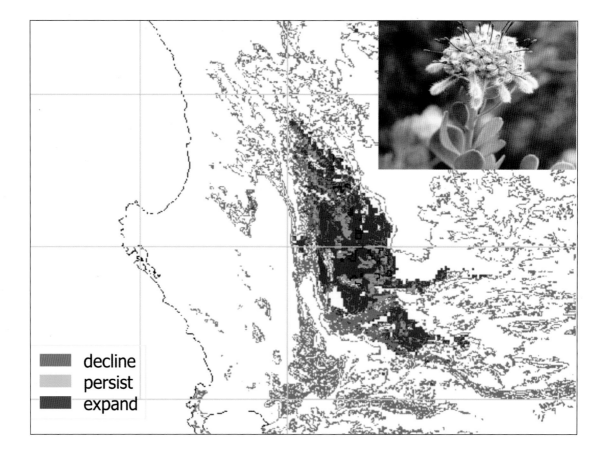

Plate 8. Range shift model of *Vexatorella amoena* (inset), a member of the Protea family whose range is found in the mountains above Cape Town, South Africa. Green denotes range retained in 2050, red denotes range projected to be lost in 2050, and blue denotes areas in which newly suitable climatic conditions appear for the species in 2050.

Populations and species with specific characteristics may be disproportionately threatened. Species with high climatic tolerances, (which can occur in a wide variety of habitats) that occupy widespread habitats, and that disperse readily seem likely to prosper, whereas sedentary habitat specialists might be expected to suffer, failing to track the changing distribution of suitable conditions. Within the British butterfly fauna, habitat specialists and sedentary species have continued to decline over the past 20 to 30 years, through ongoing land use changes, whereas half of the more mobile species and habitat generalists have extended their ranges northward (Warren et al. 2001). The average characteristics of species is already changing in response to the combined effects of climate change and landscape modification.

THREATS TO GENETIC DIVERSITY FROM CLIMATE WARMING

Biodiversity implies all levels of genetic variation: genetic diversity within populations, differences among populations, and differences among geographic races and subspecies, as well as genetic diversity among species (see Chapter 1). Changes to variation within populations and the extinction of geographic races and subspecies represent potential evolutionary responses to climate change. To date, research on threats to genetic variation in conservation biology has concentrated on the effects of habitat destruction and hybridization (e.g., Rhymer and Simberloff 1996; Avise et al. 1997; Saccheri et al. 1998; Frankham 1999). Climate warming is probably a much greater threat than either of these, with climate change disproportionately causing extinction of the parts of species' ranges that contain the greatest genetic diversity. This section concentrates on the north temperate zone, but many of the same general arguments

might also apply in the tropics, with some modification.

In the last glacial period, many temperate-zone animals and plants occurred at lower latitudes, shifting their ranges to higher latitudes after the retreat of the ice sheets (e.g., Coope 1978; Davis 1983; Huntley 1991; Hewitt 1999; Huntley, this volume). Most such species exhibit reduced genetic variation within populations and are divided into fewer subspecies and races at higher latitudes (e.g., Hewitt 1993, 1996, 1999; Stone and Sunnucks 1993; Cooper, Ibrahim, and Hewitt 1995; Green et al. 1996; Merilä, Björklund, and Baker 1997; Tolman 1997; Bernatchez and Wilson 1998; Broyles 1998; Stauffer et al. 1999; Descimon et al. 2001). Colonization northward by small numbers of individuals would have resulted in reduced levels of genetic variation being present within the new populations, and only a subset of the distinct races or subspecies present at lower latitudes managed to expand their distributions in response to the warmer climate: either another subspecies got there first, or they were unable to surmount some major geographical barrier (reviewed by Hewitt 1999; Hewitt and Nichols, this volume; Huntley, this volume).

Climate warming in the north temperate zone is already resulting in population-level extinctions and range retraction in the southern parts of the distributions of many species, and further expansions are taking place at many northern range boundaries. Genetically diverse populations and subspecies toward the warm margins of species ranges are, therefore, becoming extinct, whereas replacement populations are derived (through further founder events) from already genetically depauperate populations at the cool margins of species ranges. A loss of genetic variation is expected within many species (Descimon et al. 2001).

So far, the limited evidence that is available is consistent with these predicted patterns. For example, the expanding northern margin of the brown argus butterfly, *Aricia agestis*, that has been established in the past 15 years contains a subset of both the neutral mitochondrial DNA (I. Wynne, personal communication) and adaptive genetic variation (egg-laying preferences) present in the longer-established parts of the distribution from which the colonists were derived (Thomas et al. 2001). Meanwhile, in Mexico and southern California, the southernmost subspecies of Edith's checkerspot, *Euphydryas editha*, has recently disappeared from most of its remaining "suitable" habitats, most plausibly because of climate change; the entire subspecies may be doomed to extinction (Parmesan 1996; C. Parmesan, personal communication). Extremely localized high-elevation remnant populations of the bog fritillary butterfly, *Proclossiana eunomia*, in the Pyrenees and Cantabrian Mountains in southwestern Europe must also be seriously threatened: at 10 loci examined, these populations contained four alleles not known to occur anywhere else within the entire distribution of the species (Nève 1996; Descimon et al. 2001).

The ultimate extent of climate change will be critical. Existing interpretations usually argue that there is a zone of geographic overlap between glacial and interglacial distributions, situated in the low-latitude parts of the current species ranges (e.g., Hewitt 1993, 1999). This is where most genetic diversity is preserved. The crucial question is whether anthropogenic climate warming over the next 100-plus years will cause species distributions to disappear completely in these zones of overlap. If so, large amounts of genetic variation could be lost from many temperate-zone species, including the potential loss of many low-latitude geographic races and subspecies.

A priority is to establish whether analogous predictions will apply in tropical regions, where (less certain) changes in moisture gradients and seasonality are likely to be as or more important than changes in temperature.

SUMMARY

Current species assemblages, individual distributions, and local adaptations of populations reflect the cooler conditions that have predominated over the last million years (Overpeck et al., this volume). The effects of new global patterns of temperature and rainfall, combined with all the new interactions between species in new biological assemblages, will apply major selection pressures on almost every terrestrial species. Directly or indirectly, climate change is likely to dominate the evolutionary process over the next century and more. Species and populations that remain where they are will experience different climates and will live in biological communities with altered compositions. Species that shift their distributions will undergo evolutionary changes as they move, and they will experience new selection pressures as they enter new regions and encounter new species assemblages.

Ecological changes in phenology, abundances, and distributions are taking place almost everywhere one looks (Root and Hughes, this volume; Parmesan, this volume). It seems likely that evolutionary responses to changes in the climate and to climate-caused changes in community composition will *already* be equally widespread. The many climate-related extinctions and population collapses that have been observed in populations and metapopulations (Parmesan 1996; Thomas, Singer, and Boughton 1996; Hughes 2000; Walther et al. 2002; Parmesan, this volume) imply that selection must frequently be strong and directional. The limited number of clear examples of evolutionary responses to recent climate change simply reflects the large amount of detailed work required, first, to demonstrate that phenotypic changes have taken place and, second, to identify the extent to which genetic factors underlie these changes.

It is interesting to contemplate where all this variation comes from, and why it is maintained in existing populations. In many cases, variation is likely to be maintained by spatial variation in resources and microclimates, where variation in habitat choice, physiological responses, and dispersal capacity may have adaptive value within populations and metapopulations. However, a more intriguing possibility is that some of the variation might actually be left over from previous climate shifts. For example, a flight-polymorphic insect might have colonized its current geographic range by flight 10,000 years ago when the climate warmed (or approximately 5000 years ago when humans opened up new habitats). Once the new range was established, selection would then have acted against flying insects, because flight incurs reduced fecundity. Transient changes in morphological traits would be exceptionally difficult to detect in the fossil record, and brief behavioral and physiological changes could not be detected at all. Once virtually all individuals were flightless, selection against any recessive traits that permit flight would have been extremely low: residual variation is likely to have remained within large populations ever since. However, once colonization of new environments becomes possible again, colonization will be achieved by those exceptionally rare individuals that can fly, that are homozygous at some flight-related loci. This overcomes the problem normally associated with selecting for recessive traits when they are rare (i.e., interbreeding with individuals homozygous for the dominant allele), because the homozygous recessive individuals move away from the original population. In a hypothetical population where flight was governed by a single recessive allele and mating only took place after flight, all individuals in the new pop-

ulation would be able to fly (if all female and male founders would have been homozygous for flight ability); one quarter of the new population would be able to fly if it was founded by females that mated before dispersing (assuming flying females mated with males homozygous for flightlessness; this would occur in the second generation—the first generation would all be heterozygous and flightless).

Despite the potential for rapid evolutionary change, most of the observed evolutionary events represent changes within an apparently constrained phenotypic envelope, beyond which it seems difficult for a species to evolve. There are various genetic reasons why this could be so, including lack of appropriate genetic variation, and gene flow from other populations where different adaptations predominate (Butlin et al. 2003). It is likely that most populations will be showing climate- and weather-related evolutionary responses most of the time, yet the overall set of conditions that can be occupied by an entire species may remain more constant. If this paradigm holds during a period of widespread and rapid environmental change, it means that few species will manage to "evolve themselves out of trouble" in areas where the climate deteriorates beyond the range of conditions currently exploited by that species.

This contrast between population- and species-level responses to climate has practical implications for the management of biodiversity. Rapid evolutionary changes within populations and during range expansion are sufficient to generate major ecological changes. Therefore, everything from habitat management to population viability analyses may have to be reappraised in the light of evolutionary as well as ecological responses to climate. But the relative stability of the phenotypic envelope of an entire species may still permit predictions to be made of the future distribution of suitable climates for a species (given an accurate climate projection). The major caveats are that: (1) Many of the new climates will include *combinations* of temperature, precipitation, seasonality, and day length that do not currently exist anywhere on Earth (or at least in the same continent), and all these new climates will be combined with unprecedented CO_2 concentrations in the atmosphere. Something will live in these nonanalogue climates, but it is difficult to guess what. (2) Species will often live in biological assemblages for which there is no current analogue. Evolutionary changes in these new assemblages may play a critical role in determining species' future distributions and patterns of abundance.

ACKNOWLEDGMENTS

I am particularly grateful to Mike Singer, Ted Bodsworth, Adam Simmons, Niklas Wahlberg, Graham Stone, and Jane Hill, and to Godfrey Hewitt, Roger Butlin, Lee Hannah, and Tom Tregenza. NERC grants and EU TMR funding have supported this work.

REFERENCES

Ansart, A., P. Vernon, and J. Daguzan. 2001. Photoperiod is the main cue that triggers supercooling ability in the land snail, Helix aspersa (Gastropoda: Helicidae). Cryobiology 42:266–273.

Asher, J., M. Warren, R. Fox, P. Harding, G. Jeffcoate, and S. Jeffcoate. 2001. The Millennium Atlas of Butterflies in Britain and Ireland. Oxford: Oxford University Press.

Avise, J. C., P. C. Pierce, M. J. van den Avyle, M. H. Smith, W. S. Nelson, and M. A. Asmussen. 1997. Cytonuclear introgressive swamping and nuclear turnover of bass after an introduction. Journal of Heredity 88:14–20.

Bale, J. S., G. J. Masters, I. D. Hodkinson, C. Awmack, T. M. Bezemer, V. K. Brown, J. Butterfield, A. Buse, J. C. Coulson, J. Farrar, J. E. G. Good, R. Harrington, S. Hartley, T. H. Jones, R. L. Lindroth, M. C. Press, I. Symrnioudis, A. D. Watt, and J. B. Whittaker. 2002. Herbivory in global climate change research: direct effects of rising temperature on insect herbivores. Global Change Biology 8:1–16.

Bernatchez, L., and C. C. Wilson. 1998. Comparative phylogeography of nearctic and palearctic fishes. Molecular Ecology 7:431–452.

Berthold, P. 1998. Bird migration: Genetic programs with high adaptability. Zoology—Analysis of Complex System 101:235–245.

Berthold, P., A. J. Helbig, G. Mohr, and U. Querner. 1992. Rapid microevolution of migratory behaviour in a wild bird species. Nature 360:668–670.

Berthold, P., and F. Pulido. 1994. Heritability of migratory activity in a natural bird population. Proceedings of the Royal Society of London, Series B 257:311–315.

Billings, W. D. 1952. The environmental complex in relation to plant growth and distribution. Quarterly Review of Biology 27:251–265.

Blackenhorn, W. U., and D. J. Fairbairn. 1995. Life-history adaptation along a latitudinal cline in the water strider Aquarius remigis (Heteroptera, Gerridae). Journal of Evolutionary Biology 8:21–41.

Bradshaw, W. E., and C. M. Holzapfel. 2001. Genetic shift in photoperiodic response correlated with global warming. Proceedings of the National Academy of Sciences, USA 98:14509–14511.

Broyles, S. B. 1998. Postglacial migration and the loss of allozyme variation in northern populations of Asclepias exaltata (Asclepiadaceae). American Journal of Botany 85:1091–1097.

Bryant, S. R., C. D. Thomas, and J. S. Bale. 1997. Nettle-feeding nymphalid butterflies: Temperature, development and distribution. Ecological Entomology 22:390–398.

Bryant, S. R., C. D. Thomas, and J. S. Bale. 2002. The influence of thermal ecology on the distribution of three nymphalid butterflies. Journal of Applied Ecology 39:43–55.

Butlin, R. K., J. R. Bridle, and M. Kawata. 2003. Genetics and the boundaries of species' distributions. In Macroecology: Concepts and Consequences, T. Blackburn and K. J. Gaston, eds., pp. 274–295. Oxford: Blackwell Science.

Cody, M. L., and J. M. Overton. 1996. Short-term evolution of reduced dispersal in island plant populations. Journal of Ecology 84:53–61.

Coope, G. R. 1978. Constancy of insect species versus inconstancy in Quaternary environments. In: Diversity of Insect Faunas, L. A. Mound and N. Waloff, eds., pp. 176–187. Oxford: Blackwell.

Cooper, S. J. N., K. M. Ibrahim, and G. M. Hewitt. 1995. Post-glacial expansion and genome subdivision in the European grasshopper Chorthippus parallellus. Molecular Ecology 4:49–60.

Cox, P. M., R. A. Betts, C. D. Jones, S. A. Spall, and I. J. Totterdell. 2000. Acceleration of global warming due to carbon-cycle feedbacks in a coupled climate model. Nature 408:184–187.

Cwynar, L. C., and G. M. MacDonald. 1987. Geographical variation of lodgepole pine in relation to population history. American Naturalist 129:463–469.

Davis, M. B. 1983. Holocene vegetational history of the eastern United States. In: *Late Quaternary environments of the United States, vol. 2: The Holocene*, H. E. Wright, Jr., ed., pp. 166–181. Minneapolis: University of Minnesota Press.

Dennis, R. L. H. 1993. *Butterflies and Climate Change.* Manchester: Manchester University Press.

Dennis, R. L. H., and T. G. Shreeve. 1989. Butterfly wing morphology variation in the British Isles: The influence of climate, behavioural plasticity and the hostplant habitat. *Biological Journal of the Linnean Society* 38:323–348.

Descimon, H., M. Zimmermann, E. Cosson, B. Barascud, and G. Nève. 2001. Genetic variation, geographic variation and gene flow in some French butterfly species. *Genetics Selection Evolution* 33:S223–S249.

Fitter, A. H., and R. S. R. Fitter. 2002. Rapid changes in flowering time in British plants. *Science* 296:1689–1691.

Frankham, R. 1999. Resolving conceptual issues in conservation genetics: The roles of laboratory species and meta-analysis. *Hereditas* 130:195–201.

Green, D. M., T. F. Sharbel, J. Kearsley, and H. Kaiser. 1996. Postglacial range fluctuation, genetic subdivision and speciation in the western North American spotted frog complex, *Rana pretiosa*. *Evolution* 50:374–390.

Haes, E. C. M., and P. T. Harding. 1997. *Atlas of the Grasshoppers, Crickets and Allied Insects in Britain and Ireland*. London: The Stationery Office.

Hard, J. J., W. E. Bradshaw, and C. M. Holzapfel. 1993. Genetic co-ordination of demography and phenology in the pitcher-plant mosquito, *Wyeomyia smithii*. *Journal of Evolutionary Biology* 6:707–723.

Hewitt, G. M. 1993. Post-glacial distribution and species substructure: Lessons from pollen, insects and hybrid zones. In: *Evolutionary Patterns and Processes*, D. R. Lees and D. Edwards, eds. The Linnean Society of London Symposium Series, 14:97–123.

Hewitt, G. M. 1996. Some genetic consequences of ice ages, and their role in divergence and speciation. *Biological Journal of the Linnean Society* 58:247–276.

Hewitt, G. M. 1999. Post-glacial re-colonization of European biota. *Biological Journal of the Linnean Society* 68:87–112.

Hill, J. K., C. D. Thomas, and D. S. Blakeley. 1999. Evolution of flight morphology in a butterfly that has recently expanded its geographic range. *Oecologia* 121:165–170.

Hill, J. K., C. D. Thomas, and O. T. Lewis. 1999. Flight morphology in fragmented populations of a rare British butterfly, *Hesperia comma*. *Biological Conservation* 87:277–284.

Huey, R. B., G. W. Gilchrist, M. L. Carlson, D. Berri-gan, and L. Serra. 2000. Rapid evolution of a geographic cline in size in an introduced fly. *Science* 287:308–309.

Hughes, L. 2000. Biological consequences of global warming: Is the signal already apparent? *Trends in Ecology and Evolution* 15:56–61.

Huntley, B. 1991. How plants respond to climate change: Migration rates, individualism and the consequences for plant communities. *Annals of Botany* 67:15–22.

Intergovernmental Panel on Climate Change. 2001. *Climate Change 2001: The Scientific Basis*. http://www.grida.no/climate/ipcc—tar/

Karlsson, M. G., and J. W. Werner. 2002. Photoperiod and temperature affect flowering in German primrose. *Horttechnology* 12:217–219.

Knutti, R., T. F. Stocker, F. Joos, and G. K. Plattner. 2002. Constraints on radiative forcing and future climate change from observations and climate model ensembles. *Nature* 416:719–723.

Lair, K. P., W. E. Bradshaw, and C. M. Holzapfel. 1997. Evolutionary divergence of the genetic architecture underlying photoperiodism in the pitcher-plant mosquito, *Wyeomyia smithii*. *Genetics* 147:1873–1883.

Loeschcke, V., J. Bundgaard, and J. S. F. Barker. 2000. Variation in body size and life history traits in *Drosophila aldrichi* and *D. buzzatii* from a latitudinal cline in eastern Australia. *Heredity* 85:423–433.

Marshall, J. A., and E. C. M. Haes. 1988. *Grasshoppers and Allied Insects of Great Britain and Ireland*. Colchester: Harley Books.

McColl, G., and S. W. McKechnie. 1999. The *Drosophila* heat shock hsr-omega gene: An allele frequency cline detected by quantitative PCR. *Molecular Biology and Evolution* 16:1568–1574.

Merilä, J., M. Björklund, and A. J. Baker. 1997. Historical demography and present day population structure of the Greenfinch, *Carduelis chloris*—An analysis of mtDNA control-region sequences. *Evolution* 51:946–956.

Nève, G. 1996. *Dispersion chez une espèce à habitat fragmenté: Proclossiana eunomia (Lepidoptera, Nymphalidae)*. Ph.D. Dissertation. Université Catholique de Louvain, Louvain-la-Neuve.

Noda, H. 1992. Geographic variation of nymphal diapause in the small brown planthopper in Japan. *Jarq-Japan Agricultural Research Quarterly* 26:124–129.

Parmesan, C. 1996 Climate and species' range. *Nature* 382:765–766.

Peterson, A. T., M. A. Ortega-Huerta, J. Bartley, V. Sanchez-Cordero, J. Soberon, R. H. Buddemeier, and D. R. B. Stockwell. 2002. Future projections for Mexican faunas under global climate change scenarios. *Nature* 416:626–629.

Pigott, C. D. 1968. Biological flora of the British Isles: *Cirsium acaulon*. *Journal of Ecology* 56:597–612.

Pounds, J. A., M. P. L. Fogden, and J. H. Campbell. 1999. Biological response to climate change on a tropical mountain. *Nature* 398:611–615.

Pulido, F., P. Berthold, G. Mohr, and U. Querner. 2001. Heritability of the timing of autumn migration in a natural bird population. *Proceedings of the Royal Society of London, Series B.* 268:953–959.

Rhymer, J. M., and D. Simberloff. 1996. Extinction by hybridization and introgression. *Annual Review of Ecology and Systematics* 27:83–109.

Saboureau, M., M. Masson-Pevet, B. Canguilhem, and P. Pevet. 1999. Circannual reproductive rhythm in the European hamster (*Cricetus cricetus*): Demonstration of the existence of an annual phase of sensitivity to short photoperiod. *Journal of Pineal Research* 26:9–16.

Saccheri, I., M. Kuussaari, M. Kankare, P. Vikman, W. Fortelius, and I. Hanski. 1998. Inbreeding and extinction in a butterfly metapopulation. *Nature* 392:491–494.

Singer, M. C., and C. Parmesan. 1993. Sources of variation in patterns of plant-insect interaction. *Nature* 361:251–253.

Singer, M. C., R. A. Moore, and D. Ng. 1991. Genetic variation in oviposition preference between butterfly populations. *Journal of Insect Behaviour* 4:531–535.

Singer, M. C., D. Ng, and C. D. Thomas. 1988. Heritability of oviposition preference and its relationship to offspring performance within a single insect population. *Evolution* 42:977–985.

Singer, M. C., and C. D. Thomas. 1996. Evolutionary responses of a butterfly metapopulation to human and climate-caused environmental variation. *American Naturalist* 148:S9–S39.

Singer, M. C., C. D. Thomas, H. L. Billington, and C. Parmesan. 1994. Correlates of speed of evolution of host preference in a set of twelve populations of the butterfly *Euphydryas editha*. *Écoscience* 1:107–114.

Singer, M. C., C. D. Thomas, and C. Parmesan. 1993. Rapid human-induced evolution of insect-host associations. *Nature* 366:681–683.

Stauffer, C., F. Lakatos, and G. M. Hewitt. 1999. Phylogeography and postglacial colonization routes of *Ips typographus* L-(Coleoptera, Scolytidae). *Molecular Ecology* 8:763–773.

Stone, G. N., and P. J. Sunnucks. 1993. The population genetics of an invasion through a patchy environment: The cynipid gallwasp *Andricus quercuscalicis*. *Molecular Ecology* 2:251–268.

Takeda, M., and S. D. Skopik. 1997. Photoperiodic time measurement and related physiological mechanisms in insects and mites. *Annual Review of Entomology* 42:323–349.

Taylor, F. 1981. Ecology and evolution of physiological time in insects. *American Naturalist* 117:1–23.

Thomas, C. D., E. J. Bodsworth, R. J. Wilson, A. D. Simmons, Z. G. Davies, M. Musche, and L. Conradt. 2001. Ecological and evolutionary processes at expanding range margins. *Nature* 411:577–581.

Thomas, C. D., M. C. Singer, and D. A. Boughton. 1996. Catastrophic extinction of population sources in a butterfly metapopulation. *American Naturalist* 148:957–975.

Thomas, J. A. 1993. Holocene climate changes and warm man-made refugia may explain why a 6th of British butterflies possess unnatural early-successional habitats. *Ecography* 16:278–284.

Thomas, J. A., R. J. Rose, R. T. Clarke, C. D. Thomas, and N. R. Webb. 1999. Intraspecific variation in habitat availability among ectothermic animals near their climatic limits and their centres of range. *Functional Ecology* 13(Suppl. 1):55–64.

Tolman, T. 1997. *Butterflies of Britain and Europe.* London: HarperCollins.

Turnock, W. J., and G. Boivin. 1997. Inter- and intrapopulation differences in the effects of temperature on postdiapause development of *Delia radicum*. *Entomologia Experimentalis et Applicata* 84:255–265.

Walther, G. R., E. Post, P. Convey, A. Menzel, C. Parmesan, T. J. C. Beebee, J. M. Fromentin, O. Hoegh-Guldberg, and F. Bairlein. 2002. Ecological responses to recent climate change. *Nature* 416:389–395.

Warren, M. S., J. K. Hill, J. A. Thomas, J. Asher, R. Fox, B. Huntley, D. B. Roy, M. G. Telfer, S. Jeffcoate, P. Harding, G. Jeffcoate, S. G. Willis, J. N. Greatorex-Davies, D. Moss, and C. D. Thomas. 2001. Rapid responses of British butterflies to opposing forces of climate and habitat change. *Nature* 414:65–69.

White, R. R., and M. C. Singer. 1974. Geographical distribution of hostplant choice in *Euphydryas editha*. *Journal of the Lepidopterists' Society* 28:103–107.

Widgery, J. 2000. Orthoptera Recording Scheme for Britain and Ireland. *Newsletter* 26, Biological Records Centre, Huntingdon.

Zera, A. J., and R. F. Denno. 1997. Physiology and ecology of dispersal polymorphisms in insects. *Annual Review of Entomology* 42:207–230.

Zwaan, B. J., R. B. R. Azevedo, A. C. James, J. Van 'T Land, and L. Partridge. 2000. Cellular basis of wing size variation in *Drosophila melanogaster*: A comparison of latitudinal clines on two continents. *Heredity* 84:338–347.

PART III

Learning from the Past

Large, abrupt changes were observed in ice cores raised in Greenland and Antarctica during the 1990s, alerting biologists to a climatic past in which rapid change was common and often large. This substantially changed views of past climate change and biotic response. If rapid climate change was common in the past, how did biota endure with so few extinctions? Further, it changed views of the relationship between past changes and possible future change. The rate of future change is an order of magnitude faster than the average change in the transition from the last glacial, but not so dissimilar to the rate of change that occurred over millennia or decades during abrupt events interspersed over the same period. At the same time, it had become well established that previous interglacials may have been slightly warmer than the present, so both the point of departure for future change and its rate may find some analogies in the past. These small revolutions in climatology gave paleoecology new relevance in understanding the possible effects of human-induced climate change.

The frequency of rapid change in the ice core records was also striking. The past 11,000 years appeared as a relatively warm, stable plateau in these records. If the future were like the past, biologists would need to prepare conservation strategies for rapid and large climatic fluctuations. Although there are several reasons the future will not be like the past— most notably the rising effect of human greenhouse gas emissions to the atmosphere—the future offered by human influence in the climate system offers more, not less, chance of rapid change. Whatever its magnitude, the greenhouse gas–forced warming in mean global temperature alone will be one of the most rapid ever experienced on earth, and it carries with it the possibility of triggering renewed natural abrupt change, by disrupting thermohaline circulation or other climate interconnections.

The other major difference between past and future change is that future warming will take place in the context of a global climate that is in one of its warmest phases in 2 million years or more. So precise analogies to the past are unlikely—most rapid natural warming has taken place in transitions from cooler conditions.

Nonetheless, the past has many insights to offer with regard to the future. Rapid warming of the same order of magnitude as expected in the twenty-first century has occurred in the past. Conditions as warm or warmer than the present have existed in previous interglacials. Mechanisms that have triggered past natural rapid climate change may parallel processes under way due to anthropogenic change. Because the historical record of present and recent past change is so short, looking at proxies of changes further back in time is one of the best lines of evidence available for understanding the future.

The first chapter in this part provides an overview of past climatic change as context for understanding past biotic change. The following chapters review evidence for biological responses to past climate change—primarily species-level responses—broken down by various types of biological systems (temperate, tropical, freshwater, marine). The part concludes with a review of genetic signatures of climate change, which have the potential to alter species' responses. The evidence these authors present indicates the breadth and depth of insight to the future that may be presented by lessons from the past.

CHAPTER SEVEN

A "Paleoperspective" on Climate Variability and Change

JONATHAN OVERPECK, JULIA COLE,
AND PATRICK BARTLEIN

Understanding past climate change is central to understanding possible future impacts on biodiversity. For instance, it is from our understanding of past biotic response to climate change that we believe species will move independently, and not as coherent communities, in response to future change. Thus, while future change may be quite different from past changes, inferences on mechanisms of response, speed of response, and other factors may all be drawn from the study of past changes. This chapter highlights lessons learned from the paleoclimate record that are relevant to anticipating future climate change and its ecological impacts. We focus on the most recent record (the past 10,000–100,000 years) because the data are richest and best constrained in this period, but we briefly describe older changes as well.

Many aspects of climate variability can be investigated using paleoenvironmental proxies derived from sources such as tree rings, corals, ice cores, and lake and marine sediments. These proxies allow the study of decadal to millennial climate variability, documenting how the climate system works naturally in the absence of anthropogenic forcing such as increased concentrations of atmospheric trace gases or aerosols. These proxies provide the only way to observe and understand the propensity of the Earth's climate system to shift abruptly in ways that could, if repeated in the future, have large impacts on both societies and ecosystems. Finally, the paleoenvironmental perspective provides the only way to study how the climate system responds to large changes in climate forcing similar in magnitude to those that will occur over the next 100 or more years.

One important paleoclimatic approach uses fossils to infer how climates have changed in the past. The abundances of many species or genera on land, in lakes, and in the ocean vary in response to changing climate, and fossil records of these variations can be used to infer how climate changed in the past. The most widely used fossils are ones of small size, because they are often abundant in lake and marine sediments. Large numbers mean that powerful quantitative (statistical) approaches can be used. Well-defined methods exist to relate evidence of the present-day distribution and abundance of taxa (e.g., pollen, diatoms or foraminifera) to the present-day distribution of climate. Combining these relationships with known abundances of fossil taxa in an ancient sediment sample allows us to estimate the conditions (atmospheric, oceanographic, or limnological) in which the fossil assemblage lived. The quantitative methods for doing this kind of work are well developed (Bradley 1999).

Another common approach is to use the geochemistry of ancient sediment or fossils to infer past environmental conditions. The most widely used approaches include isotopic and trace element geochemistry. For example, in the hard parts of some fossils, the ratios of stable hydrogen or oxygen isotopes and the abundances of certain trace elements relate to the environment (e.g., temperature or salinity of water) in which the organisms lived. These geochemical methods are also widely used in other paleoclimatic archives, including ice cores (from polar ice caps and high-elevation glaciers at all latitudes), corals, and speleothems (cave formations that can grow slowly over thousands of years). For example, some species of coral grow continuously over several centuries, and geochemical methods can be used on both living heads and on fossil heads that date to much older intervals. Oxygen isotopes in the corals continuously track the temperature and salinity of the seawater in

which the coral grew, and can be typically resolved by the paleoclimatologist to about a monthly temporal resolution. Some trace elements also substitute for calcium in the calcium carbonate lattice of the coral as a function of temperature or of their concentration in seawater. Once these records are calibrated to modern environmental or climatic variations, it becomes possible to reconstruct sea surface conditions back hundreds to thousands of years.

Another well-known source of paleoenvironmental data is dendroclimatology, or the study of paleoclimate from tree rings. In this case, variations in the width, density, and geochemistry of annual rings provide ways to reconstruct a wide variety of past climate parameters, including seasonal temperature, precipitation, drought, and atmospheric circulation patterns. Moreover, because many trees and wood samples are preserved long after death, it is possible to extend some dendroclimatic records back many thousands of years. As with other sources of paleoenvironmental data that are dated to the year, it is straightforward to calibrate tree ring variations during the twentieth century directly against instrumental records of climate, and to use this calibrated relationship to turn past tree ring variations into quantitative estimates of past climate.

In contrast to the many sources of paleoenvironmental data that can be dated using annual rings, bands, or layers to yield chronologies that are accurate to the season or year (e.g., tree rings, corals, annually laminated sediments, selected ice cores, and some speleothems), many other types of data come from sources (e.g., nonlaminated sediments) that need to be dated using other methods. Among these methods, radiocarbon is the most widely used and understood, but it is useful only back to about 40,000 years; other methods include Uranium (U-series) dating, which is useful on certain carbonate materials that formed in the past ca. 250,000 years. Generally, the accuracy of time control tends to decrease further back in time, and available records become fewer. For this reason, and with some notable exceptions, spatial patterns of interannual to millennial climate variability are hard to map out prior to about 40,000 years ago.

The scientific literature can give the impression that there are almost as many ways to reconstruct past environmental conditions as there are paleoclimatologists. Some of the other innovative approaches include the use of ancient documentary or historical sources. In some countries (e.g., China) this approach can yield quantitative estimates of past climate back hundreds of years or more. Other scientists have studied inactive or "fossil" eolian (e.g., dune) landforms as evidence of former megadroughts (droughts that last a decade or longer). Inactive eolian landforms cover over 100,000 km^2 of the central United States, and many of these "paleodeserts," now underlying agricultural areas, were active at one or more times during the past 10,000 years (Forman et al. 2001). Glacial extent is another indicator of past climate. Just as the current worldwide retreat of alpine glaciers is a sign of ongoing global warming (e.g., Thompson et al. 2002), some paleoclimatologists have used former glacier sizes to reconstruct past climate conditions. For more about these and other innovative approaches to paleoclimatology, the reader should consult texts such as Crowley and North (1991), Parrish (1998), or Bradley (1999).

CLIMATE CHANGE IN LONG-TERM PERSPECTIVE

Global mean temperature and CO_2 have varied, often in close step, over hundreds of millions of years. These changes have been the focus of much research, because they indicate fundamental reorganizations among the diverse components of the earth's climate system that represent extreme possible climate states. The Earth has experienced major glacial ages at about 800–600 million years ago, 300 million years ago, and from 2 million years ago to the present (Crowley and North 1991). We are presently in a moderately warm phase (i.e., an interglacial period) that would be followed in thousands of years by another glacial phase if the pattern of the past 2 million years is not diverted by anthropogenic forcing. Major biologic changes have accompanied these changes in climate (Webb and Bartlein 1992), raising concern over the impacts of human-induced increases in atmospheric CO_2 and their associated climate change.

Earth history over the past 100 million years encompassed some of the largest long-term shifts in climate ever witnessed by the earth's biota. The late Cretaceous (c. 80 Ma) was thought to be ice-free, associated with a paleogeography quite different from today, and with CO_2 levels perhaps 8 times the modern atmosphere (Crowley and North 1991). Sixty-five million years ago, significant climate change and mass extinction resulted from a meteorite impact, which triggered sudden catastrophic changes (fires, tsunami, and immediate chilling) followed by more persistent climatic anomalies (Kring 2000). Since that time (Fig. 7.1), the evolution of climate included several thermal maxima. The first, at around 55 million years ago (the Late Paleocene), was apparently triggered by a rapid and large discharge of methane (a greenhouse gas) from marine sediments (Dickens et al. 1995; Zachos et al. 2001). The second, longer warm period in the early Eocene (50–53 million years ago) was associated with slower shifts in both atmospheric composition and plate tectonics (Zachos et al. 2001). Plate tectonic changes continued to drive fundamental changes in the geography of the Earth over the ensuing 50 million years, with warming and cooling resulting from changes in the configuration of oceans (e.g., opening of the

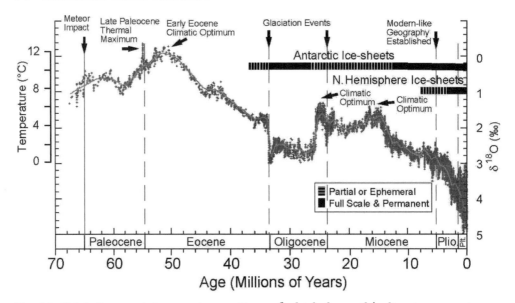

Figure 7.1. Global climate evolution over the past 65 million years from global deep-sea oxygen isotope records. The ^{18}O temperature scale is valid only for the period prior to the onset of major glaciation at ca. 35 million years, but the progressively greater glaciation after this time, first of Antarctica and then of the Northern Hemisphere, represents progressively cooler global temperatures. See text for details on the various climatic events that are highlighted. *Source:* Zachos et al. (2001). Copyright American Association for the Advancement of Science. Reprinted by permission.

Drake Passage), the uplift of mountain belts (e.g., the Tibetan Plateau), changes in the rate of seafloor spreading, and changes in atmospheric trace gas concentrations (Zachos et al. 2001). At about 5 million years ago, modern continental configurations were established. Well-understood processes led to the formation of ice sheets, first on Antarctica (at about 32 Ma) and subsequently in the northern hemisphere after about 5 million years ago.

The last significant warm period occurred about 3 million years ago, during the Pliocene, and thereafter the Earth cooled into a long period of relatively uniform glacial–interglacial oscillations from 700,000 years ago to the present (Imbrie et al. 1992, 1993). These variations define the envelope of natural climate variability familiar to most modern ecosystems, out

of which the earth's climate system is now being pushed (Fig. 7.2) (Overpeck et al. 2002). The history of climate over about the past 100 million years provides many lessons regarding long-term patterns and causes of change, but it falls short in providing strict analogs for the future because the earth's geography and climate forcing were different from those of today (Webb and Wigley 1985; Crowley 1990). The past climates most analogous to today's (i.e., the twentieth and twenty-first centuries) are those of the past 2 million years, characterized by alternation between glacial and interglacial states and a limited "envelope" of natural variability.

RAPID CLIMATE CHANGE

Although the causes of past climate change are often gradual (e.g., slow changes in radiative forcing associated with the Earth's orbital variations; Imbrie et al. 1993), paleoclimatic observations now make clear that climate often responds abruptly to these forcings (Alley et al. 2003). The concept of abrupt climate change stems directly from the discovery that major climate shifts (e.g., greater than 10°C) and ocean circulation reorganizations can occur over

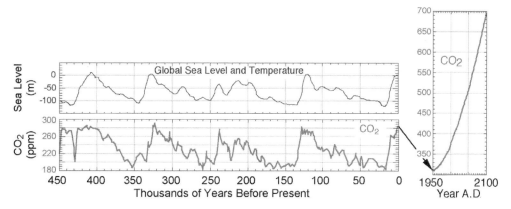

Figure 7.2. Variation in the global volume of continental ice sheets (a proxy for both global sea-level and temperature) and atmospheric CO_2 over the past 450,000 years, compared to possible CO_2 changes of the next 100 years. The amplitude of glacial to interglacial global temperature changes was ca. 5–7°C (Kutzbach et al. 1998). Glacial ice volumes were obtained using the global average deepwater $\delta^{18}O$ (Imbrie et al. 1992) scaled to a 20,000-year B.P. sea-level change equal to the observed value of 120 m (Fairbanks 1989). Past atmospheric CO_2 concentrations are from bubbles trapped in ice, as well as measured in air samples (Etheridge et al. 1996, Petit et al. 1999, IPCC 2001b).

short periods (e.g., 10 years) and have repeatedly done so in the North Atlantic region during the past 80,000 years. (Broecker 1987; Alley et al. 1993; Labeyrie et al. 2002) (Fig. 7.3).

The paleoclimatic records that appear in Figure 7.3 show several modes of variation, including, (a) the contrast between the generally warmer and less variable interglacial conditions after 11,000 years ago, and the colder, more variable glacial conditions before; (b) dozens of "millennial" time-scale variations in climate (those that take one to a few thousand years to complete) that are particularly evident during glacial times (i.e., "Dansgard-Oeschger," or "D-O," events) but are also apparent during the Holocene part of the record; and (c) several very large (in amplitude) and abrupt millennial time-scale variations, in particular the Younger-Dryas climate reversal between 13,000 and 11,500 years ago. The coherence among the records implied by the figure arises in

part from their location around the margins of the North Atlantic. Viewed globally, paleoclimatic records show evidence of both in-phase relationships between hemispheres, as for the transition between glacial and interglacial conditions, and out-of-phase or opposing relationships between hemispheres, as for the D-O oscillations (Clark et al. 2002). These differences between hemispheres in the phase or timing of abrupt variations, sometimes referred to as the "bipolar seesaw," are attributable to the variations in the relative importance of oceanic as opposed to atmospheric transmission of North Atlantic climatic variations elsewhere around the globe (Broecker et al. 1999; Stocker 2002; Morgan et al. 2002).

Although many of these abrupt shifts occurred during glacial periods, and thus likely required continental ice sheets either as a trigger or amplifier, it is clear that abrupt shifts in North Atlantic climate (and ocean circulation) are also possible during warm climates, including the future one. Several studies have suggested that in a warmer world, increased precipitation and/or Greenland Ice Sheet meltwater would reduce salinity in the North Atlantic sufficiently to significantly weaken North Atlantic thermohanline circulation (Manabe and Stouffer 1994; Stocker and Schmittner 1997; Rahmstorf 2000). The paleoperspective suggests that such an event would have a profound impact on temperature, precipitation, and perhaps

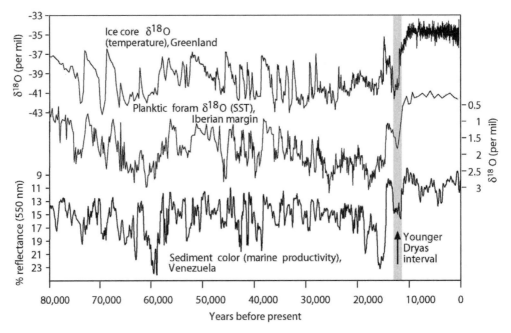

Figure 7.3. Comparison of North Atlantic millennial climate variability between 80,000 years ago and today, based on three well-known records: (a) oxygen isotopic content of ice from the GISP2 ice core, Greenland (Stuiver and Grootes 2000); (b) planktonic foraminiferal δ¹⁸O (plotted inverse) from a core off the Iberian margin, reflecting sea surface temperature—upward is warmer (Shackleton 2000): and (c) sediment color in a core from the Cariaco Basin, offshore Venezuela—darker sediments, plotted upward, indicate higher productivity based on increased nutrients from terrestrial sources (Peterson et al. 2000). The high amount of shared millennial-scale variability before 10,000 years ago is well studied but poorly understood. The last major millennial event was the "Younger-Dryas" cold event that occurred in the North Atlantic region just before the current interglacial period (the Holocene). Although the climate of Holocene was more stable than that of the glacial period in much of the North Atlantic region, this was not the case outside this region, particularly in monsoon-sensitive areas where substantial Holocene variability in precipitation is well documented.

circulation is usually accompanied by dramatic decreases in precipitation in northern South America (Hughen et al. 2000) and northwestern Africa (Gasse and van Campo 1994; Overpeck and Webb 2000). New data are emerging to support the hypothesis (Overpeck et al. 1996) that past abrupt changes in North Atlantic climate had a downstream impact as far as the summer monsoon rains in Tibet, China, and southern Asia (Porter and An 1995; Wang et al. 2001; Morrill et al. 2002; Gupta et al. 2003). The nature of this linkage (an increase in Eurasian snow cover leading to a weakened monsoon) suggests that an abrupt North Atlantic cooling in the future (e.g., IPCC 2001) could result in decreased summer monsoon rains over a significant part of Asia.

storm frequency downstream over Europe (Rahmstorf 2000; IPCC 2001b), as well as significant climatic effects in the tropics and Southern Hemisphere (Peterson et al. 2000).

A growing body of paleoclimatic data indicates that the impacts of an abrupt shift in high-latitude North Atlantic ocean

THE ENVELOPE OF NATURAL CLIMATE VARIABILITY

Paleoclimatic observations of rapid climate change and its worldwide impacts have forced reexamination of the comparative rate of human-induced climate change. Globally, the *rate* of human-induced change

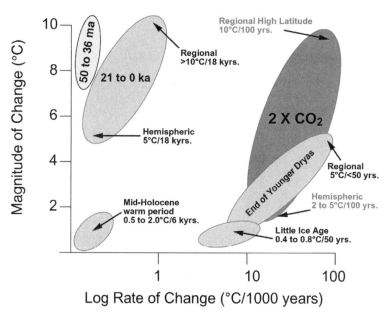

Figure 7.4. Summary comparison of the rates and magnitudes of possible future climate change (estimated in terms of mean annual temperature) with those associated with several well-known periods of past change in regions that were vegetated. The paleoperspective (see text) suggests that future anthropogenic warming might be large, and also that rates of future regional temperature change could far exceed any widespread change in the late Quaternary. *Source:* Overpeck et al. (2002).

is not likely to be an order of magnitude greater than any previous rate of natural change, as was previously believed. However, rates of change for many *regions* may exceed their past rates (see discussion below).

Absolute warming may quickly exceed levels experienced in the past 2 million years, however, indicating that there is less precedent for the magnitude of human-induced warming than for the rate of change. The most recent period that was warmer than present was probably the last interglacial period, which is known variously as the Eemian or Marine Isotope Stage 5e. Global mean temperature during the Eemian was probably little more than 1–2°C warmer than the present, but within this global figure, regional and seasonal variability was considerable. Modeling and paleodata indicate that Northern Hemisphere temperature changes were

greater, perhaps twice as high as the global mean over parts of northern Asia in summer, while Southern Hemisphere and winter changes were less or perhaps negative (Montoya et al. 2000). Associated biotic changes were consistent with these temperature changes, featuring distributional ranges in the Northern Hemisphere that often extended farther poleward than those at present.

Human-induced change may be placed in context with natural change through the concept of the "envelope of natural climate variability" (Overpeck et al. 2002). This envelope may be defined as the magnitude, rate, and "destination" of climate and atmospheric trace-gas composition change over the last several million years—the time over which many of the biotic systems on earth developed (Fig. 7.4). "Destination" refers to the configuration of the global climate forcing at a particular point in time, that is, past or future. Over the last several hundred thousand years, the magnitude, rate, and destinations of the climate system have, remarkably enough, stayed within some clear bounds (Overpeck et al. 2002). Moreover, the past 450,000 years, encompassing four major glacial-interglacial cycles, is more or less representative of the last million years of Earth history (see Fig. 7.2; Imbrie et al. 1992, 1993; Labeyrie et al. 2002). Although the exact spatial configuration of seasonal insolation and ice sheets varied through time with well-known changes in the Earth's orbit, these variations occurred within well-defined limits. Similarly, everything that is known about atmospheric trace-gas concentrations suggests that they were also tightly bounded within the natural envelope; for example, CO_2 varied naturally between about 180 and 300 ppmv (see Fig. 7.2; Petit et al. 1999) and probably stayed within these bounds much further back in time (i.e., back to 25 Ma; Zachos et al. 2001). Thus, most extant life on Earth (i.e., species, ecosystems, and biomes) has evolved within a fairly well defined envelope of climate variability and forcings.

There are clear signs (e.g., unprecedented global warmth of the late twentieth century) that anthropogenic climate forcing is driving the Earth's climate to destinations well outside this natural envelope (IPCC 2001; Robertson et al. 2001; Briffa and Osborn 2002; Hulme, this volume). In particular, important atmospheric trace-gas constituents (e.g., CO_2, CH_4, N_2O) are likely to continue increasing far beyond the natural envelope into the foreseeable future (see Fig. 7.2; IPCC 2001). At local to regional scales, species and ecosystems in some parts of the world could experience climatic conditions (e.g., high temperatures and increased potential evapotranspiration) unlike any experienced in the last million or more years (Webb and Wigley 1985; Crowley 1990). To compound the trouble this could pose to biodiversity and conservation planning, climate model intercomparison efforts (IPCC 2001b), and, more critically, comparisons between paleoclimate model simulations and data indicate that our state-of-the-art climate models do not yet provide reliable assessments of regional climate change (see Box 7.1).

Just as the future climate destinations of many regions may well lie outside the natural envelope, it is possible that the magnitudes and rates of future regional climate change will also exceed natural bounds of the last million or more years (Fig. 7.4) (Jackson and Overpeck 2000; Overpeck et al. 2002). The unprecedented rates of change likely to occur in many parts of the world are highlighted by a mapped comparison of one scenario of future mean climate change with change that occurred over much longer time scales; future change could be both large in magnitude and, for many regions, unprecedented in rate. Thus, even though much of Earth's biota has endured large and rapid climate change in the past, it may have to deal with large change in the future that occurs even more rapidly—an unnerving prospect in light of additional unprecedented, nonclimatic stresses on species and ecosystems. For example, anthropogenic deforestation and burning threaten tropical rainforests (Nepstad et al. 1999), while coastal development and rising CO_2 pose significant nonclimatic stresses to coral reefs (Kleypas et al. 1999; Wilkinson 1999).

Paleoclimatic data suggest that the issue of global sea level deserves special attention. It has been long known that glacial ice sheets grew repeatedly over large regions of North America and Eurasia, lowering global sea level substantially as they did so. Interestingly, during the past 2 million years ice sheets appear to have never grown more than it takes to lower global sea level by 130 m below today's level (see Fig. 7.2; Imbrie et al. 1992, 1993; Labeyrie et al. 2002). During the many intervening warm interglacial periods, global ice sheets almost always retreated to configurations (i.e., extent and height) similar to, or slightly smaller than, those of today. Melting interglacial Greenland and Antarctic ice sheets rarely produced more than about 6 m of sea-level rise above today's

Box 7.1. Climate Sensitivity

A key issue is how sensitive the climate system is to a specific change in radiative forcing, such as altered solar radiation or atmospheric trace-gas concentrations. Although there are many theoretical (and thus highly debated) ways to assess this climate system sensitivity, there are only two ways to place empirical constraints. The first is to wait about 100 years to see how much climate change actually occurs. The second is to use the paleorecord of climate change to unravel sensitivity from past cause and effect.

Fortunately, it is possible not only to reconstruct past climate change, but also to estimate past climate forcing (both anthropogenic and natural, specifically solar, trace-gas and volcanic; e.g., Robertson et al. 2001). Comparison of the past forcing and response (cause and effect) yields estimates of climate sensitivity. Work focused on estimating average Northern Hemisphere temperature over the last 1000 years provides a useful example. Many of the well-constrained multi-century temperature reconstructions (e.g. Mann et al. 1999; Crowley 2000; Briffa et al. 2001; Jones et al. 2001) suggest that the amplitude of temperature change over the past several centuries was relatively modest, indicative of a climate sensitivity at the low end of the IPCC range (1.4° to 5.8°, IPCC 2001a). However, as pointed out by Briffa and Osborn (2002), other recent reconstructions exhibit greater amplitude over the same time period, and thus raise the possibility that climate system sensitivity is actually closer to the high end of the same IPCC range.

Another clue that the climate system is more, rather than less, sensitive to altered climate forcing lies in the nature of the earth's regular cycling from glacial to interglacial and back every 100,000 or so years over the past 700,000 years. We have solid evidence that small variations in the earth's orbit serve as the pacemaker of the ice ages (Hays et al. 1976; Imbrie et al. 1992, 1993; Zachos et al. 2001), but one important mystery remains: How does the climate system work to turn a net global insolation anomaly of less than 0.1 percent (compared to modern; Imbrie et al. 1993) into the observed global climate variation of at least 5°C? Whatever the answer, it likely implies high climate system sensitivity.

Do the climate models being used to simulate future climate change have realistic representations of all the feedbacks that have served to amplify climate forcing in the past? The short answer, based on comparisons between simulated and observed paleoclimate change, is probably not. For example, efforts to use these models to simulate the full amount of observed hydrologic change in North Africa over the past 6000 years have repeatedly underestimated the observed amount of change associated with a North Africa that was much wetter and greener than present (Braconnot et al. 1999; Joussaume et al. 1999). Moreover, although efforts to simulate the magnitude of global temperature change over glacial-interglacial cycles have been successful using simplified and/or "tuned" models (e.g. Berger et al. 1999; Salzman 2002), the more sophisticated coupled atmosphere-ocean circulation models being used to assess future climate change have not been able to simulate the large, and often abrupt, climate shifts associated with glacial to interglacial oscillations. Nor have these models been able to simulate many regional aspects of the observed paleoclimatic change. The lesson for conservationists is not to put too much faith in simulations of future regional climate change, and also to design conservation strategies that will be robust in the face of climatic surprises.

level (Cuffey and Marshall 2000; Lambeck and Chappell 2001).

The paleoenvironmental record suggests two startling aspects of past sea-level high stands. First, the data are clear that sea level was 3–6 m above present during the peak of the last interglacial period, centered on 125,000 years ago (Esat et al. 1999; Israelson and Wohlfarth 1999; McCulloch et al. 1999; McCulloch and Esat 2000; Muhs 2002). What makes this ominous is that global average temperature was probably little more than 1–2°C warmer than today (Crowley and Kim 1994; Montoya et al. 2000). The second surprising paleoenvironmental observation is that the high rates of past sea-level rise have been at the upper end (i.e., 0.25 cm/year; Fairbanks 1989) of the range of current rise (0.1–0.325 cm/year) recently cited in the IPCC (2001) or perhaps even higher (i.e., over 2 cm/year; Esat et al. 1999; McCulloch and Esat 2000). Thus, the paleoperspective indicates that anthropogenic global warming over the next 100–200 years may be enough to trigger a sea-level rise of at least 3–6 m, primarily at the expense of a reduced Greenland Ice Sheet.

PAST CHANGES IN TELECONNECTIONS AND THEIR GLOBAL IMPACTS

None of the earth's major climate systems, including the North Atlantic, has a greater global reach than the El Niño-Southern Oscillation (ENSO). Variations between extreme states of the ENSO system in the tropical Pacific orchestrate year-to-year climate variability in many parts of the world (Kiladis and Diaz 1989; Trenberth et al. 1998). Teleconnections throughout the tropical oceans, in the western Americas, and in the Indian and African monsoon regions are among the many well-documented climate impacts of the modern ENSO system. ENSO also influences high-latitude processes, including Antarctic sea

ice extent (Simmonds and Jacka 1995; White et al. 1999), as well as aspects of Atlantic hydrography (e.g., North Atlantic SST and subtropical freshwater balance; Schmittner et al. 2000; Latif 2001). ENSO-related rainfall changes strongly influence tree mortality and fire frequency in many regions, both naturally (Swetnam and Betancourt 1998) and by abetting anthropogenic burning (Nepstad et al. 1999). ENSO has significant impacts on the state of coral reef health (Hoegh-Guldberg 1999; Wilkinson 1999; Glynn et al. 2001), human and animal disease (Linthicum et al. 1999; Pascual et al. 2000), drought (see below), and tropical storms (Gray 1984).

As is the case with North Atlantic variability, the instrumental record of ENSO provides merely a glimpse of how the tropical Pacific can behave, and how this important system influences ocean and terrestrial climate worldwide. Coral records from ENSO-sensitive regions (e.g., Tarawa and Maiana Atolls in the central Pacific; Cole et al. 1993; Urban et al. 2000) (Fig. 7.5) extend the record of ENSO back into the so-called Little Ice Age and support the idea that the recurrence intervals and durations of El Niño and La Niña can change significantly and with little warning as the mean global climate warms. The Maiana record reveals a period in the late nineteenth century (ca. 1850–1890) where the interannual variability so familiar today nearly disappears and is replaced by dominantly decadal variability. This interval of time is characterized by decadal fluctuations in other ENSO-sensitive records (Cole et al. 2000), and one decade-long La Niña appears to have been the principal cause of a prolonged drought that gripped much of the western United States between 1855 and 1865 (Cole et al. 2002).

Looking back further in the past, it is clear that the ENSO system can behave in ways that do not resemble anything we have observed in the past century of instrumental climate data (Clement et al.

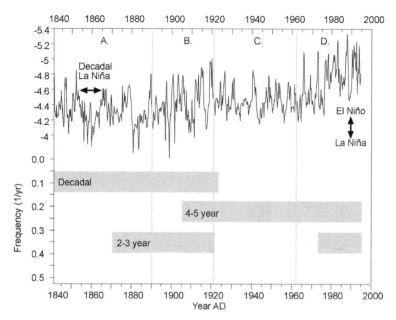

Figure 7.5. Top line shows coral $\delta^{18}O$ data that reflect ENSO activity (Urban et al. 2000); negative (upward) deviations indicate warm and wet El Niño conditions, and positive (downward) deviations denote cool and dry La Niña conditions. The decadal La Niña discussed in the text is indicated. Changes in the background state and variability of the record are also evident: A. Background is cool and dry; variability in primarily decadal. B. Background is shifting to warmer and wetter; strong interannual variability occurs at a 2–3-year period superimposed on decadal variability. C. Background is stable, and variability is weak but concentrated at a 4–5-year period. D. Interannual variability occurs at both 2–3- and 4–5-year periods, and a significant shift occurs to warmer and wetter conditions at 1976. Bars in the lower part of the figure indicate periods when variance at the indicated periods is significant; these are based on quantitative evolutive spectral analysis between ca. 1870 and 1970 (Urban et al. 2000) and estimated for the most recent and oldest intervals.

2000; Cole 2001). In the western Pacific, a coherent picture is emerging from both coral records and Australian lake sediments of greatly reduced ENSO climate variability and generally warmer conditions during the mid-Holocene, 5000–6000 years ago (Shulmeister and Lees 1995; Gagan et al. 1998; Tudhope et al. 2001). In the eastern Pacific, where ENSO's impact is strongest and should be easiest to detect, paleoclimate records are less easy to reconcile among one another.

In the absence of coral data, lake sediment data alone (Colinvaux 1972; Rodbell et al. 1999; Moy et al. 2002; Riedinger et al. 2002) suggest that El Niño activity was low before the early to middle Holocene, increasing to modern levels in the past 5000–7000 years. All these studies are consistent with the more La Niña–like state before 5000 years ago inferred from western Pacific records (Clement et al. 2000; Cole 2001). A suite of evidence from coastal South America presents a more heterogeneous picture (Sandweiss et al. 1996; DeVries 1987; Andrus et al. 2002), but still implies an ENSO system very different from today's.

Although reasonable hypotheses have been put forward using climate models to explain some of the past ENSO behavior (e.g., Clement et al. 1999; Liu et al. 2000), it appears that our understanding is still far short of explaining all of the past ENSO-related climate change. More striking is our lack of knowledge regarding how shifts, often abrupt, in ENSO variability translated into altered impacts over land outside the tropical Pacific region. The spatial pattern and strength of ENSO teleconnections clearly change through time for poorly understood reasons (Cole and Cook 1998). Although non-ENSO and regional influences likely play a key complicating role (Otto-Bliesner 1999), these influences are still generally not understood well enough to form a basis for skillful prediction, particularly when the substantial natural variability in this system and its global impacts have yet to be well simulated.

DROUGHTS, FLOOD, AND STORM FREQUENCY

Paleodata provide substantial evidence of past changes in drought, flood, and storm frequency (Gregory et al. 1995; Enzel et al. 1996; Baker 1998; Baker 2000; Liu and Fearn 1993, 2000; Knox 2000; Donnelly et al. 2001a, 2001b). The most extensive evidence has been amassed on drought frequency. Paleoclimatic evidence of megadroughts can be found worldwide (Fig. 7.6). Across northern Africa, decades- to centuries-long droughts even greater than the Sahel drought of the late twentieth century were not uncommon (Gasse 2000; Verschuren et al. 2000). Both the Middle East and southern Asia are known to have endured abrupt long-term decreases in rainfall (Heim et al. 1997; Cullen and deMenocal 2000; Weiss and

Bradley 2001; Morrill et al. 2002). Droughts in the Andes have been implicated in cultural collapse (Binford et al. 1997), and mounting evidence suggests that even the tropical lowlands of South America and Africa experienced protracted periods of lower rainfall (Ledru et al. 1998; Maley and Brenac 1998; Salgado-Labouriau et al. 1998; Sifeddine et al. 2001). So common is the evidence for decade-long droughts in places where paleoclimatologists have looked, it would be a mistake to assume any region is safe from megadrought. Moreover, even though a coherent picture of past drought fre-

Figure 7.6. Presence of substantial megadroughts in paleoclimate records worldwide: a) lake lowstands in Lake Naivasha, Kenya (Verschuren et al. 2000), b) oxygen isotope maxima in Lago Punta Laguna, Yucatan, Mexico (Hodell et al. 2001), c) lake lowstands in the southern basin of Lake Titicaca, Bolivia/Peru (Abbott et al. 1997); d) tree growth intervals in areas now submerged by lakes and streams in the eastern Sierra Nevada, California, U.S. (Stine 1994), e) precipitation in southern Nevada reconstructed from 6 tree-ring sites (Hughes and Funkhouser 1998), f) periods of dune activity that are widespread across the U.S. northern Great Plains (Forman et al. 2001), and g) periods of higher than average salinity in Moon Lake, North Dakota, U.S. (Laird et al. 1996). Note that records d and f are not derived from continuous data sources, so the absence of evidence for drought does not imply evidence of drought absence. The other records are continuous, but their resolution varies from annual (e) to decadal (b, g-top) to century-scale (a, c, g-base).

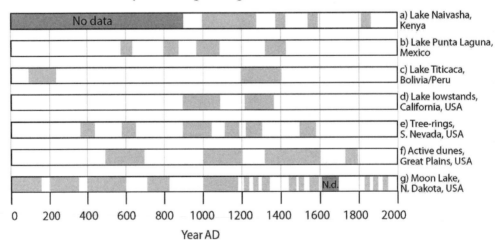

Examples of Megadrought Around the Globe

Year AD

quency and extent is emerging for some areas, the origin of megadroughts or abrupt shifts in the frequency of droughts is still mostly a mystery.

The paleorecord of floods (Knox 2000) and landfalling tropical storms (e.g., Liu and Fearn 2000), like the those of the ENSO system, also reveal changes in frequency and magnitude over the Holocene (changes that may in fact turn out to be related to changes in ENSO variability). The flood and storm records also reveal the occurrence of events in the prehistoric record that far exceed those observed during historical times, and jointly suggest that estimates of environmental risk informed only by the historical record may be underestimates.

SUMMARY

Climate variability and change during the past century have been modest relative to alterations in the climate system in the past. The past 11,000 years have been warm and stable relative to the past 2 million years. But even this period has seen abrupt climate change, making it clear that the stability of the recent past is relative. Examination of detailed proxy records of the past 200,000 years reveals many rapid, large climatic changes. Included in these are "millennial" oscillations—rapid climatic reversals that span a few thousand years or less. These and other changes may include regional or global temperature change of 10°C within a time span of as little as a decade or a few years. Some of these large, rapid changes, the D-O events, may be triggered by changes in thermohaline circulation in the North Atlantic. The timing of events in the Southern Hemisphere may be different from those in the Northern Hemisphere, perhaps one following the other. These complex paleorecords show that most of the past 2 million years has been punctuated by rapid, large global changes, raising questions

about the stability of the future, particularly as humans alter the climate system.

These observations from the past indicate that there could be three significant classes of climatic threat to biodiversity in the next 100–200 years (Overpeck et al. 2002). The first threat is of "natural" abrupt climate change. Even in the absence of significant anthropogenic climate change, abrupt changes in ocean circulation, ENSO, droughts, floods, and tropical storms all occurred in the past for poorly understood reasons. The second major threat is anthropogenic climate change, including sea-level rise. Although twentieth-century observations and models indicate that climate change associated with a doubling of atmospheric CO_2 concentrations could be small or large, the paleoclimatic record suggests that it would be a mistake to assume future change will be small. The climate system appears to have been quite sensitive to changes in climate forcing in the past, and it would be a safe bet to assume the same will be true in the future. The third, and perhaps most troubling, threat is that anthropogenic change will serve to increase the probability of abrupt climate shifts of the type described above (Alley et al. 2003). The possibility of these threats is again made more real by consideration of the paleoclimatic record. No region on earth is safe from a surprise abrupt climate change, and most regions will soon be experiencing their warmest climate in 2 million years or more.

REFERENCES

Abbott, M. B., M. W. Binford, M. Brenner, and K. R. Kelts. 1997. A 3500 C-14 yr high-resolution record of water-level changes in Lake Titicaca, Bolivia/Peru. Quaternary Research. 47:169–180.

Alley, R. B., J. Marotzke, W. D. Nordhaus, J. T. Overpeck, D. M. Peteet, R. A. Pielke, Jr., R. T. Pierrehumbert, P. B. Rhines, T. F. Stocker, L. D. Talley, and J. M. Wallace. 2003. Abrupt climate change. Science 299:2005–2010.

Alley, R. B., D. A. Meese, C. A. Shuman, A. J. Gow, K. C. Taylor, P. M. Grootes, J. W. C. White, M. Ram, E. D.

Waddington, P. A. Mayewski, and G. A. Zielinski. 1993. Abrupt increase in Greenland snow accumulation at the end of the Younger Dryas Event. *Nature* 362:527–529.

Andrus, C. F., D. E. Crowe, D. H. Sandweiss, E. J. Reitz, and C. S. Romanek. 2002. Otolith d18O record of mid-Holocene sea surface temperatures in Peru. *Science* 295:1508–1511.

Baker, V. 1998. Future prospects for past floods in India. *Memoir Geological Society of India* 41:219–228.

Baker, V. 2000. Paleoflood hydrology and the estimation of extreme floods. In *Inland Flood Hazards*, E. Wohl, ed., pp. 359–377. Cambridge: Cambridge University Press.

Berger, A., X. S. Li, and M. F. Loutre. 1999. Modelling northern hemisphere ice volume over the last 3 Ma. *Quaternary Science Reviews* 18:1–11.

Binford, M. W., A. L. Kolata, M. Brenner, J. W. Janusek, M. T. Seddon, M. Abbott, and J. H. Curtis. 1997. Climate variation and the rise and fall of an Andean civilization. *Quaternary Research* 47:235–248.

Braconnot, P., S. Joussaume, O. Marti, and N. de Noblet. 1999. Synergistic feedbacks from ocean and vegetation on the African monsoon response to mid-Holocene insolation. *Geophysical Research Letters* 26:2481–2484.

Bradley, R. 1999. *Paleoclimatology: Reconstructing Climates of the Quaternary*. San Diego: Academic.

Briffa, K. R., and T. J. Osborn. 2002. Paleoclimate—Blowing hot and cold. *Science* 295:2227–2228.

Briffa, K. R., T. J. Osborn, F. H. Schweingruber, I. C. Harris, P. D. Jones, S. G. Shiyatov, and E. A. Vaganov. 2001. Low-frequency temperature variations from a northern tree ring density network. *Journal of Geophysical Research-Atmospheres* 106:2929–2941.

Broecker, W. S. 1987. Unpleasant surprises in the greenhouse. *Nature* 328:123–126.

Broecker, W. S., S. Sutherland, and T.-H. Peng. 1999. A possible 20th century slowdown of southern ocean deep water formation. *Science* 286:1132–1135.

Clark, P. U., N. G. Pisias, T. F. Stocker, and A. J. Weaver. 2002. The role of the thermohaline circulation in abrupt climate change. *Nature* 415:863–869.

Clement, A. C., R. Seager, and M. A. Cane. 1999. Orbital controls on the El Niño/Southern Oscillation and tropical climate. *Paleoceanography* 14:441–456.

Clement, A. C., R. Seager, and M. A. Cane. 2000. Suppression of El Niño during the mid-Holocene by changes in the Earth's orbit. *Paleoceanography* 15:731–737.

Cole, J. E., J. T. Overpeck, and E. R. Cook. 2002. Multiyear La Niña events and persistent drought in the contiguous United States. *Geophysical Research Letters* Vol. 29, No. 13, 10.1029/2001GL013561, 2002 (AGU doi-type citation).

Cole, J. E. 2001. A slow dance for El Niño. *Science* 291:1496–1497.

Cole, J. E., and E. R. Cook. 1998. The changing relationship between ENSO variability and moisture balance in the continental United States. *Geophysical Research Letters* 25:4529–4532.

Cole, J. E., R. B. Dunbar, T. R. McClanahan, and N. A. Muthiga. 2000. Tropical Pacific forcing of decadal SST variability in the western Indian Ocean over the past two centuries. *Science* 287:617–619.

Cole, J. R., R. G. Fairbanks, and G. T. Shen. 1993. The spectrum of recent variability in the Southern Oscillation: Results from a Tarawa Atoll coral. *Science* 262:1790–1793.

Cole, J. E., J. T. Overpeck, and E. R. Cook. 1972. Multiyear La Niñas and prolonged US drought. *Geophysical Research Letters* 29:10.1029–2001GL013561, 25-1-4.

Colinvaux, P. A. 1972. Climate and the Galápagos Islands. *Nature* 240:17–20.

Crowley, T. 1990. Are there any satisfactory geologic analogs for a future greenhouse warming? *Journal of Climate* 3:1282–1292.

Crowley, T. J. 2000. Causes of climate change over the past 1000 years. *Science* 289:270–277.

Crowley, T. J., and K. Y. Kim. 1994. Milankovitch forcing of the last interglacial sea-level. *Science* 265:1566–1568.

Crowley, T. J., and G. R. North. 1991. *Paleoclimatology*. Oxford: Oxford University Press.

Cuffey, K., and S. Marshall. 2000. Substantial contribution to sea-level rise during the last interglacial from the Breenland ice sheet. *Nature* 404:591–552.

Cullen, H. M., and P. B. deMenocal. 2000. North Atlantic influence on Tigris-Euphrates streamflow. *International Journal of Climatology* 20:853–863.

DeVries, T. J. 1987. A review of geological evidence for ancient El Niño activity in Peru, *Journal of Geophysical Research* 92:14,471–14,479.

Dickens, G. R., J. R. Oneil, D. K. Rea, and R. M. Owen. 1995. Dissociation of oceanic methane hydrate as a cause of the carbon-isotope excursion at the end of the Paleocene. *Paleoceanography* 10:965–971.

Donnelly, J. P., S. S. Bryant, J. Butler, J. Dowling, L. Fan, N. Hausmann, P. Newby, B. Shuman, J. Stern, K. Westover, and T. Webb. 2001a. 700 yr sedimentary record of intense hurricane landfalls in southern New England. *Geological Society of America Bulletin* 113:714–727.

Donnelly, J. P., S. Roll, M. Wengren, J. Butler, R. Lederer, and T. Webb. 2001b. Sedimentary evidence of intense hurricane strikes from New Jersey. *Geology* 29:615–618.

Enzel, Y., L. Ely, P. House, and V. Baker. 1996. Magnitude and frequency of Holocene palaeofloods in the southwestern United States: A review and discussion of implications. In *Global Continental Changes: The Context of Palaeohydrology*, J. Branson, A. Brown, and K. Gregory, eds., pp. 121–137. London: Geological Society Special Publication.

Esat, T. M., M. T. McCulloch, J. Chappell, B. Pillans, and A. Omura. 1999. Rapid fluctuations in sea level recorded at Huon Peninsula during the penultimate deglaciation. *Science* 283:197–201.

Esper, J., E. R. Cook, and F. H. Schweingruber. 2002. Low-frequency signals in long tree-ring chronologies for reconstructing past temperature variability. *Science* 295:2250–2253.

Etheridge, D. M., L. P. Steele, R. L. Langenfelds, R. J. Francey, J. M. Barnola, and V. I. Morgan. 1996. Natural and anthropogenic changes in atmospheric CO2 over the last 1000 years from air in Antarctic ice and firn. *Journal of Geophysical Research-Atmospheres* 101:4115–4128.

Fairbanks, R. 1989. A 17,000-year glacio-eustatic seal level record: Influence of glacial melting rates on the Younger-Dryas event and deep-ocean circulation. *Nature* 342:617–637.

Forman, S. L., R. Oglesby, and R. S. Webb. 2001. Temporal and spatial patterns of Holocene dune activity on the Great Plains of North America: Megadroughts and climate links. *Global and Planetary Change* 29:1–29.

Gagan, M. K., L. K. Ayliffe, D. Hopley, J. A. Cali, G. E. Mortimer, J. Chappell, M. T. McCulloch, and M. J. Head. 1998. Temperature and surface-ocean water balance of the mid-Holocene tropical western Pacific. *Science* 279:1014–1018.

Gasse, F. 2000. Hydrological changes in the African tropics since the Last Glacial Maximum. *Quaternary Science Reviews* 19:189–211.

Gasse, F., and E. van Campo. 1994. Abrupt postglacial climate events in West Asia and North-Africa monsoon domains. *Earth and Planetary Science Letters* 126:435–456.

Glynn, P. W., J. L. Mate, A. C. Baker, and M. O. Calderon. 2001. Coral bleaching and mortality in panama and Ecuador during the 1997–1998 El Niño–Southern oscillation event: Spatial / temporal patterns and comparisons with the 1982–1983 event. *Bulletin of Marine Science* 69:79–109.

Gray, W. M. 1984. Atlantic seasonal hurricane frequency. 1. El Niño and 30-Mb Quasi-Biennial Oscillation influences. *Monthly Weather Review* 112:1649–1668.

Gregory, K., L. Starkel, and V. Baker. 1995. *Global Continental Palaeohydrology.* New York: John Wiley & Sons.

Gupta, A. K., D. Anderson, and J. Overpeck. 2003. Abrupt changes in the Asian southwest monsoon during the Holocene and their links to the North Atlantic Ocean. *Nature* 421:354–357.

Hays, J. D., J. Imbrie, and N. J. Shackleton. 1976. Variations in earths orbit—pacemaker of Ice Ages. *Science* 194:1121–1132.

Heim, C., N. R. Nowaczyk, J. F. W. Negendank, S. A. G. Leroy, and Z. BenAvraham. 1997. Near East desertification: Evidence from the Dead Sea. *Naturwissenschaften* 84:398–401.

Hodell, D. A., M. Brenner, J. H. Curtis, and T. Guilderson. 2001. Solar forcing of drought frequency in the Maya lowlands. *Science* 292:1367–1370.

Hodell, D. A., J. H. Curtis, and M. Brenner. 1995. Possible role of climate in the collapse of classic Maya civilization. *Nature* 375:391–394.

Hoegh-Guldberg, O. 1999. Climate change, coral bleaching and the future of the world's coral reefs. *Marine and Freshwater Research* 50:839–866.

Huang, S. P., H. N. Pollack, and P. Y. Shen. 2000. Temperature trends ever the past five centuries reconstructed from borehole temperatures. *Nature* 403:756–758.

Hughen, K. A., J. R. Southon, S. J. Lehman, and J. T. Overpeck. 2000. Synchronous radiocarbon and climate shifts during the last deglaciation. *Science* 290:1951–1954.

Hughes, M., and G. Funkhouser. 1998. Extremes of moisture availability reconstructed from tree rings for recent millennia in the Great Basin of western North America. In *The Impacts of Climate Variability on Forests*, M. Beniston and J. L. Innes, eds., pp. 109–124. Berlin: Springer Verlag.

Imbrie, J., A. Berger, E. Boyle, S. Clemens, A. Duffy, W. Howard, G. Kukla, J. Kutzbach, D. Martinson, A. McIntyre, A. Mix, B. Molfino, J. Morley, L. Peterson, N. Pisias, W. Prell, M. Raymo, N. Shackleton, and J. Toggweiler. 1993. On the structure and origin of major glaciation cycles. 2. The 100,000-year cycle. *Paleoceanography* 8:699–735.

Imbrie, J., E. Boyle, S. Clemens, A. Duffy, W. Howard, G. Kukla, J. Kutzbach, D. Martinson, A. McIntyre, A. Mix, B. Molfino, J. Morley, L. Peterson, N. Pisias, W. Prell, M. Raymo, N. Shackleton, and J. Toggweiler. 1992. On the structure and origin of major glaciation cycles. 1. Linear responses to Milankovitch forcing. *Paleoceanography* 7:701–738.

IPCC. 2001. *Climate Change 2001: The Scientific Basis. Contribution of Working Group I to the Third Assessment Report of the Intergovernmental Panel on Climate Change*, J. T. Houghton, Y. Ding, D. J. Griggs, M. Noguer, P. J. van der Linden, X. Dai, K. Maskell, and C. A. Johnson, eds., pp. 881. Cambridge: Cambridge University Press.

Israelson, C., and B. Wohlfarth. 1999. Timing of the last-interglacial high sea level on the Seychelles Islands, Indian Ocean. *Quaternary Research* 51:306–316.

Jackson, S. T., and J. T. Overpeck. 2000. Responses of plant populations and communities to environmental changes of the late Quaternary. *Paleobiology* 26:194–220.

Jones, P. D., T. J. Osborn, and K. R. Briffa. 2001. The evolution of climate over the last millennium. *Science* 292:662–667.

Joussaume, S., K. Taylor, P. Braconnot, J. Mitchell, J. Kutzbach, S. Harrison, I. Prentice, A. Broccoli,

A. Abe-Ouchi, P. Bartlein, C. Bonfils, B. Dong, J. Guiot, K. Herterich, C. Hewitt, D. Jolly, J. Kim, A. Kislov, A. Kitoh, M. Loutre, V. Masson, B. McAvaney, N. McFarlane, N. de Noblet, W. Peltier, J. Peterschmitt, D. Pollard, D. Rind, J. Royer, M. Schlesinger, J. Syktus, S. Thompson, P. Valdes, G. Vettoretti, R. Webb, and U. Wyputta. 1999. Monsoon changes for 6000 years ago: Results of 18 simulations from the paleoclimate modelling intercomparison project (PMIP). *Geophysical Research Letters* 26:859−862.

Kiladis, G., and H. F. Diaz. 1989. Global climatic anomalies associated with extremes in the Southern Oscillation. *Journal of Climate* 2:1069−1090.

Kleypas, J. A., R. W. Buddemeier, D. Archer, J. P. Gattuso, C. Langdon, and B. N. Opdyke. 1999. Geochemical consequences of increased atmospheric carbon dioxide on coral reefs. *Science* 284:118−120.

Knox, J. C. 2000. Sensitivity of modern and Holocene floods to climate change. *Quaternary Science Reviews* 19:439−457.

Kring, D. 2000. Impact events and their effect on the origin, evolution, and distribution of life. *GSA Today* 10:1−6.

Kutzbach, J., R. Gallimore, S. Harrison, P. Behling, R. Selin, and F. Laarif. 1998. Climate and biome simulations for the past 21,000 years. *Quaternary Science Reviews* 17:473−506.

Labeyrie, L., J. Cole, K. Alverson, and T. Stocker. 2002. The history of climate dynamics in the Late Quaternary. In *Paleoclimate, Global Change and the Future*, K. Alverson, R. Bradley, and T. Pedersen, eds. Berlin: Springer.

Laird, K. R., S. C. Fritz, K. A. Maasch, and B. F. Cumming. 1996. Greater drought intensity and frequency before AD 1200 in the Northern Great Plains, USA. *Nature* 384:552−554.

Lambeck, K., and J. Chappell. 2001. Sea level change through the last glacial cycle. *Science* 292:679−685.

Latif, M. 2001. Tropical Pacific/Atlantic Ocean interactions on multidecadal time scales. *Geophysical Research Letters* 28:539−542.

Ledru, M. P., M. L. Salgado-Labouriau, and M. L. Lorscheitter. 1998. Vegetation dynamics in southern and central Brazil during the last 10,000 yr BP. *Review of Palaeobotany and Palynology* 99:131−142.

Linthicum, K. J., A. Anyamba, C. J. Tucker, P. W. Kelley, M. F. Myers, and C. J. Peters. 1999. Climate and satellite indicators to forecast Rift Valley fever epidemics in Kenya. *Science* 285:397−400.

Liu, K. B., and M. L. Fearn. 1993. Lake-sediment record of Late Holocene hurricane activities from coastal Alabama. *Geology* 21:793−796.

Liu, K. B., and M. L. Fearn. 2000. Reconstruction of prehistoric landfall frequencies of catastrophic hurricanes in northwestern Florida from lake sediment records. *Quaternary Research* 54:238−245.

Liu, Z. Y., J. Kutzbach, and L. X. Wu. 2000. Modeling climate shift of El Nino variability in the Holocene. *Geophysical Research Letters* 27:2265−2268.

Maley, J., and P. Brenac. 1998. Vegetation dynamics, palaeoenvironments and climatic changes in the forests of western Cameroon during the last 28,000 years BP. *Review of Palaeobotany and Palynology* 99:157−187.

Manabe, S., and R. J. Stouffer. 1994. Multiple-century response of a coupled ocean-atmosphere model to an increase of atmospheric carbon-dioxide. *Journal of Climate* 7:5−23.

Mann, M. E., R. S. Bradley, and M. K. Hughes. 1999. Northern hemisphere temperatures during the past millennium: Inferences, uncertainties, and limitations. *Geophysical Research Letters* 26:759−762.

McCulloch, M. T., and T. Esat. 2000. The coral record of last interglacial sea levels and sea surface temperatures. *Chemical Geology* 169:107−129.

McCulloch, M. T., A. W. Tudhope, T. M. Esat, G. E. Mortimer, J. Chappell, B. Pillans, A. R. Chivas, and A. Omura. 1999. Coral record of equatorial sea-surface temperatures during the penultimate deglaciation at Huon Peninsula, *Science* 283:202−204.

Morgan, V., M. Delmotte, T. van Ommen, J. Jouzel, J. Chappellaz, S. Woon, V. Masson-Delmotte, and D. Raynaud. 2002. Relative timing of deglacial climate events in Antarctica and Greenland. *Science* 297:1862−1864.

Montoya, M., H. von Storch, and T. J. Crowley. 2000. Climate simulation for 125 kyr BP with a coupled ocean-atmosphere general circulation model. *Journal of Climate* 13:1057−1072.

Morrill, C., J. T. Overpeck, and J. E. Cole. 2002. A synthesis of abrupt changes in the Asian summer monsoon since the last deglaciation. *The Holocene*.

Moy, C. M., G. O. Seltzer, D. T. Rodbell, and D. M. Anderson. 2002. Variability of El Niño/Southern Oscillation activity at millennial timescales during the Holocene epoch. *Nature* 420:162−165.

Muhs, D. R. 2002. Evidence for the timing and duration of the last interglacial period from high-precision uranium-series ages of corals on tectonically stable coastlines. *Quaternary Research* 58:36−40.

Nepstad, D. C., A. Verissimo, A. Alencar, C. Nobre, E. Lima, P. Lefebvre, P. Schlesinger, C. Potter, P. Moutinho, E. Mendoza, M. Cochrane, and V. Brooks. 1999. Large-scale impoverishment of Amazonian forests by logging and fire. *Nature* 398:505−508.

Otto-Bliesner, B. L. 1999. El Nino La Nina and Sahel precipitation during the middle Holocene. *Geophysical Research Letters* 26:87−90.

Overpeck, J., D. Anderson, S. Trumbore, and W. Prell. 1996. The southwest Indian Monsoon over the last 18000 years. *Climate Dynamics* 12:213−225.

Overpeck, J., and R. Webb. 2000. Nonglacial rapid climate events: Past and future. *Proceedings of the National Academy of Sciences* 97:1335−1338.

Overpeck, J., C. Whitlock, and B. Huntley. 2002. Terrestrial biosphere dynamics in the climate system: Past and future. In *Paleoclimate, Global Change and the Future*, K. Alverson, R. Bradley, and T. Pedersen, eds. Berlin: Springer.

Parrish, J. T. 1998. *Interpreting Pre-Quaternary Climate from the Geologic Record*. New York: Columbia University Press.

Pascual, M., X. Rodo, S. P. Ellner, R. Colwell, and M. J. Bouma. 2000. Cholera dynamics and El Nino-Southern Oscillation. *Science* 289:1766−1769.

Peterson, L. C., G. H. Haug, K. A. Hughen, and U. Rohl. 2000. Rapid changes in the hydrologic cycle of the tropical Atlantic during the last glacial. *Science* 290:1947−1951.

Petit, J., J.-M. Barnola, I. Basile, M. Bender, J. Chappellaz, M. Davis, G. Delaygue, M. Delmotte, V. Kotlyakov, M. Legrand, V. Lipenkov, C. Lorius, L. Pepin, C. Ritz, E. Saltzman, M. Stievenard, J. Jouzel, D. Raynaud, and M. Barkov. 1999. Climate and atmospheric history of the past 420,000 years from the Vostok ice core, Antarctica. *Nature* 399:429−436.

Porter, S. C., and Z. S. An. 1995. Correlation between climate events in the North-Atlantic and China during last glaciation. *Nature* 375:305−308.

Rahmstorf, S. 2000. The thermohaline ocean circulation: A system with dangerous thresholds? An editorial comment. *Climatic Change* 46:247−256.

Riedinger, M. A., M. Steinitz-Kannan, W. M. Last, and M. Brenner. 2002. A ~6100 yr record of El Niño activity from the Galapagos Islands. *Journal of Paleolimnology* 27:1−7.

Robertson, A., J. Overpeck, D. Rind, E. Mosley-Thompson, G. Zielinski, J. Lean, D. Koch, J. Penner, I. Tegen, and R. Healy. 2001. Hypothesized climate forcing time series for the last 500 years. *Journal of Geophysical Research-Atmospheres* 106:14783−14803.

Rodbell, D. T., G. O. Seltzer, D. M. Anderson, M. B. Abbott, D. B. Enfield, and J. H. Newman. 1999. An ~15,000-year record of El Niño-driven alluviation in southwestern Ecuador. *Science* 283:516−520.

Salgado-Labouriau, M. L., M. Barberi, K. R. Ferraz-Vicentini, and M. G. Parizzi. 1998. A dry climatic event during the late Quaternary of tropical Brazil. *Review of Palaeobotany and Palynology* 99:115−129.

Salzman, B. 2002. *Dynamical Paleoclimatology: Generalize Theory of Global Climate Change*. San Diego: Academic Press.

Sandweiss, D. H., J. B. Richardson, E. J. Reitz, H. B. Rollins, and K. A. Maasch. 1996. Geoarcheological evidence from Peru for a 5000 years B.P. onset of El Niño. *Science* 273:1531−1533.

Schmittner, A., C. Appenzeller, and T. F. Stocker. 2000. Enhanced Atlantic freshwater export during El Niño. *Geophysical Research Letters* 27:1163−1166.

Shackleton, N. J. 2000. The 100,000-year ice-age cycle identified and found to lag temperature, carbon dioxide, and orbital eccentricity. *Science* 289:1897−1902.

Shulmeister, J., and B. G. Lees. 1995. Pollen evidence from tropical Australia for the onset of an ENSO-dominated climate at c. 4000 BP. *The Holocene* 5:10−18.

Sifeddine, A., L. Martin, B. Turcq, C. Volkmer-Ribeiro, F. Soubies, R. C. Cordeiro, and K. Suguio. 2001. Variations of the Amazonian rainforest environment: A sedimentological record covering 30,000 years. *Palaeogeography Palaeoclimatology Palaeoecology* 168:221−235.

Simmonds, I., and T. H. Jacka. 1995. Relationships between the interannual variability of Antarctic sea-ice and the Southern Oscillation. *Journal of Climate* 8:637−647.

Stine, S. 1994. Extreme and persistent drought in California and Patagonia during medieval time. *Nature* 369:546−549.

Stocker, T. F. 2002. Climate Change: North-South connections. *Science* 297:1814−1815.

Stocker, T. F., and A. Schmittner. 1997. Influence of CO2 emission rates on the stability of the thermohaline circulation. *Nature* 388:862−865.

Stuiver, M., and P. M. Grootes. 2000. GISP2 oxygen isotope ratios. *Quaternary Research* 53:277−283.

Swetnam, T. W., and J. L. Betancourt. 1998. Mesoscale disturbance and ecological response to decadal climatic variability in the American Southwest. *Journal of Climate* 11:3128−3147.

Thompson, L. G., E. Mosley-Thompson, M. E. Davis, K. A. Henderson, H. H. Brecher, V. S. Zagorodnov, T. A. Mashiotta, P. N. Lin, V. N. Mikhalenko, D. R. Hardy, and J. Beer. 2002. Kilimanjaro ice core records: Evidence of Holocene climate change in tropical Africa. *Science* 298:589−593.

Trenberth, K. E., G. W. Branstator, D. Karoly, A. Kumar, N.-C. Lau, and C. F. Ropelewski. 1998. Progress during TOGA in understanding and modeling global teleconnections associated with tropical sea surface temperatures. *Journal of Geophysical Research* 103:14291−14324.

Tudhope, A. W., C. P. Chilcott, M. T. McCulloch, E. R. Cook, J. Chappell, R. M. Ellam, D. W. Lea, J. M. Lough, and G. B. Shimmield. 2001. Variability in the El Niño Southern Oscillation through a glacial-interglacial cycle. *Science* 291:1511−1517.

Urban, F. E., J. E. Cole, and J. T. Overpeck. 2000. Influence of mean climate change on climate variability from a 155-year tropical Pacific coral record. *Nature* 407:989−993.

Verschuren, D., K. R. Laird, and B. F. Cumming. 2000. Rainfall and drought in equatorial east Africa during the past 1,100 years. *Nature* 403:410–414.

Vinje, T. 2001. Anomalies and trends of sea-ice extent and atmospheric circulation in the Nordic Seas during the period 1864–1998. *Journal of Climate* 14:255–267.

Vinnikov, K. Y., A. Robock, R. J. Stouffer, J. E. Walsh, C. L. Parkinson, D. J. Cavalieri, J. F. B. Mitchell, D. Garrett, and V. F. Zakharov. 1999. Global warming and northern hemisphere sea ice extent. *Science* 286:1934–1937.

Wang, Y. J., H. Cheng, R. L. Edwards, Z. S. An, J. Y. Wu, C.-C. Shen, J. A. Dorale. 2001. A high-resolution absolute-dated late Pleistocene monsoon record from Hulu Cave, China. *Science* 294:2345–2348.

Webb, T., III, and P. J. Bartlein. 1992. Global changes during the last 3 million years: Climatic controls and biotic responses. *Annual Reviews of Ecology and Systematics* 23:141–173.

Webb, T., III, and T. Wigley. 1985. What past climate climates can indicate about a warmer world. In *Projecting the Climatic Effects of Increasing Carbon Dioxide. US Department of Energy Report* pp. 237–258. Washington, D.C.

Weiss, H., and R. S. Bradley. 2001. Archaeology—What drives societal collapse? *Science* 291:609–610.

White, J. W. C., E. J. Steig, J. E. Cole, E. R. Cook, and S. J. Johnsen. 1999. Recent, annually resolved climate as recorded in stable isotope ratios in ice cores from Greenland and Antarctica. In *The ENSO Experiment Research Activities: Exploring the Linkages between the El Nino–Southern Oscillation (ENSO) and Human Health*, T. R. Karl, ed., pp. 300–302. American Meteorological Society.

Wilkinson, C. R. 1999. Global and local threats to coral reef functioning and existence: Review and predictions. *Marine and Freshwater Research* 50:867–878.

Woodhouse, C., and J. Overpeck. 1998. 2000 years of drought variability in the central United States. *Bulletin of the American Meteorological Society* 79:2693–2714.

Zachos, J., M. Pagani, L. Sloan, E. Thomas, and K. Billups. 2001. Trends, rhythms and aberrations in global climate 65 ma to present. *Science* 292:686–693.

CHAPTER EIGHT

North Temperate Responses

BRIAN HUNTLEY

The evidence available to document and understand the responses of biota to past environmental changes comes both from the fossil record, that documents past biota, and from geophysical, geochemical, geomorphological and other geological records, that document past environments (see Chapter 7). This chapter primarily draws on microfossil evidence, principally pollen and spores, and macrofossil evidence of the taxonomic composition, growth forms and phenology of past vegetation. In addition, but to a lesser extent, it draws upon fossil records of terrestrial vertebrates and invertebrates, especially those groups that have left relatively good fossil records, notably mammals, molluscs and beetles. For the purposes of this discussion, the north temperate zone is taken to include all land areas north of ca. 30°N.

For readers unfamiliar with geological time scales, the period of 65 million years spanned by the Tertiary and Quaternary geological periods may seem an unimaginably long period of time. It is important, therefore, to place this into the context of the overall history of the Earth and of life on Earth. The age of the Earth is estimated at ca. 4500 million years—in this context the period since the onset of the Tertiary represents a mere 1.4 percent of the age of the Earth, or, equating the age of the Earth to one day, the last 21 minutes of that day. Although the time of origin of single-celled organisms continues to be debated, multi-cellular organisms first appeared in the fossil record only ca. 600–700 million years ago (8.16–8.48 p.m. on the day representing the age of the Earth), whilst the land surface was apparently first colonized by plants ca. 420 million years ago (9.46

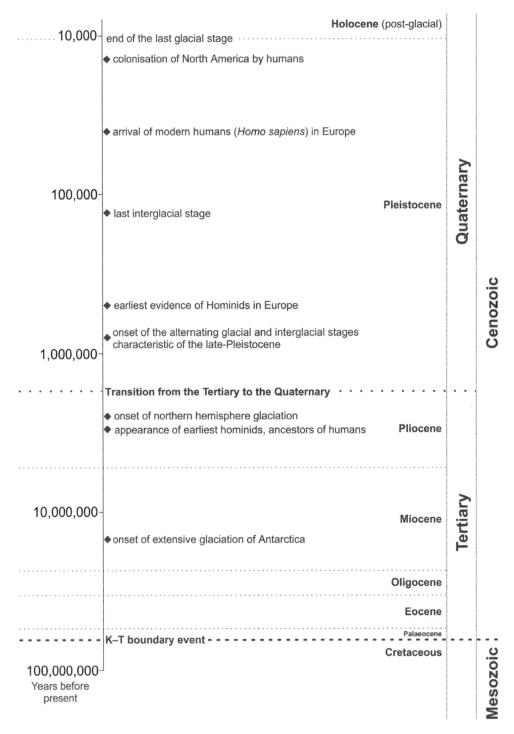

Figure 8.1. A time line for the Tertiary and Quaternary periods. The principal geological subdivisions and a few principal geological and other events are indicated. Note that the time scale is logarithmic.

p.m.) and by vertebrates ca. 70 million years later (10.08 p.m.). In contrast, early hominid ancestors of humans evolved only ca. 3 million years ago (11.59:02 p.m.), the oldest hominid fossils in Europe date from little more than half a million years ago (11.59:50 p.m.), and modern humans colonized Europe only ca. 30–40 thousand years ago (11.59:59.2 p.m.) and North America as recently as ca. 14 thousand years ago (11.59:59.7 p.m.). An outline "time line" showing major events during the last 65 million years is presented in Figure 8.1.

The sections that follow outline biological response to climate changes in the Northern Hemisphere for the past 65 million years, then provide a brief analysis of the implications of these past biotic responses in relation to efforts to conserve biodiversity.

THE TERTIARY PERIOD

The beginning of the Tertiary geological period, some 65 million years ago (11.39 p.m.), was marked by the so-called K–T (Cretaceous–Tertiary) boundary extinction event. This event was marked in particular by the demise of the dinosaurs that had dominated among terrestrial vertebrates during the preceding geological period. Geologists estimate that it took at least several million years for biodiversity to be restored to a level comparable to that before the extinction event (see e.g., Wing, Alroy, and Hickey 1995).

The world of the early Tertiary differed in important ways from that of today (Fig. 8.2). Although at least in northern temperate latitudes the principal continental units were broadly as we know them now, they were grouped much more closely together around an enclosed Arctic basin, with continuity of land connections between Alaska and eastern Asia, between northern and northwestern Europe and Greenland, and between Greenland and North America. The north pole was at that time located just to the north of the modern Bering Strait; Alaska was thus at a higher latitude than today, whereas northern Asia and Europe were as much as 10° or more lower in latitude than at present. South of these united northern continents lay the Tethys Sea, encircling the globe and connecting the modern Pacific and Atlantic basins, Mediterranean Sea and Indian Ocean. The southern continents remained isolated from those in the north: South America from North America, Africa from Europe, and India from Asia.

Palaeovegetation evidence indicates the presence of a flora with tropical Asian affinities, the London Clay Flora, in what are today the temperate latitudes of Europe. This flora extended throughout the southern parts of the northern continental regions at this time, and has been recorded as far north as Alaska and Kamchatka (Krassilov 1994). The lower than present latitude of some of these regions, coupled with their proximity to the Tethys Sea, likely account at least in part for the relatively warm nature of the climate indicated by this flora.

Further north, mixed forests of gymnosperm and angiosperm trees characteristic of a mild temperate climate were extensive, reaching even the most poleward extremities of the northern landmasses (see e.g., Basinger, Greenwood, and Sweda 1994). Although these polar forests were primarily deciduous, their diverse flora included representatives of gymnosperm families and genera that today are of restricted occurrence (e.g., Ginkgoaceae, *Metasequoia*) alongside members of gymnosperm taxa that today are widespread (e.g., Pinaceae). A similar pattern is seen among the angiosperm taxa that were present. These extensive polar forests of the early Tertiary persisted in many higher latitude regions, albeit with some changes in composition, until the onset of the pronounced global cooling that marked the transition to the Quaternary (Brown 1994).

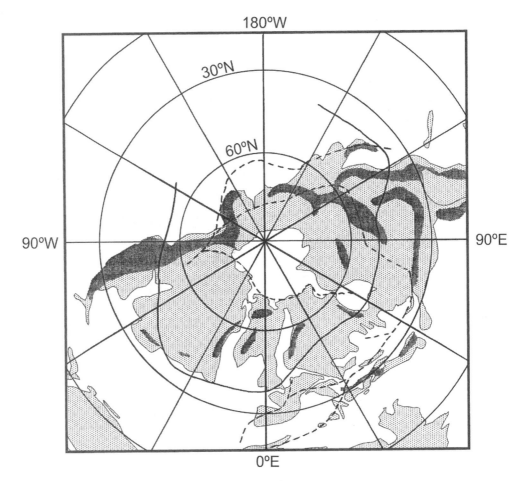

Figure 8.2. Early Tertiary palaeogeography of the northern continents. The northern continents as they were arranged in the Palaeocene. Note the isolation of the Arctic basin, the proximity of North America to northern Europe, and the continuity of the Tethys Sea in equatorial latitudes at this time. *Source*: M. C. Boulter (1994). Towards a review of Tertiary palaeobotany in the Boreal realm, fig. 1, p. 5. In *Cenozoic Plants and Climates of the Arctic*, ed. M. C. Boulter and H. C. Fisher. Berlin: Springer-Verlag. Copyright Springer-Verlag.

The presence of forest vegetation at palaeolatitudes in excess of 75° N during the Tertiary in itself indicates a very different climate regime from that of the present. The association of thermophilic vertebrate remains, including those of crocodilians, with these polar forests, is interpreted as evidence of relatively mild winter conditions without prolonged periods of sub-freezing temperatures (Estes & Hutchison 1980; Markwick 1998). In this context the relatively high concentrations of carbon dioxide in the Tertiary atmosphere are likely to be of considerable significance. Palaeoceanographic evidence, principally stable oxygen isotope ratios in calcareous marine fossils, indicates that the global oceans were substantially warmer than today during the early Tertiary, with the warming greatest at higher latitudes (Thomas, Zachos, and Bralower 2000). Although atmospheric and ocean general circulation modeling experiments have to date failed to capture the full magnitude of this warming, especially at higher latitudes, they have provided clear evidence of the importance of the higher than present atmospheric carbon dioxide concentration as a forcing factor leading to globally warmer conditions during the Tertiary than today (Bice, Sloan, and Barron 2000).

TERTIARY – QUATERNARY TRANSITION

By the onset of the Quaternary period, ca. 2.0 to 2.4 million years ago, the geography of the northern hemisphere was more or less as we know it today. One of the last significant events, happening very near the end of the Tertiary, was the formation of the land connection between North and South America at the Isthmus of Panama.

These changes were coupled with pronounced global cooling and the development of more extensive polar ice sheets in both hemispheres, and resulted in a pulse of regional and even global extinctions. On the northern continents the cooling resulted in the disappearance of the polar forest biome of the Tertiary. Tundra developed and occupied much of the northern part of the area formerly occupied by these forests, whilst to the south of the tundra biome the much less diverse boreal forests developed and persist as a circumpolar biome. Many of the tree taxa of the polar forest biome became limited to one or more isolated areas (e.g., *Liriodendron*, *Taxodium*, *Metasequoia*), while some tree taxa formerly widespread in the polar forests became globally extinct (e.g., *Glyptostrobus*).

Such extinctions were not limited to the terrestrial biota; marine molluscs and corals of the Caribbean suffered an extinction pulse at the transition from the Tertiary to the Quaternary during which more than one-fifth of marine molluscs and almost two-thirds of coral species went extinct (Jackson 1995). Just as on land this extinction pulse correlates with marked cooling, sea surface temperatures in the latitudes of the Caribbean falling by an estimated 5–6°C across the Tertiary–Quaternary transition about 2 million years ago.

EARLY AND MID-PLEISTOCENE

The Quaternary period is subdivided into only two major divisions, the Pleistocene and the Holocene, the latter comprising only the most recent 10,000 years or so. During the first ca. 1.5 million years of the Pleistocene there was a continuation of the relatively rapid general global cooling trend. Overlaid upon this were periodic fluctuations in global climate of larger magnitude than during the late Tertiary, but still without massive continental ice sheets, and with a predominantly ca. 40,000-year periodicity (Tiedemann, Sarnthein, and Shackleton 1994; Bartlein 1997). These fluctuations correspond in periodicity to the cyclic variation in one of three principal aspects of the Earth's orbit, namely the obliquity, or tilt, of the Earth's axis (see Karl and Trenberth, this volume). In this respect they contrast with the predominantly ca. 20,000-year periodicity, corresponding to that of the Earth's precession around its orbit, that is evident in some late-Tertiary records.

The early and mid-Pleistocene period was characterized by repeated climatic fluctuations that elicited changes in the predominant biomes in northern temperate latitudes. Furthermore, these repeated biotic changes also were associated with a progressive loss of diversity of tree taxa, especially at the generic level, in temperate forests. This was especially the case in Europe, where the climate and geography together resulted in greater spatial restriction of the forest biomes during cold intervals than was experienced by the equivalent biomes of North America or eastern Asia (Huntley 1993). In Britain, for example, there was an alternation between dominance of mixed forest vegetation during warm intervals and dominance of treeless vegetation of a heathland character during cold intervals (West 1961; West 1980). In the Netherlands similar palaeovegetation records for the early Pleistocene include representation of tree taxa widespread in the Tertiary but that became extinct in Europe before the mid-Pleistocene (e.g., *Carya, Magnolia, Eucommia*) (Watts 1988).

LATE PLEISTOCENE

Between about 800,000 and 1 million years ago, the "pulse" of global climatic fluctuations changed, marking the transition to the late Pleistocene (Fig. 8.3). The predominant periodicity was now ca. 100,000 years, although with a shorter cycle periodicity of ca. 20,000 years also more clearly apparent than before. These periodicities correspond to two of the principal variations in the Earth's orbit, the ca. 100,000-year periodicity in eccentricity of the Earth's orbit around the Sun and the ca. 20,000-year periodicity in the precession of the equinoxes (Hays, Imbrie, and Shackleton 1976). The magnitude of the periodic fluctuations, especially the ca. 100,000-year cycles, was much greater than previously. This late-Pleistocene period is the Quaternary "ice age." Stable oxygen isotope records from the oceans indicate the repeated development of extensive continental ice sheets during the cold, glacial stages and the reduction of these ice sheets to extreme minima comparable to the present global extent of ice sheets during the warm, interglacial stages. Global mean temperature has been estimated to have changed by 5−6°C between glacial and interglacial conditions (Schneider 1990). During interglacial stages the pattern of biomes on the northern continents was broadly similar to that which would potentially prevail today, with forest biomes occupying most of the land areas. During glacial stages, in contrast, many areas that were not ice-covered were predominated by herbaceous communities.

An important feature of this period is that the extreme conditions of glacials and interglacials occupy only a minority of the overall period, perhaps as little as one-fifth. During the majority of this period cool, intermediate conditions prevailed that were apparently closer to glacial conditions in terms of their climate and, as a result, their biota (Allen, Watts, and Huntley 2000; Huntley et al. 2003).

The glacial ice sheets of the late Pleistocene extended over large parts of the north of the Eurasian and North American continents (Denton and Hughes 1981), displacing the biota and necessitating large-scale range changes, or "migrations," by the various elements of the biota. Even where the ice sheets did not directly displace the biota, the changes in climate were extreme and the biota once again responded by extensive range changes (Huntley and Webb 1989). The biota that predominated in north temperate latitudes during interglacial stages were restricted to areas south of the ice sheets, and even within these areas were spatially restricted, especially in Europe, to very limited areas that offered the conditions they required. In Europe the tree taxa of the temperate forest biomes that predominated during interglacials were restricted to limited areas of the southern European mountains during the extremes of glacial stages. Many species' interglacial ranges extended between 1000 and 2000 km north of their glacial range limits. In a complementary manner, those species that predominated during glacials were restricted during interglacials, as they are today, to mountaintops and/or to the Arctic, their ranges extending southward by similar distances during the glacials.

The predominance of cold climatic conditions is associated with the evolution of key elements of the Arctic biota during this period. The ancestral mammoth (*Mammuthus meridionalis*), for example, was a browsing species of relatively warm climates that reached both Europe and North America during the early and mid Pleistocene, originating from a lineage that had evolved in Africa alongside the elephants

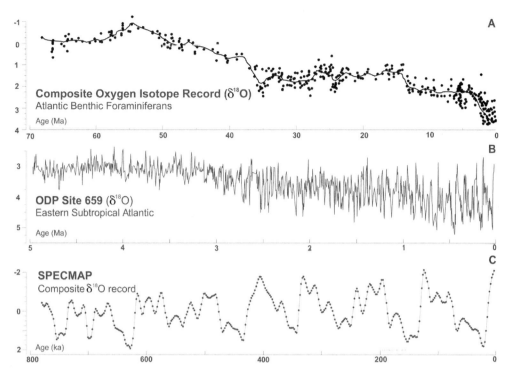

Figure 8.3. Records of Tertiary and Quaternary global climate change. Three records of stable oxygen isotope ratios (expressed as $\delta^{18}O$ 0/00) derived from the carbonate shells of Foraminiferans, single-celled ocean-dwelling organisms. These records reflect, to a first approximation, global temperatures, although those derived from planktonic species (B and C) are also influenced strongly by the global volume of ice in glaciers and ice sheets. (A) A composite record spanning the entire Tertiary. The overall cooling trend since ca. 55 Ma is clear, as are sharp downturns in temperature ca. 35 and ca. 15 Ma BP, the latter corresponding to the onset of extensive glaciation in Antarctica. (B) Record from ODP Site 659 spanning the last 5 Ma. Once again a general downward trend in global temperatures is apparent. The onset of northern hemisphere glaciation at ca. 2.4 Ma is marked by an increase in the amplitude of variability of the record. A further marked increase in the amplitude of variability, as well as the onset of prevalence of ca. 100 ka periodicity, is seen a little after 1 Ma BP; this marks the onset of the major glacial–interglacial cycles that characterize the mid- and late Pleistocene. (C) SPECMAP (Imbrie et al., 1984) composite record for the last 800 ka. This record shows no long-term trend in mean conditions but emphasizes clearly the ca. 100 ka glacial–interglacial cycles. The higher frequency orbital (Milankovitch) forcing, especially the ca. 20 ka periodicity in precession, is also apparent in this record that predominantly reflects global ice volumes over this period. *Source*: Graph A, adapted from Miller et al. (1987); graph B, adapted from Tiedmann et al. (1994); graph C, redrawn from Bartlein (1997).

(Lister and Bahn 1995). During the late Pleistocene, however, it gave rise to two separate lineages in the two continents, both of which were adapted to graze the steppe-like vegetation that was extensive during periods of intermediate climate conditions. In Eurasia this evolution progressed further, giving rise to the woolly mammoth (*M. primigenius*) (Lister and Sher 2001) that was extensive and abundant as recently as the end of the last glacial stage, persisting locally in the Russian Arctic as late as 4000 years ago (Vartanyan, Garutt, and Sher 1993). This species was strikingly cold-adapted and lived alongside other large Arctic mammals such as musk oxen (*Ovibos moschatus*) and reindeer (*Rangifer tarandus*). Along with its North American cousin, the Columbian mammoth (*M. columbi*), it was one of the many large mammals that have become extinct during the last ca. 40,000 years. A majority of these extinctions is associated either with the progressive cooling prior to and at the onset of the maximum cold interval of the last glacial stage, when Neanderthal man

(*Homo neanderthalensis*) was amongst the casualties, or with the transition from the last glacial stage to the post-glacial or Holocene, when the saber-toothed cat (*Smilodon floridanus*) of North America was one of those to be lost (Lundelius et al. 1983). In a minority of cases, including the woolly mammoth, evidence has been discovered in recent years that shows the species persisting, usually as restricted island populations (Gonzalez, Kitchener, and Lister 2000), for one to several millennia into the Holocene, before their ultimate extinction.

HOLOCENE

The Holocene is often referred to as the "post-glacial" interval, and the Quaternary ice age is considered as a phenomenon in the past. More properly, the Holocene should be considered as the present interglacial stage within the continuing Quaternary ice age. Unless some unforeseen event should occur, such as that which may have brought the Cretaceous to an end, or human interference with the climate system become so extreme as to shift it into a different state, then we can expect a future in which global climate will once again shift from its interglacial state to its intermediate state and eventually to its fully glacial state. Ice sheets will once again begin to grow and temperate biota of the northern continents will retreat from their present interglacial extreme ranges. The time scale for the onset of this global cooling, however, is likely to be some thousands of years, perhaps as much as 10,000 years into the future.

The climatic warming during the transition from the last glacial involved mean global temperature rising by ca. 5°C at rates of as much as a degree per millennium and over a period of some 5000 to 7000 years. Within this smoothed mean change, however, were many rapid fluctuations (see e.g., Stuiver, Grootes, and Bra-

ziunas 1995), the largest magnitude of which was a rapid cold snap that persisted for about 1000 years before quickly reversing itself about 11,500 years ago. This Younger Dryas cold interval is named for a plant whose fossils mark this event in the stratigraphic records from Denmark where it was first described (see Jessen 1938) (Fig 8.4). Locally, in locations sensitive to the passage of major atmospheric or oceanic fronts, the warming was much more rapid (Kroon et al. 1997) and in some regions was also of much greater magnitude. In Greenland a warming of as much as 10°C in less than a century is estimated for the end of the Younger Dryas (Dansgaard, White, and Johnsen 1989), whilst in the British Isles winter temperatures are estimated to have increased by 15–20°C at the end of the Younger Dryas, albeit more slowly than in Greenland (Atkinson, Briffa, and Coope 1987; Huntley 1994).

Not surprisingly, such rapid large magnitude changes in climate at least contributed to, if not directly caused, extinctions of some taxa. The extinction of many larger mammals at this time has already been noted. Mammals, however, were not alone in suffering extinction. In southern North America the glacial stage woodlands of the Mississippi valley and adjacent areas were characterized by the presence of a spruce, *Picea critchfieldii*, that has left numerous fossil remains. This tree disappears from the fossil record at the transition to the Holocene and was apparently an example of a plant taxon that was a casualty of the rapid climate change and became extinct at this time (Jackson and Weng 1999). Although such examples are much less frequent than are examples of extinctions of vertebrates, the occurrence at two separate localities in Scotland of seeds apparently of an extinct saxifrage (*Saxifraga* sp.) in sediments representing the last millennium of the glacial stage may represent another such example of extinction of a plant taxon at the transition to the Holo-

Figure 8.4. The Younger Dryas and other rapid temperature fluctuations during the last deglaciation. Fossil leaves (C) of the Arctic–Alpine dwarf shrub *Dryas octopetala* (Mountain Avens, shown growing in the Burren, western Ireland (B), its flowers characteristically having eight petals (A) as its epithet indicates), typify the sediments of a 1000-year-long cold interval, the Younger Dryas, at the localities in Denmark from which it was first described (Jessen 1938). This interval is but the longest and most marked of a series of climatic fluctuations that occurred during the several millennia of the transition from the last glacial stage to the Holocene. Stable oxygen isotope records from ice cores reflect the temperature over the ice sheet at the time of precipitation of the snow that eventually formed the ice being sampled. Such records from cores taken on the summit of Greenland record the many rapid climatic fluctuations during the last deglaciation, as well as the rapidity with which temperatures fell and then rose again at the onset and end of the Younger Dryas interval (D) (Stuiver et al., 1995).

cene (Birks and Mathewes 1978; Huntley 1994).

For the majority of taxa, however, the rapid climatic changes triggered extensive range changes. Many tree taxa shifted their range boundaries distances of 1000 or 2000 km during the first few millennia of the Holocene (Fig. 8.5), some achieving long-term average "migration" rates of 1 to 2 km per year and most achieving rates of at least 200 m per year (Davis 1976; Huntley and Birks 1983). Trees, moreover, were not alone in undertaking these extensive range changes; other taxonomic groups, including the terrestrial molluscs (Preece 1997), show similarly rapid and extensive range shifts. Most rapid of all in their response appear to have been the beetles (Coope 1977), at least among those groups that leave a good fossil record. The extent to which the more rapid response of beetles reflects disequilibrium between other more slowly responding groups, especially trees, and the rapidly changing climate, as opposed to a response by beetles to different aspects of climate—peak summer conditions, for example, as opposed to the overall growing season warmth or coldness of the winter—remains the subject of debate (Andersen 1993).

For the present purpose, however, the important point is that every group of terrestrial organisms we examine in the fossil record shows extensive spatial responses to the climatic warming at the onset of the Holocene at rates that are in general a challenge to account for in terms of "normal" dispersal. Such rapid rates could be maintained only by regular and reliable long distance dispersal of individuals or propagules to distances one or two orders of magnitude greater than those that can normally be observed (Collingham, Hill, and Huntley 1996).

Associated with these extensive and rapid range changes by individual taxa was a major upheaval of both the spatial patterning and structure of the biomes,

leading rapidly to the establishment of the predominant interglacial biomes and biome patterns within a millennium or so of the transition to the Holocene. Neither these patterns nor the taxonomic composition of the biomes, however, has remained static during the Holocene. These subsequent changes have occurred in response to the continuing climatic changes during the Holocene. Although these changes have been of relatively smaller magnitude, they have nonetheless resulted in, among other things, shifts of the Arctic tree line by distances of the order of 100 to 200 km (MacDonald et al. 2000), expansion and contraction of the prairie region of central North America (Grimm 1983), and westward expansion and increased abundance of drought-tolerant sclerophyll taxa in the Mediterranean region (Huntley 1988). Many of these changes have reversed direction during the Holocene, tree lines advancing northward and prairie expanding during the first few millennia of the Holocene, but then showing the reverse trends since about 6000 years ago. Others have reversed direction many times during the Holocene; many mountain tree lines, for example, have fluctuated by 100 to 200 m in elevation on millennial time scales (Karlén 1976; Karlén 1983). A few of the changes, however, have been persistently in one direction, including the expansion and increased abundance of sclerophylls in the Mediterranean.

IMPLICATIONS FOR CONSERVATION

No-analogue assemblages are indicated in many places in the Quaternary geologic record, which has led palaeoecologists to recognize that ecological communities must be recognized as no more than transient assemblages of species that together are able to tolerate the prevailing conditions (West 1964; Huntley 1996). These communities have no permanence but will be replaced by new communities with

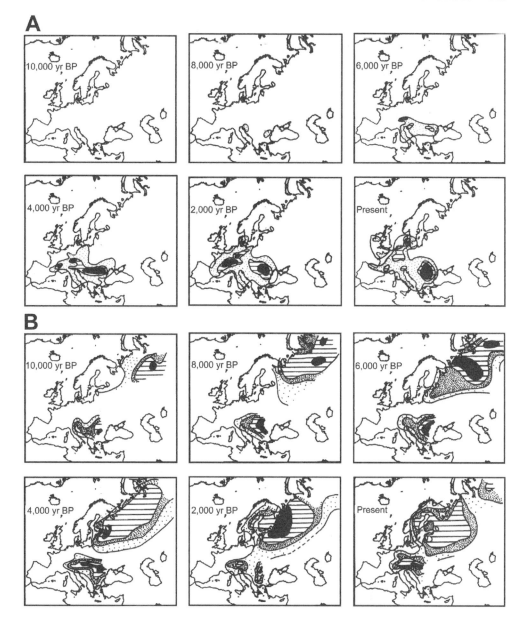

Figure 8.5. Isopoll map sequences for two European tree taxa illustrating their Holocene range changes. Isopoll map sequences at 2000 ^{14}C yr intervals illustrating the Holocene history in Europe of: (A) *Fagus* (beech) and (B) *Picea* (spruce). In each case the pollen taxon principally represents a single species within the mapped area, *F. sylvatica* and *P. abies* respectively. The density of shading indicates the relative pollen abundance; isopolls drawn at 2, 10, and 20% for *Fagus*, and at 2, 5, 10, and 20% for *Picea*. *Source:* Redrawn from Huntley (1988).

compositions often quite unlike any contemporary community as environmental conditions change. In many ways, this is simply the temporal equivalent of the Gleasonian view of spatial pattern in contemporary communities (Shimwell 1971). Both reflect the fundamental individualism of species—each species being characterized by its own unique combination of environmental tolerances and requirements, so that each responds in its own individualistic manner to spatial pattern and temporal change.

Although emphasis is often placed upon the responses of biota to climatic and other environmental changes, it is important to note that the biosphere is part of the Earth system and that, as such, it plays an active role in changes in the global environment, rather than merely passively responding to these changes (Allen et al. 1999). Thus, for example, the full magnitude of high latitude warming during the mid-Holocene cannot be accounted for without considering the positive feedback to the atmosphere resulting from boreal forest replacing tundra (Foley et al. 1994). Similarly, the early-Holocene extent and intensity of the African monsoon, and the much greater extent at that time of Lake Chad and of wetlands generally within the region of the Sahara, cannot be accounted for without the positive feedbacks to the atmosphere generated by the vegetation cover that developed in large parts of the region at that time (Street-Perrott et al. 1990). The biosphere, through its role in the global carbon cycle, is also an important source or sink for two of the most important radiatively active or "greenhouse" gases in the atmosphere: methane and carbon dioxide. Wetlands, both boreal and tropical, have been implicated in past variations in atmospheric methane concentration (Chappellaz et al. 1993, 1997), whilst soils, especially peatlands and areas of permafrost development, sequester large amounts of carbon taken up from the atmosphere as carbon dioxide by photosynthesizing green plants.

In terms of the implications of this evidence for biodiversity conservation over the next century, four key conclusions emerge:

- Adaptive genetic or evolutionary responses will, with only very few exceptions, be limited in scope to those possible by recombination of the genetic diversity already present within species' populations. In most cases this is likely to limit the scope of such responses to the current range of ecotypic variation within the species (see Thomas, this volume).

- In cases where entire biomes are threatened, then extinction of any associated specialist species is likely. Under a scenario of substantial global warming the most obviously threatened northern biomes would be tundra and polar deserts, as well as the specialized communities dependent upon polar sea ice, and also many mountain ecosystems. Associated species at risk of extinction would include Arctic vertebrates such as musk oxen (*Ovibos moschatus*) and polar bear (*Ursus maritimus*), as well as the several species of seal that breed on the Arctic sea ice.

- For most species the principal response, at least potentially, will be one of dynamic adjustment of their geographical range to maintain this in equilibrium with the changing climate. The rates of range change that can be achieved, however, will be limited intrinsically by the species' biological characteristics, as well as extrinsically by human impacts upon the extent and degree of fragmentation of suitable habitats (Collingham and Huntley 2000; Hill et al. 2001; Warren et al. 2001). It is likely that in many, if not perhaps the majority, of cases species will be unable to achieve the rates of range change required to maintain equilibrium with future climatic changes resulting from anthropogenic interference with the climate system.

- Finally, because the future combinations of climatic conditions, diurnal and seasonal solar radiation regimes, atmospheric carbon dioxide concentrations, atmospheric nitrogen deposition and UV-B fluxes will be without any current or past analogue, the future patterns, composition and structure of biomes are likely also to be without analogues, species responding individualistically, both by range change and by some degree of evolutionary adaptation, to these new combinations of conditions.

In terms of lessons from the past, the second and third of these conclusions hold the key to success in conserving biodiversity in the future. Unless the magnitude of the change in climatic conditions is constrained within bounds that do not result in loss of any biomes or major ecosystems, then some extinctions are inevitable. Unless the rate of the change in climatic conditions is constrained within bounds that permit species to achieve the range adjustments necessary to maintain dynamic equilibrium between their ranges and climate, then at the very least there will be constrictions of population size and associated loss of genetic diversity—allelic extinctions—and at the worst this too will lead to species' extinctions. Unless the approach that we take to conservation extends beyond the designation of protected areas and encompasses the wider landscape, then range change for many species will be very slow or even impossible through the often extremely impermeable managed landscapes that prevail over huge areas of the world. Failure to provide permeable landscapes that facilitate range changes will limit species' ability to respond to climatic change and increase the number that are likely to suffer extinction as a result.

SUMMARY

The geological evidence from the northern temperate latitudes indicates that extensive "migrations"—changes in the spatial extent and location of species' geographical ranges—have been the most important response of biota to past large magnitude and relatively rapid climatic changes. In addition, this evidence makes clear that rapid climatic changes of large magnitude have triggered substantial numbers of extinctions, especially among taxa adapted to biomes that have ceased to exist as a consequence of a climatic change. The record also provides evidence that taxa have evolved adaptations in response to climatic change. Substantial evolved adaptations, however, such as those that distinguished the woolly mammoth (*Mammuthus primigenius*) from its ancestor the steppe mammoth (*M. trogontherii*), are seen only in response to persistent long-term shifts in mean climatic conditions that lead to the development of a new biome and the evolution of the new biota associated with that biome. Over shorter time scales only micro-evolutionary adaptations that can be achieved by recombination of the species' preexisting genetic material take place, limiting the adaptive response to the envelope of environmental conditions to which the species' genetic diversity already enables it to adapt (Rousseau 1997).

North temperate records indicate that adaptive or genetic responses will be limited, loss of biomes is likely to be accompanied by extinctions, range adjustments will be a principal response to changing climate, and these adjustments are likely to be individualistic. The implications for conservation are that global greenhouse gas levels must be kept within constraints that avoid large-scale extinctions, at the same time that conservation strategies are made much more robust to the possible effects of changing climate.

REFERENCES

Allen, J. R. M., U. Brandt, A. Brauer, H. Hubberten, B. Huntley, J. Keller, M. Kraml, A. Mackensen, J. Mingram, J. F. W. Negendank, N. R. Nowaczyk, H. Oberhänsli, W. A. Watts, S. Wulf, and B. Zolitschka. 1999. Evidence of rapid last glacial environmental fluctuations from southern Europe. *Nature* 400, 740–743.

Allen, J. R. M., W. A. Watts, and B. Huntley. 2000. Weichselian palynostratigraphy, palaeovegetation and palaeoenvironment: The record from Lago Grande di Monticchio, southern Italy. *Quaternary International* 73/74, 91–110.

Andersen, J. 1993. Beetle remains as indicators of the climate in the Quaternary. *Journal of Biogeography* 20, 557–562.

Atkinson, T. C., K. R. Briffa, and G. R. Coope. 1987. Seasonal temperatures in Britain during the past 22,000 years reconstructed using beetle remains. *Nature* 325, 587–593.

Bartlein, P. J. 1997. Past environmental changes: Characteristic features of Quaternary climate variations. In *Past and Future Rapid Environmental Changes: The Spatial and Evolutionary Responses of Terrestrial Biota*, ed. Huntley, B., Cramer, W., Morgan, A. V., Prentice, H. C., and Allen, J. R. M., pp. 11–29. Berlin: Springer.

Basinger, J. F., D. R. Greenwood, and T. Sweda. 1994. Early tertiary vegetation of Arctic Canada and its relevance to paleoclimatic interpretation. In *Cenozoic Plants and Climates of the Arctic*, ed. Boulter, M. C., and Fisher, H. C.), pp. 175–198. Berlin: Springer-Verlag.

Bice, K. L., L. C. Sloan, and E. J. Barron. 2000. Comparison of early Eocene isotopic paleotemperatures and the three-dimensional OGCM temperature field: The potential for use of model-derived surface water $\delta^{18}O$. In *Warm Climates in Earth History*, ed. Huber, B. T., MacLeod, K. G., and Wing, S. L., pp. 79–131. Cambridge: Cambridge University Press.

Birks, H. H., and R. W. Mathewes. 1978. Studies in the vegetational history of Scotland. V. Late Devensian and Early Flandrian pollen and macrofossil stratigraphy at Abernethy Forest, Invernessshire. *New Phytologist* 80, 455–484.

Boulter, M. C. 1994. Towards a review of Tertiary palaeobotany in the Boreal realm. In *Cenozoic Plants and Climates of the Arctic*, ed. Boulter, M. C., and Fisher, H. C., pp. 1–11. Berlin: Springer-Verlag.

Brown, S. M. 1994. Migrations and evolution: Computerised maps from computerised data. In *Cenozoic Plants and Climates of the Arctic*, ed. Boulter, M. C., and Fisher, H. C., pp. 327–346. Berlin: Springer-Verlag.

Chappellaz, J., T. Blunier, S. Kints, A. Dallenbach, J. M. Barnola, J. Schwander, D. Raynaud, and B. Stauffer. 1997. Changes in the atmospheric CH_4 gradient between Greenland and Antarctica during the Holocene. *Journal of Geophysical Research-Atmospheres* 102, 15987–15997.

Chappellaz, J., T. Blunier, D. Raynaud, J. M. Barnola, J. Schwander, and B. Stauffer. 1993. Synchronous changes in atmospheric CH_4 and Greenland climate between 40-Kyr and 8-Kyr BP. *Nature* 366, 443–445.

Collingham, Y. C., M. O. Hill, and B. Huntley. 1996. The migration of sessile organisms: A simulation model with measurable parameters. *Journal of Vegetation Science* 7, 831–846.

Collingham, Y. C., and B. Huntley. 2000. Impacts of habitat fragmentation and patch size upon migration rates. *Ecological Applications* 10, 131–144.

Coope, G. R. 1977. Fossil coleopteran assemblages as sensitive indicators of climatic changes during the Devensian (Last) cold stage. *Philosophical Transactions of the Royal Society of London, Series B* 280, 313–337.

Dansgaard, W., J. W. C. White, and S. J. Johnsen. 1989. The abrupt termination of the Younger Dryas climate event. *Nature* 339, 532–534.

Davis, M. B. 1976. Pleistocene biogeography of temperate deciduous forests. *Geoscience and Man* 13, 13–26.

Denton, G. H., and T. J. Hughes, eds. 1981. *The Last Great Ice Sheets*. New York: Wiley.

Estes, R., and J. H. Hutchison. 1980. Eocene lower vertebrates from Ellesmere Island, Canadian Arctic Archipelago. *Palaeogeography Palaeoclimatology Palaeoecology* 30, 325–347.

Foley, J. A., J. E. Kutzbach, M. T. Coe, and S. Levis. 1994. Feedbacks between climate and boreal forests during the Holocene epoch. *Nature* 371, 52–54.

Frenzel, B., M. Pécsi, and A. A. Velichko, eds. 1992. *Atlas of Paleoclimates and Paleoenvironments of the Northern Hemisphere*. Budapest: Geographical Research Institute, Hungarian Academy of Sciences.

Gonzalez, S., A. C. Kitchener, and A. M. Lister. 2000. Survival of the Irish elk into the Holocene. *Nature* 405, 753–754.

Grimm, E. C. 1983. Chronology and dynamics of vegetation change in the prairie-woodland region of southern Minnesota. *New Phytologist* 93, 311–335.

Hays, J. D., J. Imbrie, and N. Shackleton. 1976. Variations in the earth's orbit: Pacemaker of the ice age. *Science* 194, 1121–1132.

Hill, J. K., Y. C. Collingham, C. D. Thomas, D. S. Blakeley, R. Fox, D. Moss, and B. Huntley. 2001. Impacts of landscape structure on butterfly range expansion. *Ecology Letters* 4, 313–321.

Huntley, B. 1996. Quaternary Palaeoecology and Ecology. *Quarternary Science Reviews* 15:591–606.

Huntley, B. 1988. Glacial and Holocene vegetation history: Europe. In *Vegetation History*, ed. Huntley, B., and Webb, T., III, pp. 341–383. Dordrecht: Kluwer Academic.

Huntley, B. 1993. Species-richness in north-temperate zone forests. *Journal of Biogeography* 20, 163–180.

Huntley, B. 1994. Late Devensian and Holocene palaeoecology and palaeoenvironments of the Morrone Birkwoods, Aberdeenshire, Scotland. *Journal of Quaternary Science* 9, 311–336.

Huntley, B., M. J. Alfano, J. R. M. Allen, D. Pollard, C. Tzedakis, J.-L. de Beaulieu, E. Grüger, and B. Watts. 2003. European vegetation during marine oxygen isotope Stage 3. *Quaternary Research* 59, 195–212.

Huntley, B., and H. J. B. Birks. 1983. *An Atlas of Past and Present Pollen Maps for Europe: 0–13000 B.P.* Cambridge: Cambridge University Press.

Huntley, B., and T. Webb, III 1989. Migration: Species' response to climatic variations caused by changes in the earth's orbit. *Journal of Biogeography* 16, 5–19.

Imbrie, J., J. D. Hays, D. G. Martinson, A. McIntyre, A. G. Mix, J. J. Morley, N. G. Pisias, W. L. Prell, and N. J. Shackleton. 1984. The orbital theory of Pleistocene climate: Support from a revised chronology of the marine δ18O record. In *Milankovitch and Climate*, ed. Berger, A., Imbrie, J., Hays, J., Kukla, G., and Saltzman, B., pp. 269–305. Dordrecht: Reidel.

Jackson, J. B. C. 1995. Constancy and change of life in the sea. In *Extinction Rates*, ed. Lawton, J. H., and May, R. M., pp. 45–54. Oxford: Oxford University Press.

Jackson, S. T., and C. Weng. 1999. Late Quaternary extinction of a tree species in eastern North America. *Proceedings of the National Academy of Science* 96, 13847–13852.

Jessen, K. 1938. Some west Baltic pollen diagrams. *Quartär* 1, 124–139.

Karlén, W. 1976. Lacustrine sediments and tree limit variations as indicators of Holocene climatic fluctuation in Lapland, northern Sweden. *Geograf. Ann.* 58a, 1–34.

Karlén, W. 1983. Holocene fluctuations of the Scandinavian alpine tree-limit. *Nordicana* 47, 55–59.

Keller, G. 2001. The end-cretaceous mass extinction in the marine realm: Year 2000 assessment. *Planetary and Space Science* 49, 817–830.

Krassilov, V. A. 1994. Tertiary climate changes in the far east based on palaeofloristic and palaeomagnetic data. In *Cenozoic Plants and Climates of the Arctic*, ed. Boulter, M. C., and Fisher, H. C., pp. 115–126. Berlin: Springer-Verlag.

Kroon, D., W. E. N. Austin, M. R. Chapman, and G. M. Ganssen. 1997. Deglacial surface circulation changes in the northeastern Atlantic: Temperature and salinity records off NW Scotland on a century scale. *Paleoceanography* 12, 755–763.

Lister, A., and P. Bahn. 1995. *Mammoths.* London: Boxtree.

Lister, A. M., and A. V. Sher. 2001. The origin and evolution of the woolly mammoth. *Science* 294, 1094–1097.

Lundelius, E. L., Jr., R. W. Graham, E. Anderson, J. Guilday, J. A. Holman, D. W. Steadman, and S. D. Webb. 1983. Terrestrial vertebrate faunas. In *The Late Pleistocene*: Vol. 1, ed. Porter, S. C., pp. 311–353. Minneapolis: University of Minnesota Press.

MacDonald, G. M., A. A. Velichko, C. V. Kremenetski, O. K. Borisova, A. A. Goleva, A. A. Andreev, L. C. Cwynar, R. T. Riding, S. L. Forman, T. W. D. Edwards, R. Aravena, D. Hammarlund, J. M. Szeicz, and V. N. Gattaulin. 2000. Holocene treeline history and climate change across northern Eurasia. *Quaternary Research* 53, 302–311.

Markwick, P. J. 1998. Fossil crocodilians as indicators of Late Cretaceous and Cenozoic climates: Implications for using palaeontological data in reconstructing palaeoclimate. *Palaeogeography Palaeoclimatology Palaeoecology* 137, 205–271.

Miller, K. G., R. G. Fairbanks, and G. S. Mountain. 1987. Tertiary oxygen isotope synthesis, sea level history, and continental margin erosion. *Paleoceanography* 2, 1–19.

Mukhopadhyay, S., K. A. Farley, and A. Montanari. 2001. A short duration of the Cretaceous–Tertiary boundary event: Evidence from extraterrestrial Helium-3. *Science* 291, 1952–1955.

Preece, R. C. 1997. The spatial response of non-marine Mollusca to past climate changes. In *Past and Future Rapid Environmental Changes: The Spatial and Evolutionary Responses of Terrestrial Biota*, ed. Huntley, B., Cramer, W., Morgan, A. V., Prentice, H. C., and Allen, J. R. M., pp. 163–177. Berlin: Springer-Verlag.

Prentice, I. C., W. Cramer, S. P. Harrison, R. Leemans, R. A. Monserud, and A. M. Solomon. 1992. A global biome model based on plant physiology and dominance, soil properties and climate. *Journal of Biogeography* 19, 117–134.

Rousseau, D.-D. 1997. The weight of internal and external constraints on *Pupilla muscorum* L. (Gastropoda: Stylommatophora) during the Quaternary in Europe. In *Past and Future Rapid Environmental Changes: The Spatial and Evolutionary Responses of Terrestrial Biota*, ed. Huntley, B., Cramer, W., Morgan, A. V., Prentice, H. C., and Allen, J. R. M., pp. 303–318. Berlin: Springer-Verlag.

Schneider, S. H. 1990. The science of climate-modelling and a perspective on the global-warming debate. In *Global Warming: The Greenpeace Report*, ed. Leggett, J., pp. 44–67. Oxford: Oxford University Press.

Shimwell, D. W. 1971. The description and classification of vegetation. Seattle: University of Washington Press.

Shukolyukov, A., and G. W. Lugmair. 1998. Isotopic evidence for the Cretaceous-Tertiary impactor and its type. *Science* 282, 927–929.

Steenbrink, J., N. van Vugt, F. J. Hilgen, J. R. Wijbrans, and J. E. Meulenkamp. 1999. Sedimentary cycles and volcanic ash beds in the Lower Pliocene lacustrine succession of Ptolemais (NW Greece):

Discrepancy between Ar-40 / Ar-39 and astronomical ages. *Palaeogeography Palaeoclimatology Palaeoecology* 152, 283–303.

Street-Perrott, F. A., J. F. B. Mitchell, D. S. Marchand, and J. S. Brunner. 1990. Milankovitch and albedo forcing of the tropical monsoons: A comparison of geological evidence and numerical simulations for 9000 yBP. *Transactions of the Royal Society of Edinburgh* 81, 407–427.

Stuiver, M., P. M. Grootes, and T. F. Braziunas. 1995. The GISP2 $\delta^{18}O$ climate record of the past 16,500 years and the role of the sun, ocean, and volcanoes. *Quaternary Research* 44, 341–354.

Thomas, E., J. C. Zachos, and T. J. Bralower. 2000. Deep-sea environments on a warm earth: Latest Paleocene-early Eocene. In *Warm Climates in Earth History*, ed. Huber, B. T., MacLeod, K. G., and Wing, S. L., pp. 132–160. Cambridge: Cambridge University Press.

Tiedemann, R., M. Sarnthein, and N. J. Shackleton. 1994. Astronomic timescale for the Pliocene Atlantic $\delta^{18}O$ and dust flux records of Ocean Drilling Program site 659. *Paleoceanography* 4, 619–639.

Vajda, V., J. I. Raine, and C. J. Hollis. 2001. Indication of global deforestation at the Cretaceous-Tertiary boundary by New Zealand fern spike. *Science* 294, 1700–1702.

van Vugt, N., J. Steenbrink, C. G. Langereis, F. J. Hilgen, and J. E. Meulenkamp. 1998. Magnetostratigraphy-based astronomical tuning of the early Pliocene lacustrine sediments of Ptolemais (NW Greece) and bed-to-bed correlation with the marine record. *Earth and Planetary Science Letters* 164, 535–551.

Vartanyan, S. L., V. E. Garutt, and A. V. Sher. 1993. Holocene Dwarf Mammoths from Wrangel-Island in the Siberian Arctic. *Nature* 362, 337–340.

Warren, M. S., J. K. Hill, J. A. Thomas, J. Asher, R. Fox, B. Huntley, D. B. Roy, M. G. Telfer, S. Jeffcoate, P. Harding, G. Jeffcoate, S. G. Willis, J. N. Greatorex-Davies, D. Moss, and C. D. Thomas. 2001. Rapid response of British butterflies to opposing forces of climate and habitat change. *Nature* 414, 65–69.

Watts, W. A. 1988. Late-Tertiary and Pleistocene vegetation history: Europe. In *Vegetation History*: Vol. 7, ed. Huntley, B., and Webb, T., III, pp. 155–192. Dordrecht: Kluwer Academic.

West, R. G. 1961. Vegetational history of the early Pleistocene of the Royal Society Borehole at Ludham, Norfolk. *Proceedings of the Royal Society of London Series B—Biological Sciences* 155, 437–453.

West, R. G. 1980. *The pre-glacial Pleistocene of the Norfolk and Suffolk coasts*. Cambridge: Cambridge University Press.

Wing, S. L., J. Alroy, and L. J. Hickey. 1995. Plant and mammal diversity in the Paleocene to early Eocene of the Bighorn Basin. *Palaeogeography Palaeoclimatology Palaeoecology* 115, 117–155.

CHAPTER NINE

Tropical Biotic Responses to Climate Change

MARK B. BUSH AND
HENRY HOOGHIEMSTRA

Although most General Circulation Models predict that global warming will induce the greatest changes in temperature at high latitudes (Chapter 1), the severest impact on biodiversity may occur in the tropics. Because moist tropical systems hold such huge diversity, and because the vast majority of those species are thought to have narrowly restricted niches (e.g. small elevational ranges, specific moisture requirements, single food plant/host), the potential exists for small climatic perturbations to have a profound effect. However, paleoecological studies reveal that the tropics have been engulfed in continual climatic change throughout the past 2 million years. To what extent can conservationists seek solace from the fact that hyperdiverse systems survived past climatic dynamism? This chapter provides an overview of some of the ways that climate change has influenced tropical biodiversity.

THE PATTERNS SET IN THE TERTIARY

The continuous movement of Earth's crustal plates has played a significant role in shaping both climatic and evolutionary patterns. As plates drifted, new continental connections provided migratory paths, some fleeting and some long-term, for biotic interchange. The new configurations also determined oceanic and atmospheric heat transport around the planet, resulting in profound climatic change. Three critical events—the rise of the Andes, the rise of the Tibetan Plateau, and the closure of the Isthmus of Panama—demonstrate the importance of ancient tectonic activity in es-

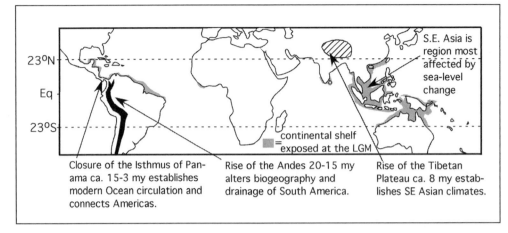

Figure 9.1. Schematic diagram showing some of the major tectonic influences shaping modern biogeographic patterns. Also shown are areas of the continental shelf exposed at the last glacial maximum.

tablishing modern patterns of biodiversity and climate (Fig. 9.1).

The uplift of the Andes started about 20 million years ago and continues to the present (Hoorn et al. 1995). The initial rise of the Andes redirected the drainage of tropical South America from a northward to an eastward pattern, thereby forming the mighty Amazon River (Hoorn et al. 1995). The easterly flow of winds and moisture across South America was interrupted by the new mountains. Air masses forced up over the mountains dumped their moisture onto the eastern Andean flank, creating some of the wettest locations on Earth. The leeward rain shadow created by this orographic effect desertified the central montane valleys, western Andean slopes, and Pacific coast of South America.

The upwarping of the Andes generated a profusion of new habitat types. Ridges and valleys with different orientation to sun and winds created a massively heterogeneous set of montane and submontane microclimates, each of which could support a novel flora and fauna. The new mountains and reconfiguration of Amazonian river basins also isolated populations, inducing allopatric speciation. Consequently, some of the most distinct phylogenetic splits among modern South American sister species of birds, insects, and mammals occur in genera riven by the Andes and major rivers (Bates eet al. 1998; Patton et al. 2000) (Fig. 9.2).

The second tectonic example is the northward drift of the Indian subcontinent and its ongoing collision with mainland Eurasia. This impact, which is often likened to a trainwreck, thrust up the Himalayas and the Tibetan Plateau. The emergence of the Tibetan Plateau provided a vast convective surface that established the modern monsoon climate of Southeast Asia about 8 million years ago. Dry seasons in Southeast Asia became less marked and prompted the spread of evergreen forest, which, since the Eocene, had been largely replaced by drier forest types (Morley 2000). This relatively recent expansion of the evergreen forests of Asia may partially explain why they are so floristically distinct at the family level from all other tropical forests (Gentry 1988).

The third plate tectonic event that ushered in modern climates was the formation of the Isthmus of Panama. Although a permanent land bridge was established only about 3 million years ago, the collision of the Caribbean Plate and the Panama–Costa Rica Arc created a significant barrier that blocked most flow between the Atlantic and Pacific basins as early as 15 million years ago (Haug and

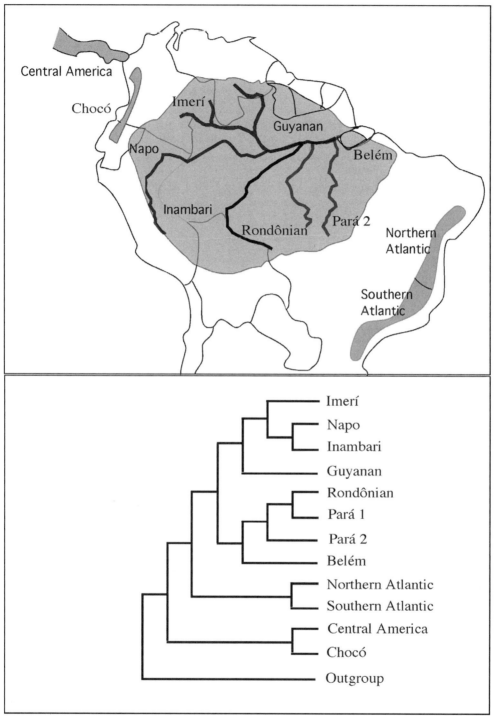

Figure 9.2. Biogeographic and phylogenetic relationships for rainforest birds in South America. (*top*) Major biogeographic regions among lowland mesic forest habitats. Note the importance of major landscape divides, mountains, intervening habitat types, and large rivers in forming the boundaries of these areas. lowland forests. (*bottom*) A composite cladogram showing the most parsimonious solution for 655 species and 1062 subspecies of passerine birds (after Bates et al. 1998). The oldest separations lie closest to the base of the tree.

Tiedemann 1998). This isthmian barrier was an essential precursor to establishing the modern oceanic thermohaline circulation system that is the principal source of heat circulation on the planet. One early consequence of the new circulation pattern was to warm the northern Atlantic, which led to increased precipitation halting the spread of grasslands and deserts. However, the North Pole was cold and the increased precipitation contributed to the formation of the modern Arctic ice cap, possibly paving the way for Quaternary ice ages.

The Panamanian land bridge, incomplete at first, provided stepping-stones for faunal and floral exchange between the Laurasian biota of North America and the Gondwanan biota of South America. This interchange was lopsided, with more species migrating south than north across the isthmus. Waves of placental mammals arrived in South America with caviomorph rodents (e.g., capybaras and guinea pigs) and primates entering when the land bridge was still very fragmented about 30–23 million years ago. Perissodactylids (e.g., tapirs and horses) and, a little later, Artiodactylids (deer, peccaries, and camels) arrived between 5 and 3 million years ago, but the largest invasion took place as the isthmus formed an intact migratory corridor about 2 million years ago (Webb 1976). The influx of North American fauna resulted in widespread extinction of endemic South American equivalent forms through intensified competition and predation. The flow of animals from south to north was more modest but included opossum, giant ground sloth, glyptodonts, toxodontids, and *Titanis*, the 3-m-tall flightless carnivorous "terror" birds. All but one species of possum went extinct from North America by the early Holocene.

Laurasian trees also spread south and some became locally important members of temperate South American floras. Two examples are *Alnus* [alders] (Betulaceae) and *Quercus* [oaks] (Fagaceae). *Alnus* arrived in Colombia about 1 million years ago and has now spread southward throughout cloud forest regions of the Andes. *Quercus* migrated south through the Isthmus of Panama, reaching Colombia about 350,000 years ago, but its southern range boundary still lies north of the equator. In Central America the modern forests of Panama and Costa Rica reflect the mixing of floras from both continents. The uplands are generally dominated by Laurasian taxa (*Quercus, Alnus, Myrica, Juglans*), whereas the lowlands were invaded en masse by the lowland forest elements of South America (e.g., Fabaceae, Annonaceae, Bombacaceae, Rubiaceae).

By about 3 million years ago, throughout the tropics, biomes with recognizable floral structure—that is, temperate deciduous forest, savanna, and tropical rain forest—were present and occupied ranges similar to those of the present, though the species composition was probably without modern analog in many areas. The great gradients of species diversity were also established. Species diversity is positively correlated with decreasing latitude, tropical altitude (trees), and increasing productivity (at least at large scales), but negatively correlated with dry season intensity and length (Gentry 1988; Lieberman et al. 1996). These gradients are the most powerful predictors of tree biodiversity and are of special importance to conservation modelers.

From this brief history it is evident that large scale changes in climate generated the patterns of modern biodiversity, induced migrations of species, and spurred extinction and speciation events. Given the power of climate to initiate such widespread biotic repercussions, future climate change can be seen as a threat to present biogeographic patterns and biodiversity.

However, balancing that view is the observation that tropical biodiversity has withstood the coming and going of about 20 ice ages in the past 2.5 million years.

Each of these events forced species to migrate, to endure rapid climate change, and to survive interactions with new competitors and predators. Although the fate of individual species in the lowland tropics during the ice-age oscillations is wholly unknown, in the intensively studied flora of North America only one species of tree, *Picea critchfieldii*, is known to have gone extinct in the last ice age (Jackson and Weng 1999). Despite our lack of species-specific knowledge, it is apparent that the coming and going of ice ages did not eliminate tropical biodiversity.

QUATERNARY CLIMATIC RIPPLES AND ICE AGES

A number of analytical tools are available to reconstruct environments based on marine, ice and lake sediment cores (Overpeck et al., this volume). One of the most important tools in the reconstruction of past vegetation types and climatic regimes is fossil pollen analysis. Once a lake becomes permanent it begins to fill with sediment and, layer upon layer, pollen is trapped in the accumulating mud. The buried pollen fossilizes and provides a record of regional vegetation, and hence climate, change through time. Tropical pollen can usually be identified to family and often to genus, and this resolution is sufficient to estimate tropical temperature change to within ca. $\pm 1.5°C$. (Weng et al. 2004). The High Plain of Bogotá has yielded the longest and most complete paleoecological record yet found in the tropics (see the case study following this chapter).

Temperature Changes

The 100,000-year rhythms of the ice ages were punctuated by shorter, sharper episodes of climate change. The complete mechanism of rapid climate change is not fully understood, but it is probable that they were induced by sudden collapses of ice sheets and massed calving of icebergs, which in turn influenced the thermohaline circulation (Bond et al. 1993) and variations in atmospheric dust content (Broecker and Denton 1989; Bond et al. 1997; Broecker 1998).

These events instigated climatic pulses that changed global temperatures $3-5°C$ in less than 10 years (Bond et al. 1997). The periods of most rapid change appear to be associated with the onset of the Holocene and more than twenty rapid cold/warm oscillations termed Dansgaard-Oeschger (D-O) cycles that took place between ca. 100,000 and 10,000 years ago. While a change of 5°C may not sound severe, it is the equivalent of swapping the climate of Atlanta for Washington, D.C., or Berlin for that of Moscow.

D-O cycles also influenced tropical climates in the Colombian Andes (Hooghiemstra eet al. 1993), and in northern Chile (Heusser et al. 1999). In tropical Africa and Southeast Asia similar records of cooling are evident, with as much as a 5°C cooling at the last glacial maximum (LGM) (Livingstone 1967; Flenley 1979). Like the neotropics, the temperature records of Africa and Southeast Asia suggest oscillations between cool and warm phases within the last glacial (e.g., Bonnefille et al. 1990). Evidence of such pronounced temperature oscillations is not discernible in the lowland tropics, possibly because the climatic response was muted (Bush et al. 2004), or perhaps simply due to a lack of suitable records.

The climatic oscillation formed by the last interglacial (ca. 115,000–105,000 years ago) and the last glacial (ca. 105,000–12,000 years ago) is of particular interest, as it is the most extreme interglacial–glacial couplet of the Quaternary. Indeed, due to its 2°C warmer-than-modern climates, the last interglacial (Marine Isotope Stage 5e, or MIS5e) may provide a climatic analog to the projected warming of Earth as a result of greenhouse gas emissions (IPCC 2001).

The most important lesson of the past 40 years of tropical climate research is that tropical systems do not have stable climates. During glacial episodes tropical temperatures in both montane and lowland environments were as much as 5°C cooler than present. It has also been learned that the temporal scale of cooling is of great importance. At a coarse scale glacials are about 5°C cooler than interglacials, but most climates are intermediate, so on finer time scales changes of lesser magnitude would be common.

Precipitation-Related Change

Considerable controversy exists over past precipitation regimes in the tropics. In part, the problem has arisen because of a desire to simplify a temporally and spatially complex system. Lake levels are often used as a proxy for precipitation, and their levels rose and fell by different amounts, at different times, across regions (Street-Perrott and Perrott 1993). The great driving forces behind tropical precipitation are sea-surface temperatures, the inter-tropical convergence zone (ITCZ), monsoons, and convection (Cerveny 1999).

Rather than trying to deal with this subject in an exhaustive way it is sufficient to observe that precipitation probably varied with the same or greater rapidity than temperature, and that in each tropical area the last ice age saw every possible permutation of cool or warm and wet or dry climates. Certainly any statement that the ice ages were "dry" or "wet" is inconsistent with available data and climate models (Bush 2002b; Bush et al. 2002).

What is of greater interest is the seasonality of precipitation, as dry season length and intensity is strongly negatively correlated with biodiversity. Far more data need to be gathered to determine seasonal responses of lowland tropical systems in the last ice age. Global circulation models suggest that in tropical areas experiencing reduced precipitation at the last glacial max-

imum, it was the wet season rains that were most reduced. In most areas, the models suggest similar or even slightly wetter dry season conditions. For example, if the cooling was the product of semi-permanent sea-surface temperatures that resemble the La Niña phase of the El Niño Southern Oscillation, the wet season would start earlier across much of Amazonia, and overall precipitation would be increased. However, the strength of this effect would be greatest in northwestern Amazonia and weakest in the southeast, adding diversity to Amazonian paleoclimates.

Another factor that would increase dry season moisture availability was the winter cold fronts that would have occurred more frequently during the ice age. Incursions of polar fronts from the northern hemisphere down to the equator and from Patagonia north to the equator would bring rain squalls during the winter (dry season). The amount of rain associated with an increased number of such cold fronts would increase dry season precipitation at the same time as reducing evapotranspirative loss.

The net effects of these precipitation changes were to leave the great expanses of tropical forests largely intact. In systems with a seasonal surplus of water, such as northern Amazonia, the Congo Basin, and most of Southeast Asia, changing wet season precipitation is of much smaller biological consequence than altering that of the dry season. In areas that are always hovering on the edge of drought deficit—for example, all tropical deserts and savannas, the Andean Altiplano, East Africa, and the Tibetan Plateau—modest reductions in wet season rains would have profound biological consequences. In ecotonal locations between forest and savanna, dry phases saw savannas replace forest, and conversely during wet phases forests expanded into savannas. The extent of these replacements in the neotropics has yet to be determined, but was probably minor (a few tens or hundreds of kilometers) when seen at the scale of forest cover. In the much drier conditions of Africa, substantial replacement of forest with savanna may have occurred for a few millennia at a time.

Atmospheric Concentration of CO_2

Other changes that appear to have tracked variation in global temperature are atmospheric concentrations of CH_4 and CO_2. Both CH_4 and CO_2 are potent greenhouse gases, but CO_2 is of special interest to biologists as it is the source of carbon for photosynthesis. During the last ice age, CO_2 concentrations tracked global temperature, and at the LGM the partial pressure of CO_2 was 170 ppm (compared with ca. 280 ppm pre-industrial and 370 ppm today). Such a substantial reduction in CO_2 availability could have important consequences for carbon budgets and productivity in any ecosystem. The best evidence for strong responses to lowered CO_2 comes from Africa (Jolly and Haxeltine 1997; Street-Perrott et al. 1997).

The effect of lowered CO_2 concentrations on plant communities is often determined, and expressed, by the level of drought stress. Plants can compensate for lowered CO_2 concentrations by having more stomata, or opening their stomata longer (Cowling and Sykes 1999). Both these responses result in increased transpirational loss and, if water is limiting, drought stress. Plants using a C4 photosynthetic pathway or that can switch between C3 and C4 pathways (CAM) would be expected to have an advantage over C3 plants under conditions of drought stress and lowered CO_2 concentrations (see Box 18.1). In Africa, where paleotemperature had been estimated based on the height of the tree line, the competitive advantage of C4 grasses over C3 trees could induce a lowering of tree line independent of temperature. As much as 40 percent of the apparent temperature signal may be attributable to the competitive superiority of C4 plants under dry, low CO_2 conditions (Jolly and Haxeltine 1997; Street-Perrott et al. 1997). However, a replacement of C3 by C4 plants is not an automatic outcome of lowering atmospheric CO_2 concentrations. C3 plants can avoid drought deficit through reducing their leaf area, thereby reducing the demand for water (Cowling and Sykes 1999), or by becoming increasingly deciduous. Such responses are going to lower growth rates and reproductive output. This reduction could be a year-round response, but is more likely to be a seasonal response coinciding with maximum dry season drought stress. In semi-deciduous forests such as Barro Colorado Island, Panama, periodic drought appears to induce predictable consequences in the degree of leaf loss, the amount of new leaves flushed, and flowers set, among both deciduous and evergreen species (Leigh 1999). How this phenology is triggered has yet to be fully understood, but the variable response of individuals and species reveals the adaptive response of tropical trees to their surroundings.

Where water is not a limiting resource, the significance of lowered CO_2 concentrations is greatly reduced. Because observations on the reduced strength of Amazonian dry seasons at the coldest times also apply to the times of lowest CO_2 concentrations, the effect of reduced CO_2 on Amazonian forests is likely to be small. Indeed, in a study conducted near the modern forest/cerrado ecotone, Pessenda and colleagues (1998) found no evidence of C3 replacement by C4 floras during the last ice age.

CONSEQUENCES OF CLIMATE CHANGE

In each of the three tropical regions the influence of climate change has reorganized floras and faunas within the past 20,000 years. However, due to the different physical attributes of those areas, the principal factor underlying those changes differed. In the wet forests of South America and Africa the lowland tropics were probably most strongly shaped by cooling, whereas in the savannas and cerrados changes in precipitation and CO_2 concentrations exerted a greater influence on ecosystems. In Southeast Asia the largest variant in the mountains was temperature, but in the great bottomlands, sea level was a far more important variable than any direct climatic factor (see below).

A further factor that can be attributed directly or indirectly to climate change was the loss of megafauna. Climate change fundamentally changed habitats, allowing humans to apply the coup de grace to many of these populations, and thereby change whole ecosystems. Modern forests throughout the Americas and Australasia may still be adjusting to the loss of an entire tier of carnivores, and all large seed predators and dispersers.

Temperature-Induced Range Migration

Paleoecological studies of North American and European ecosystems radically changed how the response of plants to major climate change is viewed. It is now well established that the great chill of the ice age caused plants to die at the northern edge of their range and for the population to extend south either through expansion of relict populations or via fresh colonization (Davis 1981; Bennett et al. 1991). For each species, the rate varied at which they died out in the north, and when and where their coverage expanded to the south. The net effect of such individualistic migration patterns was that new assemblages of plant species were forming continuously (Davis 1981; Webb 1987). The forests of the LGM contained all the species present today, but they occurred in combinations and proportions without modern analog. It was these data that laid

to rest the concept of tightly co-evolved, permanent plant communities.

The tropical forests offer some intriguing parallels, but also some stark differences to the patterns seen in temperate latitudes. First, in temperate regions virtually all the modern tundra and boreal forest environments, and even some areas of temperate forest, were icebound at the LGM. To survive, species of these habitats had to migrate south. In the tropics there was no equivalent vast ice extension. Relatively small ice caps formed on high peaks, displacing paramo/puna vegetation, but did not come lower than 3000 m elevation in most tropical regions (Clapperton 1993). Thus a much lower proportion of land was physically perturbed by ice in the tropics than in the high latitudes. Second, equatorial species could not migrate equatorward, but moved downslope to warmer climates. Downslope expansion applied only to montane species, as lowland species had nowhere to go (a 125-m drop in sea level is of little thermal consequence, roughly equivalent to 0.6°C). An Andean species moving downslope from 2000 m to maintain its climatic zone under a 5°C cooling may have needed to migrate only 30 or 40 km (Bush et al. 2004). In contrast to this, the distance that a temperate species of the Midwestern United States needed to migrate to maintain its climatic zone would have been on the order of 1500 km. Thus one could expect the response of Andean taxa to change almost in years or decades, rather than have the lags of several thousands of years seen in the Holocene reestablishment of temperate floras in Europe and North America (Davis 1981; Webb 1987; though, see Kullman 1998).

A third difference between the tropical and temperate systems is the sheer diversity of tropical species, each with subtly different niche requirements. High proportions of endemic species with distinct habitat requirements characterize areas

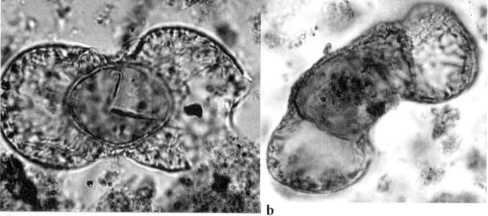

Figure 9.3. Two of the three different types of *Podocarpus* pollen recovered from ice age sediments on the Hill of Six Lakes, Brazil. *Source:* Photographs courtesy P. de Oliveira, and E. Moreno.

such as the forests of the Andean slopes, the Tanzanian Arc, and island floras of Indonesia. Though sometimes used to argue for community stability, the presence of such endemics requires only the long-term persistence of a suitable niche (not necessarily the same as their current one). The vital niche parameter may be a chemical compound shared between a number of alternate potential food plants, or a structural component such as a moist gulley or a vine tangle, or the presence of suitably sized prey rather than a fixed set of species in its food web. Clearly there are many dedicated interspecific relationships that may share a long history, for example, fig and fig wasp, *Atta* spp. and their basidiomycetes, but we should not use modern specificity to infer past community permanence (Bush 2002b).

Further hints at the impermanence of tropical communities come from the pollen records. Due to the coarse taxonomic resolution of most pollen records it is difficult to demonstrate no-analog tropical communities in the past. One of the best examples where a no-analog community does seem evident comes from a low-

land site, the Hill of Six Lakes, in Brazilian Amazonia. In that record three distinct types of *Podocarpus* pollen indicate the apparent coexistence of three species of *Podocarpus* growing alongside lowland floral elements, 20,000 years ago (Fig. 9.3). At present no species of *Podocarpus* is known from the lowlands in that region, and only two are known from all of lowland Amazonia (Colinvaux et al. 1996). The *Podocarpus* may represent extinct species, or extant species yet to be recorded in the lowlands, or they were montane species that had migrated downslope to form a community of mixed upland and lowland trees without modern analog (the last two possibilities are not mutually exclusive).

Sea Level and Biodiversity

The ca. 125-m rise of sea level associated with deglaciation led to a reorganization of littoral and coastal systems around the world. Perhaps the region most profoundly affected by these changes was the presently submerged lowlands of Southeast Asia (see Fig. 9.1). In Africa and South America the change in land area was small relative to the great lowland habitats, but in Southeast Asia more than 50 percent of the land area supporting lowland tropical forest at the LGM was flooded by 8000 years ago. Two phases of rapid, sustained sea-level rise between approximately 16,000 and 12,500 years ago, and between 11,500 and 8000 years ago (Fairbanks 1989) led to the sudden inundation of massive areas of lowland flood forest and mangrove. Harmon and colleagues (1983) estimate that the rate of shoreline advance across the Australian land bridge that connected Australia to Papua New Guinea was about 20 km per millennium, approximately 40 cm per week. Such a loss of area led to isolation of populations on the small mountainous islands of the Lesser and Greater Sundas. On first principles, a 50 percent reduction in habitat area might translate into a 10 percent extinction event when island floras and faunas equilibrate (MacArthur and Wilson 1967). Whether such extinctions have already taken place or are in the process of taking place is unclear, as these island systems are probably not equilibrial (Bush and Whittaker 1993; Whittaker 1998).

The Southeast Asian island floras and faunas are much richer than oceanic counterparts. Supersaturation of populations may have been common on most of the islands as organisms crowded into the diminishing land area. These islands still support mammal and bird faunas unusually rich in large species, for example, orangutans, Banteng cattle, tigers, gibbons, and hornbills (Whittaker 1998). Ecological relaxation undoubtedly led to the loss of larger predators from some of the smallest islands. During the 8000 years of the present sea-level highstand, population equilibria may have been approached on these islands as populations declined from supersaturation levels, leaving many species to exist as precariously small populations. Such isolated populations, made vulnerable by the unusual isolation of the Holocene high sea levels, are now threatened, some extirpated, by anthropogenic habitat destruction and hunting, for example, the Javan tiger, last seen in 1972.

Loss of Megafauna

The sudden loss of most mammals weighing more than 50 kg from neotropical forests in the late Pleistocene / early Holocene, which removed all large seed dispersers and seed predators with the exception of *Tapirus* spp., was likely due to a combination of factors, in which climate change featured prominently. This likely had long-term consequences for these forests. The important role that large mammals play as seed predators and dispersers is becoming evident (Andressen 1999; Silman et al. 2003). In a study of seed predation at Cocha Cashu, Peruvian Amazonia, Silman and colleagues (2003) demonstrated that the periodic absence of white-lipped peccary was the principal determinant in the success or failure of seedling cohorts of two of the commonest trees in the forest. With the loss of the megafauna the largest seeds in the forest may now lack a disperser (Janzen and Martin 1982). However, even medium-sized seeds may have been important dietary components for gomphotheres, giant ground sloth, giant peccary, and other extinct fauna. To a botanist, these creatures were mobile bags of seeds, capable of roaming large distances and defecating a large dollop of seeds in its own block of fertilizer. Future plant migrations will have to rely on physically smaller dispersers and a reduced variety of frugivores compared with those of the past.

Speciation and Climate Change

Repeated Quaternary climate change has been suggested as a cause of tropical speciation through a variety of mechanisms (e.g., Haffer 1969; Bush 1994; Colinvaux 1998). However, speciation is part of a continuous process of differentiation, and evidence from an increasing number of molecular studies suggests that Quaternary events do not necessarily stand out as a time of increased differentiation (e.g.,

Klicka and Zink 1999). This observation may be particularly true in the lowland neotropics, where levels of genetic divergence both between *and* within species are frequently found to exceed levels consistent with Quaternary time periods (e.g., Bates et al. 1999; Patton et al. 2000).

Considerable uncertainty surrounds how long individual species last in the fossil record, but it is believed that invertebrates generally persist longer than vertebrates (Simpson 1953). Taxonomic uncertainty and small sample sizes prevent confident statements about the duration of even well-known groups such as song birds. Klicka and Zink (1999) suggest that even among recently diverged forest songbird sister species, more than 80 percent diverged from the common ancestor more than 1 million years ago. To find that even such closely related species did not arise during the last several glacial cycles suggests that short-term climate change does not spur increases in biodiversity.

HOW RESILIENT WILL TROPICAL COMMUNITIES BE TO CLIMATE CHANGE?

Tropical paleoecological records show that deglaciation was a stop-and-go process. It is also evident from paleoecological studies that plants and animals do not migrate tidily. Some populations migrate quickly, others lag, perhaps driven by quite different ecological pressures. Yet other populations fragment, with some individuals migrating while others remain as outliers in favorable microhabitats. So long as a migratory path exists, it seems that species are adapted to withstand climate change and to survive in novel communities. Much more work needs to be done to find what determines whether isolated populations can persist for centuries or millennia, and whether they are truly genetically isolated during this time. In the short term, that populations show this mixed response to climate change probably reduces their chance of extinction.

Pleistocene climate changes did not (as far as is known) cause widespread extinctions of endemic plants, nor did the warming of the Holocene, but the effect on mammals may have been much more severe. Another bout of warming, probably similar in scale and rapidity to that at the onset of the Holocene, is under way. However, whereas the Holocene warming lay within the climatic bounds experienced within the previous 100,000 years, the projected warming will far exceed them. The apparent resilience of species to the waves of rapid climate change encountered in the Quaternary provides hope that they can ride out coming climate change. The concern is that if we do lose species, the repair process may be extremely slow. Following major extinction events in the past, such as the K–T boundary (65 million years ago), it took 400,000 years for substantial diversification of the large mammals of North America (Lillegraven and Eberle 1999), and as much as 10 million years for a full floral recovery (Wing 1998).

SUMMARY

Paleoecology is a relatively crude tool to assess extinction events, their rapidity and cause. Although megafaunal decline is evident in the Late Quaternary, ascribing those extinction events to a particular climatic change is highly speculative. Among North American trees just one species is known to have gone extinct during the period of greatest climatic perturbation in the past 100,000 years. It would appear that, in natural systems, species are resilient to climate change. Whether that same resilience should be predicted within landscapes fragmented and altered by human activities is far less certain. What is apparent is that even if species are retained, community composition will alter substantially. Species' ranges will expand or contract individualistically, and familiar communities will probably be replaced by novel species combinations. The critical factor to conserving biodiversity is not the preservation of communities (as these come and go with each climatic change), but the long-term maintenance of ecological niches.

ACKNOWLEDGMENTS

This work was supported by grant DEB-9732951 from the National Science Foundation.

REFERENCES

Andressen, E. 1999. Seed dispersal by monkeys and the fate of dispersed seeds in a Peruvian rain forest. *Biotropica* 31:145–158.

Bates, J. M., S. J. Hackett, and J. Cracraft. 1998. Area-relationships in the Neotropical lowlands: An hypothesis based on raw distributions of Passerine birds. *Journal of Biogeography* 25:783–794.

Bates, J. M., S. J. Hackett, and J. Goerck. 1999. High levels of mitochondrial DNA differentiation in two lineages of antbirds (*Drymophila* and *Hypocnemis*). *Auk* 116:1093–1106.

Bennett, K. D., P. C. Tzedakis, and K. J. Willis. 1991. Quaternary refugia of north European trees. *Journal of Biogeography* 18:103–115.

Bond, G., W. Broecker, S. Johnsen, J. McManus, L. Labeyrie, J. Jouzel, and G. Bonani. 1993. Correlations between climate records from the North Atlantic sediments and Greenland Ice. *Nature* 365:143–147.

Bond, G., W. Showers, M. Cheseby, R. Lotti, P. Almasi, P. Demenocal, P. Priore, H. Cullen, I. Hajdas, and G. Bonani. 1997. A pervasive millennial-scale cycle in North Atlantic Holocene and glacial climates. *Science* 278:1257–1266.

Bonnefille, R., J. C. Roeland, and J. Guiot. 1990. Temperature and rainfall estimates for the past 40,000 years in equatorial Africa. *Nature* 346:347–349.

Broecker, W. S. 1998. Paleocean circulation during the last deglaciation: A bipolar seesaw? *Paleoceanography* 13:119–121.

Broecker, W. S., and G. H. Denton. 1989. The role of ocean-atmosphere reorganizations in glacial cycles. *Geochimica et Cosmochimica Acta* 53:2465–2501.

Bush, M. B. 1994. Amazonian speciation: A necessarily complex model. *Journal of Biogeography* 21:5–18.

Bush, M. B. 2002a. Distributional change and conservation on the Andean flank: A paleoecological perspective. *Global Ecology and Biogeography* 11:463–473.

Bush, M. B. 2002b. On the interpretation of fossil Poaceae pollen in the humid lowland neotropics. *Palaeogeography, Palaeoclimatology, Palaeoecology* 177:5–17.

Bush, M. B., M. C. Miller, P. E. De Oliveira, and P. A. Colinvaux. 2002. Orbital-forcing signal in sediments of two Amazonian lakes. *Journal of Paleolimnology* 27:341–352.

Bush, M. B., M. R. Silman, and D. H. Urrego. 2004. 48,000 years of climate and forest change in a biodiversity hot spot. *Science* 303:827–829.

Bush, M. B., and R. J. Whittaker. 1993. Krakatau: Non-equilibration in island theory of Krakatau. *Journal of Biogeography* 20:453–457.

Cerveny, R. 1999. Present climates of South America. In *Climates of the Southern Continents*, J. E. Hobbs, J. A. Lindesay, and H. A. Bidgman, eds., pp. 107–135. Chichester: John Wiley & Sons.

Clapperton, C. W. 1993. *Quaternary Geology and Geomorphology of South America*. Amsterdam: Elsevier.

Colinvaux, P. A. 1998. A new vicariance model for Amazonian endemics. *Global Ecology and Biogeography Letters* 7:95–96.

Colinvaux, P. A., P. E. De Oliveira, J. E. Moreno, M. C. Miller, and M. B. Bush. 1996. A long pollen record from lowland Amazonia: Forest and cooling in glacial times. *Science* 274:85–88.

Cowling, S. A., and M. T. Sykes. 1999. Physiological significance of low atmospheric CO_2 for plant-climate interactions. *Quaternary Research* 52:237–242.

Davis, M. B. 1981. Quaternary history and the stability of forest communities. In *Forest Succession: Concepts and Application*, D. C. West, H. H. Shugart, and D. B. Botkin, eds., pp. 132–154. New York: Springer Verlag.

Fairbanks, R. G. 1989. A 17,000 year glacio-eustatic sea level record: Influence of glacial melting rates on Younger Dryas event and deep ocean circulation. *Nature* 342:637–642.

Flenley, J. R. 1979. *A Geological History of Tropical Rainforest*. London: Butterworths.

Gentry, A. H. 1988. Changes in plant community diversity and floristic composition on environmental and geographical gradients. *Annals of the Missouri Botanical Garden* 75:1–34.

Haffer, J. 1969. Speciation in Amazonian forest birds. *Science* 165:131–137.

Harmon, R. S., R. M. Mitterer, N. Kriausakul, N. S. Land, H. P. Schwarcz, P. Garrett, G. J. Larson, H. L. Vacher, and M. Rowe. 1983. U-series and amino-acid racemization geochronology of Bermuda: Implications for eustatic sea-level fluctuation over the past 250,000 years. *Palaeogeography, Palaeoclimatology, Palaeoecology* 44:41–70.

Haug, G. H., and R. Tiedemann. 1998. Effect of the formation of the Isthmus of Panama on Atlantic Ocean thermohaline circulation. *Nature* 393:673–676.

Heusser, L., C. Heusser, A. Kleczkowski, and S. Crowhurst. 1999. A 50,000-yr pollen record from Chile of South American millennial-scale climate instability during the last glaciation. *Quaternary Research* 52:154–158.

Hooghiemstra, H., J. L. Melice, A. Berger, and N. J. Shackleton. 1993. Frequency spectra and paleoclimatic variability of the High-resolution 30–1450 ka Funza I pollen record (Eastern Cordillera, Colombia). *Quaternary Science Review* 12:141–156.

Hoorn, C., J. Guerrero, G. A. Sarmiento, and M. A. Lorente. 1995. Andean tectonics as a cause for changing drainage patterns in Miocene northern South America. *Geology* 23:237–240.

Intergovernmental Panel on Climate Change. 2001. IPCC Summary for Policymakers. *Climate Change 2001: Impacts, Adaptation and Vulnerability.* A report of Working Group II. Cambridge: Cambridge University Press.

Jackson, S. T., and C. Weng. 1999. Late Quaternary extinction of a tree species in eastern North America. *Proceedings of the National Academy of Sciences* 96:13847–13852.

Janzen, D. H., and P. S. Martin. 1982. Neotropical anachronisms: The fruits the gomphotheres ate. *Science* 215:19–27.

Jolly, D., and A. Haxeltine. 1997. Effect of low glacial atmospheric CO_2 on tropical African montane vegetation. *Science* 276:786–788.

Klicka, J., and R. M. Zink. 1999. Pleistocene effects on North American songbird evolution. *Proceedings of the Royal Society of London* 266:695–700.

Kullman, L. 1998. Palaeoecological, biogeographical and palaeoclimatological implications of early Holocene immigration of Larix Sibirica Ledeb. into the Scandes Mountains, Sweden. *Global Ecology and Biogeography Letters* 7:181–188.

Leigh, E. G. j. 1999. Epilogue: Research on Barro Colorado Island, 1980–94. In *The Ecology of a Tropical Forest*, E. G. j. Leigh, A. S. Rand, and D. M. Windsor, eds., pp. 469–503. Washington, DC: Smithsonian Institution.

Lieberman, D., M. Lieberman, S. Peralta, and G. S. Hartshorn. 1996. Tropical forest structure and composition on a large-scale altitudinal gradient in Costa Rica. *Journal of Ecology* 84:137–152.

Lillegraven, J. A., and J. J. Eberle. 1999. Vertebrate faunal changes through Lancian and Puercan time in southern Wyoming. *Journal of Paleontology* 73:691–710.

Livingstone, D. A. 1967. Postglacial vegetation of the Ruwenzori Mountains in equatorial Africa. *Ecological Monographs* 37:25–52.

MacArthur, R. H., and E. O. Wilson. 1967. *The Theory of Island Biogeography*. Princeton: Princeton University Press.

Morley, R. J. 2000. *Origin and Evolution of Tropical Rain Forest*. Chichester: Wiley and Sons.

Patton, J. L., M. N. F. da Silva, and J. R. Malcolm. 2000. Mammals of the Rio Juruá and the evolutionary and ecological diversification of Amazonia. *Bulletin of the American Museum of Natural History* 244.

Pessenda, L. C. R., B. M. Gomes, R. Aravena, A. S. Ribeiro, and S. E. M. Gouveia. 1998. The carbon isotope record in soils along a forest-cerrado ecosystem transect: Implication for vegetation changes in Rondônia State, southwestern Brazilian Amazon region. *The Holocene* 8:631–635.

Silman, M. R., J. T. Terborgh, and R. A. Kiltie. 2003. Population regulation of a rainforest dominant tree by a major seed predator. *Ecology* 84:431–438.

Simpson, G. G. 1953. *The major features of evolution*. New York: Columbia University Press.

Street-Perrott, A. F., Y. Huang, R. A. Perrott, G. Eglinton, P. Barker, L. B. Khelifa, D. D. Harkness, and D. O. Olago. 1997. Impact of lower atmospheric carbon dioxide on tropical mountain ecosystems. *Science* 278:1422–1426.

Street-Perrott, F. A., and R. A. Perrott. 1993. Holocene variation, lake levels, and climate of Africa. In *Global Climates Since the Last Glacial Maximum*, H. E. J. Wright, J. E. Kutzbach, T. I. Webb, W. F. Ruddiman, F. A. Street-Perrott, and P. J. Bartlein, eds. Minneapolis: University of Minnesota Press.

Webb, S. D. 1976. Mammalian faunal dynamics of the great American interchange. *Paleobiology* 2:216–234.

Webb, T. I. 1987. The appearance and disappearance of major vegetational assemblages: Long term vegetational dynamics in eastern North America. *Vegetatio* 69:177–188.

Weng, C., M. B. Bush, and M. R. Silman. 2004. An analysis of modern pollen rain on an elevational gradient in southern Peru. *Journal of Tropical Ecology* 20:113–124.

Whittaker, R. J. 1998. Island biogeography, ecology, evolution and the lighthouse keeper's cat. Oxford: Oxford University Press.

Wing, S. L. 1998. Tertiary vegetation of North America as a context for mammalian evolution. In *Evolution of Tertiary Mammals of North America*, Vol. 1: *Terrestrial Carnivores, Ungulates and Ungulate-like Mammals.* C. M. Janis, K. M. Scott, and L. L. Jacobs, eds., pp. 18–30. Cambridge: Cambridge University Press.

A Record of Change from the High Plain of Bogotá

Mark B. Bush

The longest and most detailed record of paleoclimatic and paleoecological change in the Neotropics comes from a deep sedimentary basin in the Colombian Andes. In the late 1950s Thomas van der Hammen raised a long core of mud from the dry bed of an ancient lake near the modern city of Bogotá. A truck-mounted water-well drilling rig was used to recover a core of sediment more than 300 m in length. The stratigraphy of the core was described and samples were removed for pollen analysis. Pollen was separated from the sediment, examined at ×400–×1000 magnification, and identified by comparison with specimens of modern pollen.

The pollen analysis revealed cycles of vegetation change in which arboreal pollen was replaced by that of grasses and herbs (van der Hammen and González 1959). The oscillations between grassland (paramo) and forest were abrupt events, and as work continued on the core as many as 22 cold events could be seen (Fig. 9.4). Radiocarbon (^{14}C) dating could be used to establish the age of only the uppermost of these cold events, as the others took place beyond the 45,000-year range of ^{14}C dating. The grassland-dominated cold events were glacial episodes, and the brief intervening forest periods were interglacials.

Past vegetation inferred from the pollen record was used to quantify paleoclimatic change. Vegetation surrounding the modern high plain of Bogotá (2550 m elevation) is Andean forest, rich in *Quercus* (oak), *Alnus* (alder), *Myrica* (myrtle), *Podocarpus* (southern fir or yew), *Weinmannia*, and *Hedyosmum*. About 700 m upslope (3200 m elevation) from the ancient lake the Andean forest gives way to shrubby subparamo, and at about 1000 m above the lake (3500 m elevation) there is a further vegetational transition to the grassy paramo. As the cold events in the fossil record indicated that paramo had descended to some 500 m below the plain of Bogotá, an approximate temperature depression for each of these events could be calculated. If the moist air adiabatic lapse

rate is taken to have been constant at ca. 5°C of cooling per 1000 m of ascent, the 1500-m vertical change in vegetation could be translated into a temperature change, providing an inferred cooling of 7.5°C (van der Hammen 1974).

Van der Hammen and González (1959) revolutionized tropical ecology with their insight that the Andean forests migrated downslope during cold events and were replaced by cold-adapted grasslands. This record was the first to show that neotropical vegetation had been strongly influenced by the coming and going of glaciations and that the tropics did not offer steady, unchanging evolutionary environments in which species accumulated under ideal growing conditions.

In the past 40 years additional, deeper, cores have been collected and analyses have been conducted at finer temporal resolution. A detailed fossil pollen history of the past 1.4 million years has now been established for the high plain of Bogotá (Hooghiemstra 1984; 1989; Andriessen et al. 1993). These sediments have revealed a history not only of glacial cycles but also of longer-term changes in plant communities. An improved chronology of the cores has been achieved through zircon fission track dating of volcanic ash layers (Andriessen et al. 1993), wiggle-match-

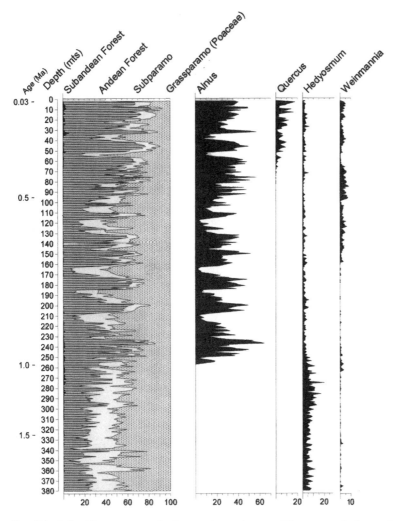

Figure 9.4. A 2-million-year record of vegetation and climate change in the high plains of the Andes. Each column in the figure represents the proportion of fossil pollen (% of total pollen recorded) from a single plant genus over time, in a sediment core taken from Lake Fuquene, Colombia. Age (million years) as represented by depth of the drill-core (meters) is shown on the Y-axis. The panel on the left provides a summary of results in which pollen types characteristic of different biomes (Andean Forest, Paramo Grassland, etc) are grouped at each time interval. Note the arrival of *Alnus* (alder) about 1 million years ago and *Quercus* (oaks) about 600,000 years ago.

ing the pollen record to that of isotopic records from ice cores and marine sediments, and tuning events to changes in Earth's orbit around the sun (Milankovitch cycles).

The revised chronology of the Funza-1 core (Van't Veer and Hooghiemstra 2000) allows insights into important biogeographic changes in Andean ecosystems. The first observation is that the vegetational changes in this system closely mirror orbital variation over the past 1.4 million years. A steady precessional forcing (19,000 and 23,000 years) is evident throughout this span. Superimposed on this regular beat, strong 100,000-year rhythms have predominated in the past 800,000 years, while the preceding 600,000 years were dominated by the 40,000-year rhythm (Hooghiemstra et al. 1993). The same patterns are evident in sediments from the North Atlantic (Imbrie et al. 1984) and ice cores from Antarctica (Petit et al. 1999). These observations demonstrate that the ecology of the highly diverse forests of the Andes are climatically linked to other regions of the globe through complex interhemispheric mechanisms of heat transport.

The Funza-1 core also offers some fascinating insights into community ecology of Andean ecosystems. The arrival of the Laurasian genus *Alnus* about 1 million years ago and *Quercus* about half a million years ago after the closing of the Panamanian land bridge to North America are clearly evident, and each of these taxa changed how the ecosystem responded to climatic change.

The arrival of *Alnus* is especially dramatic in the pollen record because this genus often grows in wet, disturbed settings and produces vast quantities of pollen. Prior to the arrival of *Alnus*, the Andean forests were rich in *Podocarpus*, *Weinmannia*, *Hedyosmum*, and *Myrica*. Although the percentage occurrence of these genera is reduced when *Alnus* arrives, it is likely that the effect on the ecology of the forest was nowhere near as profound as the effect on the pollen record suggests.

Alnus occurs along streams and as marsh forest on the high plain and may have been especially abundant around the lake. As *Alnus* is wind-pollinated and produces pollen copiously, it is also frequently overrepresented in pollen spectra. In the Funza core, *Alnus* pollen can be 1.5 times as abundant as all other pollen types combined. *Alnus*, with its abilities to disperse quickly and leave a strong signal in the pollen record, had all the attributes of an excellent palynological marker of climate change. Ever since its arrival, it is the most certain indicator of relatively warm times on the high plain of Bogotá.

Another Laurasian genus, *Quercus*, is first evident in the Bogotá pollen record ca. 600,000 BP, although it does not become locally abundant until 330,000 BP. In North America there are dozens of species of *Quercus*, but in the modern forests on the Costa Rican/Panamanian border this number has dwindled to about 13 species largely restricted to elevations above 1700 m. Only one species, *Q. humboldtii*, is currently found in Colombia. The filtering effect of a long migration with discontinuous corridors of suitable habitat is evident in this loss of diversity.

The arrival of *Quercus* at the high plain of Bogotá was not as spectacular in terms of its effect on the pollen record as that of *Alnus*, but was probably of greater ecological significance. Although wind-pollinated, *Quercus* is not as heavily overrepresented in modern pollen rain as *Alnus* (Grabandt 1985; Bush 2000), and its pollen representation provides a more accurate reflection of its importance in local floras. When *Quercus* first formed forests around the high plain of Bogotá about 330,000 BP, it was probably a dominant, or at least common, component of those forests. The paleobotanical picture that emerges from this work is thus one of wholesale changes in vegetation type (the paramo–forest transitions), coupled with more intricate movements of individual taxa (the *Alnus* and *Quercus* arrivals and associated changes)—a close analog to the effects simulated by biome and species range shift models of the future.

REFERENCES

Andriessen, P. A. M., K. F. Helmens, H. Hooghiemstra, P. A. Riezebos, and T. van der Hammen. 1993. Absolute chronology of the Pliocene-Quaternary sediment sequence of the Bogota area, Colombia. *Quaternary Science Reviews* 12:483–502.

Bush, M. B. 2000. Deriving response matrices from Central American modern pollen rain. *Quaternary Research* 54:132–143.

Grabandt, R. A. J. 1985. Pollen rain in relation to vegetation in the Colombian Cordillera Oriental. Ph.D. University of Amsterdam, Amsterdam.

Hooghiemstra, H. 1984. *Vegetational and Climatic History of the High Plain of Bogota, Colombia: A Continuous Record of the Last 3.5 Million Years*. Vaduz: Gantner Verlag.

Hooghiemstra, H. 1989. Quaternary and upper-Pliocene glaciations and forest development in the tropical Andes: Evidence from a long high-resolution pollen record from the sedimentary basin of Bogota, Colombia. *Palaeogeography, Palaeoclimatology, Palaeoecology* 72:11–26.

Hooghiemstra, H., J. L. Melice, A. Berger, and N. J. Shackleton. 1993. Frequency spectra and paleoclimatic variability of the high-resolution 30–1450 ka Funza I pollen record (Eastern Cordillera, Colombia). *Quaternary Science Reviews* 12:141–156.

Imbrie, J. D., J. Hays, D. G. Martinson, A. McIntyre, A. Mix, J. J. Morley, N. G. Pisias, W. L. Prell, and N. J. Shackleton. 1984. The orbital theory of Pleistocene climate: Support from a revised chronology of the marine ^{18}O record. In A. L. Berger, J. Imbrie, J. Hays, G. Kukla, and B. Saltzman, eds., *Milankovitch and Climate*, pp. 269–305. Dordrecht, Netherlands: Reidel.

Petit, J. R., J. Jouzel, D. Raynaud, N. I. Barkov, J.-M. Barnola, I. Basile, M. Bender, and J. Chappellaz. 1999. Climate and atmospheric history of the past 420,000 years from the Vostok ice core, Antarctica. *Nature* 399:429–436.

van der Hammen, T. 1974. The Pleistocene changes of vegetation and climate in tropical South America. *Journal of Biogeography* 1:3–26.

van der Hammen, T., and E. González. 1959. *Historia de clima y vegetación del Pleistocene Superior y del Holoceno de la Sabana de Bogotá y alrededeores.* Bogotá: Dept. Cundinamarca, Official Report of the Servicio Geologico Nacional.

Van't Veer, R., and H. Hooghiemstra. 2000. Montane forest evolution during the last 650,000 yr in Colombia: A multivariate approach based on pollen record Funza-I. *Journal of Quaternary Science* 15:329–346.

CHAPTER TEN

Southern Temperate Ecosystem Responses

VERA MARKGRAF
AND MATT MCGLONE

Southern temperate regions, while often thought of as a single biogeographic entity, are as different from one another as any of them are from comparable regions in northern temperate latitudes. Some spectacular disjunctions of taxa from one widely separated landmass to another—as exemplified by the celebrated austral *Nothofagus* (southern beech) forests—have perhaps led to an overemphasis on their similarities. However, while they share a common geological and biotic origin in the megacontinent Gondwana, 90 million years of separate geological history, climate change, waves of biotic immigrants, and different evolutionary trajectories have left them radically different in many respects.

What the austral regions do share, and what differentiates them from the northern hemisphere, are similar climatic regimes. The Southern Hemisphere is 80 percent ocean with only small landmasses at mid to high latitudes. Climates throughout are strongly modified by the thermal buffering of the ocean. Cold dry air from Antarctica is warmed and humidified by its passage over water, and the mid-latitude westerlies are both much stronger and more zonal than those of the north (Hobbs et al. 1998). Temperature seasonal variability is low without the extremes of the northern hemisphere. This results in the rarity of some functional groups such as deciduous trees, lack of cold tolerance or diapause in southern hemisphere plants and animals (Wardle 1998), and abundance of lianas, palms, and tree ferns at higher latitudes than elsewhere.

In this chapter we address the effects of history on biodiversity patterns in the

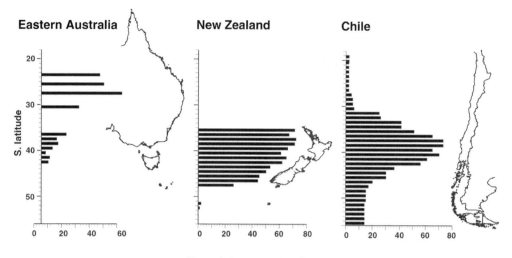

Figure 10.1. Woody plant diversity (% of total flora) in Eastern Australia, New Zealand, and Chile.

southern temperate regions in terms of their response to geological, climatic, and ecological processes in the past (Fig. 10.1). We focus primarily on plant and vegetation aspects, but also mention some instances of faunal biogeographical patterns.

BIODIVERSITY AND HISTORY

Tertiary Changes

Three major factors have controlled the biodiversity of the southern continents. First, the prolonged plate tectonic movements which split apart the megacontinent Gondwana; second, tectonic uplift and down-warping and; third, climate change. These three factors interacted throughout the past 65 million years (Cenozoic) and cannot be looked at in isolation.

The southern landmasses that currently make up Antarctica, southern South America, Australia, and New Zealand began to break apart from the southern megacontinent Gondwana approximately 100 million years ago (Fig. 10.2), and thus were interconnected at a crucial time for biotic diversification of bird, mammal, and flowering plant lineages, forming the basis for present-day biotic similarities (Cracraft 1973; Marshall and Sempere 1993; Romero 1993; Markgraf et al. 1996; Hill and Weston 2001; Lee et al. 2001). Initially, around 120 million years ago, west Gondwana lay centered on the pole, with a seasonal cool, mild, and relatively moist climate as relatively warm oceanic currents penetrated to high latitudes. Closed forests grew at the pole, and Antarctica at this time was probably a major center for angiosperm diversification (Romero 1993; Hill and Weston 2001).

From about 90 million years ago, plate tectonic movements began to shift southern landmasses equatorward by ca. 10° for South America and 20 to 30° for Australia and New Zealand, respectively, and by 38 million years ago Antarctica was separated from all southern lands (Wilford and Brown 1996). Some early interchange with the northern megacontinent of Laurasia can be assumed because of worldwide occurrences of many land mammals (Pascual 1984) and angiosperm families and genera during the early Tertiary (Burrows 1998), presumably achieved through a combination of transoceanic dispersal, movement along island chains,

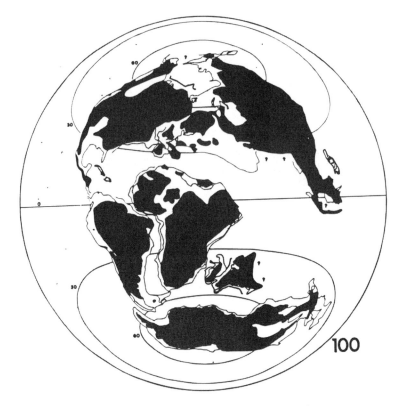

Figure 10.2. Paleogeography 100 million years ago (MA) showing Laurasia in the north, and Gondwana megacontinent before breakup. *Source:* Barron et al. (1980). Reprinted with permission from Elsevier.

and overland routes. As the four continental landmasses became gradually more isolated from one another, their biotas became more distinct. This evolutionary process was enhanced by the considerable influence of tropical biota on the formation of the southern biota, occurring at different times and from different sources in the southern landmasses. For southern South America the influx of tropical taxa began during the Paleocene (65 Ma to 54 Ma) (Romero 1993). In Australia, northward movement in the course of the Tertiary and warming climates promoted in situ evolution of tropical and subtropical taxa of Gondwanic orgin, and interchange with Southeast Asia (Macphail et al. 1994). However, even then there is little fossil evidence for a massive invasion from northern sources into Australasia; pollen fossil

records suggest most dominants have a long record extending back into the early Tertiary (ca. 60 Ma) (Kershaw et al. 1994). Early to mid-Tertiary warmth in New Zealand resulted in an influx of warm temperate taxa from both Australian and Southeast Asian sources (Lee et al. 2001).

Thus similarities of disjunct tropical and temperate forest taxa or *Nothofagus* forest avifaunas are greater between the adjacent land areas (e.g., northern Andes and southern South America (Arroyo et al. 1993) or New Guinea/New Caledonia and Australia/New Zealand (Vuillemier and Kikkawa 1991), respectively, than between Australasia and South America.

The degree of biotic isolation of the various landmasses, and the part played by vicariance (splitting of ancestral populations from formation of geographic barriers such as oceans and mountain ranges) from Gondwanic sources, versus long-distance dispersal interchange between the separating continental masses and northern tropical and subtropical sources of novelty, are under intense debate.

Molecular phylogenies and clock techniques have revolutionized our understanding of biogeography and have provided strong evidence for major long-distance dispersal events over open ocean. For instance, it is proposed that the *Lophozonia* subgenus of *Nothofagus* dispersed from Australia to New Zealand (Swenson et al. 2001); that *Laurelia* and *Nemauraon* (Atherospermataceae) dispersed from southern South America to New Zealand and New Caledonia (Renner et al. 2000), and even that some of the flightless ratite birds of New Zealand arrived subsequent to the Gondwanan breakup (Cooper et al. 2001). Pole's (1994) suggestion that the entire New Zealand flora (and perhaps a large part of the fauna) has arrived via long-distance dispersal has been strengthened by more recent work that suggests that "the floristic similarities between the Southern Andes and New Zealand mainly result from transoceanic dispersal" (Wardle et al. 2001, 92).

Many of the biotic differences between the southern temperate lands, however, have resulted from local evolutionary responses to Tertiary climate change. In addition to mountain building episodes that resulted in rain shadow effects, the crucial event was the formation of a large-scale ice sheet in Antarctica in the late Eocene / early Oligocene (38 to 36 Ma ago), triggered by the establishment of the circum-Antarctic current and cooling of the southern oceans (Ehrman and Mackensen 1992). As the Pleistocene approached, around 2 million years ago, Antarctica was more or less completely glaciated and thus no longer a source or conduit for biotic range shifts (Hill and Scriven 1995).

In consequence, throughout southern Gondwana, the closed forests of the Cretaceous and early Tertiary gave way to more open vegetation types, characterized by increasing numbers of sclerophyllous and savannah components (Romero 1993;

Archer et al. 1994), first occurrences of taxa in modern plant families (Asteraceae, Malvaceae), and increase of C4 type plants at that time, pointing to increased aridity and seasonality of precipitation (Latorre et al. 1997).

Many of these major climatic and geomorphic changes resulted in extinctions, different for plants than for animals in terms of timing and magnitude. During the early Tertiary (ca. 65 Ma to 24 Ma) in South America, land mammals were represented by marsupials and placentuals, endemic descendants from immigrants from Laurasia, which had replaced the highly peculiar Gondwanan endemic types (Pascual 1984; Marshall and Sempere 1993). Faunal turnovers occurred around 25 million years ago (late Oligocene) when Caribbean taxa immigrated, 6 million to 10 million years ago (late Miocene) with immigration from Central America and during mid- (3.4 Ma) and late (2.47 Ma) Pliocene, when an abrupt increase of Holarctic immigrants caused a great number of extinctions, reflecting the closing of the Isthmus of Panama (Great Interamerican Exchange) (Pascual 1984; Tonni et al. 1992; Marshall and Sempere 1993). Plant extinctions in southern South America apparently did not occur at the same times as the animal extinctions, and perhaps not even for the same reasons. For instance, major extinctions of taxa characteristic of the Gondwanan "*Nothofagidites* Province" (fossil floral association of late Cretaceous times composed primarily of taxa in the families of Podocarpaceae, Proteaceae, and Myrtaceae) occurred 65 million years ago simultaneous with the appearance of new taxa (Markgraf et al. 1996). A possible cause was mountain building episodes, leading to changes in seaways.

Tertiary extinction waves in Australian mammals, in contrast, appear to reflect altered climates and environments rather than influxes of immigrants—which have been confined to rodents in the latest Miocene–early Pliocene (ca. 7 to 5 Ma).

Decline in closed forest mammals in the middle Tertiary were offset by the rise in open country grazers and browsers adapted to open forest and scrub (Archer et al. 1994).

New Zealand vertebrate fossil history is poorly known, and it is conceivable that early mammals were part of the original Gondwanan inheritance, along with dinosaurs, *Sphenodon*, primitive Leiopelmatid frogs, and other ancient groups. Molecular phylogenies from extant and recently extinct groups points to an Oligocene (37 to 24 Ma) crisis when extreme reduction of the landmass to a low-lying archipelago effectively created a biotic bottleneck through which only a handful of taxa persisted but which was followed by radiation in some groups such as the ratite moa (Cooper and Cooper 1995). Plant taxa show no similar pattern, suggesting that land fragmentation was not as a critical factor as it was for vertebrates.

There have been few documented plant extinctions since the major floristic adjustment about 5 to 6 million years ago, suggesting that those plant taxa that could not survive the onset of glacial conditions were culled early on. However, several pollen types (*Dacrydium* in Australia; *Acacia* in New Zealand) vanish in the last 100,000 years (Kershaw 1984; Bussell 1992), indicating that perhaps plant taxa extinctions continued to affect the flora with stochastic losses during glacial–interglacial transitions.

Quaternary

The Quaternary is defined by the onset of marked glacial–interglacial cycles, commonly referred to as the "Ice Age," at around 1.8 million years ago. Cooling and increasing aridity had been under way since at least the mid-Miocene (ca. 15 Ma) with the establishment of the circum-Antarctic oceanic circulation, and was accelerated by the final glaciation of Antarctica in the late Pliocene. However,

although there had been cycles between warmer and cooler episodes during the Pliocene (6 to 1.8 Ma), the onset of these Quaternary cycles 1.8 million years ago marks the beginning of a new biotic era, characterized by initiation of widespread glaciation in the southern Andes, New Zealand, southern Australia, and New Guinea (Clapperton 1990).

In Antarctica, ice sheets advanced to the edge of the continental shelf in places during glacials, and the spring sea ice fringe lay 4° to 5° of latitude farther north (Burckle and Cirillii 1987). These thin glacial sea ice shelves vanished quickly with interglacial warming, and thus there is no equivalent in the south of the prolonged glacial lag of the continental ice sheets of the north. Glaciation was not extensive in Australia, being confined to limited areas of the highest mountain ranges in southeastern Australia and Tasmania. In New Zealand, the Southern Alps were extensively glaciated toward the south, with essentially a piedmont ice cap, with glaciers extending to below present sea level in the west. In southern South America the most extensive glaciation dates back to 1.2 to 1.0 Ma. During that period, south of latitude 50°S, an ice cap covered essentially all land area, from 120 m below sea level in the Chilean Channels across the Andes to the Atlantic Ocean (Mercer 1976; Clapperton 1993; Rabassa et al. 2000). Although the more recent glaciations in the southern Andes were less extensive than earlier ones, south of latitude 43°S valley glaciers descending from the Patagonian ice cap dissected, and probably eliminated, the whole temperate forest area.

In the southern hemisphere, the most biologically significant glacial episodes were the extreme climate states, be they during a glacial or an interglacial. Deglacial transitional sequences of high frequency and high amplitude climate variability imposed unique stresses on the biota. Biodiversity during Quaternary times responded first by proliferation of species with wide ecological (moisture and temperature) amplitudes, second by range shifts, and third by extinction of warm-adapted taxa (during glacials) or cold-adapted taxa (during interglacials).

All three landmasses suffered massive extinction of large vertebrate animals in the late Pleistocene, between about 40,000 and 10,000 years ago. In Australia, at least 40 species of mammals were lost, including all with a mass greater than 60 kg. New Zealand suffered loss of 40 percent of all land bird species, including all 12 moa species, and extirpation and reduction of range of land-breeding marine mammals and many reptiles and amphibians. Losses were particularly high among long-standing endemic taxa. New World mammalian faunas were also devastated, with loss of over 150 genera (Anderson 1984).

Biodiversity and Extreme Climates of the Last Glacial Maximum (LGM)

The history of the last glacial–interglacial cycle, that is, the past 20,000 years, illustrates the typical response of southern plant taxa to the range of climatic conditions experienced in the Pleistocene. The last glacial maximum (LGM: 25,000 to 17,000 calendar years before present), which presaged the late-glacial transition to the present-day interglacial, has been estimated by various means, including glacial equilibrium lines, microfossils, and isotopes in deep sea cores, to have been between 4° and 6°C cooler than the historical average over the southern continents (McGlone 1988; Harrison and Dodson 1993; McGlone et al. 1993; Hulton et al. 2002).

During glacial times—including the LGM, all pollen records from the temperate latitudes in South America, east and west of the Andes, southern Australia, and New Zealand south of latitude 37°S—show either substantially reduced proportions of tree taxa or treeless conditions.

Widespread aridity is reported for the LGM in Australia (Harrison and Dodson 1993) and southern South America in the ice-free areas north of 36°S, south of 43°S, and east of the Andes (Markgraf et al. 1992). Being a narrow oceanic archipelago, New Zealand cannot have become as arid as the southern continents, but there are indications that precipitation was reduced by at least one-third (McGlone et al. 1993). Nevertheless, despite the very large changes in vegetation communities, it appears that forest elements remained more or less in situ in sites sheltered from the full force of LGM climates, or along the coasts (McGlone et al. 1984; Hope 1994; Markgraf et al. 1995).

The area in southern South America where temperate forest components during LGM persisted—co-dominant with steppe taxa—was restricted to west of the Andes between latitudes 36° and 43°S (Markgraf 1991a). In contrast to today's high arboreal diversity in this latitudinal belt (Markgraf et al. 1995; Villagrán 1995) the dominant arboreal element was *Nothofagus* (probably *Nothofagus dombeyi*, Markgraf 1991b) accompanied only by traces of other trees (Heusser et al. 1996; Moreno 1997; Heusser et al. 1999). According to quantitative pollen-climate calibration analysis most of these mid-latitude LGM assemblages are analogous to the Patagonian steppe-woodland assemblages east of the Andes (Markgraf et al. 2002). The presence of many large mammal grazers in the Chilean lowlands (as well as east of the Andes) at that time, including horse, mastodon, camelids, deer, and ground sloth (Moreno et al. 1994; Nuñez et al. 2001), also suggests a Patagonian steppe-woodland environment. In terms of climate, LGM summer precipitation in the Chilean lowlands would have been only 30 percent of today's, winters 3°C colder, and seasonal temperature range would have been markedly higher than today.

Treeless vegetation, dominated by

grasses, herbs, and asterad shrubs, extended across most of southern Australia and down to sea level on the southern coast (Hope 1994). The central desert regions expanded, and open woodlands were widespread as far north as Queensland. Faunal records include more ecological groups at a given locality than occur at present. The mixed faunal and pollen assemblages suggest a mosaic of dense scrub to open shrublands.

New Zealand had steppe-like conditions in the far southeast, with grassland, low shrubland, or open ground predominating in the dry east, and subalpine shrubland and tussock grassland in the west (McGlone et al. 1993, 1996). Forest patches—mostly of *Nothofagus*—were part of the mix throughout, but forming an increasing component of the mosaic in the North Island (McGlone et al. 1984; Pickrill et al. 1992; Lees et al. 1998). Continuous closed forest occurred only in the northernmost Northland peninsula north of 37°S. Biogeographic considerations also support tree survival in the South Island: for instance, large disjunctions from the northern to southern South Island of all four *Nothofagus* species, one of which is typical of lower montane sites, are strong evidence for their LGM survival in far southern districts (McGlone et al. 1996).

The dominant cool climate subalpine-like vegetation that dominated during the LGM can therefore be thought of as an azonal matrix imposed by relatively infrequent cold dry episodes. Intricately woven into this matrix was a highly variable network of forest patches, ever expanding and retreating, coalescing and fragmenting in response to geomorphologic and climatic change. The similarity of vegetation and faunal response to LGM climates in the southern continents suggests a common cause. Very cold polar oceans and an extensive sea-ice belt to the south (Nelson et al. 1993; Carter et al. 2002), cooler tropical oceans, and lowered sea levels cut off major sources of atmospheric moisture, thus lowering precipitation in most places and more vigorous atmospheric circulation as a consequence of the steepened pole–equator temperature gradient increased chronic wind stress. Under these circumstances, frequent and cold Antarctic air outbreaks would have occurred (Bradbury et al. 2001).

Biodiversity and Holocene Change

Both the late-glacial interval and the Holocene (the past 10,000 years) show high environmental and climatic variability that are of special interest in context of the response of terrestrial biota to climate change.

In mid-latitudes in all southern lands, maximum climatic equability—characterized in some areas by maximum temperatures and precipitation—occurred between 9000 and 4000 years ago. There is no indication of any major extension of range of taxa at this time. Certain elements—tree ferns in New Zealand, certain moisture-loving trees and shrubs such as *Phyllocladus*, *Nothofagus*, and *Pomaderris* in Australia, seasonal drought–adapted *Nothofagus obliqua* in South America—became very much more common, suggesting local increases rather than migratory movements. In southern South America, except for the zone around 50°S, the early Holocene was characterized by high fire frequency (Markgraf and Anderson 1994; Huber and Markgraf 2003) suggesting that interannual or decadal variability of precipitation was high, as storm tracks must have been focused narrowly at 50°S (Markgraf et al. 1992; Markgraf et al. 2003).

The last few thousand years in mid-latitude temperate southern regions have been characterized by cooler, more variable climates in which drought and frost have become more prominent. El Niño/Southern Oscillation (ENSO), with its high inter-annual and decadal climatic variability, began to affect biodiversity patterns globally in late Holocene times and especially in the southern hemisphere (e.g., McGlone et al. 1992; Boninsegna and Villalba 1996; Markgraf and Diaz 2000). In the temperate forest region of South America, ENSO is expressed by greatly increased precipitation (latitude 31° to 36°) or by decreased precipitation and warmer temperatures (ca. latitude 40°S). With the onset of ENSO after about 5000 years ago, biota responded accordingly, showing increase in forest taxa diversity and cover at the lower latitudes (Markgraf 1987; Villagran and Varela 1990) and an anomalous combination of forest taxa adapted to higher moisture as well as increased drought and fire at mid-latitudes (McGlone et al. 1992; Markgraf and Diaz 2000). South of 50°S in South America, on the other hand, late Holocene climates were characterized by great stability and moisture equability, documenting again the regional differences in climate expression.

BIODIVERSITY RESPONSE TO CLIMATE

There is no good evidence for range migration as a major response to climate change in the southern continents. Forest and warm temperate species appear to have survived either within their current ranges as small patches in favourable habitat sheltered from the full impact of dry, windy, cool, and variable LGM climates, or nearby, for instance along coastlines or in complex mountainous terrain. The reverse has happened during the warm, mild interglacial peaks, with cool climate elements located in upland patches and on poor soils. However, the rigorous demands made by constant glacial–interglacial expansion and contraction, with patch density and continuity diminishing toward the cooler south and more arid interiors and north during glacial maxima, have left an imprint. Chief among these is the rapid falloff in forest species richness

from north to south, particularly noticeable in South America north of 34°S and south of 43°S (Arroyo et al. 1995; Markgraf et al. 1995; Villagrán 1995) suggesting that there is no migratory tracking of climate but simply a multi-cycle adjustment to the severity of the glacial maxima (McGlone et al. 1996).

The taxon diversity and genetic structure of the tree species reflect this LGM gradient. In situ persistence is most marked in Australia, where there is extensive local speciation of the major tree species (Burgam 2002), demonstrating old patterns not markedly affected by glacial climates. Genetic and taxon diversity of forest taxa is high in the very north of New Zealand, where forest was continuous through glacial–interglacial cycles, but is extremely low to the south where repeated in situ fragmentation and recoalescence of tree populations has occurred (Hawkins and Sweet 1989; Billington 1991; Haase 1992a,b; McGlone 2002). Analysis of the genetic characteristics of several major forest species in the rain forests of southern South America (Fitzroya, Austrocedrus, Pilgerodendron, Nothofagus) also indicates that the mid-latitude region with high present-day biodiversity was much less affected by the ice ages than were the higher latitudes (Premoli et al. 2000a,b; 2002). High genetic similarity within these mid-latitude populations, especially west of the Andes, suggests that the distances between the fragmented forests was not restricting gene flow. On the other hand, in populations east of the Andes and in the higher latitudes where forests were replaced by steppe and heath during glacial intervals, genetic variability is higher than for those to the west, which suggests that distances between the refuges were greater and that reexpansion occurred from the same refuges and not from populations west of the Andes.

LESSONS FOR FUTURE CLIMATES AND BIODIVERSITY

Global circulation modeling experiments related to increased greenhouse gases predict substantial climatic change for both temperature and precipitation (Pittock and Salinger 1982; Stouffer et al. 1989; Bridgman 1997; see also Chapter 1). Southern hemisphere mid- and high latitudes will experience a temperature increase of up to 5°C and, due to the poleward shift of the subtropical high pressure cell, especially strong in summer, precipitation will also be shifted poleward. In southern South America the seasonal rain forest region between latitudes 34° and 42°S, with its high biodiversity, would be much warmer and drier than today with most likely markedly increased fire frequency, conditions perhaps comparable to the early Holocene. Plant and animal taxa would shift by about 6° latitude and about 500-m elevation, essentially eliminating the Andean tundra zone (Arroyo et al. 1993; Contreras 1993; Rozzi et al. 1995). Similar predictions have been made for biotic changes in New Zealand and Australia (Mitchell and Williams 1996; Whetton et al. 1996).

Current biodiversity changes in the southern continents are still largely driven by anthropogenic alteration of habitat. In New Zealand and Australia, the era of widespread land clearances is now over, but the legacy of fragmentation, salinization due to excessive removal of woody cover, changed fire intensity, and introduced weeds and pests continues to have a negative impact on indigenous biodiversity. Against this continuing adjustment to massive anthropogenic change, it is difficult to distinguish the response to global warming since the beginning of the nineteenth century. For instance, there is no clear indication that New Zealand tree lines are rising, despite a 0.5°C or more increase in summer temperatures since the early years of the twentieth century (Cullen et al. 2001).

In the past, climatic variability appears to have been at least as important as climate absolutes in governing biodiversity responses in southern climates. As oceanic weather tends to be relatively aseasonal, variability assumes a major importance. The widespread treeless condition of much of the southern continents during the LGM probably owes as much to increased intra-seasonal variability of precipitation and temperature as it does to the changing means. In Australia, it has been argued that long association of the continent with persistent ENSO fluctuations in precipitation has dominated the evolutionary trajectory of the biota, with factors such as fire-resistance or tolerance, sclerophylly, and litter or clutch size in vertebrates helping responses to unpredictable climates (Nicholls 1991). Parts of southern South America are likewise under strong ENSO influences, and it would be surprising if a similar biotic tolerance had not developed. Cooler, higher latitudes are not as influenced by ENSO, but other climatic cycles and sporadic outbursts of cold high latitude air throughout the year keep variability high.

Despite massive range and population size fluctuations in the Quaternary, it can be assumed that the individual taxa that have persisted are those that can survive such fluctuations, and small population sizes in and of themselves are probably not a major threat to the vast majority of species. However, there are now two new factors. First, systematic extirpation of some taxa from major parts of their range, agricultural clearance leading to large-scale fragmentation of the forests and extensive disturbance, including fires, and introduction of pests and weeds have radically changed the context in which climatic variability operates; thus, survival of many lowland taxa cannot be assured. Second, while an increase in annual temperatures by 1 to 2°C over the pre-industrial norm is unlikely to pose any serious threat to indigenous species which have experienced such change in the relatively recent past, as temperature changes go beyond 2°C—that is, greater than experienced during previous interglacials—many species will enter new climatic territory. Some, perhaps most, will be able to survive under warmer climates, but there must be questions about interactions with exotics from warmer areas of the globes. In the remote past, whenever there have been major warmings beyond previous norms, immigration or evolution of new taxa and elimination of old, cool-adapted taxa has been the rule. There is no hint that different outcomes should be expected with anthropogenic warming, save perhaps a greater proportion of weedy immigrants due to land use change.

SUMMARY

Since the Cretaceous, the southernmost landmasses of the southern hemisphere have been in only fitful and incomplete contact with the north. Major climatic and physiographic differences, combined with this partial isolation, have led to the production of strikingly different temperate biota. Although, as we have shown, there are great differences between the southern landmasses, they have some major, historically determined factors in common: Ancient austral or paleobiotic elements have been preserved in the south. In some cases (such as with the ancient reptile lineage represented by the tuatara (*Sphenodon*) in New Zealand) this was due to isolation from novel influences such as mammals. In others, such as with the speciose Proteaceae in Australia, the ancient groups are not relictual in any sense, but represent a well-adapted radiation to unique southern conditions. When there has been massive reduction in typically southern ele-

ments—as with the Podocarpaceae and Auracariaceae during the Quaternary (1.8 Ma to present) contraction of wet forest (Hill and Brodribb 1999)—this has often been due to changing climates.

Episodic influx from tropical and subtropical elements from adjacent landmasses during the Tertiary (65 to 1.8 Ma), and increasingly from vagile northern temperate elements during the past 2 million years (Plio-Pleistocene), led to marked differences between the separate southern areas according to the source. Long-distance biotic interchange probably continued between the southern landmasses over vast stretches of ocean.

The lack of large land areas in high latitudes and the mitigating influence of ocean on climate extremes probably were responsible for the scarcity of deciduous woody taxa in the southern temperate forests and poor adaptation of fauna and flora to low temperatures.

Coolings during the glacial maxima were significantly moderated by the oceans, and the fact that glaciation was largely of the mountain valley glacier type with extensive ice caps restricted to some areas in the southern Andes. Downstream climatic effects were therefore far less strong than those known in the northern hemisphere (e.g., Whitlock et al. 2001).

Because of the southern continents' physiography and the mitigating effects of the oceans on climate extremes, the typical response of the southern rain forest taxa to climatic extremes or oscillations was not large-scale range shift, as in the northern hemisphere land areas (see also Chapter 4), but rather forest fragmentation during the glacial intervals and reexpansion during the interglacial intervals.

Plant extinction and speciation events have followed major climatic and geomorphic change. For mammals, major extinction events in South America occurred in the Pliocene (5 to 2 million years ago), timed with the closing of the Isthmus of

Panama, and during the late Pleistocene. In both cases, novel mammalian competitors, perhaps also including humans (e.g., Martin and Klein 1984; but see Markgraf 1985; Nuñez et al. 2001), played a role. In Australia, contraction of ever-wet forests appears to have resulted in major mammalian turnover in the late Tertiary, and humans caused the extinction of many species during the late Pleistocene (Roberts et al. 2001). New Zealand vertebrate extinctions probably followed reduction in land area during the Oligocene (37 to 24 Ma) inundation, and arrival of humans in the late Holocene (Anderson and McGlone 1991). Extinctions accelerated with European settlement, and newly introduced mammalian browsers and grazers effectively reduced biodiversity in the southern forests (Veblen et al. 1989).

Lessons from the past for future climates suggest that climate variability may be as important as climate absolutes in determining biotic response. Two major new factors are the degree of habitat loss to human activities and the destination of possible climate change. Habitats are heavily fragmented in most southern temperate settings now, making range adjustments to climate change difficult or impossible. Once warming exceeds about 2°C, it will exceed temperatures during most previous interglacials, raising concerns about species' abilities to survive and compete with introduced alien species in modified landscapes.

REFERENCES

Anderson, A. J., and M. S. McGlone. 1991. Living on the edge—prehistoric land and people in New Zealand. In The Native Lands: Human–Environmental Interactions in Australia and Oceania, pp. 199–241, ed. J. Dodson. Melbourne: Longmans.

Anderson, E. 1984. Who's who in the Pleistocene: A mammalian bestiary. In Quaternary Extinctions: A Prehistoric Revolution, pp. 40–89, ed. P. S. Martin and R. G. Klein. Tucson: University of Arizona Press.

Archer, M., S. J. Hand, and H. Gothelp. 1994. Patterns in the history of Australia's mammals and inferences about palaeohabitats. In History of the Australian Vegetation: Cretaceous to Recent, pp. 80–103, ed. R. S. Hill. Cambridge: Cambridge University Press.

Arroyo, M. T. K., J. J. Armesto, F. Squeo, and J. Gutierrez. 1993. Global change: Flora and vegetation of Chile. In Earth System Responses to Global Change, Contrasts between North and South America, pp. 239–264, ed. H. A. Mooney, E. R. Fuentes, and B. I. Kronberg. San Diego: Academic Press.

Arroyo, M. T. K., M. Riveros, A. Penaloza, L. Cavieres, and A. M. Faggi. 1995. Phytogeographic relationships and regional richness patterns of the cool temperate rainforest flora of southern South America. In High-Latitude Rainforests and Associated Ecosystems of the West Coast of the Americas. Climate, Hydrology, Ecology, and Conservation, ed. R. G. Lawford, P. B. Alaback, and E. Fuentes. Ecological Studies 116:134–172.

Ashworth, A. C., V. Markgraf, and C. Villagrán. 1991. Late Quaternary climatic history of the Chilean Channels based on fossil pollen and beetle analyses, with an analysis of the modern vegetation and pollen rain. Journal of Quaternary Science 6:279–291.

Bennett, K. D., S. G. Haberle, and S. H. Lumley. 2000. The last glacial—Holocene transition in southern Chile. Science 290:325–328.

Billington, H. L. 1991. Effect of population size on genetic variation in a dioecious conifer. Conservation Biology 5:115–119.

Blunier, T., J. Chappellaz, J. Schwander, A. Dallenbach, B. Stauffer, T. F. Stocker, R. Raynaud, J. Jouzel, H. B. Claussen, C. U. Hammer, and S. J. Johnson. 1998. Asynchrony of Antarctic and Greenland climate change during the last glacial period. Nature 394:739–743.

Boninsegna, J. A., and R. Villalba. 1996. Dendroclimatology in the southern hemisphere: Review and prospects. In Tree Rings, Environment and Humanity, ed. J. S. Dean, D. M. Meko, and T. W. Swetnam. Radiocarbon 38:127–141.

Bradbury, J. P., M. Grosjean, S. Stine, and F. Sylvestre. 2001. Full and late glacial lake record along the PEP 1 transect: Their role in developing interhemispheric paleoclimate interactions. In Interhemispheric Climate Linkages, pp. 265–289, ed. V. Markgraf. San Diego: Academic Press.

Bridgman, H. 1997. Future climate scenarios for the southern continents. In Climates of the Southern Continents, pp. 265–292, ed. J. E. Hobbs, J. A. Lindesay, and H. A. Bridgman. Chichester: Wiley & Sons.

Burckle, L. H., and J. Cirillii. 1987. Origin of diatom ooze belt in the Southern Ocean: Implications for late Quaternary paleoceanography. Micropaleontology 33:82–86.

Burgam, M. A. 2002. Are listed threatened plant species actually at risk? Australian Journal of Botany 50:1–13.

Burrows, C. J. 1998. Similarities between mid-Eocene floras in Germany and the modern floras of Australia, New Caledonia and New Zealand. *Records of the Canterbury Museum* 12:31–78.

Bussell, M. R. 1992. Late Pleistocene palynology of terrestrial cover beds at the type section of the Rapanui Terrace, Wanganui, New Zealand. *Journal of the Royal Society of New Zealand* 22:77–90.

Carter, L., H. L. Neil, and L. Northcote. 2002. Late Quaternary ice-rafting events in the SW Pacific Ocean, off eastern New Zealand. *Marine Geology*, 191:19–35.

Clapperton, C. 1990. Quaternary glaciations in the Southern Hemisphere. *Quaternary Science Reviews* 9:121–304.

Clapperton, C. 1993. *Quaternary Geology and Geomorphology of South America.* Amsterdam: Elsevier.

Clark, J. S., J. S. McLachlan. 2003. Stability of forest biodiversity. *Nature* 423(6940):635–638.

Contreras, L. C. 1993. Effect of global climatic change on terrestrial mammals in Chile. In *Earth System Responses to Global Change, Contrasts between North and South America*, pp. 285–294, ed. H. A. Mooney, E. R. Fuentes, and B. I. Kronberg. San Diego: Academic Press.

Cooper, A., and R. A. Cooper. 1995. The Oligocene bottleneck and New Zealand biota: Genetic record of a past environmental crisis. *Proceedings of the Royal Society of London Series B*, 261:293–302.

Cooper, A., C. Lalueza-Fox, S. Anderson, A. Rambaut, J. Austin, and R. Ward. 2001. Complete mitochondrial genome sequences of two extinct moas clarify ratite evolution. *Nature* 409:704–707.

Cracraft, J. 1973. Contenental drift, paleoclimatology, and the evolution and biogeography of birds. *Journal of Zoology*, London, 169:455–545.

Cullen, L. E., G. H. Stewart, R. P. Duncan, and J. G. Palmer. 2001. Disturbance and climate warming influences on New Zealand *Nothofagus* tree-line population dynamics. *Journal of Ecology* 89:1061–1071.

Davis, M. B., and R. G. Shaw. 2001. Range shifts and adaptive responses to Quaternary climate change. *Science* 292:673–679.

Denton, G. H., C. J. Heusser, T. V. Lowell, P. I. Moreno, B. G. Andersen, L. E. Heusser, C. Schlüchter, and D. R. Marchant. 1999. Interhemispheric linkage of paleoclimate during the last glaciation. *Geografiska Annaler* 81A:107–153.

Ehrmann, W. V., and A. Mackensen. 1992. Sedimentological evidence for the formation of an East Antarctic ice sheet in Eocene/Oligocene time. *Palaeogeography, Palaeoecology, Palaeoclimatology* 93:85–112.

Haase, P. 1992a. Isozyme variability and biogeography of *Nothofagus truncata* (Fagaceae). *New Zealand Journal of Botany* 30:315–328.

Haase, P. 1992b. Isozyme variation and genetic relationships in *Phyllocladus trichomanoides* and *P. alpinus*

(Podocarpaceae). *New Zealand Journal of Botany* 30:359–363.

Harrison, S. P., and J. Dodson. 1993. Climates of Australia and New Guinea since 18,000 yr B.P. In *Global Climates since the Last Glacial Maximum*, pp. 265–293, ed. H. E. Wright, J. E. Kutzbach, T. Webb III, W. F. Ruddiman, F. A. Street-Perrrott, and P. J. Bartlein. Minneapolis: University of Minnnesota Press.

Hawkins, B. A., and G. B. Sweet. 1989. Genetic variation in rimu—an investigation using isozyme analysis. *New Zealand Journal of Botany* 27:83–90.

Heusser, C. J., L. E. Heusser, and T. V. Lowell. 1999. Paleoecology of the southern Chilean Lake District—Isla Grande de Chiloe during middle–late Llanquihue glaciation and deglaciation. *Geografiska Annaler A*, 81:231–284.

Heusser, C. J., T. V. Lowell, L. E. Heusser, A. Hauser, B. G. Andersen, and G. H. Denton. 1996. Full-glacial–late-glacial palaeoclimate of the Southern Andes: Evidence from pollen, beetle, and glacial records. *Journal of Quaternary Science* 11:173–184.

Hill, R. S., and T. J. Brodribb. 1999. Turner Review No. 2—Southern conifers in time and space. *Australian Journal of Botany* 47:639–696.

Hill, R. S., and L. J. Scriven. 1995. The angiosperm-dominated woody vegetation of Antarctica—a review. *Review of Palaeobotany & Palynology* 86:175–198.

Hill, R. S., and P. H. Weston. 2001. Southern (Austral) ecosystems. *Encyclopedia of Biodiversity*, v. 5, pp. 361–370. San Diego: Academic Press.

Hobbs, J. E., J. A. Lindesay, and H. A. Bridgman. 1998. *Climates of the Southern Continents*, ed. J. E. Hobbs, J. A. Lindesay, and H. A. Bridgman. Chichester: Wiley & Sons.

Hope, G. S. 1994. Quaternary vegetation. In *History of the Australian vegetation: Cretaceous to Recent*, pp. 368–389, ed. R. S. Hill. Cambridge: Cambridge University Press.

Huber, U. M. 2001. Linkages between climate, vegetation and fire in Fuego-Patagonia during the late glacial and Holocene. Ph.D. Dissertation, University of Colorado, Boulder.

Huber, U. M., and V. Markgraf. 2003. Holocene moisture regime changes and fire frequency at Rio Rubens bog, southern Patagonia. In *Fire and Climatic Change in Temperate Exosystems of the Western Americas*, pp. 357–380, ed. T. W. Swetnam, T. A. Veblen, and G. Montenegro. Berlin: Springer-Verlag.

Hulton, N. R. J., R. S. Purves, R. D. McCulloch, D. E. Sugden, and M. J. Bentley. 2002. The last glacial maximum and deglaciation in southern South America. *Quaternary Science Reviews* 21:233–241.

Huntley, B., and H. J. B. Birks. 1983. *An Atlas of Past and Present Pollen Maps for Europe 0–13,000 years ago.* Cambridge: Cambridge University Press.

Kershaw, P. 1984. Late Cenozoic plant extinctions in

Australia. In *Quaternary Extinctions: A Prehistoric Revolution*, pp. 691–707, ed. P. S. Martin, and R. G. Klein. Tucson: University of Arizona Press.

Kershaw, A. P., H. A. Martin McEwen, and J. R. C. Mason. 1994. The Neogene: A period of transition. In *History of the Australian Vegetation: Cretaceous to Recent*, pp. 299–327, ed. R. S. Hill. Cambridge: Cambridge University Press.

Latorre, C., J. Quade, and W. C. McIntosh. 1997. The expansion of C4 grasses and global change in the late Miocene: Stable isotope evidence from the Americas. *Earth and Planetary Science Letters* 146:83–96.

Lee, D. E., W. G. Lee, and G. Mortimer. 2001. Where and why have all the flowers gone? Depletion and turnover in the New Zealand Cenozoic angiosperm flora in relation to palaeogeography and climate. *Australian Journal of Botany* 49:341–356.

Lees, C. M., V. E. Neall, and A. S. Palmer. 1998. Forest persistence at coastal Waikato, 24,000 years BP to present. *Journal of the Royal Society of New Zealand* 28:55–81.

Markgraf, V. 1985. Late Pleistocene faunal extinctions in Southern Patagonia. *Science* 228:1110–1112.

Markgraf, V. 1987. Paleoenvironmental changes at the northern limit of the subantarctic Nothofagus forest, lat 37°S, Argentina. *Quaternary Research* 28:119–129.

Markgraf, V. 1991a. Late Pleistocene environmental and climatic evolution in southern South America. *Bamberger Geographische Schriften* 11:271–282.

Markgraf, V. 1991b. Younger Dryas in southern South America. *Boreas* 20:63–69.

Markgraf, V. 1993. Younger Dryas in southernmost South America—an update. *Quaternary Science Reviews* 12:351–355.

Markgraf, V., and L. Anderson. 1994. Fire history of Patagonia: Climate versus human cause. *Revista Instituto Geologico Sao Paulo, Brazil* 15:35–47.

Markgraf, V., and M. M. Bianchi. 1999. Paleoenvironmental changes during the last 17,000 years in western Patagonia: Mallin Aguado, Province of Neuquen, Argentina. *Bamberger Geographische Schriften* 19:175–193.

Markgraf, V., J. P. Bradbury, A. Schwalb, S. J. Burns, Ch. Stern, D. Ariztegui, A. Gilli, F. S. Anselmetti, S. Stine, and N. Maidana. 2003. Holocene paleoclimates of southern Patagonia: Limnological and environmental history of Lago Cardiel, Argentina. *The Holocene* 13:597–607.

Markgraf, V., and H. F. Diaz. 2000. The past ENSO record: A synthesis. In *El Niño and the Southern Oscillation: Multiscale Variability and Global and Regional Impacts*, pp. 465–488, ed. H. F. Diaz and V. Markgraf. Cambridge, UK: Cambridge University Press.

Markgraf, V., J. R. Dodson, A. P. Kershaw, M. S. McGlone, and N. Nicholls. 1992. Evolution of late Pleistocene and Holocene climates in the circum-South Pacific land areas. *Climate Dynamics* 6:193–211.

Markgraf, V., and R. Kenny. 1997. Character of rapid vegetation and climate change during the late-glacial in southernmost South America. In *Past and Future Rapid Environmental Changes*, ed. B. Huntley, W. Cramer, A. V. Morgan, H. C. Prentice, and J. R. M. Allan. NATO ASI Series I 47:81–90.

Markgraf, V., M. McGlone, and G. Hope. 1995. Neogene paleoenvironmental and paleoclimatic change in southern temperate ecosystems—a southern perspective. *Trends in Ecology and Evolution* 10:143–147.

Markgraf, V. E. Romero, and C. Villagrán. 1996. History and paleoecology of South American Nothofagus forests. In *The Ecology and Biogeography of Nothofagus forests*, pp. 354–386, ed. T. T. Veblen, R. S. Hill, and J. Read. New Haven and London: Yale University Press.

Markgraf, V., R. S. Webb, K. H. Anderson, and L. Anderson. 2002. Modern pollen/climate calibration for southern South America. *Palaeogeography, Palaeoclimatology, Palaeoecology* 182:375–397.

Marshall, L. G., and T. Sempere. 1993. Evolution of the neotropical Cenozoic land mammal fauna in its geochronologic, stratigraphic, and tectonic context. In *Biological Relationships between Africa and South America*, pp. 329–392, ed. P. Goldblatt. New Haven: Yale University Press.

Martin, P. S., and R. G. Klein, Eds. 1984. *Quaternary Extinctions*. Tucson: University of Arizona Press.

McGlone, M. S. 1988. New Zealand. In *Handbook of Vegetation Science*, ed. B. Huntley and T. Webb III. Vol. 7: *Vegetation History*, pp. 558–599. London: Kluwer Academic.

McGlone, M. S. 1995. Late glacial landscape and vegetation change during the Younger Dryas climatic oscillation in New Zealand. *Quaternary Science Reviews* 14:867–881.

McGlone, M. S. 1997. The response of New Zealand forest diversity to Quaternary climates. In *Past and Future Rapid Environmental Changes: The Spatial and Evolutionary Responses of Terrestrial Biota*, pp. 73–80, ed. B. Huntley, W. Cramer, A. V. Morgan, H. C. Prentice, and J. R. M. Allen. Nato ASI Series. Series 1: Global Environmental Change, Vol. 47. Berlin: Springer-Verlag.

McGlone, M. S. 2001. A late Quaternary pollen record from marine core P69, southeastern North Island, New Zealand. *New Zealand Journal of Geology and Geophysics* 44:69–77.

McGlone, M. S. 2002. The Late Quaternary peat, vegetation and climate history of the Southern Oceanic Islands of New Zealand. *Quaternary Science Reviews* 21:683–707.

McGlone, M. S., R. Howorth, and W. A. Pullar. 1984: Late Pleistocene stratigraphy, vegetation and climate of the Bay of Plenty and Gisborne regions, New Zealand. *New Zealand Geology and Geophysics* 27:327–350.

McGlone, M. S., A. P. Kershaw, and V. Markgraf. 1992.

El Niño/Southern Oscillation climatic variability in Australasian and South American paleoenvironmental records. In *El Niño: Historical and Paleoclimatic Aspects of the Southern Oscillation*, pp. 435–462, ed. H. F. Diaz and V. Markgraf. Cambridge, U.K.: Cambridge University Press.

McGlone, M. S., D. C. Mildenhall, and M. S. Pole. 1996. History and paleoecology of New Zealand Nothofagus forests. In *The Ecology and Biogeography of Nothofagus forest*, pp. 83–130, ed. T. T. Veblen, R. S. Hill, and J. Read. New Haven: Yale University Press.

McGlone, M. S., M. J. Salinger, and N. T. Moar. 1993. Palaeovegetation studies of New Zealand's climate since the Last Glacial Maximum. In *Global Climates since the Last Glacial Maximum*, pp. 294–317, ed. H. E. Wright, J. E. Kutzbach, T. Webb III, W. F. Ruddiman, F. A. Street-Perrrott, and P. J. Bartlein. Minneapolis: University of Minnnesota Press.

McGlone, M. S., R. P. Duncan, P. B. Heenan. 2001. Endemism, species selection and the origin and distribution of the vascular plant flora of New Zealand. *Journal of Biogeography* 28:199–216.

Macphail, M. K., N. F. Alley, E. M. Truswell, and I. R. K. Sluiter. 1994. Early Tertiary vegetation: Evidence from spores and pollen. In *History of the Australian Vegetation: Cretaceous to Recent*, pp. 189–261, ed. R. S. Hill. Cambridge University Press.

Mercer, J. J. 1976. Glacial history of southernmost South America. *Quaternary Research* 6:125–166.

Mildenhall, D. C. 1995. Pleistocene palynology of the Petone and Seaview drillholes, Petone, Lower Hutt Valley, North Island, New Zealand. *Journal of the Royal Society of New Zealand* 25:207–262.

Mitchell, N. D., and J. E. Williams. 1996. The consequences for native biota of anthropogenic-induced climate change. In *Greenhouse: Coping with Climate Change*, pp. 309–323, ed. W. J. Bouma, G. I. Pearman, and M. R. Manning. Collingwood, Australia: CSIRO Publishing.

Moreno, P. I. 1997. Vegetation and climate near Lago Llanquihue in the Chilean Lake District between 20,200 and 9500 14C yr BP. *Journal Quaternary Science* 121:485–500.

Moreno, P. I., T. V. Lowell, G. L. Jacobson, Jr., and G. H. Denton. 1999. Abrupt vegetation and climate changes during the last glacial maximum and last termination in the Chilean Lake District: A case study from Canal de la Puntilla (41°S). *Geografiska Annaler* 81A:285–312.

Moreno, P. I., C. Villagrán, P. A. Marquet, and L. G. Marshall. 1994. Quaternary paleobiogeography of northern and central Chile. *Revista Chilena Historia Natural* 67:487–502.

Nelson, C. S., P. J. Cooke, C. H. Hendy, and A. M. Cuthbertson. 1993. Oceanographic and climatic changes over the past 160,000 years at Deep Sea Drilling Project Site 594 off southeastern New Zealand, southwest Pacific Ocean. *Paleoceanography* 8:435–458.

Nicholls, N. 1991. The El Niño/Southern Oscillation and Australian vegetation. *Vegetatio* 91:23–36.

Nuñez, L., M. Grosjean, and I. Cartajena. 2001. Human dimensions of late Pleistocene/Holocene arid events in southern South America. In *Interhemispheric Climate Linkages*, pp. 105–117, ed. V. Markgraf. San Diego: Academic Press.

Pascual, R. 1984. La sucesión de las edades-mamífero, de los climas y del diastrofismo sudamericanos durante el Cenozoico: Fenomenos concurrentes. *Anales Academia Nacional Ciencias Exactas Fisicas y Naturales*. Buenos Aires, 36:15–37.

Pendall, E., V. Markgraf, J. W. C. White, M. Dreier, and R. Kenny. 2001. Multiproxy record of late Pleistocene—Holocene climate and vegetation changes from a peat bog in Patagonia. *Quaternary Research* 55:168–178.

Pickrill, R. A., J. M. Fenner, and M. S. McGlone. 1992. Late Quaternary evolution of a fjord environment, Preservation Inlet, New Zealand. *Quaternary Research* 38:331–346.

Pittock, A. B., and M. J. Salinger. 1982. Southern hemisphere climate scenarios. *Climate Change* 18:205–222.

Pole, M. 1994. The New Zealand flora—Entirely long-distance dispersal? *Journal of Biogeography* 21:625–635.

Premoli, A. C., T. Kitzberger, and T. T. Veblen. 2000a. Conservation genetics of the endangered conifer Fitzroya cupressoides in Chile and Argentina. *Conservation Genetics*, 1:57–66.

Premoli, A. C., T. Kitzberger, and T. T. Veblen. 2000b. Isozyme variation and recent biogeographical history of the long-lived conifer Fitzroya cupressoides. *Journal of Biogeography* 27:251–260.

Premoli, A. C., C. P. Souto, A. E. Rovere, T. R. Allnut, and A. C. Newton. 2002. Patterns of isozyme variation as indicators of biogeographic history in Pilgerodendron uviferum (D. Don) Florin. *Diversity and Distributions* 8.

Rabassa, J., A. Coronato, G. Bujalesky, M. Salemme, C. Roig, A. Meglioli, C. Heusser, S. Gordillo, F. Roig, A. Borromei, and M. Quattrocchio. 2000. Quaternary of Tierra del Fuego, Southernmost South America: An updated review. *Quaternary International* 68–71:217–240.

Renner, S. S., D. B. Foreman, and D. Murray. 2000. Timing transantarctic disjunctions in the Atherospermataceae (Laurales): Evidence from coding and noncoding chloroplast sequences. *Systematic Biology* 49:579–591.

Roberts, R. G., T. F. Flannery, L. K. Ayliffe, H. Yoshida, J. M. Olley, G. J. Prideaux, G. M. Laslett, A. Baynes, M. A. Smith, R. Jones, and B. L. Smith. 2001. New ages for the last Australian megafauna: Continent-wide extinction about 46,000 years ago. *Science* 292:1888–1892.

Romero, E. J. 1993. South American paleofloras. In

Biological Relationships between Africa and South America, pp. 62–85, ed. P. Goldblatt. New Haven: Yale University Press.

Rozzi, R., D. Martinez, M. F. Willson, and C. Sabag. 1995. Avifauna de los bosques templados de Sudamerica. In *Ecología de los Bosques Nativos de Chile*, pp. 135–152, ed. J. J. Armesto, C. Villagrán, and M. K. Arroyo. Santiago, Chile: Editorial Universitaria.

Sandiford, A., M. Horrocks, R. Newnham, J. Ogden, and B. Alloway. 2002. Environmental change during the last glacial maximum (c. 25,000– c. 16,500 years BP) at Mt Richmond, Auckland Isthmus, New Zealand. *Journal of the Royal Society of New Zealand* 32:155–167.

Stouffer, R. J., S. Manabe, and K. Bryan. 1989. Interhemispheric asymmetry in climate response to a global increase in CO_2. *Nature* 342:660–662.

Swenson, U., R. S. Hill, and S. McLoughlin. 2001. Biogeography of *Nothofagus* supports the sequence of Gondwana break-up. *Taxon* 50:1025–1041.

Thompson, R. S., C. Whitlock, P. J. Bartlein, S. P. Harrison, and W. G. Spaulding. 1993. Climatic changes in the Western United States since 18,000 yr B.P. In *Global Climates since the Last Glacial Maximum*, pp. 468–513, ed. Wright, H. E., J. E. Kutzbach, T. Webb III, W. F. Ruddiman, F. A. Street-Perrrott, and P. J. Bartlein. Minneapolis: University of Minnesota Press.

Tonni, E. P., M. T. Alberdi, J. L. Prado, M. S. Bargo, and A. L. Cione. 1992. Changes of mammal assemblages in the Pampean region (Argentina) and their relation with the Plio-Pleistocene boundary. *Palaeogeography, Palaeoclimatology, Palaeoecology* 95:179–194.

Veblen, T. T., T. Kitzberger, R. Villalba, and J. Donnegan. 1999. Fire history in northern Patagonia: The roles of humans and climatic variation. *Ecological Monographs* 69:47–67.

Veblen, T. T., M. Mermoz, C. Martin, and E. Ramilo. 1989. Effects of exotic deer on forest regeneration and composition in northern Patagonia. *Journal of Applied Ecology* 26:711–724.

Villagrán, C. 1991. Historia de los bosques templados del sur de Chile durante el Tardiglacial y Postglacial. *Revista Chilena de Historia Natural* 64:447–460.

Villagrán, C. 1995. Quaternary history of the Mediterranean vegetation of Chile. In *Ecology and Biogeography of Mediterranean Ecosystems in Chile, California, and Australia*, ed. M. K. Arroyo, P. H. Zedler, M. D. Fox. *Ecological Studies* 108:3–20.

Villagrán, C., and J. J. Armesto. 1993. Full and late glacial paleoenvironmental scenarios for the west coast of southern South America. In *Earth System Responses to Global Change: Contrasts between North and South America*, pp. 195–208, ed. H. A. Mooney, E. R. Fuentes, and B. I. Kronberg. San Diego: Academic Press.

Villagrán, C., and J. Varela. 1990. Palynological evidence for increased aridity on the Central Chilean coast during the Holocene. *Quaternary Research* 34:198–207.

Vuillemier, F., and J. Kikkawa. 1991. Reconstructing the history of *Nothofagus* avifaunas. *Acta XX Congressus Internationlis Ornithologici* 578–586.

Wardle, P. 1998. Comparison of alpine timberlines in New Zealand and the southern Andes. In *Ecosystems, Entomology and Plants*, pp. 69–90, ed. R. Lynch. *Royal Society of New Zealand Miscellaneous Series* 48. Wellington: Royal Society of New Zealand.

Wardle, P., C. Ezcurra, C. Ramírez, and S. Wagstaff. 2001. Comparison of the flora and vegetation of the southern Andes and New Zealand. *New Zealand Journal of Botany* 39:69–108.

Weaver, P. P., E. L. Carter, and H. L. Neil. 1998. Response of surface water masses and circulation to late Quaternary climate change east of New Zealand. *Paleoceanography* 13:70–83.

Whetton, P., A. B. Mullan, and A. B. Pittock. 1996. Climate-change scenarios for Australia and New Zealand. In *Greenhouse: Coping with Climate Change*, pp. 145–168, ed. W. J. Bouma, G. I. Pearman, and M. R. Manning. Collingwood: CSIRO Publishing.

Whitlock, C., V. Markgraf, P. Bartlein, and A. Ashworth. 2001. The mid-latitudes of North and South America during the last glacial maximum and early Holocene: Similar paleoclimatic sequences despite different large-scale controls. In *Interhemispheric Climate Linkages*, pp. 391–416, ed. V. Markgraf. San Diego: Academic Press.

Wilford, G. E., and P. J. Brown. 1996. Maps of late Mesozoic–Cenozoic Gondwana break-up: Some palaeographical implications. In *History of the Australian Vegetation: Cretaceous to Recent*, pp. 5–13, ed. R. S. Hill. Cambridge: Cambridge University Press.

Microrefugia and Macroecology

Matt McGlone and Jim Clark

At the last glacial maximum (LGM), southern South America, southern Australia, and New Zealand were in largely treeless vegetation of scrub, grassland, steppe, tundra, and desert (Markgraf et al. 1995; Markgraf and McGlone, this volume). Ten thousand years later, by the beginning of the Holocene, nearly all these areas were in temperate forest cover. This rapid recolonization by forest trees and shrubs cannot easily be accommodated by the classic refugia-migration model based on Northern Hemisphere evidence. Macrorefugia—large contiguous areas where cold-sensitive trees survived the LGM—seem unlikely to have played a significant role in southern postglacial forest dynamics for several reasons.

First, southern pollen records show rapid response of tree populations to postglacial warming, but the order of arrival of species at any given site seems to be controlled more by climate and soils than by inherent spread potential. Thus, in New Zealand, tree ferns spread by light wind-dispersed spores do not precede bird-dispersed podocarp trees, and podocarp trees themselves arrive in strict order of tolerance of raw soils and stressful climates, not dispersal ability (McGlone 1997). The same pattern is seen in Chile, where, for example, heavy-seeded but pioneering *Nothofagus* spread thousands of years before light-seeded wind-dispersed *Weinmania trichosperma* (Markgraf et al. 1995; Bennett et al. 2000). Second, in southern South America, Tasmania, and New Zealand, the rugged terrain cut by short rivers flowing to the coast does not provide easy migratory pathways (McGlone 1997). This difficult terrain makes it improbable that currently highly disjunct tree populations seen, for example, in *Nothofagus* in New Zealand (McGlone et al. 1996) and *Eucalyptus* in southeastern Australia and Tasmania (Hughes et al. 1996) could have achieved their distributions by postglacial spread from macrorefugia. Finally, there are fossil indications of survival of small areas of forest in both coastal and inland areas (Lees et al. 1998; McKenzie and Kershaw 2000) and even in the coldest parts of far southern New Zealand and Chile (Pickrill et al. 1992; Markgraf et al. 1995).

The glacial vegetation of the southernmost regions of these continents has therefore long been interpreted as consisting largely of shrubland, grassland, tundra, or desert with tree microrefugia scattered throughout. Postglacial range changes occurred mainly by expansion of local populations rather than through long-distance dispersal events (McGlone 1985; Kershaw 1988; Markgraf et al. 1995). However, tree biogeography (McGlone et al. 2001) and genetic data (Haase 1992; Premoli et al. 2000; Freeman et al. 2001; McKinnon et al. 2001; Premoli et al. 2001) from these far southern regions show areas of elevated species and genetic diversity similar to those that have been used to infer macrorefugia in the Northern Hemisphere (e.g., Hewitt 1999). Are macrorefugia and microrefugia mutually exclusive—or could they be parallel processes that operate side by side?

Postglacial migration of trees has long been cited as an example of rapid spread from refuges in response to climate change (Clark 1998). However, for many taxa the palaeographic evidence for rapid postglacial migration of trees is incompatible with migration rates predicted from

157

life history characteristics (time to reproductive maturity) and seed dispersal rates, an anomaly known as Reid's Paradox (Clark et al. 1998). For a time it appeared that rare, long-distance dispersal events might explain extremely rapid migration. This interpretation seemed to agree with paleo interpretations of rapid spread, but it was based on deterministic models that did not accurately describe the long-distance dispersal process. More recent analysis has shown that model predictions of remarkably high migration potential depend not only on questionable dispersal assumptions, but also on unrealistically high net reproductive rates and short generation times (Clark et al. 2003). It now seems unlikely that many trees could have achieved their current distributions if spread from macrorefugia was the norm.

The postglacial vegetation history of the southernmost landmasses and this recent reassessment of tree migration rates suggest a revision is needed of the dominant refugia-migration hypothesis, at least for trees. For the past 2 million years, the world has been mainly in a cool, dry glacial mode, with relatively brief excursions into warm, wet interglacials. Tree populations endured long glacial periods of restricted low abundance, but many species remained capable of rapidly occupying the landscape during the shorter interglacial periods. The largest impact of glacial climates was experienced in regions of low relief. In areas with topographic diversity, such as that provided by hill slopes, ravines, incised rivers, and coastal embayments, the harsh climate regime was modified by greater insolation and cold air drainage on slopes, moist, sheltered sites, and moderation of temperature by adjacent water masses. Species that became widespread and common during interglacials may be those that survived in microrefugial environments that were dispersed more broadly than previously interpreted from paleoevidence (McLachlan and Clark 2003).

Postglacial recolonization may have proceeded by similar mechanisms in both the Northern and Southern Hemispheres, because both might have included macrorefugia and microrefugia. Taxa incapable of surviving in the dominant glacial habitats were confined either to areas where glacial–interglacial fluctuations are minimized by topography or latitude (macrorefugia), or to restricted microhabitats scattered across the landscape (microrefugia). Taxa in macrorefugia experience low extinction rates and allopatric speciation driven by climatic fragmentation, and thus macrorefugia become more diverse over time. On the other hand, taxa restricted to microrefugia may have been more exposed to stochastic extinction and go through cycles of extreme abundance and contraction, purging them of variation through allele losses during contractive phases (Premoli et al. 2001) and swamping of rare genotypes during expansion. The genetic relationships, and phylogenies revealed by chloroplast DNA, linking populations confined to macrorefugia with those currently more widely distributed may not, in most cases, be a function of postglacial spread. Instead they may reflect common ancestors that existed in pre-Pleistocene biogeographies that contributed descendents to both macrorefugia and microrefugia.

REFERENCES

Bennett, K. D., S. G. Haberle, and S. H. Lumley. 2000. The Last Glacial-Holocene transition in southern Chile. *Science* 290, 325–328.

Clark, J. S. 1998. Why trees migrate so fast: Confronting theory with dispersal biology and the paleo record. *American Naturalist* 152, 204–224.

Clark, J. S., C. Fastie, G. Hurtt, S. T. Jackson, C. Johnson, G. King, M. Lewis, J. Lynch, S. Pacala, I. C. Prentice, E. W. Schupp, T. Webb III, and P. Wyckoff. 1998. Reid's Paradox of rapid plant migration. *BioScience* 48, 13–24.

Clark, J. S., M. Lewis, J. S. McLachlan, and J. Hille Ris Lambers. 2003. Estimating population spread: What can we forecast and how well? *Ecology*, 84, 1979–1988.

Clark, J. S., and J. S. McLachlan. 2003. Stability of forest biodiversity. *Nature* 423(6940), 635–638.

Freeman, J. S., H. D. Jackson, D. A. Steane, G. E. McKinnon, G. W. Dutkowski, B. M. Potts, and R. E. Vaillancourt. 2001. Chloroplast DNA phylogeography of *Eucalyptus globulus*. *Australian Journal of Botany* 49, 585–596.

Haase, P. 1992. Isozyme variability and biogeography of *Nothofagus truncata* (Fagaceae). *New Zealand Journal of Botany* 30, 315–328.

Hewitt, G. M. 1999. Post-glacial re-colonization of European biota. *Biological Journal of the Linnean Society* 68, 87–112.

Hope, G. S. 1994. Quaternary vegetation. In *History of the Australian Vegetation: Cretaceous to Recent*, pp. 368–389, ed. R. S. Hill. Cambridge: Cambridge University Press.

Hughes, L., E. M. Cawsey, M. Westoby. 1996. Geographic and climatic range sizes of Australian eucalypts and a test of Rapoports Rule. *Global Ecology & Biogeography Letters* 5, 128–142.

Kershaw, A. P. 1988. Australasia. Handbook of Vegetation Science Vol. 7: *Vegetation History*, pp. 237–306, ed. B. Huntley, and T. Webb III. Dordrecht: Kluwer Academic.

Kot, M., M. A. Lewis, and P. van den Driessche. 1996. Dispersal data and the spread of invading organisms. *Ecology* 77, 2027–2042.

Lees, C. M., V. E. Neall, and A. S. Palmer. 1998. Forest persistence at coastal Waikato, 24,000 years BP to present. *Journal of the Royal Society of New Zealand* 28, 55–81.

Markgraf, V., M. S. McGlone, and G. S. Hope. 1995. Neogene paleoenvironmental and paleoclimatic change in southern temperate ecosystems—a southern perspective. *Tree* 10, 143–147.

McGlone, M. S. 1985. Plant biogeography and the late Cenozoic history of New Zealand. *New Zealand Journal of Botany* 23, 723–749.

McGlone, M. S. 1996. When history matters: Scale, time, climate and tree diversity. *Global Ecology and Biogeography Letters* 5, 309–314.

McGlone, M. S. 1997. The response of New Zealand forest diversity to Quaternary climates. In *Past and Future Rapid Environmental Changes: The Spatial and Evolutionary Responses of Terrestrial Biota*. Nato ASI Series. Series 1: Global Environmental Change, Vol. 47, pp. 73–80, ed. B. Huntley, W. Cramer, A. V. Morgan, H. C. Prentice, and J. R. M. Allen. Berlin: Springer-Verlag.

McGlone, M. S., R. P. Duncan, and P. B. Heenan. 2001. Endemism, species selection and the origin and distribution of the vascular plant flora of New Zealand. *Journal of Biogeography* 28, 199–216.

McGlone, M. S., D. C. Mildenhall, and M. S. Pole. 1996. History and paleoecology of New Zealand *Nothofagus* forests. In *The Ecology and Biogeography of Nothofagus Forest*, pp. 83–130, ed. T. T. Veblen, R. S. Hill, and J. Read. New Haven: Yale University Press.

McKenzie, G. M., and A. P. Kershaw. 2000. The last glacial cycle from Wyelangta, the Otway region of Victoria, Australia. *Palaeogeography Palaeoclimatology Palaeoecology* 155, 177–193.

McKinnon, G. E., R. E. Vaillancourt, H. D. Jackson, and B. M. Potts. 2001. Chloroplast sharing in the Tasmanian eucalypts. *Evolution* 55, 703–711.

Pickrill, R. A., J. M. Fenner, and M. S. McGlone. 1992. Late Quaternary evolution of a fjord environment, Preservation Inlet, New Zealand. *Quaternary Research* 38, 331–346.

Premoli, A. C., T. Kitzberger, and T. T. Veblen. 2000. Isozyme variation and recent biogeographical history of the long-lived conifer *Fitzroya cupressoides*. *Journal of Biogeography* 27, 251–260.

Premoli, A. C., C. P. Souto, T. R. Allnutt, and A. C. Newton. 2001. Effects of population disjunction on isozyme variation in the widespread *Pilgerodendron uviferum*. *Heredity* 87, 337–343.

CHAPTER ELEVEN

Responses of Marine Species and Ecosystems to Past Climate Change

KAUSTUV ROY

AND JOHN M. PANDOLFI

The fossil record is the primary source of data for understanding how marine biotas respond to changes in climatic and environmental regimes on scale of centuries to thousands to millions of years. The Phanerozoic (ca. past 570 million years) history of the planet has seen repeated changes in four major environmental variables in the sea: sea surface temperatures (SST), oceanic concentrations of CO_2, sea level, and ultra-violet radiation (Crowley and North 1991; Frakes et al. 1992). In general, the climatic history of the Phanerozoic can be divided up into alternating periods of *icehouse* (low CO_2 and temperatures) and *greenhouse* (high CO_2 and temperatures) climate (Fischer and Arthur 1977). These climatic changes played a vital role in shaping marine biodiversity patterns. For example, changes in global climate and ocean productivity may have been crucial in constraining the tempo and mode of early animal evolution (Knoll and Carroll 1999).

Similarly, many of the major extinction events such as the end-Ordovician (ca. 430 Ma) and late Devonian (ca. 367 Ma) mass extinctions, are also associated with significant changes in global temperature (Brenchley et al. 2001; McGhee 2001), although whether temperature change by itself can cause widespread extinctions of marine organisms is a subject of debate (Clarke 1993). In addition, the greenhouse world during the Cretaceous (150 to 65 Ma) or the unusually warm period during the mid-Miocene (ca. 14.5–17 Ma) provides fascinating glimpses into marine life under global climatic and oceanographic conditions that have no modern analogs. This long-term view from the fossil record is admittedly lower in resolution compared to modern or Pleistocene (past 1.8 Ma) patterns but provides the long-term perspective essential for understanding the ecological and evolutionary consequences of climate change. In contrast, the repeated glacial–interglacial changes during the Pleistocene provide a set of high-resolution records of natural experiments involving species that are mostly still living and hence insights that are directly applicable to the living biota.

In this review, we provide a general overview of how marine biotas have responded to past climatic changes and discuss the implications of past insights for predicting biological response to future global change. The focus of this chapter is on marine invertebrates, since they have the richest fossil record. Responses of tropical reef communities are discussed separately from those of other species in order to highlight some of the fundamental differences in how different systems respond to global climatic change.

CLIMATIC IMPACTS ON COMMUNITIES

Insights from the Tertiary

Temperature as well as oceanographic circulation patterns are important determinants of species distributions in the sea (Valentine 1973; Gaylord and Gaines 2000), and it is not surprising that a common biological response to climate change is a shift in the geographic range of species (Valentine and Jablonski 1993; Buzas and Culver 1994, 1998; Jablonski and Sepkoski 1996). When climate changes, co-occurring species can become separated or new species associations form, thereby changing the diversity and composition of communities. Such changes are known from a variety of taxa, from mollusks and benthic foraminifera to ostracods and bryozoans (Buzas and Culver 1994; Jablonski and Sepkoski 1996; Culver and Buzas 2000; Taylor 2000). For example, the middle Miocene (ca. 15–17 Ma BP) was a time of globally warm climates that led to northward shifts in the geographic ranges of many tropical and subtropical mollusks along both sides of the northern Pacific Ocean. In the northwestern Pacific, warm water taxa extended their range into northern Japan (e.g., Ogasawara 1994) and Kamchatka (Gladenkov and Shanster 1990), and some even made it across to Alaska (Marincovich 1988). Along the northeastern Pacific the same warming event saw shifts in the ranges of subtropical mollusks as far north as southern Washington (Addicott 1969), and tropical planktonic foraminifera are known from southern Alaska during this time (Lagoe 1983). The mid-Miocene warming was followed by gradual cooling of global climate (Zachos et al. 2001) and the disappearance of most of these tropical taxa from north temperate waters (Ogasawara 1994). A subsequent warming event during mid-Pliocene times (3–5 Ma) again brought warm water taxa into mid- and high-latitude areas (Dowsett et al. 1994).

In a regional study spanning a much larger time, Buzas and Culver (1994) examined the composition of benthic foraminiferal communities in continental shelf deposits along the Atlantic coastal plain of North America over a 55-million-year interval (Eocene–Pliocene; see Fig. 8.1). Their results show that geographic distributions of foraminiferal species changed repeatedly in response to climate and sea level fluctuations throughout this interval, resulting in considerable changes in the composition of these communities (Buzas and Culver 1994). Similarly, the Pliocene fossil record of bivalves on Niue and other central Pacific islands also reveals significant fluctuations in geographic ranges of inner reef species in response to Milankovitch-scale changes in sea level and climate (Paulay 1996).

Global climatic change can also influence species diversity and community composition in the deep sea, an environment traditionally thought to be buffered from environmental changes in shallow water (e.g., Hessler and Sanders 1967). The diversity of North Atlantic benthic ostracods fluctuated in step with Milankovitch-scale climatic fluctuations during the late Pliocene (2.85–2.40 Ma); interglacial periods saw a three- to fourfold increase in regional diversity compared to glacial times (Cronin and Raymo 1997). Available evidence suggests that during glacial times, deep sea ostracod species use the shallower bathyal environments (1000–2000 m) as refugia only to recolonize the deeper water abyssal habitats during the warmer interglacial periods (Cronin et al. 1995; Cronin and Raymo 1997). The exact mechanisms driving these changes in bathymetric distributions remain uncertain, but factors such as changes in thermohaline circulation, bottom temperature, and food availability are probably important (Cronin and Raymo 1997). A more intriguing possibility is that diversity changes in the deep sea are caused by fluctuations in food availability due to changes in surface water productivity (Cronin and Raymo 1997). Such a coupling between deep-sea diversity and surface productivity (Gage and Tyler 1991) serves to illustrate the potentially cascading and pervasive effects of global climate change on marine ecosystems.

Another example of how climate change can dramatically transform the long-term composition and nature of regional biotas comes from Antarctica. Present-day shallow water (less than 100 m) benthic communities in Antarctica are unique in that they resemble Paleozoic marine communities or deep sea assemblages; slow-moving invertebrates dominate higher trophic levels, and dense concentrations of ophiuroids (brittle stars and basket stars) and crinoids (feather stars; Fig. 11.1) result from the absence of crabs and a limited diversity of sharks and other predators (Aronson and Blake 2001). Data from the Antarctic fossil record show that prior to the onset of global cooling during the late Eocene and early Oligocene (ca. 35 Ma) Antarctic communities resembled those from lower latitudes with their full complement of shell-crushing predators (Aronson and Blake 2001). However, the cooling event led to the disappearance of crabs, sharks, and most of the predatory fish from the region. The resulting reduction in predation pressure allowed slow-moving benthic invertebrates such as crinoids to flourish in a shallow marine community that most resembles those from the deep sea (Aronson and Blake 2001).

In contrast to most marine invertebrates, tropical reefs form three-dimensional biogenic structures that typically extend significantly above the seafloor. Their growth and development depend upon the interaction of major geological forces: tectonic, climatic, and oceanographic (Rosen 1972). Climatic factors are thought to be a major control over their occurrence through geological time, and in a broad sense this is true (e.g., Stehli and Wells 1972; Veron 1995). However, recent work also suggests a different set of constraints on species diversity independent of latitude (Bellwood and Hughes 2001). Moreover, different organisms have dominated ancient reefs at different times, and many provide little in the way of analogy to modern-day reefs dominated by scleractinian corals.

Intense interest exists among paleontologists concerning the geological history of tropical reefs and associated "carbonate" environments (Newell 1971; Fagerstrom 1987; Copper 1988, 1989; Flügel and Flügel-Kahler 1992; Hallock 1997; Wood 1999; Stanley 2001). Stanley and Hardie (1998) showed a relationship between inorganic and organic carbonate production, suggesting that global changes in seawater chemistry (namely Mg/Ca ratio) have an overarching influence on develop-

Figure 11.1. Crinoids, also known as feather stars, are relatives of sea urchins and other echinoderms. Crinoid community composition in Antarctica has changed strongly in response to climate (see text). Photo courtesy of Jeff Jeffords (www.divegallery.com).

ment of reefs and associated ecosystems through geological time. Others argue that climate per se holds the key to understanding reef evolution through time (reviewed in Boucot and Gray 2001).

Recent investigation of the latitudinal distribution of tropical reefs throughout the Phanerozoic demonstrates that neither the total latitudinal range of reefs nor the width of the tropical reef zone shows a significant correlation with inferred paleo-climate (Kiessling 2001). This may be because CO_2 saturation states of seawater may be an overriding determinant of tropical reef distribution (Hoegh-Guldberg, this volume). Thus it appears that, in contrast to coral growth rates, the distribution of reefs is not a good paleoclimate indicator. It is more instructive to use independently derived estimates of paleoclimate in evaluating tropical reef response to climate change. However, these analyses are based upon reefs of varying composition, and it is well known that differences exist in the response of tropical reef organisms to fluctuations in global climate variables.

The relationship between levels of CO_2 and the ability of corals to form major

---

Box 11.1. **Calcite and Aragonite**

Calcite and aragonite are both forms of calcium carbonate, $CaCO_3$. Aragonite is a *polymorph* of calcite—it has the same chemical composition but undertakes a different crystal form. Both corals and non−coral reef builders such as rudistid bivalves secrete calcium carbonate shells or skeletons. Modern scleractinian corals secrete their skeletons in the form of aragonite.

---

structural components of reefs is still poorly known. In general, coral reefs have been less dominant in times of high CO_2 and SST (greenhouse periods), and more dominant in times of low CO_2 and SST (icehouse periods). During greenhouse periods they have been replaced as the dominant reef-builder, though not completely, by organisms that secrete calcite as part of their skeletons (i.e., rudistid bivalves in the Cretaceous), as opposed to strictly aragonite. For example, Kauffman and Johnson (1988) showed a dominance of rudistid bivalves over corals during the Cretaceous, a time of high atmospheric CO_2 concentrations that were double today's values (700 ppm). However, Paleozoic CO_2 levels were much higher than today's, and during some of this time extensive reefs were made by Anthozoans.

It remains an open question whether there may be an upper limit to SST or CO_2 concentration above which scleractinian corals are no longer able to successfully dominate tropical reefs. This appears to be counterintuitive, since corals are almost always associated with warm, tropical seas—why not the warmer the better? However, it is now known that mass bleaching of corals, resulting in high mortality, occurs as a result of summertime increase of around 2°C in SST for a period greater that 10 days (Gleeson and Strong 1995; Hoegh-Guldberg, this volume). At some point a critical temperature or CO_2 threshold is reached, and corals no longer function at their optimum during and after a transition to "greenhouse" conditions. This may have to do with a reduced surface ocean calcium carbonate saturation state associated with rising CO_2 (Gattuso et al. 1999). Rising atmospheric CO_2 causes the surface ocean water to become more acidic, making it more difficult for some organisms (including scleractinian corals) to precipitate their carbonate mineral skeletons. Thus, the important question that remains unanswered is: What is the range of temperature and CO_2 values that living representatives of coral reefs have existed under during the geological past? This is especially true for the last 1 million years of the Quaternary Period, when coral reefs much like today have arisen.

Pleistocene Insights

The Pleistocene and Holocene (last 1.8 million years) fossil record undoubtedly provides some of the best insights into how marine species and communities respond to climatic change. The multiple glacial−interglacial cycles during this time period provide a series of natural experiments, and the vast majority of species involved are still living in the world oceans (Valentine 1989; Valentine and Jablonski 1993; Pandolfi 1996, 1999; Roy et al. 1996). This high-resolution record can be used to quantify biological responses on time scales of 10^3-10^4 years (e.g., Roy et al. 1996; Pandolfi 1996, 1999) and also to gain insights into the ecological or life history characteristics that determine how individual species respond to environmental change. While considerable work has been done on quantifying the responses of marine invertebrates to Pleistocene climatic changes, except for reef corals the majority of this work has focused on extratropical species and communities; tropical assemblages still remain poorly studied. In general, the Pleistocene record shows that the most common responses to climatic change involved latitudinal shifts in range limits of individual species (Valentine and Jablonski 1993; Roy et al. 1995, 1996; Kohn and Arua 1999). During glacial times northern populations of many species went extinct, resulting in a southward shift in the northern range boundary. As temperatures warmed up during interglacials many of these regions were recolonized by individuals from the southern refugia, leading to a northward shift in the range boundary. Examples of such range shifts in response to Pleistocene climatic changes have been documented for many different groups ranging from mollusks and foraminifera to ostracods and from shallow water to the deep sea (Hazel 1970; Valentine and Jablonski 1993; Culver and Buzas 1995; Roy et al. 1995, 1996; Schmiedl et al. 1998; Cannariato et al. 1999; Cronin et al. 1999; Kitamura et al. 2000; Hellberg, Balch, and Roy 2001). A handful of studies of tropical assemblages have shown responses that are very similar, at least qualitatively, to those from temperate regions. For example, Paulay (1990) and Paulay and Spencer (1988) documented that compositions of bivalve communities inhabiting reef environments on central tropical Pacific islands changed as distributions of their component taxa shifted in response to cyclic changes in sea level. Similar changes have also been documented for molluscan assemblages from Fiji (Kohn and Arua 1999), the Aldabra Atoll (Taylor 2000), and along the Kenyan coast (Crame 1986).

While range shifts are a common response to climate change, not all species shift their ranges to the same extent or at the same time, and many species do not seem to shift their distributions at all, even when climate changes substantially. We still poorly understand the biological processes underlying such individualistic response to the same perturbation. Clearly, identifying the ecological and life history characteristics that impose or permit greater variability in the geographic ranges of some species relative to others in the same community is essential for predicting the responses of marine species to future climate change. Body size is one trait that influences almost every aspect of the biology of a species, from physiology to life history (Peters 1983), and there is evidence that size can play an important role in determining species response to climate change.

The bivalve fauna of the Californian province (28°N–34.5°N) is one of the best-studied in the world, and this region also contains extensive fossiliferous marine Pleistocene terrace deposits that preserve over 80 percent of the living bivalve species (Valentine 1961, 1989). Analyses of Pleistocene and modern distributions of these species have shown that body size appears to play an important role in mediating the responses of bivalve species to climate change. The species whose range limits shifted in response to climatic change, as a group, have significantly larger body sizes than the rest of the Pleistocene species pool (Fig. 11.2) (Roy et al. 2001, 2002). Interestingly, the same size bias toward larger size is also seen in marine bivalve species whose geographic ranges have expanded in historical times through human-mediated introductions (Roy et al. 2001, 2002). These results suggest that, in general, range limits of large bivalves are more unstable compared to smaller species (Roy et al. 2002). Whether size plays a similar role in other groups remains to be tested.

Community dynamics in response to Quaternary glacial–interglacial cycles has also been documented for reef corals, which form the three-dimensional structure of modern reefs, and these dynamics differ from those of solitary benthic invertebrates. High-resolution paleoecological records found in uplifted terraces around the world indicate that there has been little change to the species composition of coral communities during the natural variations in the Earth's climate over at least the past 500,000 years (Jackson 1992; Pandolfi 1996, 1999, 2000), even though there is general consensus that climate change is likely to adversely affect modern coral reef communities (Buddemeier and Smith 1999). Pandolfi (1996) documented persistence in coral community composition (species presence/absence and diversity) in reef coral communities that had repeatedly reassembled on the Huon Peninsula, Papua New Guinea, through a 95,000-year interval (from 125,000–30,000 years ago). Coral community structure was no different among nine different reef-building episodes, even though the communities varied spatially and existed under variable global environmental parameters (Pandolfi 1996, 1999) including marked differences in CO_2 concentration and SST. A similar trend of persistence in species composition through time (four reef-building episodes between 104,000 and 220,000 years ago), this time using coral species relative abundance, also appears to hold in the Pleistocene raised reef terraces of Barbados (Pandolfi 2000). There is, however, some evidence that individual coral species can shift their geographic range limits in response to climate change. For

example, Playford (1983) showed that the coral fauna preserved in Late Pleistocene limestones exposed on Rottnest Island was dominated by branching *Acropora* species, 350 km south of the modern range limit for the genus. Similarly, warm water coral assemblages were preserved in Late Pleistocene limestones on Cape Leeuwin, 500 km south of their modern occurrence in the Houtman-Abrolhos Islands (Kendrick et al. 1991).

What are the population-level consequences of the climate-driven changes in species range boundaries when such shifts happen? Phylogeographic analyses using mitochondrial sequences are being increasingly used to look at population-level responses of marine species to Pleistocene and Holocene climatic changes. These data nicely complement the paleontological analyses, and there is a tremendous potential for integrating these independent lines of evidence to generate a comprehensive model of species response to climate change. In general, phylogeographic data show that the current population structures of many marine species bear a legacy of the range shifts in response to late Pleistocene climatic changes (e.g., Avise 1994; Burton 1998; Dahlgren et al. 2000; Hewitt 2000; Edmands 2001; Lessios et al. 2001; Wares 2001; Wares and Cunningham 2001). Populations in the northern parts of their ranges may show little genetic differentiation compared to the southern ones largely because the northern populations went extinct during the last glacial period and these regions have been only recently recolonized (Edmands 2001; Hellberg, Balch, and Roy 2001). An im-

---

Figure 11.2. Approximate positions of the present-day range limits of extralimital bivalve species known from the Californian Pleistocene. Extralimital species are those whose Pleistocene occurrences fall at least one degree of latitude outside of their present-day range limits. Open circles represent the southern range limit of northern extralimital species; filled circles mark the northern range limit for southern extralimitals. The number beside each circle represents the number of species whose ranges end near that point. The shaded area indicates the Californian province where all the extralimital species co-existed during the Pleistocene. The inset shows the size-frequency distributions of the extralimital versus nonextralimital bivalve species. The size data were log_2-transformed following Brown (1995) and Roy et al. (2000b). The distributions are significantly different ($p = 0.01$, Kolmogorov-Smirnov test). *Source:* Adapted from Roy et al. (2001).

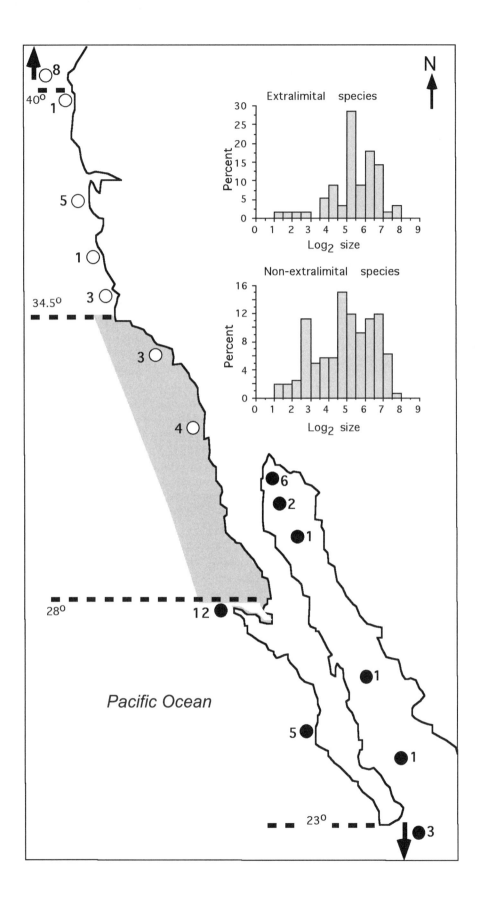

N

Extralimital species

Non-extralimital species

Pacific Ocean

40°
34.5°
28°
23°

portant implication of this pattern is that distributions of many species are still far from being in equilibrium with climate, a situation likely to be accelerated by anthropogenic warming. For coral reef taxa, current population genetic structure can also bear the signature of events associated with global climate change and sea level fluctuation during the past 1–3 my. For example, for some Indo-Pacific taxa, present patterns of genetic variation have resulted from highly pulsed dispersal events associated with range expansions during interglacial periods (Benzie 1999).

In contrast to a growing number of phylogeographic studies focusing on impacts of climate change on population structure, we still know relatively little about the phenotypic consequences of climate-driven range shifts. Results of one study strongly suggest that rapid morphological evolution may result when changing climatic conditions alter the geographic ranges of species (Hellberg, Balch, and Roy 2001). Morphological measurements from extant and Pleistocene populations of an intertidal gastropod (*Acanthinucella spirata*) show that northern populations of this species contain shell morphologies absent from southern populations or the Pleistocene fossil record. Yet these northern populations show very little genetic differentiation, a pattern consistent with a recent northward range expansion of this species (Hellberg, Balch, and Roy 2001). Whether such decoupling of genetic and morphological responses is common in other marine invertebrates is currently unknown but can be evaluated by integrating phylogeographic and paleontological data. Local or global extinction of species due to climatic shifts can also influence the phenotypic evolution of competing species through character release. For example, Pleistocene extinction of the widespread organ-pipe *Montastraea* coral led to measurable morphological and ecological effects on other, surviving species of the *Montastraea* "*annularis*" complex (Pandolfi et al. 2002).

EXTINCTION DYNAMICS AND GLOBAL CHANGE

Insights from Deep-time

Can large-scale changes in the global climate cause widespread extinctions in the sea? The fossil record reveals a complex relationship between climatic changes and extinctions of marine taxa—many past climatic events are associated with significant extinctions while others, such as the Pleistocene glacial–interglacial cycles, did not result in heightened extinctions. The major extinction events of the past where climatic change could have played a role include the late Ordovician extinction, latest Paleocene (ca. 56 Ma) extinction, Eocene-Oligocene (ca. 34 Ma) extinction, and Plio-Pleistocene (ca. 2 Ma) extinction of the tropical western Atlantic biota. Other events such as the end-Cretaceous (ca. 65 Ma) mass extinction also involved major environmental perturbations, but these were triggered by rare and unusual events such as asteroid impacts, and hence are perhaps less relevant for understanding the biotic responses to more gradual climatic shifts.

Extinction magnitudes associated with particular climatic events varied between taxa and between habitats as well as ocean basins. For planktonic and benthic foraminifera the latest Paleocene (ca. 57 Ma) extinctions mainly affected intermediate and deep-water taxa over a relatively short time period (Kennett and Stott 1991; Kaiho 1994), while the Eocene–Oligocene extinctions affected surface and deep-water species in a series of pulses spread out over a few million years (Hansen 1992; Prothero and Berggren 1992; Kaiho 1994). The Plio-Pleistocene extinction is well documented in tropical western Atlantic mollusks, foraminifera, and corals (Stanley and Campbell 1981; Stanley 1986; Allmon et al. 1993; Jackson 1995; Budd 2000), but northeastern and northwestern Pacific mollusks do not show a similar increase in extinction rates during this time (Stanley et al. 1980; Valentine and Jablonski 1991; Jackson et al. 1993).

As Clarke (1993) pointed out, past episodes of global climate change associated with increased extinction among marine ectotherms (cold-blooded animals) tend to be periods of climatic cooling (e.g., the Eocene-Oligocene and Plio-Pleistocene changes), and these events disproportionately affected warm water taxa. Beyond this correlation, very little is known about the actual mechanisms that caused the extinctions. Using basic principles of physiological ecology, Clarke (1993) argued that temperature per se is an unlikely cause of extinction of marine ectotherms; instead other ecological and environmental processes associated with temperature change (e.g., changes in seasonality, dissolved oxygen, energy flow, or oceanographic conditions) may be the causal agent. The fact that not all episodes of past climatic change saw heightened extinctions and that even the documented extinctions often showed significant ecological selectivity (e.g., Todd et al. 2002) is consistent with this view (Clarke 1993). Indeed, processes such as a decrease in dissolved oxygen concentrations as a result of climatic warming (e.g., Kaiho 1994) or changes in nutrient regimes (e.g., Todd et al. 2002) have often been proposed as the cause of extinctions during periods of climatic change.

An additional consideration here is the existence of threshold conditions as well as historical contingencies. Faunal responses to climatic events may depend on the rapidity of the climatic change. There is now considerable evidence that species and ecosystems experience changes in climate on a variety of time scales ranging from millennia to tens of millennia (Roy et al. 1996 for review), and slow cooling events such as those on Milankovitch scales may not be sufficient to cause major extinctions in a biota already adapted to very rapid fluctuations. However, if the rate of climatic cooling (or warming) is unusually rapid (e.g., Kennett and Stott 1991) or if the magnitude of change is unusually large, then many species could be vulnerable (Clarke 1993). Similarly the response of a biota to a given environmental condition may be largely contingent on its evolutionary and thermal history (Clarke 1993; Jackson 1995).

Pleistocene Insights

A surprising observation from the marine Pleistocene fossil record is that there is very little evidence for widespread global extinctions of species in the face of extreme climatic fluctuations (Valentine and Jablonski 1993; Roy et al. 1996; Pandolfi et al. 2001). For the shallow marine molluscan species of California, the extinction rate over the last million years is around 12 percent, which is not significantly different from the long-term background extinction rate for this group (Valentine 1989). Similarly for reef corals only about 8 percent (five species) extinctions are known during the past 1 million years (Budd 2000). Again this is well within the expected background rate of extinction.

This lack of extinction during the Pleistocene climate cycles is in contrast to significant extinctions seen in tropical western Atlantic mollusks, foraminifera, and corals during the latest Pliocene and early Pleistocene, which roughly coincides with significant cooling at the onset of northern hemisphere glaciation (Jackson 1995; Budd 2000). As discussed above, the Plio-Pleistocene extinction is, however, much less severe in the eastern and western Pacific faunas, a difference that highlights interoceanic and regional differences in species response to the same global climatic event. The Plio-Pleistocene extinction pulse in the tropical western Atlantic may also have served as a filter that removed the more vulnerable species, leaving the rest of the fauna relatively immune to the subsequent glacial–interglacial cycles (Jackson 1995). Associated with this filter is the higher variance associated with climate variables around 1 my ago as compared to those of the past 400,000 years.

LESSONS OF HISTORY AND THE FUTURE

Analyses of biotic responses to past climatic changes hold some of our best hopes for understanding how marine species and ecosystems respond to changing environments. In fact, many of the dynamics preserved in the fossil record are already evident today as species attempt to adapt to an increasingly warm world. For example, shifts in the geographic ranges of many species are being documented (Walther et al. 2002), and climate warming over multiple decades has led to increased abundances of warm water species in temperate marine communities (Barry et. al 1995; Southward et al. 1995; Holbrook et al. 1997; Sagarin et al. 1999). Perhaps not surprisingly, some of the species involved in such range shifts are the same ones that moved north during the last interglacial (A. J. Collins and K. Roy, unpublished). However, while the lessons from the past predict that such northward shifts in distributions of benthic species will become increasingly common under global warming, we are still far from predicting how individual species or even groups of species will respond, since the ecological and life history traits determining species response to climate change remain poorly understood. Clearly, data from the Pleistocene and Holocene records in conjunction with ecological data from living species is the best way to address this problem.

The lack of heightened extinction rates in any marine group in response to Pleistocene glacial–interglacial cycles suggests that the living biota is adapted to coping with "normal" Milankovitch-scale changes in climate, at least as far as those that have occurred over the past 2 million years; such fluctuations would change species associations and local diversity but are unlikely to lead to catastrophic extinctions. But some of the projected rates of increase in ocean temperatures for the next 100 years are 10 times those observed previously (IPCC 2001), and those for CO_2 increase are 2 times the magnitude ever experienced by coral reefs over the past 40 million years.

Another important observation is that inferences from history apply only to a world where ecosystems are healthy and habitats plentiful. Unfortunately today many marine ecosystems are far from healthy (Jackson et al. 2001) and predictions about the consequences of future global change have to take the present threats to biodiversity into account. Examination of paleontological, archaeological, historical, and ecological records of the world's oceans demonstrates historical overfishing as one of the major mechanisms for the degradation of coastal marine habitats (Jackson et al. 2001; Pandolfi et al. 2003). For example, in the Caribbean Sea, two formerly dominant species of corals, have suffered greatly reduced population densities. Although the cause of decline of these species is largely due to historical overfishing rather than climate change, bleaching due to climate change may now drive these species to extinction.

For coral reefs, further climate change and community dynamics must be placed in the context of SSTs and CO_2 concentrations. IPCC projections suggest conditions the earth would not have experienced in over 40 million years and when it did, corals were not the dominant reef-building organisms. Thus even if present reefs were in a pristine condition, one might reasonably predict shifts in the dominance patterns of tropical reefs should present trends in increasing temperature and CO_2 continue to the ultimate transition from present icehouse to greenhouse conditions.

Stress and reef community dynamics will interact. Even though reefs in the geological past have been delayed in their recovery from mass extinctions (Copper 1989), they have always survived, regardless of which organism was dominant. But today's stressed reef ecosystems may not be able to produce a dominant reef organism to be the greenhouse 3-dimensional structural counterpart to corals. Evidence for this lies in the present phase shift from calcifying scleractinian corals to noncalcifying fleshy macro-algae (seaweed) as the dominant reef organism. The possible synergy between local habitat degradation and predicted global change may push marine ecosystems into uncharted territory where even small perturbations could trigger large deleterious effects.

SUMMARY

Marine species and ecosystems show a wide range of responses to past climate change. Shifts in geographic distributions of species are a very common response to climate change in habitats ranging from the intertidal to the deep sea and in groups from mollusks to foraminifera. On the other hand, many species, including many tropical reef corals, do not show noticeable changes in their geographic distributions even in response to significant climatic perturbations such as the Pleistocene glacial–interglacial cycles. The causes of these differential responses remain poorly known, but there is increasing effort to understand the ecological and life history traits that impose or permit greater variability in the geographic ranges of some species relative to others in the same community. Shifts in geographic ranges of species are also evident today as species attempt to adapt to an increasingly warm world and understanding the biological basis of such individualistic responses is essential for predicting the consequences of future climate change. In terms of extinction, some of the past climatic events are associated with widespread extinctions of marine species, but others such as the Milankovitch-scale changes during the Pleistocene did not lead to heightened extinctions. Again the exact causes of these differences remain unclear, but threshold effects and historical contingencies almost certainly play a large role.

The past provides us with unique insights about biotic responses to climate change and the ecological and evolutionary consequences of such responses. The repeated glacial–interglacial cycles during the Pleistocene are particularly valuable since they provide replicate natural experiments involving species that are still living and hence insights directly applicable to ongoing and future climatic change. However, inferences from history apply directly only to a world where ecosystems are healthy and habitats plentiful. A key to understanding future marine responses to climate change is the possible synergistic relationship between climate change and the widespread habitat degradation and biodiversity loss due to anthropogenic activities.

ACKNOWLEDGMENTS

This work was partially supported by National Science Foundation grants to K.R. and J.M.P.

REFERENCES

Addicott, W. O. 1969. Tertiary climatic change in the marginal northeastern Pacific Ocean. *Science* 165, 583–586.

Allmon, W. D., G. Rosenberg, R. W. Portell, and K. S. Schindler. 1993. Diversity of Pliocene to Recent Atlantic coastal plain mollusks. *Science* 260, 1626–1628.

Aronson, R. B., and D. B. Blake. 2001. Global climate change and the origin of modern benthic communities in Antarctica. *Amer. Zool.* 41, 27–39.

Avise, J. C. 1994. *Molecular Markers, Natural History and Evolution.* New York: Chapman & Hall.

Barry, J. P., C. H. Baxter, R. D. Sagarin, and S. E. Gilman. 1995. Climate-related long-term faunal changes in a California rocky intertidal community. *Science* 267, 672–675.

Bellwood, D. R., and T. P. Hughes. 2001. Regional-scale assembly rules and biodiversity of coral reefs. *Science* 292, 1532–1534.

Benzie, J. A. H. 1999. Genetic structure of coral reef organisms: Ghosts of dispersal past. *Amer. Zool.* 39, 131–145.

Boucot, A. J., and J. Gray. 2001. A critique of Phanerozoic climatic models involving changes in the CO2 content of the atmosphere. *Earth-Science Rev.* 56, 1–159.

Brenchley, P. J., J. D. Marshall, and C. J. Underwood. 2001. Do all mass extinctions represent an ecological crisis? Evidence from the Late Ordovician. *Geol. Jour* 36, 329–340.

Brown, J. H. 1995. *Macroecology*. Chicago: University of Chicago Press.

Budd, A. F. 2000. Diversity and extinction in the Cenozoic history of Caribbean reefs. *Coral Reefs* 19, 25–35.

Buddemeier, R. W., and S. V. Smith. 1999. Coral adaptation and acclimatization: A most ingenious paradox. *Amer. Zool.* 39, 1–9.

Burton, R. S. 1998. Intraspecific phylogeography across the Point Conception biogeographic boundary. *Evolution* 52, 734–745.

Buzas, M. A., and S. J. Culver. 1994. Species pool and dynamics of marine paleocommunities. *Science* 264, 1439–1441.

Buzas, M. A., and S. J. Culver. 1998. Assembly, disassembly and balance in marine paleocommunities. *Palaios* 13, 263–275.

Cannariato K. G., J. P. Kennett, and R. J. Behl. 1999. Biotic response to late Quaternary rapid climate switches in Santa Barbara Basin: Ecological and evolutionary implications. *Geology* 27, 63–66.

Clarke, A. 1993. Temperature and extinction in the sea: A physiologist's view. *Paleobiology* 19, 499–518.

Copper, P. 1988. Ecological succession in Phanerozoic reef ecosystems: Is it real? *Palaios* 3, 136–152.

Copper, P. 1989. Enigmas in Phanerozoic reef development. *Mem. Assoc. Australasian Palaeontologists* 8, 371–385.

Crame, J. A. 1986. Late Pleistocene molluscan assemblages from the coral reefs of the Kenya coast. *Coral Reefs* 4:183–196.

Cronin, T. M., D. M. DeMartino, G. S. Dwyer, and J. Rodriguez-Lazaro. 1999. Deep-sea ostracod species diversity: Response to late Quaternary climate change. *Marine Micropaleo.* 37, 231–249.

Cronin, T. M., and M. E. Raymo. 1997. Orbital forcing of deep-sea benthic species diversity. *Nature* 385, 624–627.

Cronin, T. M., et al. 1995. Late Quaternary paleoceanography of the Eurasian Basin, Arctic Ocean. *Paleoceanography* 10, 259–281.

Crowley, T. J., and G. R. North. 1991. *Paleoclimatology*. Oxford Monographs on Geology and Geophysics No. 16. New York: Oxford University Press.

Culver, S. J., and M. A. Buzas. 1995. The effects of anthropogenic habitat disturbance, habitat destruction, and global warming on shallow marine benthic foraminifera. *Jour. Foram. Res.* 25, 204–211.

Culver, S. J., and M. A. Buzas. 2000. In *Biotic Response to Global Change: The Last 145 Million Years*, ed. S. J. Culver and P. F. Rawson, 122–134. Cambridge: Cambridge University Press.

Dahlgren T. G., J. R. Weinberg, and K. M. Halanych. 2000. Phylogeography of the ocean quahog (Arctica islandica): Influences of paleoclimate on genetic diversity and species range. *Marine Biol.* 137, 487–495.

Dowsett, H. J., et al. 1994. Joint investigation of the middle Pliocene climate. I: Paleoenvironmental reconstructions. *Global and Planetary Climate Change* 9, 169–195.

Edmands, S. 2001. Phylogeography of the intertidal copepod *Tigriopus californicus* reveals substantially reduced population differentiation at northern latitudes. *Mol. Ecol.* 10, 1743–1750.

Fagerstrom, J. A. 1987. *The Evolution of Reef Communities*. New York: Wiley.

Fischer, A. G., and M. A. Arthur. 1977. Secular variations in the pelagic realm. *SEPM Spec. Pub.* 25, 19–50.

Flügel, E., and E. Flügel-Kahler. 1992. Phanerozoic reef evolution: Basic questions and data base. *Facies* 26, 167–278.

Frakes, L. A., J. E. Francis, and J. I. Syktus. 1992. *Climate Modes of the Phanerozoic*. Cambridge: Cambridge University Press.

Gage, J. D., and P. A. Tyler. 1991. *Deep-sea Biology*. Cambridge: Cambridge University Press.

Gattuso, J.-P., D. Allemand, and M. Frankignouille. 1999. Photosynthesis and calcification at cellular, organismal and community levels in coral reefs: A review on interactions and control by carbonate chemistry. *Amer. Zool.* 39, 160–183.

Gaylord, B., and S. D. Gaines. 2000. Temperature or transport? Range limits in marine species mediated solely by flow. *Amer. Nat.* 155, 769–789.

Gladenkov, Y. B., and A. E. Shanster. 1990. Neogene of Kamchatka: Stratigraphy and correlation of geological events. In *Pacific Neogene Events: Their Timing, Nature and Interrelationship*, ed. R. Tsuchi. Tokyo: University of Tokyo Press.

Gleeson, M. W., and A. E. Strong. 1995. Applying MCSST to coral-reef bleaching. *Adv. Space Res.* 16, 151–154.

Hallock, P. 1997. Reefs and reef limestones in Earth history. In *Life and Death of Coral Reefs*, ed. C. Birkeland, 13–42. New York: Chapman and Hall.

Hansen, T. 1992. The patterns and causes of molluscan extinction across the Eocene/Oligocene boundary. In *Eocene-Oligocene Climatic and Biotic Evolution*, ed. D. R. Prothero, and W. A. Berggren, 341–348. Princeton: Princeton University Press.

Hazel, J. E. 1970. Atlantic continental shelf and slope of the United States: Ostracod zoogeography in the southern Nova Scotian and northern Virginian faunal provinces. *U. S. Geol. Surv. Prof. Pap.* 529-E: E1–E21.

Hellberg, M. E., D. P. Balch, and K. Roy. 2001. Climate-driven range expansion and morphological evolution in a marine gastropod. *Science* 292, 1707–1710.

Hessler, R. R., and H. L. Sanders. 1967. Faunal diversity in the deep sea. *Deep-Sea Res.* 14, 65–78.

Hewitt, G. 2000. The genetic legacy of the Quaternary ice ages. *Nature* 405, 907–913.

Holbrook S. J., R. J. Schmitt, and J. S. Stephens. 1997. Changes in an assemblage of temperate reef fishes associated with a climate shift. *Ecol. Appl.* 7, 1299–1310.

Hughes T. P. 1994. Catastrophes, phase shifts, and large-scale degradation of a Caribbean coral reef. *Science* 265, 1547–1551.

Hughes, T. P., A. H. Baird, D. R. Bellwood, M. Card, S. R. Connolly, C. Folke, R. Grosberg, O. Hoegh-Guldberg, J. B. C. Jackson, J. Kleypas, J. M. Lough, P. Marshall, M. Nystrom, S. R. Palumbi, J. M. Pandolfi, B. Rosen, and J. Roughgarden. 2003. Climate change, human impacts, and the resilience of coral reefs. *Science* 301, 929–933.

IPCC 2001. *Climate Change 2001: The Scientific Basis.* Contribution of Working Group I to the Third Assessment Report of the Intergovernmental Panel on Climate Change (eds. J. T. Houghton, Y. Ding, D. J. Griggs, M. Noguer, P. J. van der Linden, X. Dai, K. Maskell, and C. A. Johnson). Cambridge, U.K.: Cambridge University Press.

Jablonski, D., and J. J. Sepkoski, Jr. 1996. Paleobiology, community ecology, and scales of ecological pattern. *Ecology* 77, 1367–1378.

Jackson, J. B. C. 1992. Pleistocene perspectives on coral reef community structure. *Amer. Zool.* 32, 719–731.

Jackson, J. B. C. 1995. Constancy and change of life in the sea. In *Extinction Rates*, ed. J. H. Lawton, and R. M. May, 45–54. Oxford: Oxford University Press.

Jackson, J. B. C., P. Jung, A. G. Coates, and L. S. Collins. 1993. Diversity and extinction of tropical American mollusks and the emergence of the Isthmus of Panama. *Science* 260, 1624–1626.

Jackson, J. B. C., M. X. Kirby, W. H. Berger, K. A. Bjorndal, L. W. Botsford, B. J. Bourque, R. Bradbury, R. Cooke, J. Erlandson, J. A. Estes, T. P. Hughes, S. M. Kidwell, C. B. Lange, H. S. Lenihan, J. M. Pandolfi, C. H. Peterson, R. S. Steneck, M. J. Tegner, and R. Warner. 2001. Historical overfishing and the recent collapse of coastal ecosystems. *Science* 293, 629–638.

Kaiho, K. 1994. Planktonic and benthic foraminiferal extinction events during the last 100 m.y. *Palaeogeog. Palaeoclimat. Palaeoecol.* 111, 45–71.

Kauffman, E. G., and C. C. Johnson. 1988. The morphological and ecological evolution of Middle and Upper Cretaceous reef-building rudistids. *Palaios* 3, 194–216.

Kendrick, G. W., K-H. Wyrwoll, B. J. Szabo. 1991. Pliocene-Pleistocene coastal events and history along the western margin of Australia. *Quaternary Science Reviews* 10:419–439.

Kennett, J. P., and L. D. Stott. 1991. Abrupt deep-sea warming, paleoceanographic changes an benthic extinctions at the end of the Paleocene. *Nature* 353, 225–229.

Kiessling, W. 2001. Paleoclimatic significance of Phanerozoic reefs. *Geology* 29, 751–754.

Kitamura, A., H. Omote, and M. Oda. 2000. Molluscan response to early Pleistocene rapid warming in the Sea of Japan. *Geology* 28, 723–726.

Knoll, A. H., and S. B. Carroll. 1999. Early animal evolution: Emerging views from comparative biology and evolution. *Science* 284, 2129–2137.

Kohn, A. J., and I. Arua. 1999. An Early Pleistocene molluscan assemblage from Fiji: Gastropod faunal composition, paleoecology and biogeography. *Palaeogeog. Palaeoclimat. Palaeoecol.* 146, 99–145.

Lagoe, M. 1983. Oligocene through Pliocene foraminifera from the Yakataga Reef section, Gulf of Alaska Tertiary province, Alaska. *Micropaleontol.* 29, 202–222.

Lessios H. A., B. D. Kessing, and J. S. Pearse. 2001. Population structure and speciation in tropical seas: Global phylogeography of the sea urchin Diadema. *Evolution* 55, 955–975.

Marincovich, L., Jr. 1988. Recognition of an earliest middle Miocene warm-water event in a southwestern Alaskan molluscan fauna. In *Professor Tamio Kotaka Commemorative Volume on Molluscan Paleontology*, ed. J. A. Grant-Mackie, K. Masuda, K. Mori, and K. Ogasawara, 1–31. Saito Ho-on Kai Special Publication.

McGhee, G. R., Jr. 2001. The 'multiple impacts hypothesis' for mass extinction: A comparison of the Late Devonian and the late Eocene. *Palaeogeog. Palaeoclimat. Palaeoecol.* 176, 47–58.

Newell, N. D. 1971. An outline history of tropical organic reefs. *Amer. Mus. Novitiates* 2465, 1–37.

Ogasawara, K. 1994. Neogene paleogeography and marine climate of the Japanese Islands based on shallow-marine molluscs. *Palaeogeog. Palaeoclimat. Palaeoecol.* 108, 335–351.

Pandolfi, J. M. 1996. Limited membership in Pleistocene reef coral assemblages from the Huon Peninsula, Papua New Guinea: Constancy during global change. *Paleobiology* 22, 152–176.

Pandolfi, J. M. 1999. Response of Pleistocene coral reefs to environmental change over long temporal scales. *Amer. Zool.* 39, 113–130.

Pandolfi, J. M. 2000. Persistence in Caribbean coral communities over broad spatial and temporal scales. IX International Coral Reef Congress, Bali, October 23–27.

Pandolfi, J. M., R. H. Bradbury, E. Sala, T. P. Hughes, K. A. Bjorndal, R. G. Cooke, D. McArdle, L. McClenachan, M. J. H. Newman, G. Paredes, R. R. Warner, and J. B. C. Jackson. 2003. Global trajectories of the long-term decline of coral reef systems. *Science* 301, 955–958.

Pandolfi, J. M., J. B. C. Jackson, and J. Geister. 2001. Geologically sudden natural extinction of two widespread Late Pleistocene Caribbean reef corals. In *Evolutionary Patterns: Growth, Form and Tempo in the Fossil Record*, ed. J. B. C. Jackson, S. Lidgard, and F. K. McKinney, 120–158. Chicago: University of Chicago Press.

Pandolfi, J. M., C. E. Lovelock, and A. F. Budd. 2002. Character release following extinction in a Caribbean reef coral species complex. *Evolution* 56, 479–501.

Paulay, G. 1990. Effects of late Cenozoic sea-level fluctuations on the bivalve faunas of tropical oceanic islands. *Paleobiology* 16:415–434.

Paulay, G. 1996. Dynamic clams: Changes in the bivalve fauna of Pacific islands as a result of sea-level fluctuations. *Amer. Malacol. Bull.* 12, 45–57.

Paulay, G., and T. Spencer. 1992. Nine Island: Geologic and faunatic history of a Pliocene atoll. *Pacific Science Association Information Bulletin* 44(3–4):21–23.

Peters, R. H. 1983. *The Ecological Implications of Body Size.* Cambridge: Cambridge University Press.

Petit, J. R., J. Jouzel, D. Raynaud, N. I. Barkov, J.-M. Barnola, I. Basile, M. Benders, J. Chappellaz, M. Davis, G. Delayque, M. Delmotte, V. M. Kotlyakov, M. Legrand, V. Y. Lipenkov, C. Lorius, L. Pépin, C. Ritz, E. Saltzman, and M. Stievenard. 1999. Climate and atmospheric history of the past 420,000 years from the Vostok ice core, Antarctica. *Nature* 399, 429–436.

Playford, P. E. 1983. Geological research on Rottnest Island. *Journal of the Royal Society of Western Australia* 66(1–2):10–15.

Prothero, D. R., and W. A. Berggren, Eds. 1992. *Eocene-Oligocene Climatic and Biotic Evolution.* Princeton: Princeton University Press.

Rosen, B. R. 1984. Reef coral biogeography and climate through the late Cenozoic: Just islands in the sun or a critical pattern of islands? In Brenchley, P. J., ed., *Fossils and Climate.* New York: Wiley.

Roy, K., D. Jablonski, and J. W. Valentine. 1995. Thermally anomalous assemblages revisited: Patterns in the extraprovincial range shifts of Pleistocene marine mollusks. *Geology* 23, 1071–1074.

Roy, K., D. Jablonski, and J. W. Valentine. 2001. Climate change, species range limits and body size in marine bivalves. *Ecol. Lett.* 4, 366–370.

Roy, K., D. Jablonski, and J. W. Valentine. 2002. Body size and invasion success in marine bivalves. *Ecol. Lett.* 5, 163–167.

Roy, K., J. W. Valentine, D. Jablonski, and S. M. Kidwell. 1996. Scales of climatic variability and time averaging in Pleistocene biotas: Implications for ecology and evolution. *Trends Ecol. Evol.* 11, 458–463.

Sagarin, R. D., J. P. Barry, S. E. Gilman, and C. H. Baxter. 1999. Climate-related change in an intertidal community over short and long time scales. *Ecol. Monog.* 69, 465–490.

Schmiedl, G., C. Hemleben, J. Keller, and M. Segl. 1998. Impact of climatic changes on the benthic foraminiferal fauna in the Ionian Sea during the last 330,000 years. *Paleoceanography* 13, 447–458.

Southward A. J., S. J. Hawkins, M. T. Burrows. 1995. 70 years observations of changes in distribution and abundance of zooplankton and intertidal organisms in the western English-Channel in relation to rising sea temperature. *J. Therm. Biol.* 20, 127–155.

Stanley, G. D., ed. 2001. The history and sedimentology of ancient reef systems. *Topics in Geobiology* 17.

Stanley, S. M. 1986. Anatomy of a regional mass extinction: Plio-Pleistocene decimation of the western Atlantic bivalve fauna. *Palaios* 1, 17–36.

Stanley, S. M., W. O. Addicott, and K. Chinzei. 1980. Lyellian curves in paleontology: Possibilities and limitations. *Geology* 8, 422–426.

Stanley, S. M., and L. D. Campbell. 1981. Neogene mass extinction of western Atlantic molluscs. *Nature* 293, 457–459.

Stanley, S. M., and L. A. Hardie. 1998. Secular oscillations in the carbonate mineralogy of reef-building and sediment-producing organisms driven by tectonically forced shifts in seawater chemistry. *Palaeogeog. Palaeoclimat. Palaeoecol.* 144, 3–19.

Stehli, F. G., and J. W. Wells. 1971. Diversity and age patterns in hermatypic corals. *Syst. Zool.* 20:115–126.

Taylor, J. D. 1978. Faunal response to the instability of reef habitats: Pleistocene molluscan assemblages of Aldabra Atoll. *Palaeontology* 21:1–30.

Taylor, P. D. 2000. Origin of the modern bryozoan fauna. In *Biotic Response to Global Change: The Last 145 Million Years*, ed. S. J. Culver and P. F. Rawson, 195–206. Cambridge: Cambridge University Press.

Todd, J. A., J. B. C. Jackson, K. G. Johnson, H. M. Fortunato, A. Heitz, M. Alvarez, and P. Jung. 2002. The ecology of extinction: Molluscan feeding and faunal turnover in the Caribbean Neogene. *Proc. Roy. Soc. London B* 269, 571–577.

Valentine, J. W. 1961. Paleoecologic molluscan geography of the Californian Pleistocene. *Univ. Calif. Pub. Geol. Sci.* 34, 309–442.

Valentine, J. W. 1973. *Evolutionary Paleoecology of the Marine Biosphere*. Englewood Cliffs, N.J.: Prentice-Hall.

Valentine, J. W. 1989. How good was the fossil record? Clues from the Californian Pleistocene. *Paleobiology* 15, 83–94.

Valentine, J. W., and D. Jablonski. 1991. Biotic effects of sea level change: The Pleistocene test. *J. Geophys. Res.* 96, 6873–6878.

Valentine, J. W., and D. Jablonski. 1993. Fossil communities: Compositional variation at many time scales. In *Species Diversity in Ecological Communities*, ed. R. E. Ricklefs and D. Schluter, 341–349. Chicago: University of Chicago Press.

Veron, J. E. N. 1995. *Corals in Space and Time: The Biogeography and Evolution of the Scleractinia*. Sydney: University of New South Wales Press.

Walther, G.-R., et al. 2002. Ecological responses to recent climate change. *Nature* 416, 389–395.

Wares J. P. 2001. Biogeography of Asterias: North Atlantic climate change and speciation. *Biol. Bull.* 201, 95–103.

Wares J. P., and C. W. Cunningham. 2001. Phylogeography and historical ecology of the North Atlantic intertidal. *Evolution* 55, 2455–2469.

Wood, R. 1999. *Reef Evolution*. Oxford: Oxford University Press.

Zachos, J., M. Pagani, L. Sloan, E. Thomas, and K. Billups. 2001. Trends, rhythms, and aberrations in global climate 65 Ma to Present. *Science* 292, 686–693.

Zinsmeister, W. J. 1982. Review of the Upper Cretaceous–Lower Tertiary sequence on Seymour Island, Antarctica. *Jour. Geol. Soc. Lond.* 139, 779–786.

CHAPTER TWELVE

Genetic and Evolutionary Impacts of Climate Change

GODFREY M. HEWITT
AND RICHARD A. NICHOLS

ADAPT, MOVE, OR DIE

Our continually changing climate means that populations of organisms adapt to new conditions, change their range following suitable environments, or go extinct. These processes involve genetic changes (Hewitt 1993). Adaptation and hence reproductive survival is achieved through natural selection of genotypes produced by mutation and recombination, and spread by gene flow. An organism's means of dispersal, which produce gene flow and allow it to reach new suitable places, are genetically determined and the products of considerable evolution from ancient to modern times. When these two processes fail, the population dies and its unique genotypes are lost, thereby changing the genetic structure and potential of the regional metapopulation, the geographic race, and the species. The fields of molecular genetics and palaeoclimatology have made great strides in the last few years, providing much information on the genetic structure of populations and species and on the detailed patterns of past climates. Together they are providing new insights into the effects of climate changes on the evolution of organisms, particularly over the last 3 million years.

This chapter outlines the effect of climate on species distributions and then considers the genetic consequences of the accompanying episodes of population fragmentation and expansion. Finally, the implications for the future of genetic diversity and species richness are explored.

Box 12.1. **DNA Markers**

Studies to describe the spatial genetic structure of species—and hence deduce past events such as population demography, gene flow, and range shifts—may now use a large and ever growing array of molecular markers (Sunnocks 2000). These may for simplicity be divided into simple sequence and allele frequency categories, although there is some overlap.

Animal mitochondrial (mt)DNA has been foremost in the great advances made in phylogeography (Avise 2000) due to molecular and genetic properties. As a simple sequence, unclouded by recombination and biparental inheritance in most species, it allows the deduction of genealogy and lineages of present-day *haplotypes* or alleles, from which we may see genes in space and time (Hewitt 2000). Its average rate of divergence is relatively quick, some 2% per million years, with the D-loop being notably faster. This produces population differences across a species range but does not saturate with recurrent mutations over the 2 million years of Pleistocene climate changes. Other simple DNA sequences are used but in general have slower rates of divergence, more suitable for phylogenies deeper in time (nuclear gene introns average -0.7% per million years, plant chloroplast (cp)DNA—$0.02-0.01\%$ per million years).

Microsatellites are chromosomal DNA segments that do not code for functional proteins. They have much higher mutation rates ($\times 100$), many alleles of different length, and much population variation. Their high variability with thousands of loci scattered through the genome makes them very useful for population genetics and to elucidate changes that have occurred in the late Pleistocene and Holocene.

Thus for studies of the effect of climate changes on genetic structure of species through the Pleistocene and Holocene a combination of more and less variable DNA markers is desirable, but this is expensive. A useful and achievable set might comprise an mtDNA sequence and 10 moderately variable microsatellite loci. For detailed population genetic studies at a local scale and concerned with possible changes over the past century, 20 or more microsatellite loci might be used. With rapid advances in genomics, other markers are being developed (e.g., SNPs, single nucleotide polymorphisms) along with methods for their automation; but they will not be cheap. The range change studies discussed herein mostly use sequence data, and while earlier population genetic data came from the study of proteins, most recent studies employ microsatellites. Species are made of many genes, and sequences from several genes would produce a better picture of their population genetic history, but this requires considerable effort and funding.

References: Avise JC, 2000, *Phylogeography: The History and Formation of Species.* Cambridge, Mass.: Harvard University Press.

Hewitt GM, 2000, The genetic legacy of the Quaternary ice ages. *Nature* 405: 907–913.

Sunnocks P, 2000, Efficient genetic markers for population biology. *Trends in Ecology and Evolution* 15: 199–203.

RANGE CHANGES AND SHIFTING NICHES

Dated fossil remains, especially from pollen cores and beetles, provide evidence of how individual species have responded to climate changes in different places (Huntley and Birks 1983; Webb and Bartlein 1992; Coope 1994; Bennett 1997; Chapters 7–11 in this volume). Such responses comprise a number of processes that affect the genetic structure of the populations and species.

The most striking effect is a change in a species' range that tracks its shifting niche. This involves colonization and population growth in some parts and population decrease and extinction in others (Hewitt 1996). It can be remarkably rapid for many organisms, with, for example, beetles and beech trees expanding their range at over 1 km per year in the postglacial warming of Europe and North America. In modern times, natural and human-assisted invasions such as the collared dove and corn borer moth illustrate this facility (Hewitt 1993).

The repeated population bottlenecks produced by small numbers of founding individuals in such a rapid pioneer-type colonization can greatly reduce the genetic variation at the leading edge and in its descendant populations, and this can still be detected in current populations (Hewitt 1993) (Fig. 12.1). The effect is nicely demonstrated by reduced nucleotide diversity in populations of North American fishes from previously glaciated areas (Bernatchez and Wilson 1998). Popula-

tions at the trailing edge of a range shift will become fragmented and bottlenecked before becoming extinct, unless the climate reverses and they become the expanding leading edge.

What genetic properties determine the species' environmental limits, its adaptive niche, and the edge of its range is a very demanding question (Hewitt 1993; Kirkpatrick and Barton 1997; Peterson et al., this volume). It is, however, clear that responses will differ for species with distinct

Figure 12.1. A climate change causes a range shift with expansion at the leading edge and fragmentation at the trailing edge. Five genotypes are illustrated. Genetic loss occurs at the leading edge of expansion, compared with greater retention at the trailing edge. Hybrid zones form, acting as barriers to gene flow, but selected alleles can spread. After establishment, little diffusion occurs between genomic blocks. As populations are reduced at the trailing edge different genotypes may survive locally, preserving diversity over longer periods. This may become the leading edge with a climate reversal.

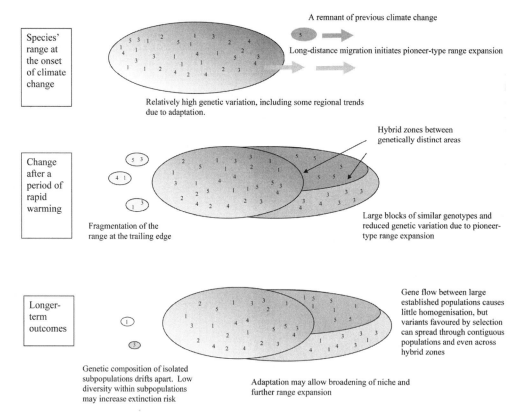

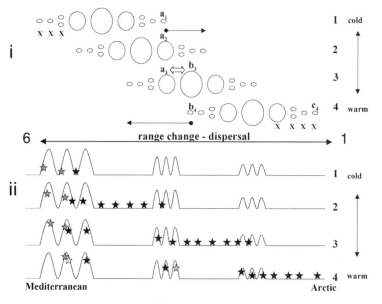

Figure 12.2. Diagrammatic range shifts due to climate changes between cold and warm conditions 1–4. In (i) the species populations are large in the center of its range and smaller at the N and S edges. Between times 1 and 2 population a expands while those at the rear go extinct, and so on as climate warms, until maximum warming and shift is reached at time 4. When climate reverses the process is reversed, as shown following time 4 with b leading. Population c is at the end of a chain of expansion and survives only in the thermal maximum. Population a_{1-3} survives though all but the warmest time, but gene flow between a_3 and b_3 would be important for transferring genetic variation and adaptation in such very warm periods. In (ii) a particular topographical case in Europe is shown. Genomes survive in the southern mountains, and one expands north as climate warms across plains and hills, which may pose temporary barriers. At the warmest time, 4, the classic Arctic–Alpine distribution occurs. The populations to the north go extinct when it cools again, including mutants and adaptations. Mutants and adaptations in the southern mountains survive and accumulate by climbing up and down.

adaptations and life histories, and inhabiting different niches in the boreal, temperate, desert, and tropical zones, on mountains and in seas. With oscillating climate over the past 2 million years, the roles of leading and trailing edge have been reversed many times (Fig. 12.2). When a large complete shift in range occurs, as in some boreal species, there is no continuously occupied region, while in large trop-

ical mountain blocks, movement over short distances in waves (called phalanx-type shifts) conserve the species and its genomic diversity (Hewitt 2000). These two extreme forms have been characterized as "pioneer" and "phalanx" modes of colonization (Nichols and Hewitt 1994).

During the last glacial maximum, at the northern limits of a temperate species' refugial area, it is likely that some species would have survived as small populations in favourable habitat pockets. This would depend on their specific niche and geographic constraints, rather like the northern and montane tree lines today. As climate warmed, some of these pockets could be the source of advances, which would produce a pattern akin to the pioneer mode, as shown in Figure 12.2. With oscillating climate these frontal outposts would flicker on and off (Hewitt 1996). These changes would leave distinct genetic patterns, which would differ in flatlands and mountains. Table 12.1 summarizes some of these patterns.

In simulations of leading edge expansion, small distinct genomic patches are created by long distance founders (Ibrahim et al. 1996), and a range of patchwork patterns can be produced using various

Table 12.1. Major climatic changes produce shifts in species ranges

Parts of Shifting Range (see Figs. 18.1 and 18.2)	Plains (e.g., Europe, North America)	Mountains (e.g., Andes, Annamites)
Leading edge	Rapid colonization, long effective dispersal. Pioneer expansion, loss of alleles. Few genotypes cover large areas. New adaptations.	Slow colonization, short effective dispersal. Phalanx advance, large populations retain more alleles and lineages.
Center (may disappear with full range shift)	Populations survive, retain genetic diversity. Some diffusion gene flow and adaptation.	Populations survive, retain more genetic diversity. More diffusion gene flow and adaptation.
Trailing edge	Rapid fragmentation, individual population drift. More allelic diversity retained among populations, before extinction.	Slower retreat, less fragmentation. Retain more genetic variability before extinction. Topography creates pockets.

Note: These can modify the genetic structure of populations, and some effects may be seen today as "footprints" or "fingerprints." These effects vary with latitude, precipitation, and life form, and here two major systems—flatlands and mountains—are contrasted.

dispersal and population growth parameters. This has recently been explored in the context of cytotype (chloroplast DNA variants) patches seen in European oaks that colonized after the last ice age (Petit et al. 2001). The detailed parapatric haplotype (mitochondrial DNA variants) pattern seen in European hedgehogs (Fig. 12.3) is also explicable in such terms (Seddon et al. 2001). Establishing the likely mode of production of any specific genomic pattern would seem to require a combination of a detailed fossil record and a full simulation analysis of expansion parameters (dispersal, growth rate) for the species and its habitat distribution through time.

Particular mention of topographically variable regions, such as mountains, is called for, since species may often be able to track their habitat up and down through many major climatic cycles (see Fig. 12.2). This probably involves colonization over short distances, which consequently maintains larger population size and hence genetic variation. Mountains in southern Europe, western United States, China, South America, and the tropics are considered important in this respect and may therefore act as refugia for the species

and their genetic variation through climate changes (Hewitt 1996, 2000).

The extensive network of pollen cores across Europe and North America shows that many species responded individually to climate changes, each tracking its own particular set of environmental conditions as it changed its range. This produced transitory mixtures of species with geographic distributions that were different from today (Huntley 1990, 1999). Such findings have been examined in terms of ecological niche theory, where the range and combination of environmental variables in which a species can survive do not alter in unison with climate change, leading to the concept of the nomadic niche (Jackson and Overpeck 2000). An important genetic corollary of such species mixing is spatially and temporally changing selection due to their interactions. This is compounded by the considerable intraspecific genomic subdivision revealed by recent molecular genetic studies (e.g. Fig. 12.3), which can be used to show that postglacial biotas are mixtures of colonist species from different glacial refugia in which considerable genomic divergence occurred (Hewitt and Ibrahim 2001). What were previously believed to be tight co-

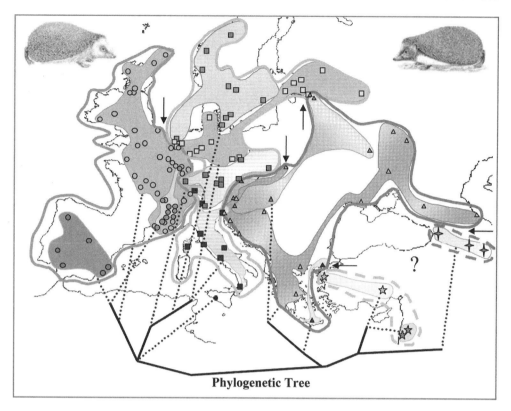

Phylogenetic Tree

Figure 12.3. The divergent genomic substructure of European hedgehogs, *Erinaceus europaeus* and *E. concolor*. The phylogenetic tree at the base connects 5 major clades that occupy distinct adjacent areas of Europe, with zones of contact (arrows indicate some of these). Within each major clade are minor clades, largely parapatrically distributed. The DNA difference among these clades indicates that they diverged 5–1 My BP, and within minor clade regions, divergence has occurred over the latter ice ages. This diversity survived glacial maxima in southern refugia and expanded over Europe after the last ice age, 10 ky BP, to produce the present complex cryptic genetic structure.

evolutionary relationships must be reviewed in light of the often transitory nature of species assemblages.

A NORTHERN HEMISPHERE EXAMPLE: THE GRASSHOPPER, THE HEDGEHOG, AND THE BEAR

Most species are subdivided geographically into distinct genomes, which are only sometimes correlated with morpho-

logical differences. In Europe, it is now possible using DNA similarity to show from which refugia particular genomes emerged to cover their present distribution. A few species have been studied using suitably variable and discriminating sequences in a large enough number of samples from across their range. Such cases have been collated, and three broad paradigm patterns of postglacial colonization routes are evident—the grasshopper, the hedgehog, and the bear (Hewitt 1999). These are illustrated in Figure 12.4.

Chorthippus parallelus, the common meadow grasshopper, has a genome divided into at least five main geographic regions, and clearly shows a postglacial expansion from a Balkan refugium (Cooper et al. 1995). Hybrid zones have been described between the Spanish and French genomes and the Italian and French/Austrian genomes along the Pyrenees and Alps, where the expanding genomes met (Flanagan et al. 1999). The divergence be-

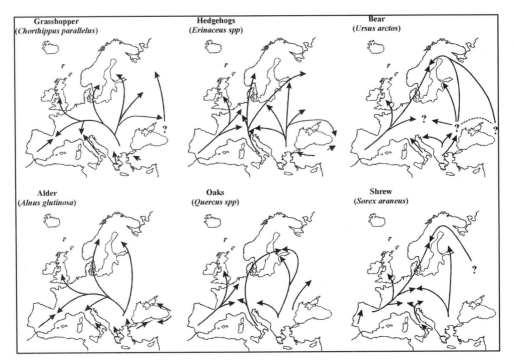

Figure 12.4. Postglacial colonization patterns for species from southern European refugia. The grasshopper, the hedgehog, and the bear are three paradigm patterns emanating from different combinations of Iberia, Italy, and the Balkans. The alder, the oak, and the shrew are similar to them.

tween genomes can be examined in detail in such hybrid ones, which provide natural laboratories for evolutionary studies (Hewitt 1988, 1993).

Erinaceus europeus and E. concolor, the European hedgehogs, show considerable DNA divergence, and surprisingly reveal distinct clades within each. The depth of divergence between these lineages suggests separations several million years ago, possibly at the beginning of the Quaternary. The hedgehog genome across Europe is divided into three principal north/south strips, with a fourth clade in Turkey and Israel (Santucci et al. 1998; Seddon et al. 2001). This pattern is quite different from that of the grasshopper, indicating colonization of northern Europe from three glacial refugia: a western/Spanish one, a central/Italian one, and an eastern/Balkan one. Interestingly, the Balkan lineage has also colonized the north side of the Caucasus, while a Turkish one has colonized from the south (Seddon et al. 2001).

Ursus arctos, the brown bear, has had its broad European distribution reduced by humans to a few isolates in the west, with larger populations in Scandinavia, Eastern Europe, and Russia. Consequently it is the object of conservation efforts. Analysis of mtDNA has revealed distinct eastern and western lineages, and U. arctos appears to have colonized most of Europe from an Iberian and a Caucasian/Carpathian refuge. These expansions met in central Sweden and formed a hybrid zone (Taberlet et al. 1998). Studies on other species are emerging, and while most broadly fit these patterns, some show somewhat different patterns. This is so for several freshwater fish, such as the barbell, chub, and perch, where large river systems like the Danube and Dnieper formed major colonization routes from Black Sea refugia (Durand et al. 1999; Nesbo et al. 1999; Englbrecht et al. 2000). Organisms with unusual life histories or ecological requirements may well show more individual patterns.

It appears that the Balkans were a source for all species in the east and for many species in the west; the grasshopper/ alder/beach/newt pattern would seem to be common. Conversely, Italian genomes rarely populated northern Europe, although exceptions include the hedgehog and the oak. It seems that the ice-capped Alps were an initial barrier to their northward expansion, while the Pyrenees apparently acted as a barrier in fewer cases, as exemplified by the grasshopper, alder, and beech (Hewitt 1999) (see Fig. 12.4).

When these expanding genomes met, they formed hybrid zones, clustering in the Alps and central Europe and to some extent in the north Balkans and Pyrenees. Such clusters are called "suture zones" (Hewitt 1988). The final melting of the Scandinavian ice cap, which allowed eastern and western colonists to meet, probably produced the neat cluster of zones in central Sweden (Jaarola et al. 1999). The Balkans appear genomically diverse and dissected, and are the source of most northward postglacial expansions. Britain received oaks, shrews, hedgehogs, and bears from Spain, and grasshoppers, alder, beech, and newts from the Balkans. The mixture in Sweden is even more complex. This causes reconsideration of community structures in terms of their evolution and stability. Furthermore, there is fossil evidence for very different communities in refugia and during the many range changes.

Partly because of these glacially induced range changes, northern Europe has less genetic variety than southern Europe in terms of numbers of species, subspecific divisions, and allelic diversity: "southern richness to northern purity," so to speak (Hewitt 1999). Within many species genetic diversity was lost during postglacial colonization into temperate regions, while recent studies on southern European populations of species show them to contain several distinct lineages (Paulo et al. 2001). This distribution of genetic diversity has significant conservation implications.

FRAGMENTATION AND GENETICS

It is clear that climate change remodels population distribution and size, and some of these changes affect the genetic composition of populations through chance and selection. It is important to know if these demographically induced genetic changes have substantial effects on the viability and dynamics of populations.

Populations of many species will fragment into smaller and more isolated subpopulations at the trailing edge of the range. In other species, the entire range may be affected in this way. This change will not necessarily accelerate the loss of genetic variation from the species as a whole. Simple models of a subdivided population suggest the rate of loss will be inversely proportional to $ND + D/4m$, where N is the size of each subpopulation, D is the number of subpopulations, and m is the migration rate (Wright 1969; Wakeley 1999). Consequently, populations with large and numerous subpopulations but low migration rates will lose the least genetic diversity. Analogous results can be found in a range of more realistic models incorporating subpopulation extinction (Nichols et al. 2001). When the migration rate is sufficiently large (e.g., $4Nm > 1$) this term is close to the total population size of the species (ND). On the other hand, when the species range is fragmented, the gene flow between populations can be much reduced (i.e., m is reduced), with the consequence that genetic impoverishment is actually retarded. For example, along the coast of Portugal, the southernmost part of the range of the lizard *Lacerta shreiberi* has become fragmented (Paulo et al. 2001). The isolates are separated by up to 100 km of unsuitable habitat, and gene flow can safely be assumed to be negligible. Because they no longer evolve in concert, different fragments have retained different genetic variants (mitochondrial DNA haplotypes) so that the region's diversity is maintained, although it is reduced in each individual population. Indeed, the more continuous northern region has lower genetic variation, probably as a consequence of a recent range expansion of the type outlined above. In many cases like this, species' genetic diversity may actually prove quite resilient to episodes of range contraction, whereas it may be lost during periods of range expansion (see Fig. 12.1).

DEMOGRAPHY AND INBREEDING DEPRESSION

Although species-wide genetic diversity may not be substantially eroded by fragmentation, less benign effects are expected at the level of the subpopulation. Small subpopulations will lose genetic variation and, of more immediate consequence, experience inbreeding depression. This loss of vigor is readily seen in inbred laboratory populations (Ralls and Ballou 1986), but it is more difficult to demonstrate in natural populations. It is harder still to demonstrate that reduced performance of individuals (e.g., low viability, fertility, or fecundity) has tangible effects on the persistence of the populations.

Mavraganis and Eckert (2001) provide an example of a detrimental effect of small subpopulation size on individual productivity in the columbine, *Aquilegia canadensis*, on small islands in the St. Lawrence River. On the other hand, Eisto and colleagues (2000) found no such pattern in the rare bellflower *Campanula cervicaria* in Finland, nor did Lammi et al. (1999) in the rare perennial *Lychnis viscaria*. Less direct evidence comes from attempts to reverse the effects of inbreeding in vertebrates. Genetic restoration appears to have been successful in the case of endangered populations of adders, *Vipera berus* (Madsen et al. 1999), and the Florida panther, *Felis concolor coryi* (Lotz et al. 2000). In both examples the populations had low genetic diversity and appeared in danger of extinction, but showed marked improvement after the introduction of new blood by deliberate transplants. Saccheri and colleagues (1998) were able to link low genetic diversity directly to extinction by following populations of the butterfly *Melitaea cinxia*, which are prone to extinction and had been recorded over many generations. They found that low heterozygosity, in addition to ecological factors such as low subpopulation size and remoteness from other subpopulations, predisposed subpopulations to extinction.

However, such subpopulation extinctions are not always a cause for concern. Indeed, the metapopulation of *Melitaea cinxia* is flourishing because colonization and recolonization of vacant habitat counterbalance the extinctions. On the other hand, whole-species extinction does sometimes appear to be related to small population size and low genetic diversity. Island populations of birds, for example, have a higher rate of extinction than comparable mainland species, especially island endemics (Frankham 1998). The genetic diversity at marker loci is less in island species than in related mainland birds and is particularly reduced in endemic species and those with reduced dispersal (Frankham 1997). The reduction in diversity of genetic markers was sufficiently severe to indicate that the population sizes had been small enough to induce inbreeding depression.

ISOLATION AND VULNERABILITY

Most cases of island extinctions have been precipitated by human arrival, most likely due to habitat destruction and the introduction of exotic species. Climate change may have analogous effects, by rapid change in the distribution of suitable habitat (Thomas et al. 2004)—changes that require rapid adaptive responses and changes in the assemblages of species that co-occur.

Olivieri and Vitalis (2001) have pointed out that fragmented populations may be drawn into what they call an "evolutionary vortex." Short-term selection for reduced dispersal and local conditions thereby produces populations that are unable to adapt and spread in the longer term. The cliff-dwelling plant *Centaurea corymbosa* (Colas et al. 1997) may be trapped in this way: there is unfilled suitable habitat very near to its surviving populations, but seed dispersal is over only a few meters.

Movement between populations may ameliorate the effects of inbreeding or otherwise introduce favorable alleles in much the same way as the deliberate effects at genetic restoration. Rapid spread of immigrant genes is seen in the case of mice on the Isle of May (Berry et al. 1991) and in laboratory models such as butterflies (Saccheri and Brakefield 2002) and daphnia (Ebert et al. 2002). Yet dispersal is a two-edged sword. It can impede adaptation to local conditions and bring in genotypes that are better adapted to dispersal than local persistence.

Boulding and Hay (2001) modeled adaptation to sudden change in the face of dispersal from unaffected populations. They found that the adverse effect of immigrant maladaptive genes was sufficient to outweigh the contribution of migrants to the size of the population. This conclusion contrasts with the results of ecological metapopulation models in which migration provides a source of individuals to rescue populations from extinction. Each outcome clearly applies to the particular details of the models, but the contrast emphasizes that gene flow does not necessarily equip a population to respond better to rapid environmental change.

DISPERSAL AND ADAPTATION

There are many cases of local, population-level genetic adaptations in species that range over altitude, latitude, and environments. For example, temperate plants that spread from south to north postglacially have reproductive biologies now adapted to their present light (photoperiod) and temperature regimes (Gray 1997). Transplant experiments have been employed to demonstrate genetic adaptations to different altitude and latitude in several pine species in Europe and North America (Matyas 1997; Davis and Shaw 2001). Particularly interesting demonstrations occur in insect colonizations, such as the native pitcher-plant mosquito, *Wyeomyia smithii*, and the recent invasion by the corn-borer moth, *Ostrinia nubialis*, in eastern North America (Showers 1981; Armbruster et al. 1998). Genetic adaptations include life history traits such as photoperiod response and development time, which appear to rapidly evolve due to strong selective pressure.

Since range shifts involve dispersal, there will be selection during expansion for genotypes more able in this respect. A number of studies demonstrate this response (Thomas, this volume). The lodgepole pine *Pinus contorta* expanded its range postglacially northward into the Yukon, and the dispersal ability of the seed was increased through reduced wing loading (Cwynar and MacDonald 1987). More recently it was shown that in the European ground beetle *Pterostichus melinarius* colonizing Canada, long-winged genotypes were in higher frequency at the leading edge than in established populations (Niemela and Spence 1999). Further detailed examples are described in the bush crickets *Conocephalus discolor* and *Metrioptera roeselii*, where populations established in the past 20 years have increased frequency of long-winged forms (Thomas et al. 2001). Such differences are accompanied by changes in morphology and physiology, as in the butterfly *Pararge aegeria* (Hill et al. 2001), and will likely involve genetic trade-off with other components of fitness. Many insect species show long- and short-winged polymorphism (Ritchie et al. 1987), which is probably a product of such interacting selective pressures. The genomes of organisms surviving many climatic cycles are likely to retain the ability to respond.

DISPERSAL AND GENE FLOW

Dispersal is the mechanism of expansion in range change and of allele exchange among established populations within the species range. The gene flow produced varies greatly with the frequency and distance of dispersers, and the density and patchiness of population structures. It also depends on the compatibility of the genotypes brought together. Rapid pioneer expansion involves long distance dispersal, with the colonies produced showing genetic distinctness among and homogeneity within themselves. Large areas may be filled by similar genotypes (Ibrahim et al. 1996) (see Fig. 12.1).

Once dense populations have become established, dispersal is a diffusion process, and migrants make little difference to the resident genome for most genes. But when a favored allele or genotype arrives, it should spread relatively quickly depending on its selective advantage (Barton and Hewitt 1989). These may be dispersing genotypes from more central regions that are better adapted or produce fitter recombinants. This gene flow and adaptation may occur through the body of the range, where there are no major barriers to dispersal. Such selective sweeps of alleles through populations can carry linked portions of the genotype with them until separated by recombination. On the other hand, if the heterozygote or hybrid between the migrant genotype and the resident is less fit than they are, this will prevent gene flow. This pattern of selection may generate a hybrid zone, which can act as a partial barrier to gene flow across the genome (Barton and Hewitt 1989), permeable to selected alleles, but severely retarding others (see Fig. 12.1). While dispersal and adaptation to local conditions occur during a climatically induced range change, their interactions are various (Hewitt 1993; Davis and Shaw 2001).

Dispersal and gene flow among populations might seem crucial to adaptation to climatically induced range change, but this does not necessarily mean that genotypes adapted to a particular part of the species range have also migrated in concert with the changing environment. Some evidence to the contrary comes from a study of pocket gophers (*Thomomys talpoides*). Hadly and colleagues (1998) obtained ancient mitochondrial DNA from remains in Lamar Cave, Yellowstone Park, Wyoming, dating back 2400 years. The ancient haplotypes (mtDNA variants) provided evidence of genetic continuity with the present-day populations, while haplotypes from six extant populations in the surrounding 200-mile region were all distinct from those at Lamar Cave. Over the same time period, the fossil morphology at Lamar Cave has changed in step with the climatic record, with gophers being smaller during the medieval warm period. The population seems to have persisted at Lamar Cave without significant immigration, and adaptation may well have occurred within the resident genome.

Conversely, it is argued that gene flow from the center of the species range may maintain suboptimal genotypes in much of the range of *Pinus contorta* (Rehfeldt et al. 1999). Any such mismatch between the population and the environment will reduce a population's viability, but it can nevertheless persist. However, it does seem reasonable that populations with lower genetic variation will be less able to adapt and therefore at greater risk. This may be critical when habitats change or shift.

LIMITS TO THE RATE OF CHANGE

Rapid climate change may require rates of adaptation so demanding that the species may fail to meet them, even in more persistent parts of the species range. Population genetics theory can be used to calculate limits to the rate of change in quantitative characters (see Barton and Partridge 2000). However, the consequences for the population if a trait fails to track the environment are less obvious.

Some populations appear to be able to adapt very rapidly. Collections of pitcher-plant mosquitoes (*Wycomyia smithii*) collected in 1988 and 1993 showed evidence of a genetic change in photoperiodic response to the current climate change (Bradshaw and Holzapfel 2001). They appear to have responded swiftly and appropriately to the recent warming.

On the other hand, Etterson and Shaw (2001) followed the fate of transplanted individuals of the legume *Chamaecrista fasciculata* and the response of their offspring to the consequent selection. In plants transplanted to warmer, drier climates, the change over one generation in traits such as leaf thickness and number did not appear fast enough to match the predicted rate of climate change. There was genetic variance for the traits in question, but antagonistic correlations between traits are predicted to retard the response. Extrapolating from these results, they predict that the rate of future environmental change may exceed that of the populations' evolutionary response.

Although range changes can deplete genetic variation over wide areas of expansion or in small fragmented populations, it is less clear that traits with important effects on fitness will be as severely affected. Reed and Frankham (2001) found no correlation between measures of genetic variation from molecular markers and those for life history traits. Merila and Sheldon (2000) found that fitness-related traits retain genetic variation despite the expected action of selection to rapidly select the optimal variants. This may reflect the large mutational target, because so much of the genome can affect these traits, and their similarly wide range of environmental influences.

Even in the laboratory, it takes carefully designed experiments to demonstrate that loss of variation has an adverse effect because of reduced evolutionary potential rather than inbreeding depression. Frankham and colleagues (1999) provide a persuasive example, where *Drosophila* populations that had been put through bottlenecks of differing severity were then challenged to adapt to an increase in salt concentration. The genetically depleted populations succumbed earlier.

The importance of such considerations and experiments to understanding this interplay of demography, genetic diversity, selection, and survival has been highlighted by a recent meta-analysis of butterflies across Europe (Schmitt and Hewitt 2004). A strong correlation exists between lower genetic diversity and recent population decline in species that colonized postglacially to the north and west from southern and eastern ice age refugia. This suggests that the phylogeographic signature may contribute to the present pattern of threat and loss. Clearly, more research on the detailed causes and wider surveys of other organisms are needed.

REFUGIA AND SPECIATION

The recent combination of fossil series and DNA sequences has allowed the identification of refugial areas and postglacial colonization routes for distinct genomes within several species. This is particularly clear for Europe and North America but is being applied in other parts of the world (Hewitt 2000). Refugial areas are often rich in species, genomes, and alleles compared with colonized regions. Furthermore these distinct intraspecific genomes have evolved different genetic adaptations, so their potential for future change may well vary also.

Tropical rain forests are among the richest biomes and were reduced in overall area during the last glacial maximum. Notably, it was argued that Amazonian lowland forests were fragmented by savannah into small isolated refugia as a result of ice age aridity. However, while few suitable pollen cores are available, these indicate that lowland forests were not so fragmented in the ice ages, with lowland and some montane species mixed together (Colinvaux et al. 2000). Likewise few genetic studies exist in this region, but these show that genome divergence is often older than the Pleistocene, and possibly due to geological activity in front of the rising Andes (Patton and da Silva 1998). Clearer evidence, both palaeobotanic and genetic, for ice age refugia and fragmentation is available for the Australian wet tropics (Moritz et al. 2000).

Such refugial areas, where genomes survive and diverge over several major oscillations, are likely to allow speciation and the accumulation of diversity. These rich areas appear to be those that can sustain populations throughout the climatic fluctuations, rather than just at the extremes. Mountainous and tropical areas appear particularly successful (see Fig. 12.2). Different locations retain different genomes and thereby a diversity of adaptive potential. Divergence and speciation appear from genetic evidence to have been proceeding continually through the Pleistocene (Hewitt 1996; Avise et al. 1998) and seem to accumulate in refugia.

Indeed, certain tropical mountain regions in Africa and South America have been considered as engines of speciation (Fjeldsa and Lovett 1997). The mountainous southern peninsulas of Europe, the western mountains of the United States, and the upland blocks of Queensland rain forest all show species, subspecies, and genomes with DNA divergence that dates from the Pliocene through to the late Pleistocene (Schneider et al. 1998; Hewitt 1999; Knowles 2000) (e.g., Fig. 12.3). Recent work shows that speciation can be quite rapid, even postglacial (Schluter 1998), but many such products go extinct with glacial oscillations.

PAST, PRESENT, AND FUTURE

In the short term local populations of a species may survive customary environmental changes through their individual plasticity. Nonetheless selection continues to modify the genotype to changed conditions, but genetic correlations among traits may impede response to selection from rapid climate change. When drastic switches in climate occur, dispersal and range shifts are probably the species' primary means of survival, with local adaptation under selection proceeding effectively only in relatively stable times. Given the changing climate, shifting niches, and species mixtures, chance and selection are constantly modifying the population and species genome. There are repeated major cycles, and the lineages that endure in the long term have the attributes to survive such climatic change. So why should it be any worse this time around?

The Holocene temperature changes of a few degrees induce small range shifts in most organisms, with dispersal, expansion, and adaptation in front, complemented by fragmentation, inbreeding, and loss at the rear. Temperature is currently rising, which may be particularly critical for species in warmer parts where there is much habitat loss, and much genetic diversity may be lost in the tropics. Organisms with fair-sized ranges should survive such events, but the increasing fragmentation of habitat and ranges will exacerbate problems of isolation, particularly for localized species. The preservation of the remaining corridors for dispersal is therefore a clear priority. In the longer term, larger and more rapid changes, such as millennial oscillations, produce larger range shifts and are made even more problematic because of fragmentation caused by humans. While populations of species dispersed and went extinct over considerable areas in the past, now ranges are more dissected, with dispersal more difficult and extinction more likely.

Furthermore, many of the refugial areas themselves have increasingly fragmented and modified habitats. Consequently, for many species the structure and survival of the source populations and their genomic legacy may be endangered. Fossil and genetic data have localized a number of these refugial areas for glacial periods, and it is increasingly important that their conservation be addressed. Their long-term conservation and genetic relevance will be placed in doubt, however, as climate moves beyond past envelopes of natural variability (Overpeck et al., this volume).

SUMMARY

Rapid changes to species' ranges have occurred as they tracked suitable habitat during past climate change. Changes in species' distribution have dramatic consequences for the genetic diversity in different parts of the species' range. Rather counterintuitively, areas where a population has invaded and grown rapidly may have lowest genetic diversity. Conversely, in areas where the range has shrunk and fragmented, yet not shifted dramatically, genetic diversity can persist—although component patches may be genetically impoverished.

These founder events and population bottlenecks can be demonstrated to have deleterious effects on individual fitness. However, evidence of longer-term effects on population persistence in natural populations is rarer and sometimes contradictory. The importance of standing genetic diversity to evolutionary flexibility and

population persistence is even more elusive and requires further study.

Future climate change may be particularly rapid. Reconstruction of similar episodes from the past suggests that rapid change led to transitory mismatches between genomes and the local environment, especially in newly colonized areas where selection may have been for effective dispersal. Again, there is limited and equivocal evidence of the consequences of this mismatch for the persistence of local populations and genomes. Some studies suggest that climate change may outstrip a population's rate of adaptation; other lines of evidence indicate that local genomes have persisted through past climatic fluctuations.

Species diversity, like genetic diversity, appears high in those regions where species can persist locally throughout climatic cycles. Species and genetic diversity in other regions was generated during the last climatic cycle and may be particularly vulnerable to change. A more disturbing possibility is that loss of habitat and dispersal routes has made the critical refugial areas more vulnerable too.

REFERENCES

Alley, R. B. 2000. Ice-core evidence of abrupt climate changes. *Proceedings of the National Academy of Sciences USA* 97:1331–1334.

Armbruster, P., W. Bradshaw, and C. Holzapfel. 1998. Effects of postglacial range expansion on allozyme and quantitative genetic variation of the pitcher-plant mosquito *Wyeomyia smithii*. *Evolution* 52: 1697–1704.

Avise, J. C., D. Walker, and G. C. Johns. 1998. Speciation durations and Pleistocene effects on vertebrate phylogeography. *Proceedings of the Royal Society of London B*, 265: 1707–1712.

Barton, N. H., and G. M. Hewitt. 1989. Adaptation, speciation and hybrid zones. *Nature* 341: 497–503.

Barton, N., and L. Partridge 2000. Limits to natural selection. *Bioessays* 22: 1075–1084.

Bennett, K. D. 1997. *Evolution and Ecology: The Pace of Life*. Cambridge: Cambridge University Press.

Bernatchez, L., and C. C. Wilson. 1998. Comparative

phylogeography of nearctic and palearctic fishes. *Molecular Ecology* 7: 431−452.

Berry, R. J., G. S. Triggs, P. King, H. R. Nash, and L. R. Noble. 1991. Hybridization and gene flow in house mice introduced into an existing population on an island. *Journal of Zoology* 225: 615−632.

Boulding, E.G., and T. Hay. 2001. Genetic and demographic parameters determining population persistence after a discrete change in the environment. *Heredity* 86: 313−324.

Bradshaw, W. E., and C. M. Holzapfel. 2001. Genetic shifts in photoperiodic response correlated with global warming. *Proceedings of the National Academy of Sciences USA* 98:14509−14511.

Colas, B., I. Olivieri, and M. Riba. 1997. Centaurea corymbosa, a cliff-dwelling species tottering on the brink of extinction: A demographic and genetic study. *Proceedings of the National Academy of Sciences USA* 94: 3471−3476.

Colinvaux, P. A., P. E. De Oliveira, and M. B. Bush. 2000. Amazonian and neotropical communities on glacial time-scales: The failure of the aridity and refuge hypotheses. *Quaternary Science Reviews* 19: 141−169.

Coope, G. R. 1994. The response of insect faunas to glacial-interglacial climatic fluctuations. *Philosophical Transactions of the Royal Society, London B* 344: 19−26.

Cooper, S. J. B., K. M. Ibrahim, and G. M. Hewitt. 1995. Postglacial expansion and subdivision of the grasshopper Chorthippus parallelus in Europe. *Molecular Ecology* 4:49−60.

Cwynar, L. C., and G. M. MacDonald. 1987. Geographical variation of lodgepole pine in relation to poulation history. *The American Naturalist* 129: 463−469.

Dansgaard, W., S. J. Johnsen, H. B. Clausen, D. Dahl-Jensen, N. S. Gundestrup, G. U. Hammer, C. S. Hvidberg, J. P. Steffensen, A. E. Sveinbjornsdottir, J. Jouzel, and G. Bond. 1993. Evidence for general instability of past climate from a 250-kyr ice-core record. *Nature* 364: 218−220.

Davis, M. B., and R. G. Shaw. 2001. Range shifts and adaptive responses to Quaternary climate change. *Science* 292: 673−679.

Durand, J. D., H. Persat, and Y. Bouvet. 1999. Phylogeography and postglacial dispersion of the chub (Leucicus cephalus) in Europe. *Molecular Ecology*, 8, 989−997.

Ebert, D., C. Haag, M. Kirkpatrick, M. Riek, J. W. Hottinger, and V. I. Pajunen. 2002. A selective advantage to immigrant genes in a Daphnia metapopulation. *Science* 295: 485−488.

Eisto, A. K., M. Kuitunen, A. Lammi, V. Saari, J. Suhonen, S. Syrjasuo, and P. M. Tikka. 2000. Population persistence and offspring fitness in the rare bellflower Campanula cervicaria in relation to population size and habitat quality. *Conservation Biology* 14: 1413−1421.

Englbrecht, C. C., J. Freyhof, A. Nolte, K. Rassmann, U. Schliewen, and D. Tautz. 2000. Phylogeography of the bullhead Cottus gubio (Pisces : Teleosteil : Cottidae) suggests a pre-Pleistocene origin of the major central European populations. *Molecular Ecology* 9:709−722.

Etterson, J. R., and R. G. Shaw. 2001. Constraint to adaptive evolution in response to global warming. *Science* 294:151−154.

Fjeldsa, J., and J. C. Lovett. 1997. Geographical patterns of old and young species in African forest biota: The significance of specific montane areas as evolutionary centres. *Biodiversity and Conservation* 6: 323−344.

Flanagan, N. S., P. L. Mason, J. Gosalvez, and G. M. Hewitt. 1999. Chromosomal differentiation through an Alpine hybrid zone in the grasshopper Chorthippus parallelus. *J. Evol. Biol.* 12:577−585.

Frankham, R. 1997. Do island populations have less genetic variation than mainland populations? *Heredity* 78: 311−327.

Frankham R. 1998. Inbreeding and extinction: Island populations. *Conservation Biology* 12: 665−675.

Frankham, R., K. Lees, M. E. Montgomery, P. R. England, E. Lowe, and D. A. Briscoe. 1999. Do population bottlenecks reduce evolutonary potential? *Animal Conservation* 2: 255−260.

Gray, A. J. 1997. Climate change and the reproductive biology of higher plants. In Huntley, B., et al., eds., *Past and Future Rapid Environmental Changes*, 371−380. NATO ASI Series Vol. I 47. Berlin: Springer-Verlag.

Hadly, E. A., M. H. Kohn, J. A. Leonard, and R. K. Wayne. 1998. A genetic record of population isolation in pocket gophers during Holocene climatic change. *Proceedings of the National Academy of Sciences USA* 95: 6893−6896.

Hewitt, G. M. 1988. Hybrid zones. *Trends in Ecology and Evolution.* 3:158−167.

Hewitt, G. M. 1993. Postglacial distribution and species substructure: Lessons from pollen, insects and hybrid zones. In D. R. Lees, and D. Edwards, eds., *Evolutionary Patterns and Processes*. Linnean Society Symposium Series 14: 97−123. London: Academic Press.

Hewitt, G. M. 1996. Some genetic consequences of ice ages, and their role in divergence and speciation. *Biological Journal of the Linnean Society* 58: 247−276.

Hewitt, G. M. 1999. Post-glacial recolonization of European Biota. *Biological Journal of the Linnean Society* 68: 87−112.

Hewitt, G. M. 2000. The genetic legacy of the Quaternary ice ages. *Nature* 405: 907−913.

Hewitt, G. M. 2001. Speciation, hybrid zones and phylogeography—or seeing genes in space and time. *Molecular Ecology* 10: 537−549.

Hewitt, G. M., and K. M. Ibrahim. 2001. Inferring glacial refugia and historical migrations with molecular phylogenies. In J. Silvertown and J. Antonovics, eds., *Integrating Ecology and Evolution in a Spatial Context*, 271–294. BES symposium volume, Oxford: Blackwell.

Hill, J. K., Y. C. Collingham, C. D. Thomas, D. S. Blakeley, R. Fox, D. Moss, and B. Huntley. 2001. Impacts of landscape structure on butterfly range expansion. *Ecology Letters* 4: 313–321.

Huntley, B. 1990. European vegetation history: Palaeovegetation maps from pollen data—1300 yr BP to present. *Journal of Quaternary Science* 5: 103–122.

Huntley, B. 1999. The dynamic response of plants to environmental change and the resulting risks of extinction. In G. M. Mace, A. Balmford, and J. R. Ginsberg, eds., *Conservation in a Changing World*, 69–85. Cambridge: Cambridge University Press.

Huntley, B., and H. J. B. Birks. 1983. *An Atlas of Past and Present Pollen Maps for Europe*. Cambridge: Cambridge University Press.

Ibrahim, K., R. A. Nichols, and G. M. Hewitt. 1996. Spatial patterns of genetic variation generated by different forms of dispersal during range expansion. *Heredity* 77: 282–291.

Jaarola, M., H. Tegelstrom, and K. Fredga. 1999. Colonization history in Fenno-scandian rodents. *Biological Journal of the Linnean Society* 68:113–127.

Jackson, S. T., and J. T. Overpeck. 2000. Responses of plant populations and communities to environmental changes of the late Quaternary. *Paleobiology* 26: 194–220 Suppl. S.

Jouzel, J. 1999. Calibrating the isotopic paleothermometer. *Science* 286: 910–911.

Kirkpatrick, M., and N. H. Barton. 1997. Evolution of a species' range. *American Naturalist* 150: 1–23.

Knowles, L. L. 2000. Tests of Pleistocene speciation in montane grasshoppers from the sky islands of western North America (*Genus Melanoplus*). *Evolution* 54: 1337–1348.

Lammi, A., P. Siikamaki, and K. Mustajarvi. 1999. Genetic diversity, population size, and fitness in central and peripheral populations of a rare plant *Lychnis viscaria*. *Conservation Biology* 13: 1069–1078.

Lotz, M., D. Shindle, D. Land, J. Osborne, M. Alvarado, K. Al-Muhyee, S. Bass, and D. Jansen. 2000. Florida Panther Genetic Restoration Progress Report January–June 2000 [online] Available: http://www.atlantic.net/~oldfla/panther/gene30.html [Accessed 21/6/02].

Madsen, T., R. Shine, M. Olsson, and H. Wittzell. 1999. Conservation biology—Restoration of an inbred adder population. *Nature* 402: 34–35.

Matyas, C. 1997. Genetics and adaptation to climate change: A case study of trees. In Huntley, B., et al., eds., *Past and Future Rapid Environmental Changes*, 357–370. NATO ASI Series Vol. I 47. Berlin: Springer-Verlag.

Mavraganis, K., and C. G. Eckert. 2001. Effects of population size and isolation on reproductive output in *Aquilegia canadensis* (Ranunculaceae). *Oikos* 95: 300–310.

Merila, J., and B. C. Sheldon. 2000. Lifetime reproductive success and heritability in nature. *American Naturalist* 155: 301–310.

Moritz, C., J. L. Patton, C. J. Schneider, and T. B. Smith. 2000. Diversification of rainforest faunas: An integrated molecular approach. *Annual Reviews of Ecology and Systematics* 31: 533–563.

Nesbo, C. L., T. Fossheim, L. A. Vollestad, and K. S. Jakobsen. 1999. Genetic divergence and phylogeographic relationships among European perch (*Perca fluviatilis*) populations reflect glacial refugia and postglacial colonization. *Molecular Ecology* 8, 1387–1404.

Nichols, R. 2001. Gene trees and species trees are not the same. *Trends in Ecology and Evolution* 16: 358–364.

Nichols, R. A., M. W. Bruford, and J. J. Groombridge. 2001. Sustaining genetic variation in a small population: Evidence from the Mauritius kestrel. *Molecular Ecology* 10: 593–602.

Nichols, R. A., and G. M. Hewitt. 1994. The genetic consequences of long distance dispersal during colonization. *Heredity* 72: 312–317.

Niemela, J., and J. R. Spence. 1999. Dynamics of local expansion by an introduced species: *Pterostichus melanarius* Ill. (Coleoptera: Carabidae) in Alberta, Canada. *Diversity and Distributions* 5: 121–127.

Olivieri, I., and R. Vitalis. 2001. La Biologie des extinctions. *M S-Medicine Sciences* 17: 63–69.

Patton, J. L., and M. N. F. da Silva. 1998. Rivers, refuges and ridges: The geography of speciation of Amazonian mammals. In Howard, D. J., and S. H. Berlocher, eds., *Endless Forms: Species and Speciation*, 114–129. Oxford: Oxford University Press.

Paulo, O. S., C. Dias, M. W. Bruford, W. C. Jordan, and R. A. Nichols. 2001. The persistence of Pliocene populations through the Pleistocene climatic cycles: Evidence from the phylogeography of an Iberian lizard. *Proceedings of the Royal Society of London B* 268: 1625–1630.

Petit, R. J., R. Bialozyt, S. Brewer, R. Cheddadi, and B. Comps. 2001. From spatial patterns of genetic diversity to postglacial migration processes in forest trees. In Silvertown J., Antonovics J., eds., *Integrating Ecology and Evolution in a Spatial Context*. BES symposium volume. Oxford: Blackwell.

Ralls, K., and J. Ballou. 1986. Captive breeding programs for populations with a small number of founders. *Trends in Ecology and Evolution* 1: 9–22.

Reed, D. H., and R. Frankham. 2001. How closely correlated are molecular and quantitative measures of genetic variation? A meta-analysis. *Evolution* 55: 1095–1103.

Rehfeldt, G. E., C. C. Ying, D. L. Spittlehouse, and D. A. Hamilton. 1999. Genetic responses to climate change in *Pinus contorta*: Niche breadth, climate change and reforestation. *Ecological Monographs* 69:375–407.

Ritchie, M. G., R. K. Butlin, and G. M. Hewitt. 1987.

Morphology, causation and fitness effects of macropterism in *Chorthippus parallelus* (Orthoptera: Acrididae). *Ecol. Entomol.* 12: 209–218.

Saccheri, I. J., and P. M. Brakefield. 2002. Rapid spread of immigrant genomes into inbred populations. *Proceedings of the Royal Society of London B* 269: 1073–1078.

Saccheri, I., M. Kuussaari, M. Kankare, P. Vikman, W. Fortelius, and I. Hanski. 1998. Inbreeding and extinction in a butterfly metapopulation. *Nature* 392: 491–494.

Santucci, F., B. C. Emerson, and G. M. Hewitt. 1998. Mitochondrial DNA phylogeography of European hedgehogs. *Molecular Ecology* 7:1163–1172.

Schluter, D. 1998. Ecological causes of speciation. In Howard, D. J., Berlocher, S. H., eds., *Endless Forms: Species and Speciation*, 114–129. Oxford: Oxford University Press.

Schmitt, T., and G. M. Hewitt. 2004. The genetic pattern of population threat and loss: A case study of butterflies. *Molecular Ecology* 13: 21–31.

Schneider, C. J., M. Cunningham, and C. Moritz. 1998. Comparative phylogeography and the history of endemic vertebrates in the wet tropics rainforests of Australia. *Molecular Ecology* 7: 487–498.

Seddon, J. M., F. Santucci, N. J. Reeve, and G. M. Hewitt. 2001. DNA footprints of European hedgehogs, Erinaceus europaeus and E. concolor: Pleistocene refugia, postglacial expansion and colonisation routes. *Molecular Ecology* 10: 2187–2198.

Shackleton, N. 2001. Paleoclimate–climate change across the hemispheres. *Science* 291: 58–59.

Showers, W. B. 1981. Geographic variation of the diapause response in the European corn borer. In Denno, R. F., and Dingle, H., eds., *Insect Life History Patterns*, 97–111. New York: Springer-Verlag.

Templeton, A. R. 2001. Using phylogeographic analyses of gene trees to test species status and processes. *Molecular Ecology* 10:779–792.

Thomas, C. D., E. J. Bodsworth, R. J. Wilson, A. D. Simmons, Z. G. Davies, M. Munsche, and L. Conradt. 2001. Ecological and evolutionary processes at expanding range margins. *Nature* 411:577–581.

Thomas, C. D., A. Cameron, R. E. Green, M. Bakkenes, L. J. Beaumont, Y. C. Collingham, B. F. N. Erasmus, M. Ferreira de Siqueira, A. Grainger, L. Hannah, L. Hughes, B. Huntley, A. S. Van Jaarsveld, G. E. Midgely, L. Miles, M. A. Ortega-Huerta, A. T. Peterson, O. L. Phillips, and S. E. Williams. 2004. Extinction risk from climate change. *Nature* 427: 145–148.

Wakeley, J. 1999. Nonequilibrium migration in human history. *Genetics* 153: 1863–1871.

Webb, T., and P. J. Bartlein. 1992. Global changes during the last 3 million years: Climatic controls and biotic responses. *Annual Reviews of Ecology and Systematics* 23: 141–173.

Williams, D., D. Dunkerley, P. DeDecker, P. Kershaw, and M. Chappell. 1998. *Quaternary Environments*. London: Arnold.

Wright, S. 1969. *Evolution and the Genetics of Populations: A Treatise in Four Volumes. Vol. 2, The Theory of Gene Frequencies.* Chicago: University of Chicago Press.

Phylogeny and Distribution of Fishes of the *Characidium lauroi* Group as Indicators of Climate Change in Southeastern Brazil

Paulo A. Buckup and Marcelo R. S. Melo

The genus *Characidium* includes small (typically less than 10 cm) fishes of the South American family Crenuchidae. The *C. lauroi* group of species inhabits cool water streams in mountain ranges and plateaus of southeastern Brazil, being largely absent from nearby warmer lowland streams. As freshwater organisms, they are particularly sensitive to long-term climate changes because, unlike terrestrial vertebrates, they are bound to their aquatic environment and cannot disperse across land barriers when climatic conditions shift optimal conditions over the geographic landscape.

The group includes five species: *Characidium lauroi* inhabits hillside tributaries of the upper and upper-middle reaches of the Paraíba do Sul River in the mountains of the coastal Serra da Bocaina and the inland Serra da Mantiqueira. *Characidium japuhybense* occurs as disjunct populations in coastal streams between the Ribeira de Iguape drainage in the west and the freshwater tributaries to the Ilha Grande Bay in the east. *Characidium schubarti* occurs on both sides of the coastal mountain range, including the Serra de Paranapiacaba. *Characidium* sp.1 (voucher MNRJ 21687) is restricted to right bank tributaries of the Paraíba do Sul River draining the high mountains of the Serra dos Órgãos, a section of the coastal mountain range located north of Rio de Janeiro with elevations above 2000 m, isolated by relatively low elevations (less than 500 m). *Characidium* sp.2 (voucher MNRJ 21542) inhabits tributaries of the Iguaçu River (Paraná Basin) in the southeastern slope of the Serra da Esperança, and the headwaters of the Ribeira River, between the coastal Serra do Mar and the inland Serra Geral.

Comparing the phylogeny of the group (Fig. 12.5) with the known geographic distribution of the species illuminates the evolutionary history of this group of fishes. If the geographic distribution of all populations of the *C. lauroi* group is considered (Fig. 12.6), a pattern of disjunct ranges emerges. Except for the partial overlap of the two oldest lineages, separate species and clusters of conspecific populations are widely separated by large areas where members of the *C. lauroi* group are absent. These gaps are likely to represent real distributional gaps, not sampling biases, because this southeastern region of Brazil is relatively well represented in museum collections.

Within the currently accepted models of biological speciation, discontinuity among geographic ranges of populations belonging to monophyletic lineages is best explained by hypotheses of extinction within a previously continuous distribution. In the case of the *C. lauroi* group, the emergence of species range discontinuity can be explained by environmental alterations associated with climatic changes—sea level rise in the case of the coastal clade, and temperature increase in the case of the inland clade. These findings have considerable relevance for predicting the impacts of future climate change.

The coastal clade of the *C. lauroi* group comprises populations of *C. japuhybense* that are currently isolated in independent coastal streams. Steep coastal mountain divides that sometimes abut against the Atlantic Ocean isolate these watersheds. In

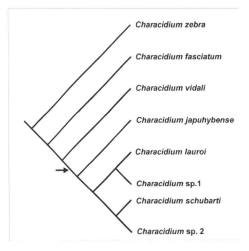

Figure 12.5. Most parsimonious cladogram of species relationships based on 18 morphological characters, requiring 29 evolutionary transformations, overall character consistency index of 0.80, and retention index of 0.79. Arrow indicates clade comprising the species of the C. lauroi group.

one instance, a population was found in a coastal island stream isolated from the continent by several kilometers of seawater. The low morphological divergence among populations of C. japuhybense is indicative of a relatively recent history of isolation by sea level changes. We hypothe-

size that the currently isolated ranges of C. japuhybense were once interconnected through a coastal system of lowland waterways during the Pleistocene (2–0.01 Ma) when sea levels were lower. Continued sea level rise associated with establishment of the current warm spell has all but obliterated the coastal lowlands, resulting in the present isolation of the coastal taxa.

The inland clade comprising C. lauroi, C. schubarti, Characidium sp.1, and Characidium sp.2 also exhibits a pattern of discontinuous geographic ranges, which may be explained by past climate change. When we plot the occurrence of members of this clade in a map, a pattern of strong association between isolated clusters of populations and high mountain ranges emerges. Such pattern is correlated with the habitat requirements of these fishes, which are restricted to clear cold water streams. As all species show this marked habitat requirement, it can be most parsimoniously assumed that their common ancestor also lived in cold mountain streams. As small water-bound vertebrates are incapable of

Figure 12.6. Geographic distribution of species of Characidium lauroi group in relation to altitude in southeastern Brazil. Thick lines represent state boundaries.

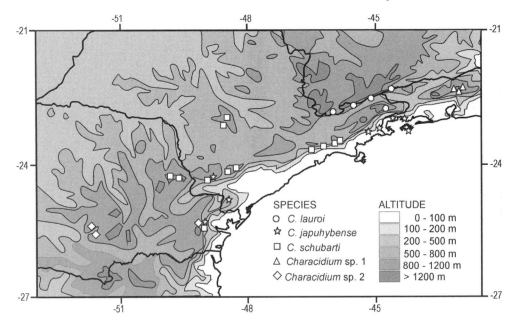

long distance dispersal we must also assume that cold water habitats were more prevalent in southeastern Brazilian highlands during the initial evolutionary history of the C. lauroi group, thus permitting a former contiguity of populations. Colder climatic conditions prevailing during late Pleistocene glacial periods were likely to have expanded cold water stream regimes in these highlands. As a warmer climate developed during the past few thousand years, the original ranges were progressively reduced at low altitudes. In the Paraíba do Sul, populations of the Serra dos Órgãos became isolated from those of tributaries that drain the Serra da Mantiqueira and the coastal Serra da Bocaina. These areas are separated by elevations below 500 m, where current climate conditions are not compatible with the maintenance of viable populations of the C. lauroi group. Likewise the formerly widespread populations in the Paraná tributaries became fragmented into several areas of high elevation where conditions are compatible with the requirements of a cold water species. This progressive fragmentation of habitats produced speciation events within the Paraíba do Sul and the Paraná Basin. These events led to the establishment of C. lauroi and Characidium sp.1 in the Paraíba do Sul, and of C. schubarti and Characidium sp.2 in the Paraná Basin.

One of the challenges faced by scientists studying current trends in global warming is the need to produce reliable estimates of future impacts of climatic change. Most predictive models try to foresee changes in biodiversity based on current prevailing conditions (e.g., Peterson et al. 2002). The results of our study suggest that information on phylogenetic history of a group may provide an additional predictive tool. In the case of the C. lauroi group of species, the perception that this group of fishes has a common history that is best explained by past climate change is possible only be-cause we have appropriate phylogenetic data. If we extend our interpretation of history of the C. lauroi group into the future, we can predict changes in species ranges that would result from expected global warming trends. Future warming is likely to produce vertical shifts in the distribution of cold water streams in southeastern Brazilian mountains, which may lead to further range reduction and even extinction of populations and species of the C. lauroi group. Such prediction is possible only because we have phylogenetic data indicating that the distribution of these species is the result of interaction between climatic change and historical evolutionary constraints.

REFERENCES

Ab'Sáber, A. N. 1957. O problema das conexões antigas e da separação da drenagem do Paraíba e do Tietê. Bol. Paulista de Geografia 26: 38–49.

Buckup, P. A. 1992. Redescription of Characidium fasciatum, type species of the Characidiinae (Teleostei, Characiformes). Copeia 4: 1066–1073.

Buckup, P. A. 1993. Phylogenetic interrelationships and reductive evolution in neotropical Characidiin fishes (Characiformes, Ostariophysi). Cladistics 9: 305–341.

Malabarba, M. C. S. L. 1998. Phylogeny of fossil Characiformes and paleobiogeography of the Tremembé Formation, São Paulo, Brazil, pp. 68–84. In Malabarba, L. R., Reis, R. E., Vari, R. P., Lucena, Z. M. S., Lucena, C. A. S., eds., Phylogeny and Classification of Neotropical Fishes. Porto Alegre: EDIPUCRS.

Melo, M. R. S. 2001. Sistemática, filogenia e biogeografia do grupo Characidium lauroi Travassos, 1949 (Characiformes, Crenuchidae). Rio de Janeiro: UFRJ, Museu Nacional (M.Sc. Dissertation).

Peterson, A. T., M. A. Ortega-Huerta, J. Bartley, V. Sánchez-Cordero, J. Soberón, R. H. Buddemeier, and D. R. B. Stockwell. 2002. Future projections from Mexican faunas under global climate change scenarios. Nature 416: 626–629.

Wiley, E. O. 1981. Phylogenetics: The Theory and Practice of Phylogenetic Systematics. New York: John Wiley and Sons.

PART IV

Understanding the Future

Uncertainty is the central issue in managing the biological consequences of future climate change. There is uncertainty in the climate models, uncertainty in tuning climate projections to regional scales useful for ecological analysis, and uncertainty in the conceptual, statistical, and mathematical models of future biological responses. With all these sources of uncertainty, what is the prospect for true understanding?

As the preceding chapters have indicated, past and present changes provide insights which can serve as the foundation for understanding of the future. Large-scale shifts in vegetation growth form (plant functional type) can be expected, leading to dramatic changes such as the replacement of forests by grasslands, or the reverse. Within these large-scale vegetation limits, species range shifts will be independent of one another (individualistic). Rapid change may be accommodated by long-distance dispersal or expansion from small micropockets of suitable habitat. Changes in species composition and biomass may add or remove CO_2 from the atmosphere, resulting in feedbacks to climate. All of these responses and others may be modeled, providing projections of future changes in biodiversity that are spatially significant and highly relevant to furthering understanding, in spite of their uncertainty.

Biologists, like climate policymakers, must view uncertainty in a framework of risk management. The probability of climate change affecting biodiversity is high, even if the probability of any individual projection corresponding exactly to future conditions is low. Business decisions in the insurance industry are commonly based on probabilities similar to those associated with climate projections or modeled species' responses. If conservationists are in the business of preserving biodiversity, then such probabilities must be the basis for management decisions in the biological realm as well.

This section begins with an explanation of models of global climate, and how these are scaled-down for regional ecological analyses. Subsequent chapters look at terrestrial modeling techniques at different scales—single species to global, and different levels of complexity—static to dynamic. From this terrestrial focus, the section moves to chapters on marine and freshwater responses. Finally, the difficult issues of synergies between multiple climate change effects are addressed.

Chapter 18, on synergistic effects, is the product of a multi-author collaboration. It is the synthesis of a companion volume, *Climate Change and Biodiversity: Synergistic Impacts*, published by Conservation International (Hannah and Lovejoy 2003). This chapter focuses on some of the most important issues identified in select contributions to that publication. Copies of the companion volume are available in limited quantities to purchasers of this book, and may be requested by contacting Conservation International.

Many of the tools described in the chapters in this section—from regional climate models to species range shift models—are available for and operable on personal computers. Thus the realm of modeling is moving, from academia and institutions with large computing facilities, to the level of conservation managers and regional researchers. This shift in the scale of research tools promises to provide many opportunities for interactive management and research. Model results will be able to be developed in or near individual protected areas, and results of monitoring in the field will be able to be used to revise models. This "decentralization" of modeling will help establish its utility as a management tool, in spite of multiple sources of uncertainty. Ultimately, reducing uncertainty through interaction between modeling and the field will provide the best hope of understanding the future.

CHAPTER THIRTEEN

Climate Change Projections and Models

SARAH C. B. RAPER
AND FILIPPO GIORGI

Assessing the impact of future climate change on biodiversity requires projections of possible future climates. One approach is to use past climates and climate changes as a proxy for the future (see Part III). Another approach, and that most widely used in impact assessment, is to create a mathematical model of the Earth's climate system. Such models are necessarily very complex and are usually run on supercomputers or their equivalent. In this chapter the structure and application of these models is described and some results are summarized.

The term *scenario* refers to a plausible and often simplified description of how the future may develop. It is not possible to make exact predictions for the driving forces behind future emissions, but plausible future emissions scenarios can be constructed that are based on a coherent and internally consistent set of assumptions. The resultant evolution of the climate is called a climate change projection, the term *projection* implying one of many possibilities for the future. It is synonymous with the term *climate change scenario*, but the latter often refers to the climate at a particular time in the future.

This chapter describes the complementary use of simple and complex climate models in making future climate projections. Simple climate models can be used to give global average temperature projections for the full range of emissions scenarios. To assess changes on a regional scale and for projections of other climate parameters, complex 3-D climate models are used. Finally, various methods for assessing changes on a smaller spatial scale are discussed.

A HIERARCHY OF MODELS

There is a hierarchy of models for the investigation of climate change. These models vary in their comprehensiveness and complexity. They range from global scale with coarse resolution down to regional models which may be run at very high resolutions (a few kilometers to a few meters). Because of the trade-off between resolution and computing power, models with resolutions of only a few meters can be run only for a very limited area. It is regional models, usually at moderate resolution and embedded in a general circulation model (GCM), which are most useful in understanding biodiversity impacts.

Complex mathematical models that simulate global climate are called general circulation models. Coupled atmosphere ocean general circulation models (AOGCMs) combine an atmospheric GCM with a model of ocean processes that are important to climate. These models consider the atmosphere and ocean in three dimensions, but still some important physics—for example the physics of cloud formation—take place at a finer resolution than the models are able to represent (sub-grid scale) and have to be represented schematically. AOGCMs generally do not include a treatment of the gas cycles (for instance carbon cycles between the biosphere and atmosphere), so that greenhouse gas concentrations must be provided as an input to the model. However, climate models which include gas cycles are now becoming available. These models are able to include in detail such potentially important feedbacks as the effect of climate change on the carbon cycle. For instance, warming may increase release of carbon (as CO_2) from decomposing organic material in soils, which in turn contributes to further warming. This next generation of models is called earth system models.

A growing category of models that operate at scales finer than those of AOGCMs is that of regional climate models (RCMs).

RCMs use mathematical descriptions of physical processes similar to those of AOGCMs. However, they cover not the whole globe but only limited areas of interest. Because of this, they can reach horizontal resolutions much higher than those of AOGCMs. Present-day RCMs are used for a wide range of applications, from seasonal prediction to regional simulations of climatic changes and paleoclimates. RCMs can vary in horizontal resolution from several tens of kilometers to on the order of a few kilometers. Because these models can reach high resolutions, they can provide fine-scale climate change information which can be very important for many biodiversity applications.

GENERAL CIRCULATION MODELS

Since we are interested in climate change on the time scale of decades, it is necessary to consider the atmosphere, the land, the cryosphere (ice caps), and the oceans. The oceans in particular have a very large thermal inertia, which tends to slow climate change. Thus it is necessary to couple an atmosphere general circulation model (AGCM) to an ocean general circulation model (OGCM) for climate change projections. The resulting coupled model is the AOGCM. The coupling task is not simple, for example, because of the difference in the time scales involved in atmospheric and oceanic processes.

In AOGCMs the atmosphere and oceans are represented in three dimensions in one of two ways. Gridded GCMs divide the atmosphere and oceans into a number of horizontal and vertical grid boxes, the number being dependent upon the spatial resolution of the GCM. Spectral GCMs represent the horizontal and vertical dimensions by a number of mathematical functions; the larger the number of functions, the higher the spatial resolution. The horizontal spatial resolution of current GCMs ranges from about 1° to 10° of latitude or

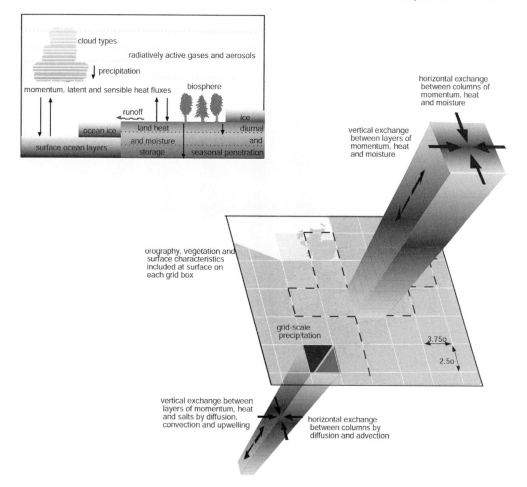

Figure 13.1. A schematic diagram illustrating an AOGCM. The inset gives examples of processes that are parameterized. *Source*: Courtesy of David Viner.

longitude (96 km to 960 km at the equator), the number of vertical layers range from one to 45 (Fig. 13.1 and Table 13.1). GCMs represent the physical processes of the atmosphere and oceans by solving a series of fundamental equations describing the conservation of momentum, mass, and energy in each grid box at finite time steps.

The grid or spectral resolution of a GCM determines the space and time scales of processes that can be modeled explicitly (see Henderson-Sellers and McGuffie 1987). Processes that operate on space or time scales too small to be modeled ex-

plicitly are called sub-grid-scale processes and must be parameterized. Parmeterization is a process by which sub-grid-scale processes are described in terms of scaled variables, estimated using physically based and/or semi-empirical formulations. It is important that processes are parameterized in a mutually consistent way; for example, if two processes produce a feedback on each other they should be resolved to a similar accuracy. An example of a parameterized process is convection within a grid cell.

Changes in land cover can have a significant impact on regional-scale climate change (IPCC 2001). For instance, conversion of forest to pasture can alter precipitation at regional levels, due to convection changes associated with differences in sur-

Table 13.1. Some Models Utilized by the IPCC Third Assessment Report

Model Name	Center	Reference	Atmospheric Resolution	Ocean Resolution	Land Surface	Flux Adjustment
CGCM2	CCCma	Flato and Boer 2001	T32(3.8×3.8)L10	1.8 ×1.8 L29	M, BB	H, W
CSIRO Mk2	CSIRO	Gordon and O'Farrell 1997	R21(3.2×5.6)L9	3.2 × 5.6 L21	C	H, W, M
CSM 1.0	NCAR	Boville and Gent 1998	T42(2.8×2.8)L18	2.0 ×2.4 L45*	C	—
DOE PCM	NCAR	Washington et al. 2000	T42(2.8×2.8)L18	0.67 × 0.67 L32	C	—
GFDL_R15_a	GFDL	Manabe et al. 1991; Manabe and Stouffer 1996	R15(4.5×7.5)L9	4.5 ×3.7 L12	B	H, W
GFDL_R30_c	GFDL	Knutson et al. 1999	R30(2.25×3.75)L14	1.875 × 2.25 L18	B	H, W
ECHAM3/LSG	DKRZ	Cubasch et al. 1997; Voss et al. 1998	T21(5.6×5.6)L19	4.0 ×4.0 L11	C	H, W, M
ECHAM4/OPYC	DKRZ	Roeckner et al. 1996	T42(2.8×2.8)L19	2.8 × 2.8 L11*	C	H, W
HadCM3	UKMO	Gordon et al. 2000	2.5 × 3.75 L19	1.25 × 1.25 L20	C	—
HadCM2	UKMO	Johns 1996; Johns et al. 1997	2.5 × 3.75 L19	2.5 × 3.75 L20	C	H, W

Center: CCCma Canadian Centre for Climate (Modelling and Analysis) (Canada); CSIRO Commonwealth Scientific and Industrial Research Organisation (Australia); NCAR National Center for Atmospheric Research (USA); GFDL Geophysical Fluid Dynamics Laboratory (USA); DKRZ Deutsche KlimaRechenZentrum (Germany); UKMO United Kingdom Met Office (UK). *Atmospheric resolution:* Horizontal and vertical resolution. The former is expressed either as degrees latitude 3 longitude or as a spectral truncation with a rough translation to degrees latitude 3 longitude. Vertical resolution is expressed as "Lmm," where mm is the number of vertical levels. *Ocean resolution:* Horizontal and vertical resolution. The former is expressed as degrees Latitude 3 longitude, while the later is expressed as "Lmm," where mm is the number of vertical levels. * indicates increased resolution at the equator. *Land surface scheme:* B = standard bucket hydrology scheme; BB = modified bucket scheme with spatially varying soil moisture capacity and/or a surface resistance; M = multi-layer temperature scheme; C = a complex land surface scheme usually including multi-soil layers for temperature and soil moisture, and an explicit representation of canopy processes. *Source:* IPCC 2001.

face heating and evapotranspiration. Land cover changes involve feedbacks that could significantly influence the rate of change of atmospheric CO_2, the climate system response, and the response of the biosphere to global change on time scales of decades to centuries. These effects are currently being studied with coupled models. Inclusion of models of the biosphere may change the global-scale response to increasing CO_2, especially due to effects of soil carbon stores, and they certainly significantly affect the simulations of local- and regional-scale change (IPCC 2001; Betts and Shugart, this volume). AOGCMs have various land surface schemes (see Table 13.1).

The IPCC (2001) has evaluated AOGCMs against observations and compared the differences between models. As well as looking at present-day climatology, the performance of the models in simulating twentieth-century climate change and selected palaeoclimates was assessed. It was concluded that in general the models provide credible simulations of climate, at least down to sub-continental scales and over temporal scales from seasonal to decadal. They state that no single model can be considered "best" and that it is important to utilize results from a range of coupled models.

GLOBAL PATTERNS OF CLIMATE CHANGE FROM COMPLEX CLIMATE MODELS

Running a single AOGCM for a 200-year simulation may take several months on a

---

**Box 13.1.   Emissions Scenarios of the Special Report
on Emissions Scenarios (SRES)**

A set of emissions scenarios approved by the IPCC are described in Nakićenović et al. (2000). There are 35 scenarios with data on the full range of gases needed for climate modeling. These scenarios arose from four different storylines about how the world might develop, providing consistent relationships between emission driving forces. Each scenario represents a quantification of one of the four storylines by a specific economic modeling group. Six illustrative scenarios were chosen as representative of the full set.

---

super computer. It would therefore be very costly and time consuming to run an AOGCM to assess the many emissions scenarios to choose from. It is, however, possible to tune a simpler (often only one-dimensional) climate model to a particular AOGCM in order to reproduce a good estimate of the global mean temperature response simulated by the AOGCM. The IPCC (Cubasch et al. 2001) used such simple models to explore the range of possible future temperature response to the 35 SRES emissions scenarios (see Box 13.1) using seven different AOGCMs. The tuning process involved the specification of the AOGCMs climate sensitivity and the adjustment of parameters to give the correct heat flux into the ocean. The results of such numerous simple model projections also act as an aid to identifying which scenarios would be most useful for more detailed assessment with an AOGCM.

The temperature projections from the simple climate model are shown in Figure 13.2 for the six illustrative SRES scenarios and for full SRES scenario envelopes. The individual time series and inner envelope (darker shading) are the average results for the seven model tunings. The average climate sensitivity (change in global mean temperature produced by a model in response to a doubling of atmospheric CO_2) used is 2.8°C and the range is 1.7 to 4.2°C. The range of global mean temperature change from 1990 to 2100 for the full set of SRES scenarios with the seven model tunings is 1.4 to 5.8°C. The wide range of emissions scenarios and the uncertainty in the AOGCMs (as represented by the climate sensitivity) are very important sources of uncertainty in the projections. However, the range of 1.4 to 5.8°C is not the extreme range of possibilities because, for example, some AOGCMs have climate sensitivities outside the range considered.

AOGCMs model climate in enough detail that any individual run has its own "weather" and natural climatic variability, in addition to global climate change. To separate out the global climate change element, several runs must be made and the variability averaged out. Several AOGCM simulations run with the same forcing but starting from different times will result in different internally generated climate variability in each simulation. Such a group of simulations is called an ensemble. Averaging over the ensemble members to give the ensemble mean reduces the "noise" of natural variability and gives an improved estimate of the model's forced or deterministic climate change. The idea can be taken a step further by taking the results from several AOGCMs and producing multi-model ensembles (Cubasch et al. 2001).

Ensemble studies indicate GCM agreement on maximum warming in the high latitudes of the northern hemisphere and minimum warming in the Southern Ocean (due to ocean heat uptake). Generally the warming over the land is greater

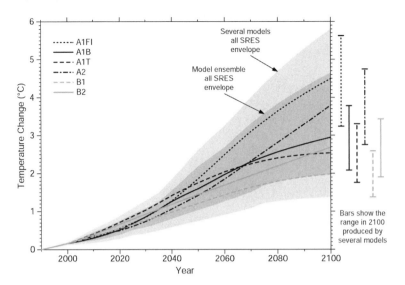

Figure 13.2. Future global mean temperature changes for the 6 illustrative SRES scenarios (see box 13.1) using a simple climate model tuned to 7 AOGCMs. The dark shading represents the envelope of the full set of 35 SRES scenarios using the simple model ensemble mean results. The light envelope is based on the GFDL—R15—and DOE PCM parameter settings. The bars show the range of simple model results in 2100 for the 7 AOGCM model tunings. *Source:* IPCC (2001). Reprinted with permission.

than over the ocean, implying that average warming over the land will be greater than that of the global mean. As climate warms, northern hemisphere snow cover and sea-ice extent decrease. Projections indicate an increase in globally averaged water vapor, evaporation, and precipitation. It is also noted that increases in the variability of precipitation are likely to accompany increases in mean precipitation. Most AOGCM projections show a weakening of the thermo-haline circulation (THC) accompanying the global warming. The weakening of the THC in the North Atlantic results in a reduction of the poleward heat transport, which leads to a minimum of the surface warming in the North Atlantic region extending to the extreme northeastern part of North America and northwestern Europe. However, the effect is not strong enough to cancel the warm-

ing, so that all the integrations show a net warming in northwestern Europe.

ENSO contributes a large part of the inter-annual variability of climate in many parts of the world. Many AOGCMs indicate a shift to a more El Nino–like regime with global warming. In particular, the central and eastern equatorial Pacific sea surface temperatures are projected to warm more than the western equatorial Pacific, with a corresponding eastward shift of precipitation.

Changes in the statistics of climate variability are of considerable interest in their own right and affect the occurrence of extremes. For the study of variability it is therefore important to analyze the individual climate experiment results before any differencing or averaging has been done. Projections of future changes in extreme events are summarized in Table 13.2, reproduced from the assessment of Cubasch and colleagues (2001). Information on changes in the extremes of other phenomena are not forthcoming, since it was concluded that there is insufficient information about recent trends and confidence in the models, and scientific understanding is still inadequate. For example, there is little agreement yet among models about future changes in mid-latitude storm intensity, frequency, and variability

Table 13.2. Estimates of Confidence in Observed and Projected Changes in Extreme Weather and Climate Events

Confidence in Observed Changes (latter half of the 20th century)	Changes in Phenomenon	Confidence in Projected Changes (during the 21st century)
Likely	Higher maximum temperatures and more hot days[a] over nearly all land areas	Very likely
Very likely	Higher minimum temperatures, fewer cold days and frost days over nearly all land areas	Very likely
Very likely	Reduced diurnal temperature range over most land areas	Very likely
Likely, over many areas	Increase of heat index[b] over land areas	Very likely, over most areas
Likely, over many northern hemisphere mid- to high-latitude land areas	More intense precipitation events[c]	Very likely, over many areas
Likely, in a few areas latitude	Increased summer continental drying and associated risk of drought	Likely, over most continental interiors (lack of consistent projections in other areas)
Not observed in the few analyses available	Increase in tropical cyclone peak wind intensities[d]	Likely, over some areas
Insufficient data for assessment	Increase in tropical cyclone mean and peak precipitation intensitiesd	Likely, over some areas

[a]Hot days refers to a day whose maximum temperature reaches or exceeds some temperature that is considered a critical threshold for impacts on human and natural systems. Actual thresholds vary regionally, but typical values include 32°C, 35°C or 40°C. [b]Heat index refers to a combination of temperature and humidity that measures effects on human comfort. [c]For other areas, there are either insufficient data or conflicting analyses. [d]Past and future changes in tropical cyclone location and frequency are uncertain. *Source:* IPCC 2001.

—though a growing number of studies are addressing this issue (Carnell and Senior 2001).

REGIONAL CHANGE

Climate change information at the regional to local scale is needed to assess the impacts of climate change on human activities and the environment and thus to develop suitable policies to react to such impacts. To date, the uncertainties on regional climate change information are still very high, because of the complexity of the processes involved and the limitations of current models. In fact, regional climates are affected not only by large-scale processes (e.g., the response of the general circulation to greenhouse gas forcing) but also by local processes related, for example, to local topography, land use, and pollutant emissions.

Current coupled AOGCMs are still characterized by horizontal spatial resolutions of a few hundred kilometers, which prevent them from adequately describing climate patterns at regional to local scales. For this reason, in recent years a number of approaches, generally referred to as "regionalization" techniques, have been developed to enhance the regional information of AOGCMs and to provide fine-scale climate information. The approaches can be divided into three categories:

1. High-resolution and variable-resolution "time slice" atmospheric GCM (or AGCM) experiments

2. Nested regional climate models (RCMs)

3. Statistical downscaling methods

Time slice experiments allow production of higher-resolution results without massive investment in computer time. They do this by limiting high-resolution modeling to a few specific "time slices" taken from a longer, coarser-scale AOGCM run. Within each time period (or "slice") selected from a transient AOGCM, climate change is simulated with a higher-resolution global atmospheric model (either of uniform resolution or with increasing resolution over a given area of interest). Typically, time slices are of a few decades in length (e.g., 1960–1990 to represent present-day climate and 2070–2100 to represent future climate) and the AOGCM grid interval reaches ~50 km with variable resolution and ~120 km with uniform resolution.

Nested RCMs are limited area atmospheric models that are run over a selected region of the globe. The meteorological initial and lateral boundary conditions necessary to run the regional model, as well as the sea surface temperature and greenhouse gas/aerosol forcing, are obtained from a corresponding AOGCM simulation. Because the regional model covers only a limited domain it can be run at a much higher resolution than the global model, and current RCM experiments have reached horizontal grid intervals of less than 10 km.

Statistical downscaling techniques essentially apply incremental change from GCM projections and "drape" them over present climate. They consist of the development of statistical relationships between large-scale predictors (say, 500 mb height) and regional- to local-scale predictands (say, temperature or precipitation). These can be obtained either by using analyses of local weather station observations or from GCM simulations of present-day climate. The statistical relationships are then applied to the output of AOGCM simulations to infer regional- to local-scale information for different time periods in the future.

All these regionalization techniques have their advantages and limitations, and the choice of any of them may depend on the particular application (Giorgi et al. 2002). The development of these regionalization techniques is still relatively recent, so that a comprehensive picture of regional climate changes based on their use is still not available. Rather, a multitude of studies for individual regions have been carried out, most of them process and sensitivity studies, not actual regional climate projections. These studies have clearly indicated that complex topographical and land use features at sub-GCM grid scale can substantially affect the climate change signal over many regions of the world.

For biodiversity studies, use of regionalization tools, such as regional climate models or downscaling, may provide information critical to determining biotic response on relevant scales (e.g., Giorgi 1995). For instance, in a typical GCM, the mountains of western North America, from the Sierra Nevada of California to the Rockies of Colorado, may be represented as a single mountainous block. Clearly, reasonable studies of biodiversity change in the inter-mountain west require finer-scale modeling of regional processes such as orographic rainfall and temperature variation. Similar considerations hold for many other mountainous areas that are of particular importance to biodiversity (Hannah 2001).

Today a wide range of high- and variable-resolution AGCMs, RCMs, and statistical downscaling models are available, which have been used for many different applications (Giorgi et al. 2001). Although much of the effort in the past decade has been on the development and understanding of these regionalization techniques, they are increasingly being applied to climate change simulations, so that in the next few years more fine-scale simulations of climate change will become available for use in biodiversity applications.

On the other hand, at present there is still a paucity of comprehensive studies with regional models, and therefore much of the regional climate change information currently available is derived from the AOGCM simulations. As such, this information is reliable only at the broad regional or sub-continental scale. Consistent patterns of climate change over different regions are emerging from different AOGCM simulations. These are illustrated in Figure 13.3A and B, which analyze the consistency across 9 AOGCM simulations in temperature and precipitation changes for the A2 and B2 emission scenarios over 22 broad regions of the world.

The results show model agreement in simulating warming much greater than the global average not only in high-latitude northern hemisphere regions but also over the Mediterranean, central Asia, and Tibet regions in summer. Inconsistency of regional warming results across models is found mostly in some tropical and sub-tropical regions. The results also indicate increases in both summer and winter precipitation over high-latitude regions. In winter, increases in precipitation are seen over northern mid-latitudes, tropical Africa and Antarctica, and in summer over southern and eastern Asia. Australia, central America, and southern Africa show consistent decreases in winter rainfall. It is also evident that many more regions of inconsistent results are found for precipitation than for temperature.

In summary, current AOGCM simulations are suggesting a number of possible patterns of change at the broad regional scale, and regionalization techniques have shown that sub-AOGCM-scale factors (e.g., topography and land use) can strongly modulate these patterns at finer scale. This represents a substantial advancement compared to previous IPCC assessments, in which few or no regional statements were presented (except for broad latitudinal bands). On the other hand, much more research is needed to fully evaluate the uncertainties associated with the simulation of regional climatic changes and the contribution of the different physical processes involved. This will require a more comprehensive and coordinated use of different AOGCMs and regionalization techniques.

SUMMARY

This chapter describes the structure and application of the hierarchy of climate models used in impact assessments. Simple models and complex coupled atmosphere-ocean general circulation models (AOGCMs) can be used in a complementary way to produce a range of possible future global climate change. Regional climate models use input from the AOGCM projections together with information about the local terrain to give finer-scale climate output. Because no single model can be considered "best," this summary is based on the comparison of results from several AOGCMs.

Human interference with the climate system is projected to cause a global mean temperature change from 1990 to 2100 in the range 1.4 to 5.8°C. Assigning probabilities to this range is at present problematic. The biggest uncertainties are in the emissions scenarios and in the climate sensitivity. The warming is expected to be generally greater over the land than over the ocean, with greatest warming over the land in high latitudes of the northern hemisphere. Snow cover and sea-ice extent are projected to decrease, whereas globally averaged water vapor, evaporation, and precipitation are expected to increase.

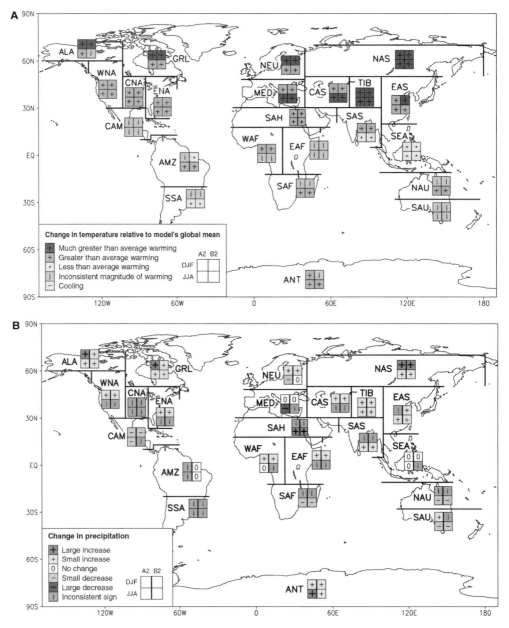

Figure 13.3. (A) Inter-model agreement in regional temperature change relative to each model's global mean annual change for the A2 and B2 scenarios. Nine AOGCMs are used in the analysis as listed by Giorgi et al. (2001). Regions are classified as showing either agreement on warming in excess of 40% of the global average ("much greater than average warming"), agreement on warming greater than the global average ("greater than average warming"), agreement on warming less than the global average ("less than average warming"), disagreement among model experiments on the magnitude of regional relative warming ("inconsistent magnitude of warming"), or agreement in cooling ("cooling"). The global mean annual change in the simulations spans 1.2–4.5°C for A2 and 0.9–3.4°C for B2, and therefore a regional amplification of 40% represents warming ranges of 1.7–6.3°C for A2 and 1.3–4.7°C for B2. (B) Inter-model agreement in regional precipitation change for the A2 and B2 scenarios. Nine AOGCMs are used in the analysis as listed by Giorgi et al. (2001). Regions are classified as showing either agreement on increase with an average regional change greater than 20% (of present-day values) ("large increase"), agreement on increase with an average regional change between 5% and 20% ("small increase"), agreement on a change between −5% and 5% ("no change"), agreement on decrease with an average regional change between −5% and −20% ("small decrease"), agreement on decrease with an average regional change of less than −20% ("large decrease"), or disagreement ("inconsistent sign"). Agreement in panels A and B is defined by a consistent result from at least 7 out of 9 model experiments. *Source:* IPCC (2001). Reprinted with permission.

Associated with these changes in average conditions some likely changes in extreme events can be identified. For example, it is very likely that there will be higher maximum and higher minimum temperatures resulting in more hot days and fewer cold and frost days over nearly all land areas. Many areas will also very likely experience more intense precipitation events.

The broad patterns of climate change which emerge from the AOGCM simulations can be strongly affected by local factors such as topography and land use change. Regionalization techniques, such as RCMs and statistical downscaling, are powerful tools that can be used to provide the fine-scale information needed in many biodiversity applications. More comprehensive and coordinated research into the use of different AOGCMs, and regionalization techniques is needed to fully evaluate the uncertainties associated with the simulation of regional climate changes.

REFERENCES

Boville, B. A., and P. R. Gent, 1998. The NCAR climate system model, version one. *J. Climate*, 11, 1115–1130.

Carnell, R. E., and C. A. Senior, 2001. Changes in mid-latitude variability due to increasing greenhouse gases and sulphate aerosols. *Clim. Dyn.*, 14, 369–383.

Cubasch, U., G. A. Meehl, G. J. Boer, R. J. Stouffer, M. Dix, A. Noda, C. A. Senior, S. Raper and K. S. Yap, 2001. Projections of future climate change. In *Climate Change 2001: The Scientific Basis. Contribution of Working Group I to the Third Assessment Report of the Intergovernmental Panel on Climate Change*, Houghton, J. T., Y. Ding, D. J. Griggs, M. Noguer, P. van der Linden, X. Dai, K. Maskell, C. I. Johnson, eds. New York: Cambridge University Press, 525–582.

Cubasch, U., R. Voss, G. C. Hegerl, J. Waszkewitz, and T. J. Crowley, 1997. Simulation of the influence of solar radiation variations on the global climate with an ocean-atmosphere general circulation model. *Clim. Dyn.*, 13, 757–767.

Flato, G. M., and G. J. Boer, 2001. Warming asymmetry in climate change experiments. *Geophys. Res. Lett.*, 28, 195–198.

Giorgi, F., 1995. Perspectives for regional earth system modeling. *Global and Planetary Change*, 10, 23–42.

Giorgi, F., B. Hewitson, J. Christensen, M. Hulme, H. Von Storch, P. Whetton, R. Jones, L. Merns, and C. Fu, 2002. Regional climate information—Evaluation and projections. In *Climate Change 2001: The Scientific Basis. Contribution of Working Group I to the Third Assessment Report of the Intergovernmental Panel on Climate Change*, Houghton, J. T., Y. Ding, D. J. Griggs, M. Noguer, P. van der Linden, X. Dai, K. Maskell, C. I. Johnson, eds. New York: Cambridge University Press, 583–638.

Gordon, C., C. Cooper, C. A. Senior, H. T. Banks, J. M. Gregory, T. C. Johns, J. F. B. Mitchell, and R. A. Wood, 2000. The simulation of SST, sea ice extents and ocean heat transports in a version of the Hadley Centre coupled model without flux adjustments. *Clim. Dyn.*, 16, 147–168.

Gordon, H. B., and S. P. O'Farrell, 1997. Transient climate change in the CSIRO coupled model with dynamic sea ice. *Mon. Wea. Rev.*, 125, 875–907.

Hannah, L., 2001. The role of a global protected areas system in conserving biodiversity in the face of climate change. In *Climate Change and Protected Areas*, Visconti, G., and M. Balaban, eds. Dordrecht: Kluwer.

Henderson-Sellers, A., and K. McGuffie, 1987. *A Climate Modelling Primer*. Chichester: John Wiley & Sons.

IPCC, 2001. *Climate Change 2001: The Scientific Basis. Contribution of Working Group I to the Third Assessment Report of the Intergovernmental Panel on Climate Change*, Houghton, J. T., Y. Ding, D. J. Griggs, M. Noguer, P. van der Linden, X. Dai, K. Maskell, and C. I. Johnson, eds. New York: Cambridge University Press.

Johns, T. C., 1996. A description of the Second Hadley Centre Coupled Model (HadCM2). Climate Research Technical Note 71, Hadley Centre, United Kingdom Meteorological Office, Bracknell Berkshire RG12 2SY, United Kingdom.

Johns, T. C., R. E. Carnell, J. F. Crossley, J. M. Gregory, J. F. B. Mitchell, C. A. Senior, S. F. B. Tett, and R. A. Wood, 1997. The second Hadley Centre coupled atmosphere-ocean GCM: Model description, spinup and validation. *Clim. Dyn.*, 13, 103–134.

Knutson, T. R., T. L. Delworth, K. W. Dixon, and R. J. Stouffer, 1999. Model assessment of regional surface temperature trends (1949–1997). *J. Geophys. Res.*, 104, 30981–30996.

Manabe, S. J., and R. J. Stouffer, 1996. Low-frequency variability of surface air temperature in a 1000-year integration of a coupled atmosphere-ocean-land model. *J. Climate*, 9, 376–393.

Manabe, S., R. J. Stouffer, M. J. Spelman, and K. Bryan, 1991. Transient responses of a coupled ocean-atmosphere model to gradual changes of atmospheric CO2. Part I: Annual mean response. *J. Climate*, 4, 785–818.

Nakićenović, N., J. Alcamo, G. Davis, B. de Vries, J. Fenhann, S. Gaffin, K. Gregory, A. Grübler, T. Y. Jung, T. Kram, E. L. La Rovere, L. Michaelis, S. Mori, T. Morita, W. Pepper, H. Pitcher, L. Price, K. Raihi, A. Roehrl, H.-H. Rogner, A. Sankovski, M. Schlesinger, P. Shukla, S. Smith, R. Swart, S. van Rooijen, N. Victor, and Z. Dadi, 2000. *IPCC Special Report on Emissions Scenarios*. New York: Cambridge University Press.

Roeckner, E., J. M. Oberhuber, A. Bacher, M. Christoph, and I. Kirchner, 1996. ENSO variability and atmospheric response in a global coupled atmosphere-ocean GCM. *Clim. Dyn.*, 12, 737–754.

Voss, R., R. Sausen, and U. Cubasch, 1998. Periodically synchronously coupled integrations with the atmosphere-ocean general circulation model ECHAM3 / LSG. *Climate Dyn.* 14, 249–266.

Washington, W. M., et al., 2000. Parallel climate model (PCM): Control and transient simulations. *Clim. Dyn.*, 16, 755–774.

CHAPTER FOURTEEN

Modeling Distributional Shifts of Individual Species and Biomes

A. TOWNSEND PETERSON, HANQIN TIAN,
ENRIQUE MARTÍNEZ-MEYER,
JORGE SOBERÓN,
VÍCTOR SÁNCHEZ-CORDERO,
AND BRIAN HUNTLEY

The observation of rapidly changing climates on global and local scales (Karl et al. 1996; Magnuson 2001) demands investigation of the magnitude of consequences to be expected for species and natural communities. In general, theoretical considerations lead to a three-way expectation for climate change effects on species' distributions (Holt 1990). If species can track appropriate conditions spatially, we can term this response niche-tracking. The alternative to moving is *adapting*—evolution in situ. Finally, failing both movement and adaptation, by definition, a species finds itself outside of the conditions that constitute its niche, and *extinction* occurs. This three-way reasoning provides a framework for understanding climate change effects on species' geographic distributions. Modeling of species' range shifts is one way to explore the first of these alternatives.

SINGLE-SPECIES ECOLOGICAL NICHE MODELING

Grinnell (1917, 1924) originally defined the ecological niche of a species as the suite of environmental factors that determine its geographic distribution. Later, the concept was redefined either to mean the "role" of a species in a community (Elton 1927) or to include interactions with other species (Hutchinson 1957; MacArthur 1972). However, because of our focus on geographic dimensions of species' distributions, the most useful focus for ecological niche modeling is on a broad-scale view most akin to a Grinnellian niche.

Ecological niche or "climate envelope"

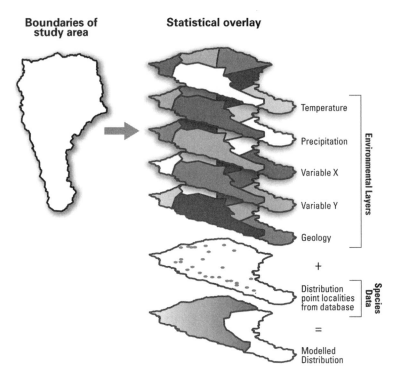

Boundaries of study area

Statistical overlay

Temperature

Precipitation

Variable X

Variable Y

Geology

Environmental Layers

+

Distribution point localities from database

Species Data

=

Modelled Distribution

Figure 14.1. Schematic diagram of a single-species range shift model. Known occurrence points for the species are correlated with various climatic and environmental variables to produce a simulated distribution. *Source: Lischke et al. (1998). In View from the Alps,* MIT Press.

models relate occurrence data to ecological-environmental characteristics of a landscape under present-day conditions, and can be projected to future change scenarios to identify potential distributional areas. The underlying principle of almost all models of this type is the description of present-day relationships between a species' geographic distribution and multiple environmental variables (Fig. 14.1). Models of this type include BIOCLIM (Nix 1986), DOMAIN (Carpenter et al. 1993), and FLORAMAP (Jones and Gladkov 1999), among many others. Still other approaches use completely different inferential procedures, and may be based more on physiological models than on geographic correlations (Sykes et al. 1996).

These diverse approaches to ecological niche modeling differ in several characteristics. First, they may use either point data indicating presences at which the species has been recorded (e.g., Nix and Switzer 1991) or gridded data indicating the presence *and absence* of the species across the region of interest (e.g., Huntley et al. 1995), although the latter approach is clearly based to a large degree on known presences and assumptions regarding absences. Second, niche modeling methods may use either a broad selection of environmental dimensions or a preselected set of dimensions that are expected to be particularly relevant ("bioclimatic variables"; Huntley et al. 1995), which of course involves assumptions regarding which environmental dimensions will be most relevant to limiting species' distributions. Finally, niche modeling approaches may fit a model using some predetermined statistical approach that yields a "global" fit (e.g., Yee and Mitchell 1991), may relax assumptions regarding the form of the relationship (e.g., Huntley et al. 1995), or may avoid any assumptions whatsoever about the form of the relationships involved (e.g., Stockwell and Noble 1992). These differences among methods affect the characteristics of the predictions in resulting models.

More complex applications have been developed that generate multidimensional or heterogeneous sets of decision rules (e.g., neural networks, genetic algorithms). Such models have proven accurate in tests under diverse circumstances (e.g., Peterson et al. 2002d), and many of the following examples will be drawn from work using GARP, a genetic algorithm which draws on multiple rule-sets (Stockwell and Noble 1992; Stockwell 1999; Stockwell and Peters 1999). Many software packets permitting development of niche models are now available for use on personal computers and may be obtained over the Internet (e.g., CLIMEX, DIVA, ARP).

TESTS OF THE SINGLE-SPECIES APPROACH

An appropriate starting point in assessing the ecological niche modeling approach to understanding species' potential distributions in response to climate change is by examining its precepts. Here, the principal question is whether ecological niches constitute stable, long-term constraints on species' potential geographic distributions. Some investigators have argued that rearrangements of interactions in ecological communities in the face of changing climates will have effects much more profound than those caused directly by climate change on species' tolerances (Davis et al. 1998). For example, single-species models cannot address the effects of interspecific competition on species' distributions, and they assume that present distributions are in equilibrium with climate, which may not always be the case. If such limitations were dominant, the predictivity of future distributional phenomena might prove nil, so this question is central to this contribution.

Given the yet-to-be-observed nature of future climate change effects, the only approach available for testing model predictivity is that of postprediction of past events; several scenarios permit rich quantitative tests of climate change effects on species. First, on a most proximate time scale, the too-frequent phenomenon of species' invasions (Lawton and Brown 1986; Carlton 1996) allows numerous tests of ecological niches as constraints on species' distributional possibilities when placed in very different community contexts. The question is whether conditions modeled as ecological niches of species on native distributional areas can predict distributional possibilities on a very different landscape (in this case, a different continent).

Initial explorations of the possibilities for such predictions of invasions were promising (Higgins et al. 1999). Indeed, quantitative tests indicated that species obey the same ecological "rules" even when transplanted to completely different geographic and ecological situations (Peterson and Vieglais 2001). Subsequent tests with diverse taxa (e.g., Fig. 14.2) have confirmed high levels of predictivity of species' invasions essentially without exception (Iguchi et al. 2005; Papes and Peterson, submitted; Peterson 2003; Peterson et al. 2003a; Peterson and Robins 2003; Peterson et al. 2003b; Roura-Pascual et al. 2005).

A second line of evidence for stable distributional constraint based on ecological niches is provided by recent analyses of ecological niches and geographic distributions of 23 mammal species across the transition from the last glacial maximum to present (Martínez-Meyer et al. 2004). Here, ecological niches modeled based on present ranges predicted Pleistocene distributions, and vice versa; for 18 of 23 species, at least one of the reciprocal predictions (Pleistocene−Present or Present−Pleistocene) proved statistically significant (Fig. 14.3). The general picture is one of species' geographic distributions tracking

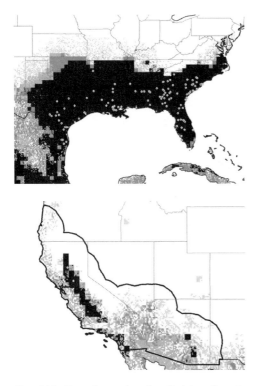

Figure 14.2. Exemplar results of predictivity of species invasions, showing (top) the native range of the glassy-winged sharpshooter (Homalodisca coagulata) and (bottom) projection of the ecological model developed on the native range to the invaded range in California. Known occurrence points on both ranges are shown, and the highly significant prediction of the geographic distribution of invasive populations is evident. Source: Peterson et al. (2004).

particular climate regimes closely, and of modeled ecological niches providing substantial predictive power regarding distributions under diverse climate regimes.

A final—and longest-term—line of evidence for stable predictivity of distributions from niche models is that of predicting geographic distributions among closely related species. This approach, pioneered in general comparisons (Huntley et al. 1989) and developed in detail more recently (Peterson et al. 1999), asks the question of whether ecological niches have diverged over moderate periods of evolutionary time. This question may be addressed by using ecological niche models developed for each of a pair or set of

closely related species. In analyses focused on sister species pairs (birds, mammals, butterflies) separated by the Isthmus of Tehuantepec, in southern Mexico (Fig. 14.4), all of 37 such comparisons were statistically significant, whereas comparisons with randomly selected confamilial species were generally not significant (Peterson et al. 1999). This result suggested that niche conservatism is generally maintained at least at the level of closely related species, although exceptions have been encountered (Peterson and Holt 2003).

Thus, several independent lines of evidence indicate that species' modeled ecological niches do predict geographic distributions: in spite of different ecological and community contexts in the case of invasive species (Peterson and Vieglais 2001); very different climates and geography in the case of the Pleistocene–recent transition (Martínez-Meyer et al. 2004); and speciation events, changed climates, and moderate periods of evolutionary time in the interpredictivity analyses of related species (Peterson et al. 1999). Hence, the premise that ecological niches represent stable constraints on species' distributional possibilities, and that these constraints can be modeled with high degrees of predictive accuracy, is strongly supported, despite theoretical and experimental arguments to the contrary (e.g., Davis et al. 1998). This result suggests that it is informative to apply the results of projections of such models under different future climate change scenarios.

PREDICTIONS FOR SPECIES' DISTRIBUTIONS UNDER CHANGED CLIMATES

Ecological niche modeling has blossomed rapidly over the past two to three decades into a field replete with analytical tools: predictive models for individual species are developed using bioclimatic envelopes (Midgley et al. 2002), generalized linear

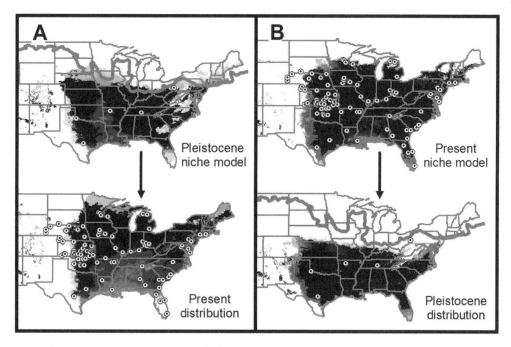

Figure 14.3. Pleistocene–recent comparisons for *Scalopus aquaticus* (eastern mole), modeled in both directions. (A) Pleistocene occurrences were used to create a Pleistocene niche model (*top*), projected onto present climate maps to produce a prediction of present distributional patterns, which is shown with known present occurrences overlain (*bottom*). (B) Present-day known occurrences were used to develop a present-day niche model (*top*), projected onto Pleistocene climate maps to produce a prediction of Pleistocene distributional patterns, which is shown with known Pleistocene occurrences overlain (*bottom*). *Source:* Courtesy of Enrique Martínez-Meyer.

and additive models (Moisen and Edwards 1999), Bayesian probability approaches (Aspinall 1992), and genetic algorithms (Stockwell and Peters 1999), among many other approaches (Guisan and Zimmerman 2000), as described above. These models all share the common architecture of building on known occurrences and geographic coverages describing the ecological landscape to produce models of species' ecological niches; these models, when projected onto changed climate scenarios, can provide predictions of species' future distributional potential.

Some implementations of this approach for questions of climate change have not relied on GCM projections. For example, Crumpacker and colleagues (2001) used climate envelope models to project potential distributions of trees and shrubs in Florida across artificial scenarios of increased temperature (+1 to +2°C) and changed precipitation (−20 to +10 percent). These approaches are valuable as

Figure 14.4. Predictivity of geographic distributions from ecological characteristics of sister species pairs: geographic distributions of *Atthis heloisa* (circles) and *A. ellioti* (squares), each overlain on geographic predictions (dark gray) based on the ecological characteristics of occurrence points of the other species (line divides the predictions and points of one species from the other. *Source:* Peterson et al. (1999).

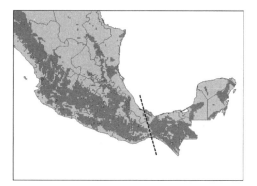

sensitivity analyses, but their real-world predictability is limited by the artificial nature of the changed-climate scenarios employed.

Most applications have used GCM projections, focusing on 2050 as the future climate scenario, since this is far enough into the future to show substantial effects but carries less uncertainty than longer-term (e.g. 2100) scenarios. For example, Texeira and Arntzen (2002) used Hadley Centre HadCM2 general circulation model results for 2050 and generalized linear models of ecological niches to develop predictive equations regarding climate responses of a salamander Chioglossa lusitanica endemic to the Iberian Peninsula. Projection of these models onto present-day climates yielded a predicted distributional area of 22,850 km^2 for the species, but projection to 2050 climates yielded a potential distributional area 15.1 percent smaller, suggesting that climate change would act to reduce this species' geographic range significantly. Similar studies have been applied in many other parts of the world.

Australia is a region that has been particularly well-studied using single-species niche models. Several modeling techniques have been developed in Australia, and applied to a variety of plant, invertebrate, and vertebrate species. In general, these studies have shown range fragmentation or even total loss of range with climate change of as little as 0.5 to 1.0°C (Hughes 2003). In a study of 92 endemic plant species, most were found to lose range, and 28 percent to lose all range with only a 0.5°C warming (Pouliquen-Young and Newman 2000). Acacia species were found to be more robust but were still predicted to lose all climate space with a 2.0°C warming. Studies of vertebrates have included both threatened taxa (Brereton et al. 1995; Pouliquen-Young and Newman 2000) which were found to be dominated by losses of climate space, and common species (Chapman and Milne

1998), some of which were projected to lose range, while others showed potential range expansion.

Europe and North America have been the focus of perhaps the greatest number of single-species studies, as well as other modeling efforts. Higher plants in Europe were modeled by Huntley and colleagues (1995), while more recent multi-taxa modeling has focused on both Great Britain and the continent (Berry et al. 2002). In North America, multiple studies have focused on plants (e.g., Crumpacker et al. 2001), vertebrates (e.g., Peterson 2003b), and invasive plants and insect pests (e.g., Logan et al. 2003).

The Alps have been the focus of a number of studies on montane effects. Modeling of alpine plant species was put in context of climate change and other modeling methods by Lischke and colleagues (1998). A particularly elegant alpine example is a fine-scale, spatially explicit modeling exercise that focused on distributions of high montane plant species in Austria (Gottfried et al. 1999). This study used a network of sample sites and fine-resolution digital elevation models to assess species' ecological niches, and then, in a GIS environment, forecast future potential distribution based on microclimates that are a function of topography (Fig. 14.5). Their results showed the expected reductions of distributional areas as a function of climate change (moving a distribution up the side of a cone will necessitate reduction in distributional area), but also unexpected range expansions in alpine and subnival (below snowline) species, as species colonized particular topographic regions of the mountain slopes.

Among high-biodiversity areas and hotspots, South Africa's Cape Floristic Region has been the focus of intensive modeling efforts for endemic plants, particularly proteas (Erasmus et al. 2000; Midgley et al. 2002), and Mexico has received considerable attention (Peterson et al. 2001;

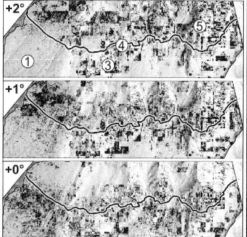

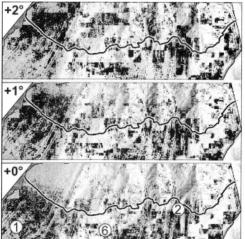

Erigeron uniflorus   *Sedum alpestre*

Figure 14.5. Exemplar of a modeled climate–mediated range shift for an alpine species. Predicted distribution patterns of *Erigeron uniflorus* and *Sedum alpestre*. Bottom: current distribution; *center*: +1.0°C scenario; *top* +2.0°C scenario. *Source*: Gottfried et al. (1999). Copyright Blackwell Scientific Publishers. Reprinted with permission.

Peterson et al. 2002b). A major modeling initiative has recently been undertaken for the Mediterranean (Araujo et al., in press). Modeling for other biodiversity hotspots is an important research priority.

The role of species' dispersal ability in determining the viability of potentially suitable future climate space has also been addressed (e.g., Iverson et al. 1999). Forecast exercises have used a variety of assumptions regarding dispersal abilities, ranging from no dispersal through limited dispersal scenarios to universal capabilities for movement (Carey and Brown 1994; Johnston and Schmitz 1997; Kadmon and Heller 1998; Hill et al. 1999; Iverson et al. 1999; Erasmus et al. 2000; Porter et al. 2000; Price 2000; Peterson et al. 2001; Xu and Yan 2001).

Many such predictive applications have required detailed physiological information on species (Johnston and Schmitz 1997), or customized data collection (Gottfried et al. 1999). For this reason, few efforts have been able to cross diverse

species groups with varied biological characteristics (Price 2000; Peterson et al. 2002b). A first set of fauna-wide predictions (all birds, mammals, and butterflies in Mexico) showed striking emergent properties (Peterson et al. 2002b); whereas only a few species (~3 percent) actually were likely to go extinct globally, rearrangements (local colonizations and extinctions) of many species' distributions in some regions were predicted to affect more than 40 percent of species in local communities. In contrast, other regions were predicted to remain relatively immune to climate change effects (Fig. 14.6A–D).

Such cross-taxon, flora, or fauna-wide assessments are now being implemented for a number of regions worldwide (e.g., Hughes et al. 1996; Bakkenes et al. 2002; Erasmus et al. 2002) and are resulting in similar predictions of broad species-level reorganizations across landscapes. Indeed, some generalizations are becoming possible. For example, the idea of predicting the behavior of entire species assemblages under climate change scenarios based on community characteristics (McDonald and Brown 1992; Sala et al. 2000) appears untenable: species frequently show idiosyncratic responses to climate change (Berry et al. 2002; Peterson et al. 2002b;

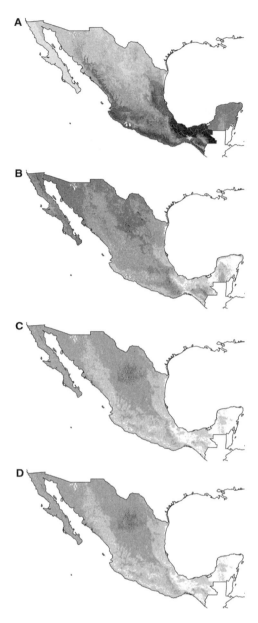

Figure 14.6. Modeled species turnover in biological communities (1870 species) across Mexico, including modeled current species richness (white <155 species, shades of gray 155–610 species, black 611–763 species), local extirpations (white <29 species, gray 29–112 species, black 113–140 species), colonizations (white <25 species, shades of gray 25–95 species, black 96–119 species), and species turnover (white <10%, shades of gray 10–40%, black >40). The southern quarter of these maps, however, may be subject to some bias, and thus should be interpreted with precaution. *Source:* Reprinted from Peterson et al. (2002b).

Midgley et al. 2003), with some migrating to higher latitudes, others simply reducing habitable distributional areas, and so on.

Moreover, contrary to the customary expectations of climate change effects being most severe in montane areas owing to area effects (McDonald and Brown 1992), montane areas appear to offer better buffering against large-scale horizontal translations, to which flatland areas are much more vulnerable (Midgley et al. 2002; Peterson 2003b). Extinction threat estimated from the results of an array of such studies shows an overall trend of habitat loss, resulting in double-digit extinction risk estimates even for mid-range, 2050 climate change projections (Thomas et al. 2004).

MODELING BIOME DISTRIBUTIONAL SHIFTS OWING TO CLIMATE CHANGE

Climate change has the potential to affect the structure, function, and geographical distribution of terrestrial ecosystems (Melillo et al. 1996). Evidence from the pollen record reveals that plant distributions have changed in both distribution and abundance across regions and continents over recent millennia (Prentice and Webb III 1998). Current projections of the response of the terrestrial biosphere to global climate change indicate potentially large expansions of tropical forests, and as much as 50 to 90 percent spatial displacement of extratropical biomes (Neilson et al. 1992; Neilson et al. 1998). Reliable assessments of potential impacts of future climate change on biome distribution rely greatly on biogeography models (Prentice and Webb 1989; Melillo et al. 1996; Tian et al. 1998a).

Biogeography models have evolved from statistically based climate-vegetation classification approaches to process-based biome models. The Holdridge Life Zone system is the best-known climate-vegetation classification. Here, broad-scale vegetation distributions (or life zones) are determined primarily by three bioclimate variables: biotemperature, mean annual precipitation, and the ratio of potential evapotranspiration to mean annual precipitation. The Holdridge Life Zone system includes 38 life zones, defined by biotemperature and plant moisture demand. Biotemperature is defined as the mean value of daily mean temperature above 0°C, divided by 365. Biotemperature, which is closely related to growing degree days, gives a measure of heat available during the growing season and is likely to be more directly related to plant growth than is mean annual temperature. Plant demand for moisture is expressed through the ratio of mean annual precipitation to the potential evapotranspiration (PET) ratio. This approach has been criticized for ignoring the dynamics of individual species and vegetation succession, as well as seasonality and other factors (Tian et al. 1998a; Goudriaan et al. 1999). To avoid the shortcomings of climate-vegetation classification approach, Box (1981) used the idea of plant functional types (PFT) to express the relationship between macroclimate and plant life forms (characteristic species types). Each PFT represents a set of plant species characterized by physiognomic features, morphological traits, and response to climate (Box 1981).

The Holdridge Life Zone system and Box's PFT approach may be used in projections of vegetation shifts with climate change. Studies using this approach find that both generally agree on the direction of mid- to high-latitude vegetation changes under climate change (doubled CO_2 scenarios) but disagree on the sign of the change in the tropics. Tropical forests expand under the Holdridge model but contract under the Box model (King and Leemans 1990). The accuracy with which these approaches predict current vegetation ranges from less than 50 percent to about 77 percent in both cases (Prentice et al. 1992) is based on statistical correlation with climate variables. These empirical models, however, have difficulty in predicting patterns of future redistribution of vegetation, in which climate, CO_2 levels, and other factors may differ from present conditions. Ideally, models for extrapolation in space and time would be based on ecological processes.

Such newer biome models are called process-based biogeography models, as opposed to those based on statistical correlations in empirical data. Three process-based biogeography models—BIOME (Prentice et al. 1992), MAPSS (Neilson et al. 1992), and DOLY (Woodward et al. 1998)—have been used to predict equilibrium responses of potential vegetation to climate change at regional and global scales. Their objective is to identify ecophysiological and ecological constraints on distributions of plant functional types under different climates at equilibrium. In phase 1 of the vegetation/ecosystem modeling and analysis project (VEMAP) (VEMAP Members 1995), for example, two of the biogeography models, MAPSS (Neilson 1995) and BIOME 3 (Haxeltine and Prentice 1996), were also used to provide estimates of global vegetation redistribution under climate change scenarios. Both MAPSS and BIOME3 produced large shifts of cold-limited vegetation boundaries to higher latitudes and elevations, although the water-controlled boundaries could exhibit any direction of change. Both models predicted that tundra would increase by as much as 33 to 66 percent of its present extent under all future climate scenarios. Many of these simulations indicated potentially large expansions of tropical and, in some cases, temperate forests as well.

Simulation of vegetation redistribution in biogeography models is essentially a static/equilibrium view—it reflects modeling of a given CO_2 level at a given time in the future, rather than a more realistic buildup of CO_2 over time. Static/equilibrium models provide useful "snapshots" of what a terrestrial biosphere in equilibrium with a particular climate might look like, but can provide only inferential information about how the biosphere will make transitions from one condition to another (Neilson et al. 1998). These models do not simulate processes of plant growth, competition, mortality, or changes in ecosystem structure that control vegetation dynamics and ecosystem functioning. To address these limitations requires a new modeling approach integrating vegetation dynamics with ecosystem function, known as dynamic global vegetation models (DGVMs; Prentice and Webb 1989). Interest in developing such models has grown considerably (Foley et al. 1996; Friend et al. 1997; Woodward et al. 1998; Bachelet et al. 2001); an overview of DGVMs and their applications is provided in Chapter 15.

Both equilibrium and dynamic vegetation models have a major limitation for modeling biodiversity response to climate change—they represent only PFT, not species. Thus, there is a fundamental barrier to comparison of species and biogeography models. The former is denominated in species but does not reflect competition; the latter is denominated in plant functional types, so its modeling of competition cannot be directly translated into effects on species composition.

TESTS OF BIOGEOGRAPHY MODELS

A critical need for developing reliable biogeography models is model validation. Compared with the effort expended in building biogeography models, relatively little effort has been put into validating or verifying results derived from them. Many efforts have been made to reconstruct terrestrial vegetation based on paleoecological and paleoclimatological data, which could potentially be used for evaluating biogeography models. Overpeck (1995) argued that the most useful assessments of possible future change would be based on

approaches with demonstrated ability to simulate past environmental changes. Paleoecological approaches thus provide the only way to determine how well biogeography models simulate the response of vegetation/biomes to changing climatic conditions. For example, Williams and colleagues (1998) adapted a biomization method, developed for European pollen data, for use with pollen data from eastern North America, and compared its estimated biomes with biome maps (present-day and 6000 years ago from NCAR CCM1) simulated by BIOME1. Only fair agreement was observed, and significant offsets existed in placement of biomes by BIOME1. For the present day, BIOME1 simulated the boundary between temperate deciduous and cool mixed forests too far south, and the steppe-forest boundary too far west. These biases are also evident in the simulations of 6000 years ago, despite the fact that CCM1 simulates warmer-than-present temperatures in the central United States. To the north, however, BIOME1 at 6000 years ago correctly simulates the cool mixed forest–taiga boundary farther northwestward than at present. Although biogeography models cannot be fully validated with data currently available (Oreskes et al. 1994; Rastetter 1996), validation studies force examination of interactions among data, model structure, parameter sets, and predictive uncertainty (Tian et al. 1998a).

RESEARCH NEEDS

Especially key in understanding the results of single-species modeling exercises will be improved understanding of the likelihood of movement versus evolutionary adaptation as potential responses to climate change. A recent synthesis in theory regarding evolutionary adaptation of ecological niches indicates that niches should show stability and conservatism in most situations (Holt and Gaines 1992; Holt

1996; Holt and Gomulkiewicz 1997), a prediction that has seen ample empirical support (Peterson et al. 1999; Peterson 2003a). To the extent that such conservatism holds true broadly across many species and across moderate evolutionary time periods, the three possible responses by species to climate change (movement, adaptation, extinction) would reduce to just two for many of them (movement and extinction), and the situation would become more tractable (but see Thomas, this volume).

This evaluation, however, is based on preliminary evidence. Empirical information is largely limited to observations of broad-scale conservatism of niche characteristics. Only two studies (Etterson and Shaw 2001; Nadkarni and Solano 2002) have measured evolutionary potential of ecological niche characteristics in the face of climate change—one finding that antagonistic genetic correlations would likely retard evolutionary change to such a point that population persistence without movement would be unlikely (Etterson and Shaw 2001); the other, testing experimentally deleterious effects on epiphytes when transplanted to trees at lower elevations that are naturally exposed to less cloud water (Nadkarni and Solano 2002). Extension of the results of these studies to broader geographic areas and development of additional examples would constitute critical "next steps" toward integration of evolutionary and ecological models.

Single-species models can be used in aggregate to predict community shifts due to climatic change only by assuming that species behave roughly independently of one another (Feria and Peterson 2002). Biome models predict distributions of whole assemblages of plant functional types, ignoring the details of species composition. Both approaches therefore omit consideration of species' interactions and otherwise resort to very different modeling approaches. Therefore, a very interesting avenue of research would be to classify plant species according to functional types, and then to compare predictions of community changes with the single-species model predictions for functional type distributions.

Numerous additional research frontiers remain. For example, models of species' invasions have yet to be integrated with climate change effects to provide a view of the changing opportunities for invasions. Additional retrospective tests of the predictivity of species' distributions in the face of change (e.g., plant distributions in the Pleistocene tested with palynological data) will provide more and better evaluations of levels of predictivity than may be expected. Quantitative evolutionary models need to be explored in greater detail, providing a much-improved theoretical framework regarding evolutionary change in ecological niches, which can then be interpreted in terms of geographic possibilities that open or not for species as a result of evolutionary change.

Finally, modeling of species in the tropics, where most biodiversity lies, has received much less attention than modeling of temperate species, a discrepancy that needs to be addressed, particularly for biodiversity hotspots.

SUMMARY

The field of modeling climate change effects on biodiversity has followed a curious path: two distinct and independent lines of models have each led to series of predictions regarding the phenomenon. The two approaches, however, do not in reality compete: rather, they take independent routes to evaluate two separate manifestations of biodiversity (species composition versus ecosystem distributions). These two approaches remain separate, distinct, and not evaluated in tandem for any system to permit comparisons.

Some generalizations do begin to

emerge. From the single-species models, the best generalization is that few generalizations exist—species show idiosyncratic responses to climate change, and there are many variants and exceptions to the expected "rules" of moving northward and upward in elevation. Although the broadest trend would indeed mirror those rules, individual species often deviate from them quite significantly. Generalization may be possible at regional levels, though, with mountain systems offering more buffering against climate change effects than flatlands systems (Peterson 2003b).

Biome models in general indicate poleward shifting of vegetation types. While this is not directly translatable into losses of biodiversity, it is reasonable to assume that longer distance displacements and more extreme transitions (e.g., forest to grassland) in biome will have the most negative effects on biodiversity (see Malcolm et al., this volume). In regions where changes are most dramatic, biome models may be useful tools in planning for biodiversity conservation (see Scott, this volume).

Results of single-species modeling studies are more directly applicable to issues in biodiversity conservation, and indicate that many species may lose large amounts or all of their range within decades due to climate change. Other species may expand their ranges. Evidence for projected species range losses has been most pronounced in higher-latitude studies, primarily in the southern hemisphere (South Africa, Australia) and for restricted-range species. More subtle effects have been observed at middle latitudes (e.g., Mexico) and in more broadly distributed species. Studies within the tropics and biodiversity hotspots have been relatively few, leaving understanding the weakest precisely where the most biodiversity is at risk.

ACKNOWLEDGMENTS

We thank numerous individuals for their collaboration and assistance in the development of these ideas, in particular Dave Vieglais, Ricardo Schachetti-Pereira, and David Stockwell for critical technological innovations, and Robert Holt and Ed Wiley for helpful discussion of biological aspects. This work has been funded by the U.S. National Science Foundation and the U.S. Environmental Protection Agency.

REFERENCES

Aber, J., R. P. Neilson, S. McNulty, J. M. Lenihan, D. Bachelet, and R. J. Drapek. 2001. Forest processes and global environmental change: Predicting the effects of individual and multiple stressors. *BioScience* 51:735–751.

Anderson, R. P., M. Laverde, and A. T. Peterson. 2002a. Geographical distributions of spiny pocket mice in South America: Insights from predictive models. *Global Ecology and Biogeography* 11:131–141.

Anderson, R. P., M. Laverde, and A. T. Peterson. 2002b. Using niche-based GIS modeling to test geographic predictions of competitive exclusion and competitive release in South American pocket mice. *Oikos* 93:3–16.

Anderson, R. P., D. Lew, and A. T. Peterson. 2003. Evaluating predictive models of species' distributions: Criteria for selecting optimal models. *Ecological Modelling* 162:211–232.

Araujo, M. B., W. Thuiller, and S. Lavorel. In press. Assessing vulnerability of biodiversity to climate change in Europe. *Global Change Biology*.

Aspinall, R. 1992. An inductive modeling procedure based on Bayes' theorem for analysis of pattern in spatial data. *International Journal of Geographic Information Systems* 6:105–121.

Austin, M. P., A. O. Nicholls, and C. R. Margules. 1990. Measurement of the realized qualitative niche: Environmental niches of five *Eucalyptus* species. *Ecological Monographs* 60:161–177.

Bachelet, D., R. P. Neilson, J. M. Lenihan, and R. J. Drapek. 2001. Climate change effects on vegetation distribution and carbon budget in the U.S. *Ecosystems* 4:164–185.

Bakkenes, M., J. R. Alkemade, F. Ihle, R. Leemans, and J. B. Latour. 2002. Assessing effects of forecasted climate change on the diversity and distribution of European higher plants for 2050. *Global Change Biology* 8:390–407.

Beaumont, L., and L. Hughes. 2002. Potential

changes in the distributions of latitudinally restricted Australian butterflies in response to climate change. *Global Change Biology* 8: 954–971.

Berry, P. M., T. P. Dawson, P. A. Harrison, and R. G. Pearson. 2002. Modeling potential impacts of climate change on the bioclimatic envelope of species in Britain and Ireland. *Global Ecology and Biogeography* 11:453–462.

Box, E. O. 1981. *Macroclimate and Plant Forms: An Introduction to Predictive Modeling in Phytogeography.* The Hague, Netherlands: Junk.

Brereton R., S. Bennett, and I. Mansergh. 1995. Enhanced greenhouse climate change and its potential effect on selected fauna of south-eastern Australia: A trend analysis. *Biol. Conserv.* 72:339–354.

Brown, J. H., T. J. Valone, and C. G. Curtin. 1997. Reorganization of an arid ecosystem in response to recent climate change. *Proceedings of the National Academy of Sciences USA* 94:9729–9733.

Carey, P. D., and N. J. Brown. 1994. The use of GIS to identify sites that will become suitable for a rare orchid, *Himantoglossum hircinum* L., in a future changed climate. *Biodiversity Letters* 2:117–123.

Carlton, J. T. 1996. Pattern, process, and prediction in marine invasion ecology. *Biological Conservation* 78:97–106.

Carpenter, G., A. N. Gillison, and J. Winter. 1993. DOMAIN: A flexible modeling procedure for mapping potential distributions of animals and plants. *Biodiversity and Conservation* 2:667–680.

Carson, D. J. 1999. Climate modeling: Achievements and prospects. *Quarterly Journal of the Royal Meteorological Society* 125:1–27.

Chapman, A. D., and D. J. Milne. 1998. *The Impact of Global Warming on the Distribution of Selected Australian Plant and Animal Species in Relation to Soils and Vegetation.* Canberra: ERIN Unit, Environment Australia.

Clark, J. S., S. R. Carpenter, M. Barber, S. Collins, A. Dobson, J. A. Foley, D. M. Lodge, M. Pascual, R. Pielke Jr., W. Pizer, C. Pringle, W. V. Reid, K. A. Rose, O. E. Sala, W. H. Schlesinger, D. H. Wall, and D. Wear. 2001. Ecological forecasts: An emerging imperative. *Science* 293:657–660.

Cramer, W., A. Bondeau, F. I. Woodward, I. C. Prentice, R. A. Betts, V. Brovkin, P. M. Cox, V. Fisher, J. A. Foley, A. D. Friend, C. Kucharik, M. R. Lomas, N. Ramankutty, S. Sitch, B. Smith, A. White, and C. Young-Molling. 2001. Global response of terrestrial ecosystem structure and function to CO_2 and climate change: Results from six dynamic global vegetation models. *Global Change Biology* 7:357–373.

Crumpacker, D. W., E. O. Box, and E. D. Hardin. 2001. Implications of climate warming for conservation of native trees and shrubs in Florida. *Conservation Biology* 15:1008–1020.

Dale, V. H., L. A. Joyce, S. McNulty, R. P. Neilson, M. P. Ayres, M. D. Flannigan, P. J. Hanson, L. C. Irland, A. E. Lugo, C. J. Peterson, D. Simberloff, F. J. Swanson, B. J. Stocks, and B. M. Wotton. 2001. Climate change and forest disturbances. *BioScience* 51:723–734.

Davis, A. J., L. S. Jenkinson, J. H. Lawton, B. Shorrocks, and S. Wood. 1998. Making mistakes when predicting shifts in species range in response to global warming. *Nature* 391:783–786.

DeStasio, B. T., Jr. 1996. Potential effects of global climate change on small north-temperate lakes: Physics, fish, and plankton. *Limnology and Oceanography* 41:1136–1149.

Elton, C. S. 1927. *Animal Ecology.* London: Sidgwich and Jackson.

Erasmus, B. F., M. Kshatriya, M. W. Mansell, S. L. Chown, and A. S. van Jaarsveld. 2000. A modeling approach to antlion (Neuroptera: Myrmeleontidae) distribution patterns. *African Entomology* 8:157–168.

Erasmus, B. F., A. S. van Jaarsveld, S. L. Chown, M. Kshatriya, and K. J. Wessels. 2002. Vulnerability of South African animal taxa to climate change. *Global Change Biology* 8:679–693.

Etterson, J. R., and R. G. Shaw. 2001. Constraint to adaptive evolution in response to global warming. *Science* 294:151–153.

Feria, T. P., and A. T. Peterson. 2002. Using point occurrence data and inferential algorithms to predict local communities of birds. *Diversity and Distributions* 8:49–56.

Foley, J. A., I. C. Prentice, N. Ramankutty, S. Levis, D. Pollard, S. Sitch, and A. Haxeltine. 1996. An integrated biosphere model of land surface processes, terrestrial carbon balance and vegetation dynamics. *Global Biogeochemical Cycles* 10:603–628.

Foster, P. 2001. The potential negative impacts of global climate change on tropical montane cloud forests. *Earth-Science Reviews* 55:73–106.

Friend, A. D., A. K. Stevens, R. G. Knox, and M. G. R. Cannell. 1997. A process-based, terrestrial biosphere model of ecosystem dynamics (Hybrid v3.0). *Ecological Modelling* 95:249–287.

Gottfried, M., H. Pauli, K. Reiter, and G. Grabherr. 1999. A fine-scaled predictive model for changes in species distribution patterns of high mountain plants induced by climate warming. *Diversity and Distributions* 5:241–251.

Goudriaan, J., H. H. Shugart, H. Bugmann, W. Cramer, A. Bondeau, R. H. Gardner, L. A. Hunt, W. K. Lauenroth, J. J. Landsberg, S. Lindner, I. R. Noble, W. J. Parton, L. F. Pitelka, M. Stafford Smith, R. W. Sutherst, C. Valentin, and F. I. Woodward. 1999. Use of models in global change studies, pp. 106–140. In B. Walker, W. Steffen, J. Canadell, and J. Ingram, eds., *The Terrestrial Biosphere and Global Change, Implications for Natural and Managed Ecosystems.* Cambridge: Cambridge University Press.

Grinnell, J. 1917. Field tests of theories concerning distributional control. *American Naturalist* 51:115–128.

Grinnell, J. 1924. Geography and evolution. *Ecology* 5:225–229.

Guisan, A., and N. E. Zimmermann. 2000. Predictive habitat distribution models in ecology. *Ecological Modelling* 135:147–186.

Haxeltine, A., and I. C. Prentice. 1996. BIOME3: An equilibrium terrestrial biosphere model based on ecophysiological constraints, resource availability and competition among plant functional types. *Global Biogeochemical Cycles* 10:693–709.

Hay, S. I., J. Cox, D. J. Rogers, S. E. Randolph, D. I. Stern, G. D. Shanks, M. F. Myers, and R. W. Snow. 2002. Climate change and the resurgence of malaria in the East Africa highlands. *Nature* 415:905–909.

Higgins, S. I., D. M. Richardson, R. M. Cowling, and T. H. Trinder-Smith. 1999. Predicting the landscape-scale distribution of alien plants and their threat to plant diversity. *Conservation Biology* 13:303–313.

Hill, J. K., C. D. Thomas, and B. Huntley. 1999. Climate and habitat availability determine 20th century changes in a butterfly's range margin. *Proceedings of the Royal Society of London B* 266:1197–1206.

Hoffman, M. H. 2001. The distribution of *Senecio vulgaris*: Capacity of climatic range models for predicting adventitious ranges. *Flora* 196/5:395–403.

Holmgren, M., M. Scheffer, E. Ezcurra, J. R. Gutiérrez, and G. M. J. Mohren. 2001. El Niño effects on the dynamics of terrestrial ecosystems. *Trends in Ecology and Evolution* 16:89–94.

Holt, R. D. 1990. The microevolutionary consequences of climate change. *Trends in Ecology and Evolution* 5.

Holt, R. D. 1996. Demographic constraints in evolution: Towards unifying the evolutionary theories of senescence and niche conservatism. *Evolutionary Ecology* 10:1–11.

Holt, R. D., and M. S. Gaines. 1992. Analysis of adaptation in heterogeneous landscapes: Implications for the evolution of fundamental niches. *Evolutionary Ecology* 6:433–447.

Holt, R. D., and R. Gomulkiewicz. 1997. The evolution of species' niches: A population dynamic perspective, pp. 23–46. In H. G. Othmer, F. R. Adler, M. A. Lewis, and J. D. Dalton, eds., *Case Studies in Mathematical Modeling—Ecology, Physiology, and Cell Biology.*

Hubbell, S. T. 2001. *The Unified Neutral Theory of Biodiversity and Biogeography.* Princeton, N.J.: Princeton University Press.

Hughes, L. 2003. Climate change and Australia: Trends, projections and impacts. *Austral. Ecology* 28(4):423–443.

Hughes, L., E. M. Cawsey, and M. Westoby. 1996. Climatic range sizes of *Eucalyptus* species in relation to future climate change. *Global Ecology and Biogeography Letters* 5:23–29.

Hulme, M., E. M. Barrown, N. W. Arnell, P. A. Harrison, T. C. Johns, and T. E. Downing. 1999a. Relative impacts of human-induced climate change and natural climate variability. *Nature* 397:688–691.

Hulme, M., J. Mitchell, W. Ingram, J. Lowe, T. Johns, M. New, and D. Viner. 1999b. Climate change scenarios for global impacts studies. *Global Environmental Change* 9:S3–S19.

Huntley, B., P. J. Bartlein, and I. C. Prentice. 1989. Climatic control of the distribution and abundance of Beech (*Fagus* L.) in Europe and North America. *Journal of Biogeography* 16:551–560.

Huntley, B., P. M. Berry, W. Cramer, and A. P. McDonald. 1995. Modeling present and potential future ranges of some European higher plants using climate response surfaces. *Journal of Biogeography* 22:967–1001.

Huston, M. A., L. W. Aarssen, M. P. Austin, B. S. Cade, J. D. Fridley, E. Garnier, J. P. Grime, J. Hodgson, W. K. Lauenroth, K. Thompson, J. H. Vandermeer, D. A. Wardle, A. Hector, B. Schmid, C. Beierkuhnlein, M. C. Caldeira, M. Diemer, P. G. Dimitrakopoulos, J. A. Finn, H. Freitas, P. S. Giller, J. Good, R. Harris, P. Högberg, K. Huss-Danell, J. Joshi, A. Jumpponen, C. Körner, P. W. Leadley, M. Loreau, A. Minns, C. P. H. Mulder, G. O'Donovan, S. J. Otway, J. S. Pereira, A. Prinz, D. J. Read, M. Scherer-Lorenzen, E.-D. Schulze, A.-S. D. Siamantzioudas, E. Spehn, A. C. Terry, A. Y. Troumbis, F. I. Woodward, S. Yachi, and J. H. Lawton. 2000. No consistent effect of plant diversity on productivity. *Science* 289:1255.

Hutchinson, G. E. 1957. Concluding remarks. *Gold Spring Harbor Symposia on Quantitative Biology* 22:415–427.

Iguchi, K., K. Matsuura, K. McNyset, A. T. Peterson, R. Scachetti-Pereira, K. A. Powers, D. A. Vieglais, E. O. Wiley, and T. Yodo. 2005. Predicting invasions of North American basses in Japan using native range data and a genetic algorithm. *Transactions of the American Fisheries Society.*

Inouye, D. W., B. Barr, K. B. Armitage, and B. D. Inouye. 2000. Climate change is affecting altitudinal migrants and hibernating species. *Proceedings of the National Academy of Sciences USA* 97:1630–1633.

Iverson, L. R., A. Prasad, and M. W. Schwartz. 1999. Modeling potential future individual tree-species distributions in the eastern United States under a climate change scenario: A case study with *Pinus virginuana. Ecological Modelling* 115:77–93.

Johnston, K. M., and O. J. Schmitz. 1997. Wildlife and climate change: Assessing the sensitivity of selected species to simulated doubling of atmospheric CO_2. *Global Change Biology* 3:531–544.

Jones, P. G., and A. Gladkov. 1999. FloraMap: A computer tool for predicting the distribution of plants and other organisms in the wild. Cali, Colombia: Centro Internacional de Agricultura Tropical.

Kadmon, R., and J. Heller. 1998. Modeling faunal responses to climatic gradients with GIS: Land
snails as a case study. *Journal of Biogeography* 25:527–
539.

Kaiser, J. 2000. Rift over biodiversity divides ecologists. *Science* 289:1282–1283.

Karl, T. R., R. W. Knight, D. R. Easterling, and R. G.
Quayle. 1996. Indices of climate change for the
United States. *Bulletin of the American Meteorological So-
ciety* 77:279–292.

King, G. A., and R. Leemans. 1990. Effects of global
climate change on global vegetation. In G. A.
King, J. K. Winjun, P. K. Dixon, and L. Y. Arnaut,
eds., *Response and Feedback of Forest Systems to Global Climate Change.* Corvallis, Ore.: Environmental Protection Agency.

Lavorel, S. 1999. Global change effects on landscape
and regional patterns of plant diversity. *Diversity
and Distributions* 5:239–240.

Lawton, J. H., and K. C. Brown. 1986. The population and community ecology of invading insects.
Philosophical Transactions of the Royal Society of London B
314:607–617.

Lischke, H., A. Guisan, A. Fischlin, J. Williams, and
H. Bugmann. 1998. Vegetation responses to cli-
mate change in the Alps: Modeling studies. In
P. Cebron, U. Dahinden, H. C. Davies, H. Imboden,
and C. C. Jager, eds., *Views from the Alps.* Cambridge,
Mass.: MIT Press.

Logan, J. A., J., Regniere, and J. A. Powell. 2003. Assessing the impacts of global warming on forest
pest dynamics. *Frontiers in Ecology and the Environment*
1:130–137.

Loreau, M., S. Naeem, P. Inchausti, J. Bengtsson, J. P.
Grime, A. Hector, D. U. Hooper, M. A. Huston,
D. Raffaelli, B. Schmid, D. Tilman, and D. A. War-
dle. 2001. Biodiversity and ecosystem function-
ing: Current knowledge and future challenges.
Science 294:804–808.

MacArthur, R. 1972. *Geographical Ecology.* Princeton,
N.J.: Princeton University Press.

Magnuson, J. 2001. 150-year global ice record reveals major warming trend. *Inter-American Institute
for Global Change Research* 24:22–25.

Martin, T. E. 2001. Abiotic vs. biotic influences on
habitat selection of coexisting species: Climate
change impacts? *Ecology* 82:175–188.

Martínez-Meyer, E., A. T. Peterson, and W. W. Hargrove. 2004. Ecological niches as stable distributional constraints on mammal species, with implications for Pleistocene extinctions. *Global Ecology
and Biogeography.*

McDonald, K. A., and J. H. Brown. 1992. Using
montane mammals to model extinctions due to
global change. *Conservation Biology* 6:409–415.

Melillo, J. M., I. C. Prentice, G. D. Farquhar, E.-D.
Schulze, and O. E. Sala. 1996. Terrestrial biotic responses to environmental change and feedbacks
to climate, pp. 444–481. In J. T. Houghton, L. G.

Meira Filho, B. A. Callander, N. Harris, A. Katten-
berg, and K. Maskell, eds., *Climate Change 1995: The
Science of Climate Change.* Cambridge: Cambridge
University Press.

Midgley, G. F., L. Hannah, D. Millar, M. C. Rutherford, and L. W. Powrie. 2002. Assessing the vulnerability of species richness to anthropogenic
climate change in a biodiversity hotspot. *Global
Ecology and Biogeography* 11:445–452.

Midgley, G. F., L. Hannah, D. Millar, W. Thuiller, and
A. Booth. 2003. Developing regional and species-
level assessments of climatic change impacts on
biodiversity in the Cape Floristic. *Biological Conservation* 112:87–97.

Moisen, G. G., and T. C. Edwards, Jr. 1999. Use of
generalized linear models and digital data in a
forest inventory of northern Utah. *Journal of Agricultural, Biological, and Environmental Statistics* 4.372–390.

Mortsch, L. D., and F. H. Quinn. 1996. Climate
change scenarios for Great Lakes Basin ecosystem
studies. *Limnology and Oceanography* 41:903–911.

Nadkarni, N. M., and R. Solano. 2002. Potential
effects of climate change on canopy communities
in a tropical cloud forest: An experimental approach. *Oecologia* 131:580–586.

Neilson, R. P. 1995. A model for predicting continental-scale vegetation distribution and water
balance. *Ecological Applications* 5:362–385.

Neilson, R. P., G. A. King, and G. Koerper. 1992. Toward a rule-based biome model. *Landscape Ecology*
7:27–43.

Neilson, R. P., I. C. Prentice, and B. Smith. 1998. Simulated changes in vegetation distribution under
global warming, pp. 439–456. In R. T. Watson,
M. C. Zinyowera, R. H. Moss, and D. J. Dokken,
eds., *The Regional Impacts of Climate Change: An Assessment
of Vulnerability.* Cambridge: Cambridge University
Press.

Newton, P.C.D., H. Clark, G. R. Edwards, and D. J.
Ross. 2000. Experimental confirmation of
ecosystem model predictions comparing transient and equilibrium plant responses to elevated
atmospheric CO_2. *Ecology Letters* 4:344–347.

Nix, H. A. 1986. A biogeographic analysis of Australian elapid snakes, pp. 4–15. In R. Longmore,
ed., *Atlas of Elapid Snakes of Australia.* Canberra: Australian Government Publishing Service.

Nix, H. A., and M. A. Switzer, eds. 1991. *Rainforest An-
imals: Atlas of Vertebrates Endemic to Australia's Wet Tropics.*
Canberra: Australian National Parks and Wildlife
Service.

Oreskes, N., K. Shrader-Frechette, and K. Belitz.
1994. Verification, validation and confirmation of
numerical models in the earth sciences. *Science*
263:641–646.

Overpeck, J. T. 1995. Paleoclimatology and climate
system dynamics. *Review of Geophysiology Supplement*
863–871.

Papes, M., and A. T. Peterson. 2003. Predicting the

potential invasive distribution for *Eupatorium adenophorum* Spreng. in China. *Journal of Wuhan Botanical Research* 21:137–142.

Parmesan, C. 1996. Climate and species' range. *Nature* 382:765–766.

Parmesan, C., N. Ryrholm, C. Stefanescu, J. K. Hill, C. D. Thomas, H. Descimon, B. Huntley, L. Kaila, J. Kullberg, T. Tammaru, J. Tennent, J. A. Thomas, and M. Warren. 1999. Poleward shift of butterfly species' ranges associated with regional warming. *Nature* 399:579–583.

Parson, E. A., R. W. Corell, E. J. Barron, V. Burkett, A. Janetos, L. Joyce, T. R. Karl, M. C. MacCracken, J. M. Melillo, M. G. Morgan, D. Schimel, and T. Wilbanks. 2003. Understanding climatic impacts, vulnerabilities, and adaptation in the United States: Building a capacity for assessment. *Climatic Change* 57:9–42.

Payette, S., M.-J. Fortin, and I. Gamache. 2001. The subarctic forest-tundra: The structure of a biome in a changing climate. *BioScience* 51:709–718.

Peterson, A. T. 2001. Predicting species' geographic distributions based on ecological niche modeling. *Condor* 103:599–605.

Peterson, A. T. 2003a. Predicting the geography of species' invasions via ecological niche modeling. *Quarterly Review of Biology* 78:419–433.

Peterson, A. T. 2003b. Projected climate change effects on Rocky Mountain and Great Plains birds: Generalities of biodiversity consequences. *Global Change Biology* 9:647–655.

Peterson, A. T., L. G. Ball, and K. C. Cohoon. 2002a. Predicting distributions of tropical birds. *Ibis* 144:e27–e32.

Peterson, A. T., and K. C. Cohoon. 1999. Sensitivity of distributional prediction algorithms to geographic data completeness. *Ecological Modelling* 117:159–164.

Peterson, A. T., and R. D. Holt. 2003. Niche differentiation in Mexican birds: Using point occurrences to detect ecological innovation. *Ecology Letters* 6:774–782.

Peterson, A. T., M. A. Ortega-Huerta, J. Bartley, V. Sanchez-Cordero, J. Soberon, R. H. Buddemeier, and D.R.B. Stockwell. 2002b. Future projections for Mexican faunas under global climate change scenarios. *Nature* 416:626–629.

Peterson, A. T., M. Papes, and D. A. Kluza. 2003. Predicting the potential invasive distributions of four alien plant species in North America. *Weed Science* 51:863–868.

Peterson, A. T., and C. R. Robins. 2003. Using ecological-niche modeling to predict Barred Owl invasions with implications for Spotted Owl conservation. *Conservation Biology* 17:1161–1165.

Peterson, A. T., V. Sanchez-Cordero, C. B. Beard, and J. M. Ramsey. 2002c. Ecologic niche modeling and potential reservoirs for Chagas disease, Mexico. *Emerging Infectious Diseases* 8:662–667.

Peterson, A. T., V. Sanchez-Cordero, J. Soberon, J. Bartley, R. H. Buddemeier, and A. G. Navarro-Siguenza. 2001. Effects of global climate change on geographic distributions of Mexican Cracidae. *Ecological Modelling* 144:21–30.

Peterson, A. T., R. Scachetti-Pereira, and W. W. Hargrove. 2004. Potential geographic distribution of *Anoplophora glabripennis* (Coleoptera: Cerambycidae) in North America. *American Midland Naturalist* 151:170–178.

Peterson, A. T., R. Scachetti-Pereira, and D. A. Kluza. 2003b. Assessment of invasive potential of *Homalodisca coagulata* in western North America and South America. *Biota Neotropica* 3: Online journal: http://www.biotaneotropica.org.br/v3n1/pt/abstract?article+BN00703012003.

Peterson, A. T., J. Soberon, and V. Sanchez-Cordero. 1999. Conservatism of ecological niches in evolutionary time. *Science* 285:1265–1267.

Peterson, A. T., D.R.B. Stockwell, and D. A. Kluza. 2002d. Distributional prediction based on ecological niche modeling of primary occurrence data, pp. 617–623. In J. M. Scott, P. J. Heglund, and M. L. Morrison, eds., *Predicting Species Occurrences: Issues of Scale and Accuracy*. Washington, D.C.: Island Press.

Peterson, A. T., and D. A. Vieglais. 2001. Predicting species invasions using ecological niche modeling. *BioScience* 51:363–371.

Poiani, K. A., W. C. Johnson, G. A. Swanson, and T. C. Winter. 1996. Climate change and northern prairie wetlands: Simulations of long-term dynamics. *Limnology and Oceanography* 41:871–881.

Porter, W. P., S. Budaraju, and N. Ramankutty. 2000. Calculating climate effects on birds and mammals: Impacts on biodiversity, conservation, population parameters, and global community structure. *American Zoologist* 40:597.

Pouliquen-Young, O., and P. Newman. 2000. *The Implications of Climate Change for Land-Based Nature Conservation Strategies*. Final Report 96/1306. Australian Greenhouse Office, Environment Australia, Canberra, and Institute for Sustainability and Technology Policy, Murdoch University, Perth.

Prentice, I. C., W. Cramer, S. P. Harrison, R. Leemans, R. A. Monserud, and A. M. Solomon. 1992. A global biome model based on plant physiology and dominance, soil properties and climate. *Journal of Biogeography* 19:117–134.

Prentice, I. C., and N. R. Webb. 1989. Developing a global vegetation dynamics model: Results of an IIASA summer workshop. RR-89-7, International Institute for Applied Systems Analysis, Laxenburg, Austria.

Prentice, I. C., and T. Webb III. 1998. BIOME 6000: Reconstructing global mid-Holocene vegetation patterns from paleoecological records. *Journal of Biogeography* 25:997–1005.

Price, J. 2000. Modeling the potential impacts of climate change on the summer distributions of Massachusetts passerines. *Bird Observer* 28:224–230.

Rastetter, E. B. 1996. Validating models of ecosystem response to global change. *BioScience* 46:190–198.

Roura-Pascual, N., A. Suarez, C. Gómez, P. Pons, Y. Touyama, A. L. Wild, and A. T. Peterson. 2005. Geographic potential of Argentine ants (*Linepithema humile* Mayr) in the face of global climate change. *Proceedings of the Royal Society of London B.*

Sagarin, R. D., J. P. Barry, S. E. Gilman, and C. H. Baxter. 1999. Climate-related change in an intertidal community over short and long time scales. *Ecological Monographs* 69:465–490.

Sala, O. E., F.S.I. Chapin, J. J. Armesto, E. Berlow, J. Bloonfield, R. Dirzo, E. Huber-Sanwald, L. F. Huenneke, R. B. Jackson, A. Kinzig, R. Leemans, D. M. Lodge, H. A. Mooney, M. Oesterheld, N. L. Poff, M.T. Sykes, B. H. Walker, M. Walker, and D. H. Wall. 2000. Global biodiversity scenarios for the year 2100. *Science* 287:1770–1773.

Shaver, G. R., J. Canadell, F.S.I. Chapin, J. Gurevitch, J. Harte, G. Henry, P. Ineson, S. Jonasson, J. M. Melillo, L. Pitelka, and L. Rustad. 2000. Global warming and terrestrial ecosystems: A conceptual framework for analysis. *BioScience* 50:871–882.

Still, C. J., P. N. Foster, and S. H. Schneider. 1999. Simulating the effects of climate change on tropical montane cloud forests. *Nature* 398:608–610.

Stockwell, D.R.B. 1999. Genetic algorithms II, pp. 123–144. In A. H. Fielding, ed., *Machine Learning Methods for Ecological Applications.* Boston: Kluwer Academic.

Stockwell, D.R.B., and I. R. Noble. 1992. Induction of sets of rules from animal distribution data: A robust and informative method of analysis. *Mathematics and Computers in Simulation* 33:385–390.

Stockwell, D.R.B., and D. P. Peters. 1999. The GARP modeling system: Problems and solutions to automated spatial prediction. *International Journal of Geographic Information Systems* 13:143–158.

Stockwell, D.R.B., and A.T. Peterson. 2002a. Controlling bias in biodiversity data, pp. 537–546. In J. M. Scott, P. J. Heglund, and M. L. Morrison, eds., *Predicting Species Occurrences: Issues of Scale and Accuracy.* Washington, D.C.: Island Press.

Stockwell, D.R.B., and A. T. Peterson. 2002b. Effects of sample size on accuracy of species distribution models. *Ecological Modelling* 148:1–13.

Sykes, M. T., I. C. Prentice, and W. Cramer. 1996. A bioclimatic model for the potential distributions of north European tree species under present and future climates. *Journal of Biogeography* 23:203–233.

Texeira, J., and J. W. Arntzen. 2002. Potential impact of climate warming on the distribution of the golden-striped salamander, *Chioglossa lusitanica*, on the Iberian Peninsula. *Biodiversity and Conservation* 11:2167–2176.

Thomas, C. D., A. Cameron, R. E. Green, M. Bak-

kenes, L. J. Beaumont, Y. C. Collingham, B.F.N. Erasmus, M. Ferreira de Siqueira, A. Grainger, L. Hannah, L. Hughes, B. Huntley, A. S. Van Jaarsveld, G. E. Midgely, L. Miles, M. A. Ortega-Huerta, A. T. Peterson, O. L. Phillips, and S. E. Williams. 2004. Extinction risk from climate change. *Nature* 427:145–148.

Tian, H. 2002. Dynamics of the terrestrial biosphere in changing global environments: Data, models and validation. *Journal of Geographical Science* 12:86–91.

Tian, H., C. Hall, and Y. Qi. 1998a. Modeling primary productivity of the terrestrial biosphere in changing environments: Toward a dynamic biosphere model. *Critical Reviews in Plant Sciences* 15:541–557.

Tian, H., J. M. Melillo, D. W. Kicklighter, A. D. McGuire, and J. Helfrich. 1999. The sensitivity of terrestrial carbon storage to historical atmospheric CO_2 and climate variability in the United States. *Tellus* 51B:414–452.

Tian, H., J. M. Melillo, D. W. Kicklighter, D. McGuire, J. Helfrich, B. Moore III, and C. J. Vörösmarty. 1998b. Effect of interannual climate variability on carbon storage in Amazonian ecosystems. *Nature* 396:664–667.

Tilman, D., D. Wedin, and J. Knops. 1996. Productivity and sustainability influenced by biodiversity in grassland ecosystems. *Nature* 379:718–720.

VEMAP. 1995. Vegetation/ecosystem modeling and analysis project (VEMAP): Comparing biogeography and biogeochemistry models in a continental-scale study of terrestrial ecosystems to climate change and CO_2 doubling. *Global Biogeochemical Cycles* 9:407–438.

Visser, M. E., A. J. van Noordwijk, J. M. Tinbergen, and C. M. Lessells. 1998. Warmer springs lead to mistimed reproduction in great tits (*Parus major*). *Proceedings of the Royal Society B* 265:1867–1870.

Walther, G.-R., E. Post, P. Convey, A. Menzel, C. Parmesan, T.J.C. Beebee, J.-M. Fromentin, O. Hoegh-Guldberg, and F. Fairlein. 2002. Ecological responses to recent climate change. *Nature* 416:389–395.

Welk, E., K. Schubert, and M. H. Hoffmann. 2002. Present and potential distribution of invasive garlic mustard (*Alliaria petiolata*) in North America. *Diversity and Distributions* 8:219–233.

White, A., M.G.R. Cannell, and A. D. Friend. 1999. Climate change impacts on ecosystems and the terrestrial carbon sink: A new assessment. *Global Environmental Change* 9:S21–S30.

Williams, J.W., R. L. Summers, and T. Webb III. 1998. Applying plant functional types to construct biome maps from eastern North American pollen data: Comparisons with model results. *Quaternary Science Reviews* 17:607–628.

Woodward, F. I., M. R. Lomas, and R. A. Betts. 1998. Vegetation-climate feedback in a greenhouse world. *Philosphical Transactions of the Royal Society of London B* 353:29–39.

Yee, T. W., and N. D. Mitchell. 1991. Generalized additive models in plant ecology. *J. Veg. Sci.* 2:587–602.

Xu, D., and H. Yan. 2001. A study of the impacts of climate change on the geographic distribution of *Pinus koraiensis* in China. *Environment International* 27:201–205.

Modeling Species Range Shifts in Two Biodiversity Hotspots

Guy F. Midgley and Dinah Millar

Two adjacent biodiversity hotspots in southern Africa, the Cape Floristic Region and Succulent Karoo, are projected to undergo significant changes in both temperature and rainfall over the next 50 years (Rutherford et al. 1999). Both hotspots have a temperate Mediterranean-type climate and have extraordinary levels of plant species richness, with more than 8000 species in the Fynbos biome typical of the Cape, and more than 5000 in the Succulent Karoo. These biomes are characterized by distinct plant growth forms (proteas, ericas, and restioids dominate the fire-prone Fynbos, while leaf-succulent forms dominate the Succulent Karoo). Will future anthropogenic climate change alter the geographic distributions of these biomes and their constituent species ranges?

Previous modeling results had suggested significant reductions in surface area suitable for persistence of both biomes (Rutherford et al. 1999). The Fynbos biome was projected to lose roughly 50 percent of its surface area by 2030 (for a rise in atmospheric CO_2 to 475 ppm), and 72 percent for doubled CO_2 (projected for 2050). The Succulent Karoo is even more negatively affected by climate change, with a loss of 89 percent of its current range by 2030, and 99 percent by 2050. By 2050 the biome persists mainly in the southern Cape, in areas already transformed by human activity.

To test the possible impacts of these biome shifts on biodiversity, range shifts of 343 species of protea and 20 leaf succulent species (Mesembryanthemaceae) endemic to the Succulent Karoo were simulated using a generalized additive modeling approach (Midgley et al. 2002). Species distribution data were obtained from the Protea Atlas Database and from field records. A climatic database at a spatial resolution of 1 × 1 minute (approximately 1.5 × 1.8 km at this latitude) provided temperature and soil water availability data (Rutherford et al. 1999). Anthropogenic climate change was simulated by the general circulation model HadCM2, which projects temperature increases of between 1.3 and 2.5°C for the southern Cape, and potential reductions of up to 25 percent in winter rainfall by 2050.

Total range loss was projected for nearly a quarter of the protea species by 2050. Another 10 to 20 percent of the remaining protea species had no overlap between present and future (2050) range. Most range shifts in the proteas were to the south and upslope, with some upslope shifts trending northward near the southern edge of the continent. The range size of some species expanded, a phenomenon noted in other multi-species modeling studies. Figure 14.7 shows illustrative potential range changes for two protea species, one losing roughly 40 percent of its range, and one which almost doubles its range. Both show a southward (poleward) range shift and increasing range fragmentation.

Dispersal of many Fynbos species is limited to periods immediately post-fire (Cowling 1992). Fire frequencies in this vegetation type are on the order of 10 to 15 years, implying limited temporal windows for dispersal. Management of Fynbos for climate change will depend on a well-developed understanding of post-fire dispersal.

Results for Succulent Karoo species are derived from 20 endemic species with large distributions, due to limited data for restricted range species. The median range loss was 60 percent by 2050. Three species

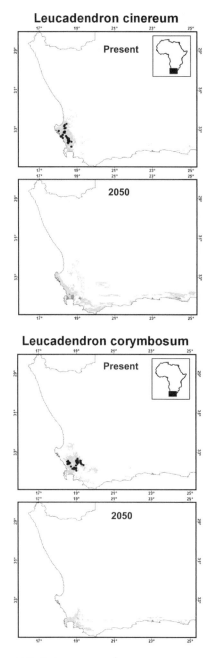

Figure 14.7. Proteaceae examples.

themaceae are endemic to individual quartz outcrops). The larger range sizes of endemic Mesembryanthemaceae modeled appear to reduce the predicted range losses relative to Proteaceae, but suggest significant range size reductions nonetheless. The relationship between small range size and high displacement noted in the proteas raises the possibility that the many highly range-restricted Mesembryanthemaceae may be at high risk.

Illustrative spatial shifts for two Succulent Karoo species (Fig. 14.8) show one that retains most of its range, and another which experiences significant range loss. Southward migration and range fragmentation is evident in both species, echoing the Proteaceae results. Spatial shifts predicted for *Ruschia caroli* are likely representative of regional endemics with smaller range sizes, with significant range contraction in the northern region of Namaqualand, and potential colonization of novel range in the southern Cape.

Range shifts in the narrow endemics of the Succulent Karoo therefore appear critical for their survival of climate change, but are not linked to discreet disturbance events as in the Fynbos. Mesembryanthemaceae dispersal is thought to be highly limited because seeds are shed from persistent capsules during rainfall events and are passively transported. However, it has been suggested that these small seeds could be transported by wind to achieve significant long-distance dispersal (Midgley et al. 2002)—lack of understanding of dispersal is therefore also a strong constraint to the estimation of extinction risk for these species in the Succulent Karoo.

Projections of habitat loss derived from biome-level modeling may over-estimate climate change impacts but are supported in location and direction of change by species-level modeling. Current range size seems an important predictor of range loss under climate change, with larger-range endemics suffering lower range losses.

showed range increases, two species lost more than 80 percent of their range, but no extinctions were predicted. These results are less robust than those of the proteas, due to limited sample size and difficulty in finding species with ranges large enough to model (many Mesembryan-

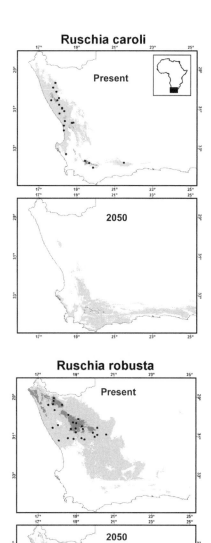

Figure 14.8. Mesembryanthemaceae examples. Modeled range in gray. Dark gray indicates modeled range of upper elevations.

Model results showing range shifts upslope and poleward (in this southern hemisphere case, southward) are consistent with theoretical predictions and field observations of range shifts due to climate change. Understanding and managing dispersal is of major importance to the long-term conservation of these two biodiversity hotspots as climate changes.

REFERENCES

Cowling, R. M. (1992) *The Ecology of Fynbos: Nutrients, Fire and Diversity.* Oxford University Press, Cape Town.

Midgley, G. F., L. Hannah, D. Millar, M. C. Rutherford, and L. W. Powrie. (2002) Assessing the vulnerability of species richness to anthropogenic climate change in a biodiversity hotspot. *Global Ecology & Biogeography*, 11, 445–451.

Rutherford, M. C., G. F. Midgley, W. J. Bond, L. W. Powrie, R. Roberts, and J. Allsop. (1999) *South African Country Study on Climate Change—Plant Biodiversity:Vulnerability and Adaptation Assessment.* Department of Environmental Affairs and Tourism, Pretoria, South Africa.

CHAPTER FIFTEEN

Dynamic Ecosystem and Earth System Models

RICHARD A. BETTS
AND HERMAN H. SHUGART

Species range shifts and biome redistribution will not occur instantaneously. The responses of ecosystems to climatic changes are constrained by the growth and development rates of individuals, and mortality rates in populations as a whole. Furthermore, feedbacks from ecosystems onto environmental conditions can either dampen or magnify rates of change, making simple cause-effect assumptions inapplicable to the problem of prediction. Assessment of potential rates of change in ecosystems requires models which include time-dependent ecosystem processes and interactions between ecosystems and the environment.

All models necessarily include a number of approximations and assumptions, in order to meet with computational constraints and provide results which can be interpreted and understood at the appropriate level. Different types of model will make these approximations and assumptions in different areas, depending on the specific questions being addressed. For example, a model of interactions between particular species of plants will explicitly consider the detailed physiological processes at a level at which differences between and within species are significant, and may make simplifications in their treatment of two-way interactions with the global environment. In contrast, a model of the interactions between large-scale vegetation patterns and atmospheric processes may neglect inter-species differences and group like forms in a small number of functional types, while including detailed mechanistic representations of the global physical environment. Here we describe a range of models which are applied to the study of the role of dynamic ecosystems and interactions with the climate system, from the patch scale to the globe.

GAP MODELS

Gap models stress the dynamic changes in the structure and the composition of the vegetation at local scales, in response to changes in climate and disturbance regimes. The models typically simulate successional dynamics by accounting for the birth, growth, death, and interactions with the environment for hundreds of individual plants of many species living on a small plot of land. The predictions of multiple plots are then combined to obtain a prediction of the change in species composition and biomass across ecological landscapes (tens of kilometers). Because they simulate the fates of each of the millions of plants involved in landscape succession, these models are called "individual-based models."

Gap models vary in their inclusion of processes, which may be important in the dynamics of particular sites being simulated (e.g., hurricane disturbance, flooding, formation of permafrost, etc.), but share a common set of characteristics. These latter characteristics involve an emphasis on the demography and natural history of plant species, relatively general rules for physiological trade-offs among species, and an emphasis on the understanding of plant growth and reproductive processes. Each individual plant is simulated as an independent entity with respect to the processes of establishment, growth, and mortality typical of the species. This feature is common to most individual-tree-based forest models and provides sufficient information to allow computation of species- and size-specific demographic effects. Gap model structure emphasizes two features important to a dynamic description of vegetation pattern: (1) the response of the individual plants and species to the prevailing environmental conditions, and (2) how the individual modifies those environmental conditions. The models are hierarchical in that the higher-level patterns observed (i.e., population, community, and ecosystem) are the integration of plant responses to the environmental constraints defined at the level of the individuals. The individual-based approach provides a useful link between more detailed physiological models and the larger-scale responses. Gap models have been tested in terms of their ability to reproduce important features of a wide variety of forests and other ecosystems (Table 15.1), including their ability to reproduce past vegetation patterns (Fig. 15.1).

A key feature of individual-based models is that two implicit assumptions associated with traditional modeling approaches are not necessary. These are the assumptions that (1) the unique features of individuals are sufficiently unimportant to the degree that individuals are assumed to be identical, and (2) the population is "perfectly mixed" so that there are no local spatial interactions of any importance. With gap models, each individual tree (or other plant in the versions of the models that simulate grassland or savanna) is simulated. The population geometry in the vertical dimension (which plant shades which other plant) is modeled explicitly in all gap models, and often the spatial position of each plant is considered in the determination of competition and access to resources.

As is the case with many of the earlier individual-based models used in forestry, gap models simulate the establishment, diameter growth, and mortality of each tree in a given area. Calculations are on a weekly to annual time step. At least initially, gap models were developed for plot of a fixed size, thus giving them their name. Many of the models focus on a size unit (ca. 0.1 ha) that approximates a forest canopy gap (Shugart and West 1980). Gap models feature relatively simple protocols for estimating the model parameters (Botkin et al. 1972; Shugart 1984). For many of the more common temperate and boreal forest trees, there is a considerable body of information on the performance of individual trees (growth rates, establishment requirements, height–diameter relations) that can be used directly in estimating the parameters of gap models. The models have simple rules for interactions among individuals (e.g., shading, competition for limiting resources, etc.) and for birth, death, and growth of individuals.

What are the implications of gap models and other individual-based models for the response of vegetation to climatic change? If one considers the dynamics of living material (biomass) on a single-canopy-sized piece of a forest over multiple generations (Fig. 15.2a), the expected changes in biomass are quasi-cyclical or in the form of a saw-toothed curve (Shugart 1998). Bormann and Likens (1979a, b) considered the dynamic response of a watershed in New Hampshire following clear-cutting to be a summation of several such saw-toothed curves initially synchro-

nized by the clear-cutting application at time zero. The distances between the "teeth" in the saw-toothed, small-scale biomass curve are determined by how long a particular tree lives and how much time is required for a new tree to grow to dominate a canopy gap. Thus, the curves for several small plots of land eventually become desynchronized from chance differences in the timing of the death of a particular tree. The summation of several of these curves for local biomass dynamics can be used to predict the biomass of the entire mosaic. This is the dynamic response of interest for the change that one might expect from deforested land being restored to a forest condition in an effort to increase the regional storage of organic carbon.

The larger-scale biomass dynamics (Fig. 15.2) is a simple statistical consequence of summing the dynamics of the parts of the

Figure 15.1. Comparison of gap model paleoclimate simulation of species composition with an actual pollen record over 4000 years; (top) pollen diagram constructed from stratigraphy of the Soppensee and (bottom) pollen reconstruction using the FORCLIM gap model. Source: Litschke et al. (1998). In View from the Alps, MIT Press.

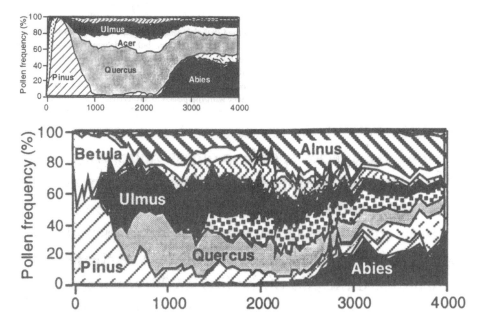

Table 15.1 Examples of Model Tests Applied to Different Classes of Dynamic Ecosystem Models

Model Class	Type of Test	Example Case
Gap model	With a priori parameter estimation for species, predict forest-level features (total biomass, leaf area, stem density, average tree diameter, etc.).	Prediction of biomass for floodplain forests (Phipps 1977; Pearlstine et al. 1985). Predict effects of timber harvest on nutrient regimes in northern (U.S.) hardwood forests (Aber et al. 1978).
Gap model	Run model for a long period of time; test results against mature forests preserved in a region.	Predict structure and composition of relic mature forests in Sweden (Leemans and Prentice 1987).
Gap model	Calibrate model on stands of a given age; predict stand structure on stands of a different age.	Predict the structure and composition of old-growth forests in the Great Smoky Mountains National Park (U.S.); test by independently predicting composition and structure of forest clear cut 40 and 70 years before (Busing and Clebsch 1987).
Gap model	Use introduction of disease or change in conditions (e.g., fire frequency, etc.) as a "natural" experiment.	Predict the composition of mature forests in eastern Tennessee (U.S.) before the introduced chestnut blight eliminated one of the major tree species (Shugart and West 1977).
Gap model	Predict independent tree diameter increment data.	Predict the growth and diameter increment for subtropical rain forest trees (van Daalen and Shugart 1989).
Gap model	Predict forestry yield tables.	Test model on its ability to reproduce Swedish yield tables (Leemans and Prentice 1987).
Gap model	Predict forest composition change to single environmental gradients.	Predict change from deciduous to coniferous forest in the mountains of New England (Botkin et al. 1972).
Gap model	Predict forest composition response to multiple environmental gradients.	Predict composition and structure of boreal forests on north and south facing slopes and for different ages since wildfire in Fairbanks, Alaska, region (Bonan 1989).
Gap model	Reconstruct composition of vegetation under past climates (Paleo-reconstruction).	Reproduce 16,000-year record of forest change based on fossil pollen chronology from eastern Tennessee (Solomon et al. 1980).
DGVM	Reproduce actual global vegetation patterns.	Taking present-day climate as input, reproduce present-day global vegetation patterns (Cramer et al. 2001).
DGVM	Reproduce observed surface water budget.	Taking present-day climate as input, reproduce present-day runoff and evapotranspiration (Foley et al. 1996).
DGVM	Reproduce observed NPP.	Taking present-day climate as input, reproduce present-day NPP (Foley et al. 1996).
GCM-DGVM	Produce realistic, stable atmospheric CO_2 concentrations and internally consistent set of global carbon pools in simulated climate system with no constraints on CO_2.	Initializing at pre-industrial CO_2, climate, and vegetation, allow climate-carbon model to run free and test for stable atmospheric CO_2 and land and ocean carbon (Cox et al. 2000).

continued

Table 15.1 Continued

Model Class	Type of Test	Example Case
GCM-DGVM	Reproduce observed variability in atmospheric CO_2 associated with emergent internal climate variability.	Compare relationship between observed and modeled El Nino / Southern Oscillation and atmospheric CO_2 (Jones et al. 2001).
GCM-DGVM	Reproduce observed atmospheric CO_2 anomalies associated with major volcanic eruptions.	Imposing aerosol loading due to Mount Pinatubo eruption, reproduce observed atmospheric CO_2 anomaly in subsequent months (Jones et al. 2001).
GCM-DGVM	Reproduce decadal-scale trends in atmospheric CO_2.	Driving model with all radiative forcings (as opposed to just CO_2 alone) reproduce observed 20th-century CO_2 record (Jones et al. 2001).
EMIC	Reproduce known changes in regional-scale vegetation cover.	Driving model with orbital forcing, reproduce mid-Holocene sparsening of west African vegetation (Claussen et al. 1999).

mosaic. If there has been a synchronizing event, such as a clear-cutting, one would expect the mosaic biomass curve to rise as all the parts are simultaneously covered with growing trees (point a in Fig. 15.2b). Eventually, some patches have trees of sufficient size to dominate the local area, and there is a point in the forest development when the local drops in biomass are balanced by the continued growth of large trees at other locations and the mosaic biomass curve levels out (point b in Fig. 15.2b). If the trees over the area have relatively similar longevities, there is also a subsequent period when several (perhaps the majority) of the pieces that constitute the forest mosaic all have deaths of the canopy dominant trees (point c in Fig. 15.2b). Over time, the local biomass dynamics become desynchronized and the biomass curve varies about some level (point d in Fig. 15.2). The implications of this for using more or less natural landscapes to store carbon are that the forest area should initially be a sink for carbon, that one would then expect the landscape to become a source, and finally that the

landscape should settle into a condition where it is, on average, a neutral for carbon fluxes. At this time one would need to protect the forested landscape from change that would make it a source.

What is the structure of a mature forest system at the end of this process? Whittaker (1953) reviewed the Watt (1947) pattern and process concept to redefine the climax concept of Clements (1916). Similar ideas were developed by Bormann and Likens (1979a, b) in their "shifting-mosaic steady-state concept of the ecosystem" as well as what Shugart (1984) called the "quasi-equilibrium" landscape. In these views, the structure of a mature forest (at the scale of several hectares) is as a heterogeneous mixture of patches in different phases or stages of gap-phase replacement. The mature forest should have patches with all stages of gap-phase dynamics, and the proportions of each should reflect the proportional duration of the different gap-replacement stages. The occurrence of such patterns has been documented for several different mature forest systems, and the presence of shade-in-

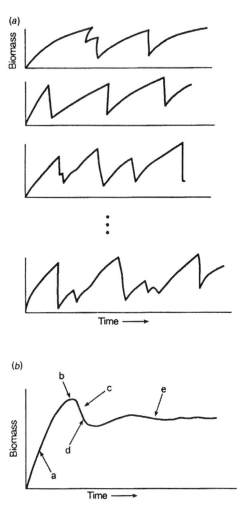

(a)

(b)

Figure 15.2. Biomass dynamics for an idealized landscape. The response is from a relatively large, homogeneous area composed of small patches with gap-phase biomass dynamics. The upper panel of the figure indicates the individual dynamics of the patches that are summed to produce the landscape biomass dynamics. The landscape biomass dynamics curve in the lower panel has 4 sections indicated: (a) Increasing landscape biomass curve rising as all of the patches are simultaneously covered with growing trees. (b) Local drops in biomass are balanced by the continued growth of large trees at other locations. The landscape biomass curve levels out. (c) If the trees have relatively similar longevities, there is a period when several (perhaps the majority) of the patches that comprise the forest mosaic all have deaths of the canopy dominant trees. (d) The local biomass dynamics become desynchronized and the landscape biomass curve varies about some level.

tolerant trees in mature undisturbed forest in patches is one observation consistent with the mosaic dynamics of mature forests (tropical rain forests: Aubréville 1938, Jones 1955–1956, Whitmore 1975, Knight 1975, Hartshorn 1978; temperate forests: Jones 1945, Raup 1964, White 1979, Oliver 1981, Peterken and Jones 1987; see Whitmore 1982 for review and discussion).

Gap models are applied to examine the response of forested systems to climate changes in both predictive and reconstructive (see Fig. 15.1) modes. As an example of the application of a gap model to assess the effects of a regional climate change, the effects of several different climate change cases were inspected for forests growing on sites in the North American boreal forest at a site near Fairbanks, Alaska. The model was developed and tested on its ability to simulate the stand composition, recovery after fire, and response to slope and aspect of the site. Fairbanks is in the zone of discontinuous permafrost. North-facing slopes and poorly drained sites have permafrost, a persistent presence of an ice layer in the soil, and are dominated by black spruce (*Picea marina*), one of the few tree species that can grow under such conditions. South-facing slopes have no permafrost, and the recovery of the system after a wildfire features a succession from birch and aspen (*Betula* and *Populus*) to a forest dominated by white spruce (*Picea glauca*).

Using the boreal forest gap model to assess climate change effects, Bonan and colleagues (1990) and Smith and colleagues (1995) investigated the responses of the forest by taking several different climate change projections from general circulation models. The models were run for several centuries on 100 plots simulating conditions associated with both north-facing and south-facing slopes. The climate conditions used in the models were then "transitioned" to those associated with a "greenhouse warming" over a 50-year time step. The results of this model experiment were to produce conditions in which the cold forests of black spruce growing on north-facing slopes were largely unaffected by the climatic warming, but the warmer, white spruce forests of the south-facing slopes were strongly affected by the change. Conditions on the south-facing slopes were outside the ecological conditions under which the common tree species near Fairbanks are known to survive. For white spruce the limiting condition appeared to be moisture stress brought on by the temperature change. Subsequent tree-ring studies investigating a run of warmer than usual temperatures near Fairbanks have found evidence for the same moisture-stress-mediated effect in white spruce, matching the prediction of the gap model a decade before (Barber et al. 2000). This field validation of a priori predictions from a gap model is reassuring for those who would apply the models to assessing climate change.

The use of gap models to assess climate change has been greatest in the Northern Hemisphere, where climate projections and data relating to tree growth, mortality, and other parameters are more reliable. The perspective provided by these studies is limited in part because gap models address localities or landscapes, while temperate range shifts due to climate change may occur on a larger scale. Where range shifts exceed tens of kilometers, they may

not be captured even in landscape-scale composites of multiple gap models. Nonetheless, range shifts can be inferred from changes in species composition, which are well-represented in gap models.

These results confirm the poleward and upslope movements predicted by theory and observed in past and current biological responses to climate change. Gap models show increases in species with southern and lowland affinities at more northerly and montane sites. These responses increase in response to greater climate forcing. Figure 15.3 illustrates changes in vegetation simulated by a gap model in response to a change in climate at a montane site in the Alps.

The responses of gap models to climate change in the north temperate zones have been used to reconstruct paleoecological changes in the vegetation, particularly since the end of the Pleistocene (Solomon and Webb 1985). Most present-day tests involve monitoring the models' ability to reproduce vegetation composition in different locations with different climates and particularly along altitudinal gradients (Shugart 1984, 1998, 2003). One significant response of the models along simulated altitudinal gradients has been the oc-

Figure 15.3. Projected response of vegetation to a step-change in climate simulated by the FORCLIM model, using climatic change downscaled from a transient IPCC business-as-usual GCM run. Source: Litschke et al. (1998). In *View from the Alps*, MIT Press.

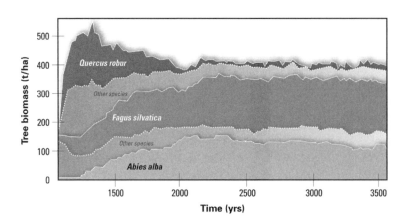

currence of hysteretic, or state-dependent, responses to climate change along the gradients (Shugart et al. 1980). In this situation, the vegetation present at a given point along the gradient depends upon history. Parts of the vegetation gradient can have more than one stable vegetation type. The vegetation in the multi-stable state zone depends on the climate history.

The models have been used less regularly for future assessments of climate change in other regions. In the subtropical and tropical settings, less is known about the climate response of species and many of the evaluations of change have involved inspections of the consequences of changes in disturbance regimes (which are strongly mediated by climatic conditions). Many of these have involved assessing the changes in vegetation to altered fire regimes (Shugart and Noble 1981) or changes in hurricanes (Doyle 1981; O'Brien et al. 1992) or drought frequency (Desanker 1996).

Studies of the effects of climate change and variation on forest composition and structure include several early large-scale applications of gap models to assess continental and subcontinental forest system responses (Shugart 1984). In some cases the growth of individuals was uniformly altered. These mostly theoretical applications often assessed the sensitivity of the forest structural dynamics to elevated-CO_2 growth enhancement. Shugart and Emanuel (1985) investigated the effects of increased plant growth in five different gap models including such diverse types of forest as subtropical rain forest, Australian *Eucalyptus* forest, and temperate mixed forests. The principal result of the findings was that an increase in tree growth was not translated directly into an equivalent forest growth (or carbon storage). Forests demonstrated an increased rate of gap filling following a tree death, so that forests with high disturbance rates were more responsive to whatever positive effects an increase in CO_2 might manifest at the individual tree level. In other cases

the performance of species growth was uniformly altered. These include growth enhancement and reduction model experiments with pollution effects on forests and climate changes. The implication of this work is that the age (or stage of development) of a forest can have a profound effect on its response to environmental change. There can be profound effects on tree populations in response to relatively small (ca. 5 to 10 percent) changes in annual growth rates. In some instances, tree species subjected to small decreases in their rate of growth can be totally eliminated from the simulated forest. In other instances, species that are subjected to equally small effects can actually increase as their competitors are diminished by growth reductions. The response of the ecosystem in toto under relatively small changes in tree growth rates can be greater than that inferred from simply averaging the reduction in growth across all the species that constitute the stand.

Probably the most significant outcome of a number of gap model–based evaluations of climate change is their depiction of significant inertial effects in the responses of forests to change. Established, vigorously growing trees are difficult to kill by stress in the model simulations. Since a wide variety of trees can survive well out of their ranges when nurtured in arboreta, this may not be an unrealistic result. Forests under stress tend to initially resist change and transition abruptly as the dominant trees reach senescence and die. The age and stand structure of the forest have a pronounced effect on this response, its timing, and its magnitude.

Solomon (1986a) used the FORENA model to simulate the response of forests to CO_2-induced climate changes across eastern North America. Changing temperature and precipitation at a constant rate resulted in a distinctive dieback of extant trees at most locations. Transient responses in species composition and carbon storage continued for as long as 300 years after simulated climate changes ceased. This was an early demonstration that forest tree population dynamics can influence large spatial-scale stand composition and carbon dynamics. FORENA computes the effects of soil moisture on tree growth, but the simulations of Solomon (1986b) used a very deep mesic silt-loam soil at all sites that tended to minimize the effects of precipitation deficits. Even more drastic forest changes are predicted under CO_2-induced climate change if shallower, coarser textured soils are used in the simulations. Similar model exercises using ZELIG to assess warming in the southeastern United States (Urban and Shugart 1989) with soils with less water holding capacity predicted increased evaporative demand with increased temperatures. The resultant increase in droughtiness severely restricted tree growth in greenhouse warming climate change scenarios, even if precipitation increased. In analogous simulations for the Great Lakes region, forests dropped to 23 to 54 percent of the current biomass (Botkin et al. 1989). On poor sites in both regions, forests in these simulations could be converted to grassland or savanna with very low productivity and carbon storage in biomass.

Climate change simulations with similar models also show changes in species composition, biomass, or both. In some cases species composition shifts with little change in stand biomass, as is the case for simulations of forests in the Pacific Northwest (Dale and Franklin 1989). Simulations of Wisconsin and Quebec forests (Overpeck et al. 1990) showed that forest productivity changed with climate, but resulted in small changes in species composition. Other model simulation results for the Great Lakes region (Pastor and Post 1988; Botkin et al. 1989) found significant changes in species composition without additional disturbances. As in simulations for the southeastern United States, dry sites could lose the capability to grow most tree species and be converted to oak savannas or even prairies with a 2° to 4° warming.

Several climate model results indicate increased rates of forest disturbance as a result of weather that is more likely to cause forest fires, convective windstorms, coastal flooding, and hurricanes. When increased disturbance frequency is added to the simulations, then significant changes in forest composition as well as changes in biomass are projected by the simulations.

Gap models can be used to make detailed predictions about specific ecosystems and species. They can explain or predict local responses to environmental changes and make use of detailed site-specific information. Because they include time-dependent processes and therefore represent successional changes within the ecosystem, they can be used to examine different stages of ecosystem response to a given climate change. However, they do not map regional-scale change, so that their application in mapping species range shifts is limited. In theory, an array of patch models could be used to map range shifts, including competition, but this application is in the developmental stage.

DYNAMIC GLOBAL VEGETATION MODELS

Dynamic global vegetation models (DGVMs) combine the large-scale applicability of biogeography (biome) models (Chapter 16) with the time-dependent vegetation dynamics of gap models. Terrestrial vegetation is modeled at the global scale including the dynamics of competition and succession. Plant physiological processes such as photosynthesis, respiration, and transpiration are modeled mechanistically from physical principles. Taking climatic variables such as temperature, precipitation, humidity, and sunshine hours as input, DGVMs calculate the carbon balance and development of each individual plant within the modeled ecosystem. This includes the uptake of carbon by the plant through photosynthesis (e.g. Farquhar et al. 1980; Collatz et al. 1991, 1992), the release of carbon by the plant through respiration (e.g. Lloyd and Taylor 1994), and the loss of water to the atmosphere via transpiration (e.g. Monteith 1981). The net uptake of carbon by the plant, the net primary productivity, gives the carbon available for growth, and this is allocated to different parts of the plant such as roots, leaves, and woody biomass. Observed relationships with environmental factors such as temperature, soil moisture availability, and day length are used to determine stages in plant life cycles such as bud burst and leaf-drop. Other physiological processes affecting growth and development, such as frost damage and drought-induced embolism, may also be represented. Species are represented as groups with similar physiological and structural properties, termed plant functional types (PFTs). Groupings for PFTs typically distinguish between evergreen and deciduous types, broadleaf and needleleaf woody vegetation, and the two different photosynthetic pathways (C3 and C4) observed in herbaceous vegetation.

The calculations are applied to large-scale ecosystems as entities (e.g. at global climate model resolutions of approximately 3°) rather than specific individuals. DGVMs take time-dependent input data—such as monthly climatic variables for many years or decades—and output large-scale ecosystem structure, which again varies in time. The inclusion of rates of growth and spreading therefore makes DGVMs suitable for time-dependent modeling of vegetation responses to transient climate change.

Different modeling groups employ a range of different approaches to representing vegetation dynamics at large scales. All DGVMs use plant functional types (PFTs), as individual species would be too numerous to model explicitly, but the number of PFTs identified can vary among models.

For instance, the VECODE model (Brovkin et al. 1997) identifies only evergreen and deciduous trees and grass, whereas other models (LPJ (Sitch 2000) and IBIS (Foley et al. 1996)) distinguish dry and cold deciduousness, tropical, temperate and boreal trees, C3 and C4 grasses, and shrubs. Competition between PFTs is represented either in terms of individuals over several patches and scaled up to the grid cell (HYBRID; Friend et al. 1997) or in terms of competition between populations (TRIFFID; Cox 2001).

DGVMs are concerned with the general functional character of vegetation rather than the behavior and distribution of particular species, so they cannot directly model biodiversity. However, DGVMs can be used to investigate a number of processes of relevance to climate change and biodiversity. Changes in the general viability of a PFT under a climate change will imply potential impacts on species within that PFT. Furthermore, changes in the abundance of species making up a particular PFT can impinge on the resources available to the species of another PFT, so changing competitive balances between PFTs suggests changes in ecosystem structure which have implications for biodiversity. If a PFT is simulated to disappear in a particular location, this may be taken to imply a threat of local extinction of all species within that PFT. However, it should be noted that the PFT by definition does not represent the diversity within a group of similar species, some of which may be more resilient than others.

Cramer and colleagues (2001) used six DGVMs to simulate the responses of global vegetation to a climate change scenario produced with the HadCM2 general circulation model (GCM). The CO_2 concentration was prescribed to increase to 750 ppmv by 2100 in both the GCM and the DGVMs. In the GCM, average temperatures over land increased by 5°C, and many regions received more rainfall on average over the year. The DGVMs showed qualitatively similar responses of global vegetation to these CO_2 and climate changes, although the responses of the different models varied in magnitude (see Plate 6). Vegetation cover generally increased over much of the land surface, either due to the regional climatic changes or as a direct result of higher CO_2 concentrations enhancing photosynthesis. Tree cover became more dense in several savanna regions, and the warming climate also led to increased tree cover in the high latitudes. As well as a general increase in tree cover, the temperate and boreal forests saw a shift in the ratio of different tree types, with broadleaf tree cover being particularly favored by the warming climate and expanding more than needleleaf trees.

However, some regions experienced a major loss of vegetation cover by 2100, largely as a result of shifting precipitation patterns in the GCM climate simulation. Although many regions received more rainfall under the warmer climate, some locations saw their rainfall reduced dramatically. Southwestern Africa received less rainfall and the vegetation cover became more sparse, and in Amazonia a severe reduction in simulated rainfall led to major decrease in forest cover. In particu-

lar, a reduction in rainfall was simulated by the GCM in Amazonia. Drying in the Amazon basin under global warming is a feature of several (but not all) GCMs, and appears to be associated with an El Niño–like pattern of sea-surface temperature change in the Pacific (Cox et al. 2004). Teleconnections between Pacific surface temperatures and Amazonian precipitation in El Niño events have been shown to be well-simulated in the Hadley Centre GCM (Spencer and Slingo 2003), and the ENSO-related interannual variability in the global carbon cycle has also been shown to be well-simulated by HadCM3LC (Jones et al. 2001). In additional simulations neglecting CO_2 fertilization, in order to directly compare the model responses to climate change, the drying led to a decrease in forest cover and biomass in all six DGVMs (Fig. 15.4). However, the magnitude of

Figure 15.4. Time series of Amazon forest biomass in 6 different dynamic global vegetation models (DGVMs), driven by a climate projection from the HadCM2 climate model (Cox et al, in press). These particular simulations do not account for the direct fertilization effect of rising atmospheric CO_2; other simulations with the same models including CO_2 fertilization showed smaller losses of biomass or even small increases. Figure by kind permission of the MetOffice, British Crown Copyright 2003.

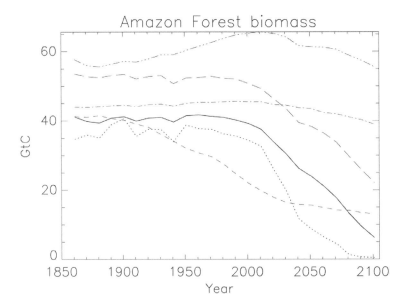

this forest dieback varied among the models. In the most extreme case, the dieback extended over most of the Amazon basin, while in the least extreme case it was limited to the northeastern quadrant.

All six DGVMs suggested that the current increase in terrestrial carbon storage may not be maintained indefinitely, due to saturation of photosynthesis at high CO_2 concentrations, loss of forest in Amazonia, and an increase in CO_2 release by soil microbial respiration with temperature. Some DGVMs simulated the global land carbon store to stop increasing by 2100; others simulated a switch to a decreasing carbon store before that time.

EARTH SYSTEM MODELS: COUPLING THE BIOTA TO ITS GLOBAL ENVIRONMENT

Changes in global ecosystems under climate change may themselves exert further effects on climate. These feedbacks could be both global and local in nature. Changes in total carbon storage of an ecosystem will modify the concentration of CO_2 in the local atmosphere, and since CO_2 has a long residence time and is well mixed, this change will be spread throughout the global atmosphere. Changes in vegetation structure may also modify regional climates through the surface energy and water budgets. In many cases such feedbacks will be predominantly local, mainly affecting the environment of the ecosystem providing the feedback. In other cases, biotic feedbacks can exert significant remote effects through teleconnections, sometimes extending across the globe.

Given the potential for major feedbacks from ecosystems, it is clear that predictions of future climate change should consider ecosystem responses and their effect on climate. This has led to the development of "earth system models," which couple models of the physical climate system (the atmosphere and oceans) to models of the terrestrial and marine biosphere (Foley et al. 1998; Cox et al. 2000; Ganapolski et al. 1998). The physical and biological models interact via biogeochemical cycles and through the impact of life on the physical properties of the Earth's surface. Some current models feature a DGVM and/or an interactive carbon cycle included within an existing GCM to provide detailed simulations with as much mechanistic process representation as possible. Other models use approximations in the more complex parts of the Earth system (e.g. atmospheric dynamics) in order to increase computational efficiency and facilitate a greater number of simulations—such models are termed earth-system models of intermediate complexity (EMICs).

The inclusion of DGVMs in GCMs allows climate prediction simulations to take into account feedbacks from ecosystems responding to climatic changes at global and regional scales (Cox et al. 2001; Betts et al. 2004). Coupled GCM-DGVMs are therefore potentially valuable for understanding and predicting synergistic responses of ecosystems to climate change over time scales of centuries and spatial scales of hundred of kilometers.

For example, the Hadley Centre climate model HadCM3 has been further developed to include terrestrial and marine ecosystems and their interactions with climate. Standard practice for such simulations is to drive the GCM with a scenario of future greenhouse gas (GHG) concentrations derived from a carbon cycle model beforehand; however, this procedure neglects the feedback effects of the climate change on the GHG concentrations themselves. The new version of HadCM3 rectifies this for CO_2 by calculating CO_2 concentrations within the climate model itself, allowing ecosystems to respond to climate change and hence alter their exchange of carbon with the atmosphere and oceans. Land ecosystems also influence the simulated climate directly by modifying the character of the land surface.

Vegetation dynamics are modeled with the "TRIFFID" DGVM (Cox et al. 2001), which represents global vegetation in terms of five plant functional types (broadleaf and needleleaf tree, C3 and C4 grass, and shrub). TRIFFID simulates the carbon fluxes between the vegetation, soil, and atmosphere, and allows the changes in vegetation to modify the physical characteristics of the land surface such as albedo and water availability. The "HadOCC" ocean carbon cycle model (Palmer and Totterdell 2001) simulates the physical and biological flows of carbon within the ocean and between the ocean and atmosphere. Simulations of climate change with this new model, HadCM3LC, therefore include feedbacks from land and ocean ecosystems via the carbon cycle (Cox et al. 2000), and also feedbacks from changes in land surface characteristics (Betts et al. 2004).

The impact of global-scale synergisms acting through the carbon cycle was investigated with HadCM3LC by comparing vegetation changes in simulations with and without climate–carbon cycle feedbacks. In the first simulation, atmospheric CO_2 was prescribed to observed concentrations from 1860 to 2000, then prescribed to the IPCC's IS92a concentration scenario from 2000 to 2100. This follows standard practice in GCM climate change simulations. Vegetation was allowed to respond to the CO_2 rise and climate change, but the changes in ecosystem carbon stocks under climate change did not feed back to atmospheric CO_2.

In the second simulation, the atmospheric CO_2 concentration was calculated within the model in response to changing carbon fluxes at the land and ocean surfaces. Anthropogenic emissions were prescribed from the IS92a scenario, but unlike the previous simulation, CO_2

concentrations were further modified by the response of global ecosystems and oceanic carbon uptake to climate change itself. The inclusion of an interactive carbon cycle within the model therefore introduced a global synergism between ecosystem responses via the carbon cycle.

Both simulations showed notable changes in global vegetation cover as a result of climate change. While the general global patterns of change were similar, the magnitude of vegetation changes at regional scales was shown to depend greatly on whether climate–carbon cycle feedbacks were included (Betts et al. 2004).

In the simulation with prescribed CO_2 concentrations, the patterns of vegetation change were broadly similar to those seen above, with DGVMs driven with one-way coupling to the climate (Cramer et al. 2001). The boreal forests increased in

density and the Arctic tree line moved poleward, while large areas of forest in Amazonia were lost as a result of reduced precipitation in that region (Fig. 15.5a). However, the inclusion of biogeophysical vegetation feedbacks had amplified some of the changes. In particular, the thinning of Amazonian forest cover had an important impact on climate and hence on the rate of forest loss itself (Betts et al. 2004). Much of the rainfall in the Amazon basin relies on water transported from over the oceans through the repeated cycling of water through rainfall and evaporation. Trees significantly increase evaporation by drawing up soil moisture and transpiring it through their leaves, so reduced forest cover leads to decreased evaporation and hence a diminished supply of water for further rainfall. In the model, drought-induced dieback of the forest in eastern Amazonia caused further rainfall reductions in the west, accelerating the westward expansion of forest loss. Although the higher CO_2 concentration made the trees more water-efficient and

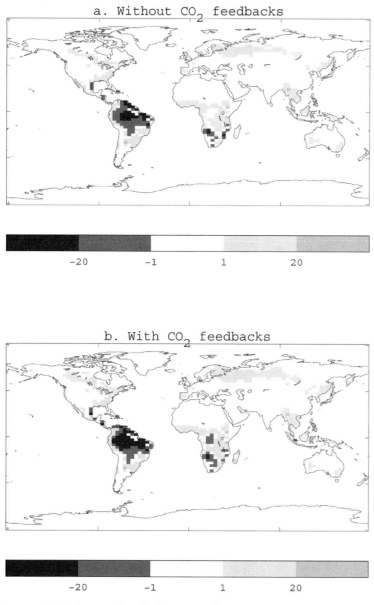

Figure 15.5. (a) Change in broadleaf tree cover from 1860–2100 simulated by the HadCM3LC coupled climate–carbon cycle model (Cox et al. 2000) without CO_2-climate feedbacks. (b) Further changes when CO_2-climate feedbacks are included.

hence increased their resilience to drier conditions, this could only partly offset the effects of drought and the overall effect was a major loss of forest cover biting into Amazonia from the northeast.

In the simulation with an interactive carbon cycle, the changes in global ecosystems exerted a major positive feedback on rising CO_2 and global warming (Betts et al. 2004). Although more vigorous growth in the boreal forests increased their uptake of carbon, dieback of the Amazonian forests was associated with large quantities of carbon being released. The rising temperatures also increased microbial respiration throughout the world's soils, leading to a further and even greater loss of carbon from the terrestrial biosphere. The CO_2 rise was accelerated, reaching 1000 ppmv by 2100, and average land temperatures increased by 8K. Warming over land was therefore increased by more than 50 percent by the inclusion of climate–carbon cycle feedbacks. Temperatures previously projected for the end of the twenty-first century were now simulated to arrive around 2060.

In this new simulation with its more rapid climate change, ecosystem changes were generally similar in character to those in the previous simulation but greater in magnitude (Fig. 15.5b). Forest expansion was more rapid in the boreal forests, as was the shift from needleleaf to broadleaf tree cover. In the Amazon basin, the rainfall reduction was magnified by the more rapid global warming, and the forest dieback was therefore accelerated. The equatorial forests of central Africa also began to become disadvantaged by the more extreme climate change in this simulation.

Although biodiversity cannot be directly derived from information on plant functional type distribution, it seems reasonable to assume that greater changes in PFTs imply greater changes in biodiversity. These results can therefore be interpreted as indicating increased impacts on biodiversity through synergisms between anthropogenic greenhouse gas emissions and carbon cycle feedbacks from ecosystem changes.

EMICs use simpler representations of the atmosphere and biosphere to allow simulations over longer periods than those currently possible with GCMs. They can therefore be used to explore long-time-scale global synergies. Their high computational efficiency allows multiple simulations exploring many combinations of drivers and internal feedback processes. Claussen and colleagues (1999) used the CLIMBER EMIC to simulate the response of global vegetation to gradual changes in insolation over the past 9000 years. The

model results suggested that the abrupt desertification of the Saharan region 4000 to 6000 years ago may have been a consequence of synergies between the atmosphere circulation, vegetation, and ocean surface temperatures. Gradual changes in global mean temperatures reduced precipitation over the Sahara, initiating a reduction in vegetation cover which further reduced precipitation. Devegetation became rapid, despite the insolation forcing remaining smooth (Fig. 15.6a–d). The timing of the abrupt desertification was found to depend on sea-surface temperatures. Simulations of increasing CO_2 concentration and climate change with CLIMBER suggest some potential for northward expansion of Sahelian grasslands into the Sahara, as a consequence of an increase in summer precipitation enhanced by vegetation–atmosphere feedbacks (Claussen et al. 2003).

SUMMARY

For conservation ecology and climate change, models are computer-accessible tools for experimenting with modern theories of vegetation interaction with climatic change. The range of models now in existence can provide guidance on potential ecosystem changes on scales ranging from a few meters to the globe, with interactions between individuals, ecosystems, and the physical environment taken into account.

Gap model insights have considerable relevance to conservation biology. Gap models support the notions of poleward and upslope range shifts, as implied by changes in relative abundance of species. More complex ecosystem dynamics predicted by these tools include multiple-equilibria dynamics and potential surprises with abrupt change in the response to environmental change. These latter dynamics imply that the simple monitoring of ecosystem change may not provide the

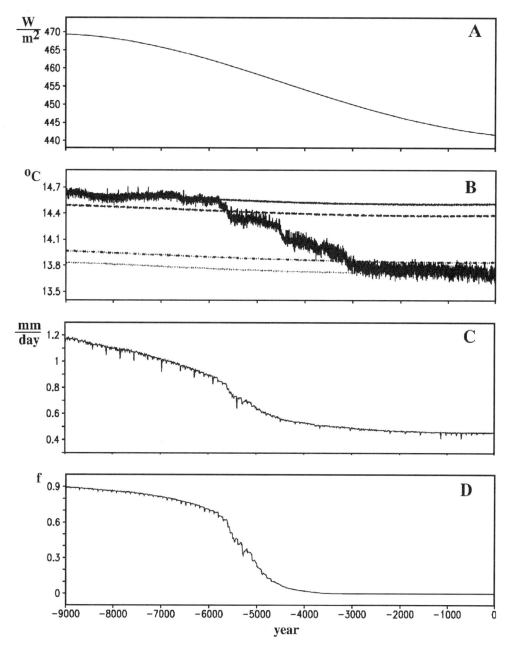

Figure 15.6. Timeseries of (A) solar irradiance (Wm^{-2}), and associated change in (B) global mean surface temperature (°C), (C) annual mean Saharan precipitation (mm day^{-1}), and (D) fractional coverage of vegetation in the Sahara.

information to project the consequences of climate change. What is required is an understanding of the dynamics of ecosystem change, including the dynamic change in the structure and composition of ecosystems.

DGVMs have shown that vegetation change may be significant at the global scale, with major transitions in plant functional type, and some regions experiencing increased growth and biomass while in other regions the vegetation may become more sparse. Coupled GCM-DGVMs indicate that such ecosystem responses may exert major feedbacks on the rates of global and regional climate change through impacts on the carbon and water cycles. In the Hadley Centre model, under the warming climate, the pattern of surface warming in the Pacific becomes El Niño–like and rainfall is reduced in Amazonia. Thinning of Amazonian forest cover hampers the regional water cycle and leads to further dieback, and associated carbon emissions contribute to an enhanced CO_2 increase and further global warming. Even more carbon is released from global soils, more than offsetting the uptake by vegetation elsewhere in the world. The overall response of each ecosystem to climate change therefore also depends on the responses of ecosystems elsewhere.

Current work on dynamic ecosystem models, at all scales, tends to focus on the direct relationships between climate and the ecosystem. For example, the modeled ecosystem changes generally result from physiological responses to meteorological or hydrological variables such as temperature, soil moisture, or atmospheric CO_2.

However, indirect relationships are also potentially important, and currently receive comparatively little attention. Disturbances such as fire or insect attack may be climatically dependent and exert major influences on ecosystem structure and carbon storage. While a number of current dynamic ecosystem models include some representation of certain disturbances (Woodward and Lomas 2004; Thonicke et al. 2001), these are arguably less advanced than the models of physiology and ecosystem dynamics. Successful models of long-term ecosystem–environment dynamics are likely to require full consideration of the role of disturbance.

With computational power expanding all the time, the inclusion of more detailed processes at finer spatial and temporal scales over wider areas becomes increasingly possible. Gap modeling approaches may poten-tially make use of this by providing individual-based representations of ecosystems over larger areas, while DGVMs applied at the global scale may increase their number of PFTs to better represent diversity. Earth system models may move to higher resolutions and incorporate additional processes which are currently missing. Simple models will always be important for understanding processes and exploring uncertainties with multiple simulations. However, regionally specific predictions of climate change and its impacts on particular species or groups of species will benefit greatly from the increasing computational power available to modelers.

REFERENCES

Aber, J. D., D. B. Botkin, and J. M. Melillo. 1978. Predicting the effects of differing harvest regimes on forest floor dynamics in northern hardwoods. *Canadian Journal of Forest Research* 8:306–315.

Aubréville, A. 1938. La forêt colonaile: les forêts de l'Afrique occidentale française. *Annales Academie Sciences Colonaile* 9:1–245. (Translated by S. R. Eyre. 1991. Regeneration patterns in the closed forest of Ivory Coast, pp. 41–55. In S. R. Eyre, ed., *World Vegetation Types*. MacMillan, London.)

Barber, V. A., G. P. Juday, and B. P. Finney. 2000. Reduced growth of Alaskan white spruce in the twentieth century from temperature-induced drought stress. *Nature* 405:668–672.

Betts, R. A. 1999. Self-beneficial effects of vegetation on climate in an Ocean-Atmosphere General Circulation Model. *Geophysical Research Letters* 26(10):1457–1460.

Betts, R. A., P. M. Cox, M. Collins, P. P. Harris, C. Huntingford, and C. D. Jones. 2004. The role of ecosystem-atmosphere interactions in simulated Amazonian precipitation decrease and forest dieback under global climate warming. *Theoretical and Applied Climatology*.

Bonan, G. B. 1989. Environmental factors and ecological processes controlling vegetation patterns in boreal forests. *Landscape Ecology* 3:111–130.

Bonan, G. B., H. H. Shugart, and D. L. Urban. 1990. The sensitivity of some high latitude boreal forests to climatic parameters. *Climatic Change* 16:9–31.

Bormann, F. H., and G. E. Likens. 1979a. Catastrophic disturbance and the steady state in northern hardwood forests. *American Scientist* 67:660–669.

Bormann, F. H., and G. E. Likens. 1979b. *Pattern and Process in a Forested Ecosystem*. New York: Springer-Verlag.

Botkin, D. B., J. F. Janak, and J. R. Wallis. 1972. Some ecological consequences of a computer model of forest growth. *J. Ecol.* 60:849–872.

Botkin, D. A., R. A. Nisbet, and T. E. Reynales. 1989. Effects of climate change on forests of the Great Lake states, pp. 2-1 to 2-31. In J. B. Smith and D. A. Tirpak, eds., *The Potential Effects of Global Climate Change on the United States: Appendix D—Forests*. Washington, D.C.: Office of Policy, Planning, and Evaluation, U.S. Environmental Protection Agency.

Brovkin, V., A. Ganopolski, and Y. Shvirezhev. 1997. A continuous climate-vegetation classification for use in climate-biosphere studies. *Ecological Modelling* 101:251–261.

Busing, R. T., and E. E. C. Clebsch. 1987. Application of a spruce-fir forest canopy gap model. *Forest Ecology and Management* 20:151–169.

Claussen, M., V. Brovkin, A. Ganapolski, C. Kubatski, and V. Petoukhov. 2003. Climate change in northern Africa: The past is not the future. *Climatic Change* 57:99–118.

Claussen, M., C. Kubatzki, V. Brovkin, A. Ganopolski, P. Hoelzmann, and H.-J. Pachur. 1999. Simulation of an abrupt change in Saharan vegetation in the mid-holocene. *Geophysical Research Letters* 26:2037–2040.

Clements, F. E. 1916. *Plant Succession: An Analysis of the Development of Vegetation*. Washington, D.C.: Carnegie Inst. Pub. 242.

Collatz, G. J., J. T. Ball, C. Grivet, and J. A. Berry. 1991. Physiological and environmental regulation of stomatal conductance, photosynthesis and transpiration: A model that includes a laminar boundary layer. *Agricultural and Forest Meteorology* 54:107–136.

Collatz, G. J., M. Ribas-Carbo, and J. A. Berry. 1992. A coupled photosynthesis-stomatal conductance model for leaves of C_4 plants. *Australian Journal of Plant Physiology* 19:519–538.

Cox, P. M. 2001. Description of the TRIFFID dynamic global vegetation model. Technical Note 24, Hadley Centre, Met Office. Available online at www.hadleycentre.com

Cox, P. M., R. A. Betts, M. Collins, P. P. Harris, C. Huntingford, and C. D. Jones. 2004. Amazon dieback under climate–carbon cycle projections for the 21st Century. *Theoretical and Applied Climatology*.

Cox, P. M., R. A. Betts, C. D. Jones, S. A. Spall, and I. J. Totterdell. 2000. Acceleration of global warming due to carbon-cycle feedbacks in a coupled climate model. *Nature* 408:184–187.

Cox, P. M., R. A. Betts, C. D. Jones, S. A. Spall, and I. J. Totterdell. 2001. Modelling vegetation and the carbon cycle as interactive elements of the climate system. In R. Pearce, ed., *Meteorology at the Millennium*. London: Academic Press.

Cramer, W., A. Bondeau, F. I. Woodward, I. C. Prentice, R. A. Betts, V. Brovkin, P. M. Cox, V. Fisher, J. A. Foley, A. D. Friend, C. Kucharik, M. R. Lomas, N. Ramankutty, S. Stitch, B. Smith, A. White, and C. Young-Molling. 2001. Global response of terrestrial ecosystem structure and function to CO_2 and climate change: Results from six dynamic global vegetation models. *Global Change Biology* 7(4):357–374.

Dale, V. H., and J. F. Franklin. 1989. Potential effects of climate change on stand development in the Pacific Northwest. *Canadian Journal of Forest Research* 19:1581–1590.

Davis, M. B., and D. B. Botkin. 1985. Sensitivity of cool-temperate forests and their fossil pollen record to rapid temperature change. *Quarternary Research* 23:327–340.

Desanker, P. V. 1996. Development of a MIOMBO woodland dynamics model in Zambesian Africa using Malawi as a case study. *Climatic Change* 34:279–288.

Doyle, T. W. 1981. The role of disturbance in the gap dynamics of a montane rain forest: An application of a tropical forest succession model, pp. 56–73. In D. C. West, H. H. Shugart, and D. B. Botkin, eds., *Forest Succession: Concepts and Application*. New York: Springer-Verlag.

Farquhar, G. D., S. von Caemmerer, and J. A. Berry. 1980. A biogeochemical model of photosynthetic CO_2 assimilation in leaves of C3 species. *Planta* 149:78–90.

Foley, J. A., S. Levis, I. C. Prentice, D. Pollard, and S. L. Thompson. 1998. Coupling dynamic models of climate and vegetation. *Global Change Biology* 4(5):561–579.

Foley, J. A., I. C. Prentice, N. Ramankutty, S. Levis, D. Pollard, S. Sitch, A. Haxeltine. 1996. An integrated biosphere model of land surface processes, terrestrial carbon balance and vegetation dynamics. *Global Biogeochemical Cycles* 10(4):603–628.

Friend, A. D., H. H. Shugart, and S. W. Running. 1993. A physiology-based model of forest dynamics. *Ecology* 74:792–797.

Friend, A. D., A. K. Stevens, R. G. Knox, and M. G. R. Cannell. 1997. A process-based, terrestrial biosphere model of ecosystem dynamics (Hybrid v3.0). *Ecological Modelling* 95:249–287.

Ganapolski, A., C. Kubatzki, M. Claussen, V. Brovkin, and V. Petoukhov. 1998. The influence of vegetation-atmosphere-ocean interactions on climate during the mid-Holocene. *Science* 280: 1916–1919.

Hartshorn, G. S. 1978. Tree falls and tropical forest dynamics, pp. 617–638. In P. B. Tomlinson and M. H. Zimmermann, eds., *Tropical Trees as Living Systems*. Cambridge: Cambridge University Press.

Hojer, R., M. Bayley, C. F. Damgaard, and M. Holm-strup M. 2001. Stress synergy between drought and a common environmental contaminant: Studies with the collembolan Folsomia candida. *Global Change Biology* 7:485–494.

Jones, C. D., M. Collins, P. M. Cox, and S. A. Spall. 2001. The carbon cycle response to ENSO: A coupled climate-carbon cycle model study. *Journal of Climate* 14(21):4113–4129.

Jones, C. D., and P. M. Cox. 2001. Modelling the volcanic signal in the atmospheric CO2 record. *Global Biogeochemical Cycles* 15(2):453–466.

Jones, E. W. 1945. The structure and reproduction of the virgin forests of the north temperate zone. *New Phytologist* 44:130–148.

Jones, E. W. 1955–1956. Ecological studies on the rain forest of southern Nigeria. IV. The plateau forest of the Okumu forest reserve. *J. Ecol.* 43:564–594; 44:83–117.

Kercher, J. R., and M. C. Axelrod. 1984. A process model of fire ecology and succession in a mixed-conifer forest. *Ecology* 65:1725–1742.

Knight, D. H. 1975. A phytosociological analysis of species-rich tropical forest on Barro Colorado Island, Panama. *Ecological Monographs* 45:259–284.

Leemans, R., and I. C. Prentice. 1987. Description and simulation of tree-layer composition and size distributions in a primaeval *Picea-Pinus* forest. *Vegetatio* 69:147–156.

Litschke, H., A. Guisan, A. Fischlin, J. Williams, H. Bugmann. 1998. Vegetation responses to climate change in the Alps—Modeling studies, pp. 309–350. In P. D. Cebon, U. Dahinden, H. Davies, D. Imboden, and C. Jaeger, eds. *A View from the Alps: Regional Perspectives on Climate Change*. Boston: MIT Press.

Lloyd, J., and J. A. Taylor. 1994. On the temperature dependence of soil respiration. *Functional Ecology* 8(3):315–323.

Monteith, J. L. 1981. Evaporation and surface temperature. *Quarterly Journal of the Royal Meteorological Society* 107(651):1–27.

O'Brien, S. T., B. P. Hayden, and H. H. Shugart. 1992. Global change, hurricanes, and a tropical forest. *Climatic Change* 22:175–190.

Oliver, C. D. 1981. Forest development in North America. *Forest Ecology and Management* 3:153–168.

Overpeck, J. T., D. Rind, and R. Goldberg. 1990. Climate-induced changes in forest disturbance and vegetation. *Nature* 343:51–53.

Palmer, J. R., and I. J. Totterdell. 2001. Production and export in a global ocean ecosystem model. *Deep-Sea Research* 48(5):1169–1198.

Pastor, J., and W. M. Post. 1986. Influences of climate, soil moisture, and succession on forest carbon and nitrogen cycles. *Biogeochemistry* 2:3–27.

Pastor, J., and W. M. Post. 1988. Response of northern forests to CO$_2$-induced climate change. *Nature* 334:55–58.

Pearlstine, L., H. McKellar, and W. Kitchens. 1985. Modeling the impacts of river diversion on bottomland forest communities in the Santee River Floodplain, South Carolina. *Ecological Modelling* 29:283–302.

Peterken, G. F., and E. W. Jones. 1987. Forty years of change in Lady Park Wood: The old-growth stands. *J. Ecol.* 75:477–512.

Phipps, R. L. 1977. Simulation of wetlands forest dynamics. *Ecological Modelling* 7:257–288.

Raup, H. M. 1964. Some problems in ecological theory and their relation to conservation. *J. Ecol.* 52(Suppl.):19–28.

Shugart, H. H. 1984. *A Theory of Forest Dynamics: The Ecological Implications of Forest Succession Models*. New York: Springer-Verlag.

Shugart, H. H. 1998. *Terrestrial Ecosystems in Changing Environments*. Cambridge: Cambridge University Press.

Shugart, H. H. 2003. *A Theory of Forest Dynamics: The Ecological Implications of Forest Succession Models*. Caldwell, N.J.: Blackburn Press. [A reprinting of the out-of-publication Shugart (1984) text.]

Shugart, H. H., and W. R. Emanuel. 1985. Carbon dioxide increase: The implications at the ecosystem level. *Plant, Cell and Environment* 8:381–386.

Shugart, H. H., W. R. Emanuel, and G. Shao. 1996. Models of forest structure for conditions of climatic change. *Commonwealth Forestry Review* 75:51–64.

Shugart, H. H., W. R. Emanuel, D. C. West, and D. L. DeAngelis. 1980. Environmental gradients in a beech-yellow poplar stand simulation model. *Mathematical Biosciences* 50:163–170.

Shugart, H. H., and S. B. McLaughlin. 1986. Modelling SO2 effects on forest growth and community dynamics, pp. 478–491. In W. E. Winner, H. A. Mooney, and R. A. Goldstein, eds., *Sulfur Dioxide and Vegetation: Physiology, Ecology, and Policy Issues*. Stanford, Calif.: Stanford Press.

Shugart, H. H., and I. R. Noble. 1981. A computer model of succession and fire response to the high altitude Eucalyptus forest of the Brindabella Range, Australian Capital Territory. *Australian Journal of Ecology*. 6:149–164.

Shugart, H. H., and T. M. Smith. 1996. A review of forest patch models and their application to global change research. *Climatic Change* 34:131–153.

Shugart, H. H., and D. C. West. 1977. Development of an Appalachian deciduous forest succession model and its application to assessment of the impact of the chestnut blight. *Journal of Environmental Management* 5:161–179.

Shugart, H. H., and D. C. West. 1980. Forest succession models. *Bioscience* 30:308–313.

Sitch, S. 2000. The role of vegetation dynamics in the control of atmospheric CO$_2$ content, Ph.D. thesis, Lund University, Sweden.

Smith, T. M., P. N. Halpin, H. H. Shugart, and C. M. Secrett. 1995. Global forests, pp. 146–179. In K. M. Strzepek and J. B. Smith, eds., *If Climate Changes: International Impacts of Climate Change*. Cambridge: Cambridge University Press.

Solomon, A. M. 1986a. Comparison of taxon calibrations, modern analog techniques, and forest-stand simulation models for the quantitative reconstruction of past vegetation: A critique. *Earth Surface Processes and Landforms* 11:681–685.

Solomon, A. M. 1986b. Transient response of forests to CO_2-induced climate change: Simulation experiments in eastern North America. *Oecologia* 68:567–79.

Solomon, A. M., H. R. Delcourt, D. C. West, and T. J. Blasing. 1980. Testing a simulation model for reconstruction of prehistoric forest-stand dynamics. *Quaternary Research* 14:275–293.

Solomon, A. M., and T. Webb III. 1985. Computer-aided reconstruction of late Quaternary landscape dynamics. *Annual Reviews of Ecology and Systematics* 16:63–84.

Spencer, H., and Slingo, J. 2003. The simulation of peak and delayed ENSO teleconnections. *Journal of Climate* 16(11):1757–1774.

Thonicke, K., S. Venevsky, S. Sitch, and W. Cramer. 2001. The role of fire disturbance for global vegetation dynamics: Coupling fire into a Dynamic Global Vegetation Model. *Global Ecology and Biogeography* 10(6):661–677.

Urban, D. L., and H. H. Shugart. 1989. Forest response to climate change: A simulation study for Southeastern forests, pp. 3–1 to 3–45. In J. Smith and D. Tirpak, eds., *The Potential Effects of Global Climate Change on the United States*. EPA-230-05-89-054, Washington, D.C.: U.S. Environmental Protection Agency.

van Daalen, J. C., and Shugart, H. H. 1989. OUTENIQUA—A computer model to simulate succession in the mixed evergreen forests of the southern Cape, South Africa. *Landscape Ecology* 24:255–267.

Watt, A. S. 1947. Pattern and process in the plant community. *J. Ecol.* 35:1–22.

White, P. S. 1979. Pattern, process and natural disturbance in vegetation. *Botanical Review* 45:229–299.

Whitmore, T. C. 1975. *Tropical Rain Forests of the Far East*. Oxford: Clarendon Press.

Whitmore, T. C. 1982. On pattern and process in forests, pp. 45–59. In E. I. Newman, ed., *The Plant Community as a Working Mechanism*. Special Publ. No. 1, British Ecological Society. Oxford: Blackwell Scientific.

Whittaker, R. H. 1953. A consideration of climax theory. The climax as a population and a pattern. *Ecological Monographs* 23:41–78.

Woodward, F. I., and M. Lomas. 2004. Simulating vegetation processes along the Kalahari transect. *Global Change Biology* 10(3):383–392.

Migration of Vegetation Types in a Greenhouse World

Jay R. Malcolm, Adam Markham, Ronald P. Neilson, and Michael Garaci

Projections of global warming suggest that within a century, we may see temperature increases that rival those during the recent glacial retreat, when ecosystems experienced massive change. Just as in the past, shifts in species ranges can be expected; however, because of the more rapidly changing climate, there is the possibility that species' migrations will lag behind shifting climatic zones. Such lags have potentially serious implications for ecosystems, including impacts on geochemical cycling and species richness. For example, in migration scenarios in which trees were perfectly able to keep up with warming, a 7−11 percent increase in global forest carbon was seen, whereas in another that allowed no migration, a 3−4 percent decline in forest carbon was observed (Solomon and Kirilenko 1997). Several simulation studies suggest that if migration fails to make up for warming-induced local losses of species, then a net decline in local diversity can be expected (Sykes and Prentice 1996; Kirilenko and Solomon 1998).

Despite the potential importance of migration in influencing ecosystem properties, little is known about the migration rates that global warming might require. For example, information on overall magnitudes, variation among localities, and differences among climate scenarios is lacking. Perhaps equally important, the implications of anthropogenic habitat loss for future migration are poorly understood. Simulation studies suggest that potential migration rates can decline precipitously under habitat loss (Schwartz 1992; Pitelka et al. 1997); however, implications for global warming have not been examined. Better assessments of future migration rates may assist in identifying possible ecological responses, and also may assist in identifying populations or locations that are disproportionately important in affecting future migration rates.

These issues were investigated in a study that simulated required migration rates under global warming by using 14 combinations of global climate and vegetation models (Malcolm et al. 2002). The climate models simulated climatic conditions for a given CO_2 concentration, and based on this climate the vegetation models mapped the equilibrium distributions of major vegetation types (biomes) at a resolution of one-half degree of latitude and longitude. Vegetation distributions under recent climatic conditions were compared against those under doubled-CO_2 climates. To estimate required migration rates, biome distributions were used as proxies for species distributions. Species are often restricted to one biome type or another, and hence movements of biomes may provide estimates of the migration rates that species will have to achieve. In addition, the same sorts of derived climate variables used to model climate envelopes also are useful in mapping species climatic envelopes (especially for plants). To calculate migration distances, we reasoned that the nearest possible immigration source for a map cell of a future ($2 \times CO_2$) biome type was assumed to be the nearest map cell of the same biome type in the current climate. Thus, migration distance was calculated as the distance between a future cell and its nearest same-biome-type cell in the current climate. These distances were then divided by 100 years, which is a conservative estimate of the length of the period over which the warming is expected to occur.

A second set of migration calculations assessed possible implications of anthropogenic habitat loss. Instead of calculating the nearest possible (crow fly) distances between map cell centers as above, migration distances were calculated assuming that: (1) species could travel only on land and (2) map cells with more than a critical threshold of habitat loss were off-limits to migration. Based on simulations by Turner (reported in Pitelka et al. 1997) and Schwartz (1992), two thresholds of human development were used: 55 percent or 85 percent (as judged using 1-km resolution satellite data; Loveland et al. 2000).

These calculations showed that even in the absence of anthropogenic habitat loss, some 16 percent of the terrestrial land surface had rates in excess of 1000 m/yr (Fig. 15.7). These rates are well above those recorded during postglacial times, which were typically in the range of 100–200 m/yr (Clark 1998). The high rates were consistent among scenarios, indicating that a wide range of assumptions about 2 × CO_2 climate forcing and vegetation responses all lead to expectations of high future migration rates. The required migration rates varied latitudinally, with approximately 35 percent of map cells above 60 degrees north latitude displaying rates greater than or equal to 1000 m/yr. Higher migration rates at higher latitudes are expected given the consistently greater poleward warming shown by climate models (IPCC 1996). Even within 20 degrees of the equator, however, nearly 1 map cell in 10 had rates above 1000 m per year, which is of concern given the possibility of lower intrinsic rates of migration in tropical compared to temperate zones.

The overall effect of anthropogenic habitat loss was relatively small at the global scale, but it sometimes had important regional effects. Strong regional increases in required migration rates along the poleward margins of developed areas were indicated (Fig. 15.8), supporting conclusions that increased connectivity among natural habitats within developed areas may be critical in aiding organisms to attain the necessary rates of migration.

The migration capabilities of organisms are poorly known, and climate is only one of many factors influencing species distributions; nevertheless, it is evident that climate change has potential to lead to a "weedier" future dominated by fast moving and climatically tolerant species. For invasive species and others with high dispersal capabilities, migration capabilities may easily exceed 1000 m/yr. Examples of species with such capabilities include goldenrod (*Soldago* L.), which invaded Eu-

Figure 15.7. Average required migration rates across 14 combinations of global climate and vegetation models. For a future (2 × CO_2) map cell of biome type x, the migration distance was the distance to the nearest same-biome-type cell in the current climate. Distances were divided by 100 years to yield a migration rate.

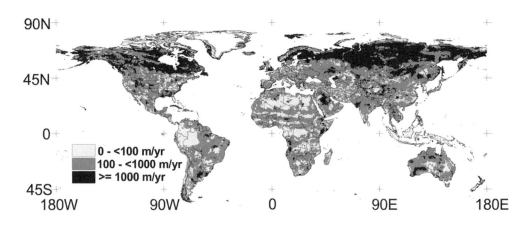

90N
45N
0
45S

0 - <100 m/yr
100 - <1000 m/yr
>= 1000 m/yr

180W   90W   0   90E   180E

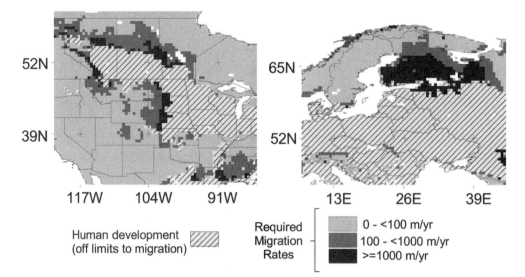

Figure 15.8. Mean migration rates calculated using shortest terrestrial paths, but assuming that highly human-modified map cells (55% or more of the underlying habitat modified by humans) were off-limits to migration (central North America on the left and northern Europe on the right). The rates shown are the differences between rates with and without human-modified habitat; hence they represent migration rates attributable to habitat loss that are over and above those imposed by global warming.

rope in the 1800s at rates close to 20,000 m/yr, and coyotes (*Canis latrans* Say), which expanded into eastern North America in the 1900s at rates above 20,000 m/yr. At the other end of the spectrum are slow moving taxa, such as many forest understory herbs and earthworms, whose migration capabilities appear to be less than 10 m/yr (references in Malcolm et al. 2002). Surprisingly, even though temperate trees are perhaps the best-studied group in this respect, their maximum migration capabilities are unknown. Populations appeared to accompany the Holocene warming without significant lags (Prentice et al. 1991); however, even these paleomigration rates are difficult to explain based on observed population parameters (Clark 1998). Added to this uncertainty concerning migration rates is the possibility that populations may persist in the face of unfavorable conditions for

decades or longer, adding a further potential lag to community change. Any filtering effect of global warming is of particular concern if it involves taxa such as trees that disproportionately modulate resource availability. Changes in overstory tree communities—including forest dieback, transitions from forested to nonforested ecosystems, or domination by early-successional taxa—all can be expected to result in marked changes in species composition and patterns of biodiversity.

REFERENCES

Clark, J. S. 1998. Why trees migrate so fast: Confronting theory with dispersal biology and the paleorecord. *American Naturalist* 152, 204–224.

IPCC. 1996. *Climate Change 1995: The Science of Climate Change*, ed. Houghton, J. T., Meira Filho, L. G., Callander, B. A., Harris, N., Kattenberg, A., and Maskell, K., p. 572. Cambridge: Cambridge University Press.

Kirilenko, A. P., and A. M. Solomon. 1998. Modeling dynamic vegetation response to rapid climate change using bioclimatic classification. *Climatic Change* 38, 15–49.

Loveland, T. R., B. C. Reed, J. F. Brown, D. O. Ohlen, Z. Zhu, L. Yang, and J. W. Merchant. 2000. Development of a global land cover characteristics database and IGBP DISCover from 1km AVHRR data. *International Journal of Remote Sensing* 21, 1303–1330.

Malcolm, J. R., A. Markham, R. P. Neilson, and

M. Garaci. 2002. Estimated migration rates under scenarios of global climate change. *Journal of Biogeography* 29, 835–849.

Pitelka, L. F. et al. 1997. Plant migration and climate change. *American Scientist* 85, 464–473.

Prentice, I. C., P. J. Bartlein, and T. Webb III. 1991. Vegetation and climate changes in eastern North America since the last glacial maximum. *Ecology* 72, 2038–2056.

Schwartz, M. W. 1992. Modelling effects of habitat fragmentation on the ability of trees to respond to climatic warming. *Biodiversity and Conservation* 2, 51–61.

Solomon, A. M., and A. P. Kirilenko. 1997. Climate change and terrestrial biomes: What if trees do not migrate? *Global Ecology and Biogeography* 6, 139–148.

Sykes, M. T., and I. C. Prentice. 1996. Climate change, tree species distributions and forest dynamics: A case study in the mixed conifer/northern hardwoods zone of northern Europe. *Climatic Change* 34, 161–177.

CHAPTER SIXTEEN

Climate Change and Marine Ecosystems

OVE HOEGH-GULDBERG

Approximately 70 percent of the earth's surface is covered by seawater. Within this dominant realm is a diversity of life forms that persist within a bewildering variety of environmental circumstances. The importance of the world's oceans increases as their role in determining global climate is appreciated along with examples of highly diverse ecosystems such as coral reefs.

Climate change is already changing the distribution and abundance of marine organisms and ecosystems. Like other organisms, marine plants and animals live within a fairly narrow set of physical and chemical conditions. Rapid changes to these environmental conditions may result in physical acclimation by the organism in the first instance, or changes in the abundance of organisms as critical limits are exceeded. Within this framework, ecological adjustments occur as changing conditions lead to varying relationships among organisms. Rates of change are also important to the capacity of a community or ecosystem to change.

Global circulation models recognize the strong physical links between ocean temperature and global climates. Beyond these broad features of the coupling of ocean and atmosphere, many other changes occur as sea temperature changes. Even minor changes to sea temperature will result in changes to the currents that flow across the earth's surface. Once changed, currents can affect the flow of heat and moisture among regions of the world. Currents are also critical to local marine conditions, and changes can have major impacts on ecosystems such as coral reefs and temperate kelp forests. There is also a wide array of more subtle influences with consequences for marine populations, such as reduced or increased genetic connectivity of marine populations as currents change.

Understanding how this global climate change will affect the earth's marine biodiversity is the focus of the present chapter. After laying the foundation for how changes in the earth heat content will affect the earth's oceans, we examine recent changes to marine ecosystems that have been strongly linked to recent climate change. In the final part of this chapter, the earth's extensive marine ecosystems will be examined with a view to understanding the type and scale of impact that is likely to occur. As will become clear, marine ecosystems are projected to undergo major changes over the next 50 years. A clear understanding of these changes is critical to the adaptive response of many human societies that are currently vulnerable through their dependence on oceanic ecosystems for either daily subsistence or industrial wealth.

THE PHYSICAL STRUCTURE AND FUNCTION OF THE EARTH'S OCEANS

There are two major transport layers within the ocean. Surface waters (down to 100 to 400 m depending on season and latitude) consist of low density seawater that is generally well-mixed, warmer, and more illuminated and oxygenated. These surface waters of the ocean are dominated by a series of currents that move under the combined influence of wind movements, the Coriolis effect and the location of landmasses such as continents (Fig. 16.1). Prominent in the surface layers of the ocean are huge gyres that circulate water within the oceanic basins of the northern and southern hemispheres. Due to the Coriolis effect, northern hemisphere gyres rotate clockwise while those of the southern hemisphere rotate counterclockwise. Smaller currents and eddies form as a consequence of these major patterns of water flow. At very high latitudes gyres tend to flow in the opposite direction.

The movement of surface waters to higher latitudes brings critical heat to latitudes that receive reduced solar input. This particular feature of oceanic circulation has important consequences for life at higher latitudes. Without the transport of warm water from the southern to northern Atlantic via the Gulf Stream, for example, the terrestrial and aquatic climates of northern Europe would be significantly colder.

Below the surface layer of the ocean are waters that are colder and denser and which move as a function of forces that act on the so-called thermohaline circulation (Fig. 16.2). The boundary between the two layers of the oceans is defined by a large scale change in seawater density known as the pycnocline.

The thermohaline circulation is a massive and long-lived flow of water from low to high latitude and from deep to shallow and back again. A major driving force of the earth's thermohaline circulation is the temperature differential between equatorial and more poleward locations. This force leads to a rapid cooling of warm saline water originating from lower latitudes. Cooling leads to an increase in density of the already more saline water, which sinks as it progresses through the high latitude seas. The sinking of oceanic waters at the poles powers a "conveyor belt"–like system in which deep waters move toward the equator while the surface components of the thermohaline circulation move poleward. The residence time of deep water can be as long as 200–500 years for the Atlantic Ocean and 1000–2000 years for the Pacific Ocean.

The interaction between the two major layers of the ocean has major impacts on productivity in the upper or photic layer of the ocean. Deep waters tend to be rich in inorganic nutrients such as phosphates, nitrates, and carbonates. As a result, areas in which deep waters are brought to the

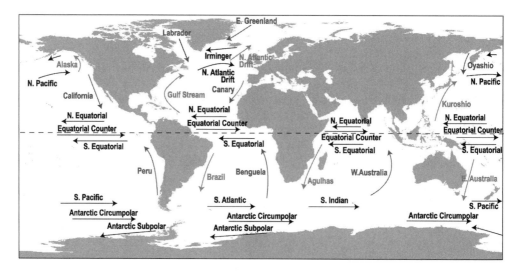

Figure 16.1. Principal water currents in the world's oceans.

surface (e.g., through Ekman Transport or upwelling) create zones of high primary productivity at the surface. These zones are critically important to fisheries and marine food webs generally.

CLIMATE CHANGE WITHIN THE OCEAN

The increase in greenhouse gases within the earth's atmosphere is set to change three fundamental variables associated with oceanic environments: the calcium carbonate saturation state, sea level, and temperature of the earth's oceans. As will be discussed, not all these changes (in isolation) are likely to have a negative impact on marine biodiversity. The combination of these changes, however, is expected to drive major changes on the distribution and abundance of marine organisms.

Figure 16.2. Global conveyor belt. Interaction between surface and deep water currents as a function of location. Colder, more saline water has a greater density than does warmer, less saline water. Water moving from the warmer, lower latitudes cools at the poles and sinks, motivating a global scale movement of oceanic waters.

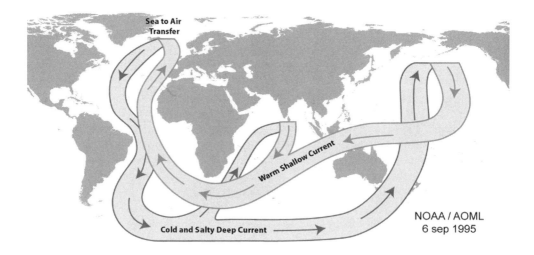

Reduced Total Carbonate Alkalinity

Total carbonate alkalinity of seawater will decrease as carbon dioxide increases within the earth's atmosphere (Gattuso et al. 1998; Kleypas et al. 1999; see Chapter 18 for additional discussion). This particular variable is expected to substantially change the acidity and carbonate ion pool (carbonate saturation state) of the global ocean. Gattuso and colleagues (1998) and Kleypas and colleagues (1999) have calculated that doubling carbon dioxide concentrations in the atmosphere will decrease the aragonite saturation state in the tropics by 30 percent by 2050 (under a doubling of carbon dioxide). Proportional changes are even higher under mid- and high-range scenarios. Conditions in which the alkalinity has been decreased experimentally by a similar amount promote lower calcification rates of a variety of plants and animals of 25 percent (range 11–40 percent; Langdon et al. 2000).

Increased Sea Level

The level of the ocean has fluctuated by more than 100 m over the past 100,000 years as ice stored on land has changed in volume. During the last ice age, sea level was 120 m below where it is today. During the transition out of this period of glaciation, sea level changed at an average rate of 10 mm/yr (some periods, rates were as high as 40 mm/yr). During the interglacial periods, rates of sea level rise have been much slower (0.1–0.2 mm/yr over the past 3000 years; Church et al. 2001). Changes in sea level have had major impacts on the abundance and particularly the distribution of both marine and terrestrial diversity.

Sea level will rise as climate change pushes planetary temperatures higher. This occurs due to the thermal expansion of ocean water (responsible for about 70 percent of the increase), the melting of glaciers, and changes to the distribution of

Greenland and Antarctic ice sheets (responsible for the remainder). The expected increase in sea level is approximately 9–29 cm over the next 40 years, or 28–98 cm by 2090 (Church et al. 2001; IPCC 2001). These changes may sound trivial but are expected to have major impacts on coastal regions and human infrastructure. Vast areas of the world's coastal regions are expected to be inundated. A 25-cm rise, for example, would displace a large number of people from the delta regions of major rivers such as the Nile, Ganges, and Yangtze as well drowning Pacific and Indian Ocean nations such as the Maldives, Kiribati, and Tuvalu (Church et al. 2001).

In concert with the direct effects of coastal inundation are the impacts of storm surge, which could result in a five-fold increase in displaced people by 2080 (Nichols et al. 1999). Impacts on marine ecosystems also vary according to proximity to coastlines; in some cases only minor changes are likely, while in others major impacts are likely. According to Nichols and colleagues (1999), sea level rise could cause the loss of up to 22 percent of the world's coastal wetlands by 2080. Combined with other human impacts, this number is likely to climb to a loss of 70 percent of the world's coastal wetlands by the end of the twenty-first century.

Sea Temperature Increase

Early models of climate change suggested that oceans would lag significantly behind atmospheric rates of warming due to the large volume and high thermal capacity of seawater (4.2 MJ m^{-3} °C^{-1} at 20°C versus 1.29 kJ m^{-3} °C^{-1} at 20°C). Modeling of potential climate change suggested that oceans would be relatively slow to respond to forcing by increases in greenhouse gases. The recent rapid responses in ocean temperature have challenged this notion. The heat content of the global ocean has increased 2.3 × 10^{23} J between the mid-

1950s and mid-1990s, representing a volume mean warming of 0.06°C. Significantly, this increase in heat content has not been distributed evenly. Substantial increases have occurred in the upper layers of the ocean, with the mean temperature increase for the upper 300 m of the global ocean over the same three decades being 0.31°C (Levitus et al. 2000). Deep oceanic warming is also occurring, and rates vary strongly with latitude (Barnett et al. 2001; Gille 2002).

Sea temperature, in turn, has an influence on a broad range of other critical features of the marine environment. Due to its direct effects on the density of seawater, changes in global temperatures can play directly upon the rates and directions of ocean water movement. Most global circulation models indicate that the thermohaline circulation of the planet, for example, is likely to weaken as greenhouse warming continues. Dickson and colleagues (2002) have produced convincing data that indicate a rapid and sustained freshening (decreased salinity) of the deep Atlantic Ocean. Though these changes may appear small (0.03 ppm salinity change over the past 40 years), they could indicate that major changes may be in store for the heat budget and functioning of the earth's oceans. Changes in the functioning of climate systems such as the El Niño Southern Oscillation (ENSO) have also been projected to occur under climate change by many greenhouse gas–driven global circulation models. Other models, however, disagree. Direct evidence of strong changes in ENSO over the past 100 years are lacking, although ENSO events in the late twentieth century were strong compared with ENSOs of previous cool (glacial) and warm (interglacial) periods (Tudhope et al. 2001).

When combined with elevated sea temperatures, changes to ocean circulation can have a major impact on local conditions and on other aspects such as the genetic connectivity of marine populations.

In this respect, variations in ocean circulation like that seen during ENSO events can have major impacts on the distribution, abundance, and health of ecosystems. For example, during the thermal anomalies of the 1997–98 ENSO disturbance, reef-building corals underwent a phenomenon called mass coral bleaching. In this event, coral bleached across most coral reef realms (Hoegh-Guldberg 1999), and an estimated 16 percent of the world's reef-building corals died (GCRMN 2000). Strong ENSO events on top of rising background sea temperatures combine to push sea temperatures above the thermal tolerance of corals. The lack of certainty as to how ENSO will change as the climate warms adds a level of uncertainty to projections of how the pattern of coral bleaching and mortality will change over the years to come. Rising background temperatures will eventually reach the thermal threshold of corals without the need for warmer than normal ENSO years (Hoegh-Guldberg 1999).

Biological Responses to Climate Change in the Ocean

Not surprisingly, organisms in the ocean respond to changes in the physical and chemical makeup of their environment. These responses may be mild, as organisms adjust their physiological processes to the new conditions (acclimation), or acute, as organisms sicken or experience higher mortality rates as their thresholds for particular types of change are exceeded. The latter may result in a shift in the genetic structure (adaptation) and/or geographic range of a population

There is strong evidence that both types of changes have already occurred among marine organisms in response to climate change. In this section, the range of different responses to climate change in marine organisms is explored. Although the discussion focuses on four major categories, a large number of other marine habitats are also showing strong signs of change as warming proceeds.

Temperate Nearshore Habitats

Coastal ecosystems line the earth's islands and continents and are generally dominated by food webs that depend on attached plants (algae or angiosperms) or waterborne microalgae. They are critically important for the flow of resources within the ocean and are the basis for more than 60 percent of the productivity of the ocean. Changes in light, temperature, salinity, and wave stress have direct impacts on their health and functioning. While diurnal and seasonal variation may be extreme, these organisms are still sensitive to shifts in environmental conditions that exceed those to which they are adapted. Consequently, it is not surprising that numerous examples of coastal ecosystems are now showing signs that recent climate change has already produced pronounced changes in the abundance and distribution of marine organisms within these habitats. Owing to space constraints, the discussion of these changes is limited to several key examples.

As with terrestrial fauna such as birds and butterflies (Parmesan et al. 2000), there has been a poleward range shift of intertidal marine species (Barry et al. 1995; Southward et al. 1995). Comparison of surveys of composition rocky intertidal communities at Monterrey on the mid-California coastline for the period 1931–33 and 1993–94 clearly indicate that species' ranges have shifted northward (Barry et al. 1995). In the study of Barry and colleagues, the geographic distributions of 45 invertebrate species were compared. The results are stark. The abundances of 8 out of 9 southern species and 5 out of 8 northern species have increased and decreased respectively over the 60-year period. These observations are largely consistent with the expectations that southern, warm-loving species should in-

vade northward as warming occurs, while northern, cold-adapted species should decline as they retract up the warming California coastline. Similar observations have been made for intertidal communities in southwestern Britain and the western English Channel, which show extensive changes in species composition and abundance over the past 70 years (Southward et al. 1995). As in California, it is the warm-water species that have replaced the colder-water ones. In at least one case, changes in the temperate intertidal communities have been matched by changes in the composition of benthic fish communities. Holbrook and colleagues (1997) have also recorded the northerly intrusion of warm-water fish in concert with long-term changes in sea temperature.

Much of the change seen among intertidal organisms is probably driven by differing tolerances for environmental extremes as well as more subtle shifts in response to environmental averages. Alleviation of freezing stress, reminiscent of the reasons underlying the poleward advance of many terrestrial species (Parmesan et al. 2000; Walther et al. 2002), may be one key reason for the poleward movement of warm-water intertidal species along the North American and European coastlines. Freezing is a severe obstacle for many species—ice formation usually has disastrous consequences for nonfreeze-tolerant organisms, with their cells literally being ripped apart as ice crystals grow with the cellular matrix (Storey and Storey 1996). Not all organisms are vulnerable, however. Intertidal animals that are freeze-tolerant can endure days or weeks of continuous freezing with 50–80 percent of total body water frozen (Aarset 1982; Storey and Storey 1988, 1992). The strategies that allow these organisms to avoid death by freezing include controlling the location and rate of ice crystal formation, regulating cell volume (and hence specific gravity), and the inclusion of specific proteins that minimize and stabilize cell structures as freezing occurs. The interplay between inherent physiological abilities, the physiological cost of certain physiological strategies, and the shift in environmental conditions are likely to yield some complex outcomes in terms of the movement of species in response to climate change.

The tolerance of elevated heat stress is also species specific. Each specific component of an organism is tailored for the thermal circumstances in which it finds itself. Enzyme systems (as most cellular components) function optimally within the temperature range typical of the environment they inhabit (e.g., Hochachka and Somero 1984). At the upper end of a thermal range, organisms cope with short periods in which thermal transients push physiologies beyond those experienced on average. For intertidal organisms, low tides during the middle of hot summer days can push temperatures to as much as 20°C beyond average temperatures experienced during other parts of the diurnal and tidal cycle.

There is a range of ways that organisms have adapted to higher temperatures, the discussion of which is beyond the scope of this chapter. However, the presence of these physiological features will clearly be important in determining which organisms invade in a poleward direction. Adaptations to temperature range from those that are behavioral to those that are more molecular in nature. Molecular chaperones, for example, which include heat-shock proteins (Hsps), are a universal feature of all organisms (Feder and Hoffmann 1999). Hsp expression is widely correlated with resistance to stress and helps stabilize proteins under increasing thermal stress. They generally increase during heat stress and help organisms cope with stress-induced denaturation of other proteins. In studies of congeneric intertidal snails (*Tegula* spp.) from California, expression and synthesis of heat-shock proteins was correlated with tidal height and emersion temperature. Snails from the highest

positions in the tidal zone had the highest rates of 77, 70, and 38 kDa Hsp, while species that were emersed for longer periods had five-fold lower expression rates (Tomanek and Somero 1999). Similar patterns have been observed with other intertidal organisms such as corals (Sharp et al. 1997).

Temperate intertidal communities are likely to change quite substantially under the projected enhanced greenhouse conditions. As the majority of marine organisms have free-swimming larval stages, the speed at which populations move poleward will be determined by the direction and strength of available currents. Species that have short-lived larval stages might also be expected to have slower rates of invasion. In some cases, increased sea temperature may not necessarily induce a poleward range shift. Other variables, such as habitat availability, light, and other factors may limit range shifts even if temperatures become more favorable.

Coral Reefs

Tropical intertidal and subtidal regions are dominated by ecosystems that are characterized by a framework of scleractinian corals. Coral reef ecosystems deserve separate mention in this chapter for two reasons. First, they stand out as a coastal ecosystem that has an order of magnitude greater biodiversity than all other coastal ecosystems (Bryant et al. 1998). Second, they have undergone major changes over the past 20 years, much of which has been associated with climate change (Hoegh-Guldberg 1999) and other stresses (Bryant et al. 1998).

The biodiversity of coral reefs is extraordinary with an estimated million species of plants, animals, and protists living in an estimated 400,000 km^2 of coral reef. To biologists starting with Charles Darwin, the biodiversity of coral reefs represented a paradox in the context of the nutrient-poor waters of the earth's tropical oceans

(Darwin 1842; Odum and Odum 1955). Despite the lack of external nutrients, these extraordinary ecosystems form rich and complex food chains that support large populations of fish, birds, turtles, and marine mammals (e.g., Kepler et al. 1994; Maragos 1994). In contrast, the open oceanic waters even a small distance away from coral reefs are usually devoid of significant populations of animal and plant life. The tight recycling of nutrients is now understood to lie at the heart of why coral reefs are so biologically diverse (Muscatine and Porter 1977; Hatcher 1988).

The distribution and abundance of reef-building corals vary as a function of latitude, with the greatest abundance of corals located closest to the equator. Light, temperature, and the carbonate alkalinity of seawater decrease in a poleward direction, making the formation of carbonate reefs more difficult at higher latitudes (Kleypas et al. 1999). Both the carbonate alkalinity and temperature of the ocean have already changed and are expected to change quite

substantially in the future. This has consequences for coral reef formation.

Coral reefs have already experienced major impacts from climate change. Tropical oceans are 0.5−1.0°C warmer than they were 100 years ago (Hoegh-Guldberg 1999). Major disturbances to coral reefs (termed mass coral bleaching) have increased dramatically over the past 30 years and have been linked irrefutably to periods of warmer than normal sea temperatures. Coral bleaching occurs when corals rapidly lose the cells and/or the pigments of the symbiotic dinoflagellates (zooxanthellae) that populate their tissues by the millions. Bleaching results in colonies turning from brown to white, often with spectacular host pigments (pocilloporins; Dove et al. 2001) (Fig. 16.3) being exposed. Reef-building corals that lose these important symbionts may experi-

Figure 16.3. Bleached coral (*Acropora* sp.) showing presence of host tissue pigments that are intensified when the brown symbiotic dinoflagellates are absent. *Source:* Photo by O. Hoegh-Guldberg, University of Queensland.

ence mortality rates that may exceed 90 percent.

Mass coral bleaching events occurred in seven major episodes from 1979 to 2002, with concomitant losses of reef-building coral cover across thousands of square miles of coral reef. The intensity and frequency of these events has increased over the past 30 years, because ENSO anomalies are probably reaching higher sea temperatures than those seen before due to the already higher background sea temperatures. The warming episodes associated with the 1997–98 ENSO event were the most widespread and intense of the century. Coral reefs in almost all parts of the planet experienced coral bleaching; in many areas, bleaching exceeded that seen before. Coincident with the majority of events was a strong thermal signal (Hoegh-Guldberg 1999), with water temperatures exceeding those recorded prior to 1997–98. Between 10 percent and 16 percent of the world's living coral reefs died during the 1997–98 mass bleaching cycle (GCRMN 2000; Hodgson and Liebeler 2001). Mortality in some areas was much higher than the global average. For instance, western Indian Ocean reef systems lost as much as 46 percent of all living reef-building corals (GCRMN 2000).

A key observation regarding the triggers for mass bleaching has been that the thermal threshold for coral bleaching is only 1–2°C above the summer maxima (for 3–5-week exposure times). The thermal threshold above which corals and their symbionts will experience heat stress and bleaching also varies geographically, indicating that corals and zooxanthellae have evolved over time to the local temperature regime (Coles et al. 1976; Hoegh-Guldberg 1999). Corals closer to the equator have thermal thresholds for bleaching that may be as high as 31°C, while those at higher latitudes may bleach at temperatures as low as 26°C. The size and length of thermal anomalies explain much of the frequency and intensity of any particular incident of mass coral bleaching (Strong et al. 2000; Hoegh-Guldberg 2001). Small anomalies over long periods may have the same impact as large anomalies over short periods. Thresholds may also vary with season at a particular site. In experiments on the central Great Barrier Reef, Berkelmans and Willis (1999) revealed that the winter maximum upper thermal limit for the ubiquitous coral Pocillopora damicornis was 1°C lower than the threshold for bleaching in summer. These shifts are evidence of thermal acclimation, a physiological adjustment that can occur in most organisms up to some upper or lower thermal limit.

Changes in reef-building coral communities are likely to have huge impacts on marine biodiversity. Corals form the essential framework within which a multitude of other species make their home. How these communities and interactions will change as coral health and abundance decline is still largely unassessed. Fish that depend on corals for food, shelter, or settlement cues may experience dramatic changes in abundance or go extinct. Thousands of other organisms are also vulnerable. For example, over 55 species of decapod crustacean are associated with living colonies of a single coral species, Pocillopora damicornis (Abele and Patton 1976; Black and Prince 1983). Nine of these are known to be obligate symbionts of living pocilloporid coral colonies. Branching corals of the genus Acropora, for example, have 20 species of obligate symbionts that depend solely on Acropora for habitat. The spacing of corals on a habitat may be critical for the reproductive success of coral associates that require sexual reproduction to proceed to the next generation. As corals become rarer (and spaced farther and farther apart), these organisms may be threatened as the chance of finding a partner or attaining successful fertilization becomes vanishingly small.

While sea temperatures experienced

during the past decade have been the highest on instrumental record (Lough 1999), and at some sites the highest sea temperatures for thousands of years (Gagan et al. 2000), the long-term patterns have still to be deciphered. There is, for example, growing evidence that sea temperatures in the Pacific decreased abruptly at the end of the eighteenth century by as much as 0.75°C (Linsley et al. 2000; Hendy et al. 2002). Decadal and even century-scale periodicity appear to play important roles in determining sea temperatures throughout the earth's tropical oceans. A full understanding of these patterns is critical to an understanding of the impacts of greenhouse forcing of global temperatures on the oceans in which corals grow.

The scale and intensity of recent mortality events on coral reefs appear unprecedented. *Acropora cervicornis*, for example, was a dominant species across the central shelf lagoon of Belize up until 20 years ago. In the 1980s, however, disease (white band disease) resulted in the complete mortality of *A. cervicornis*. Stands of the foliose (scroll-like) coral *Agaricia tenuifolia* quickly replaced *A. cervicornis* in the early 1990s but were wiped out by the high water temperatures of 1998. The latter event also caused mass coral bleaching and mortality of a wide variety of other coral species. Investigation of reef deposits reveals that the scale of these mortality events appears to have been unique in the past 3000 years (Aronson et al. 2000). The mortality of *A. cervicornis* in the 1990s left an unambiguous layer of coral branches in the sediments of reefs throughout the Caribbean. Aronson and his colleagues analyzed 38 cores from across the 375-km^2 central lagoon basin, but no similar layer could be found in sediments stretching back at least as far as 3000 years ago.

Fish Populations

Coastal fisheries are critical resources for hundreds of millions of people. Many scientists now point to the dramatic overexploitation of fisheries and the subsequent decline in fish stocks as the major factor in ecosystem change over the past two centuries (Jackson et al. 2001). As key predators and grazers within ecosystems like coral reefs, fish and other exploited organisms have a major role in determining the abundance of corals, macroalgae, and open substratum in many ecosystems.

Recent evidence has revealed that oceanographic and climatic variability may play a dominant role in fish stocks (Klyashtorin 1998; Babcock Hollowed et al. 2001; Attrill and Power 2002). The evidence is two-fold. First, fish stocks are tightly correlated with measures of climate variability such as the ENSO and North Atlantic Oscillation (NAO) indexes. Second, climate shifts (e.g., 1°C increase in sea temperature in the late 1970s) produce major changes in dominant fish stocks in areas like the North Atlantic. Exploitation of fish stocks aside, small changes in climate appear to have large effects on the abundance of fish stocks.

Klyashtorin (1998) examined major Atlantic and Pacific commercial species in relationship to the atmospheric circulation index (ACI) over most of the past century (1900–94). The atmospheric circulation index (ACI) is a measure of basic atmospheric conditions in the Atlantic-Eurasian region. Fish stocks such as Atlantic and Pacific herring, Atlantic cod, European, South African, Peruvian, Japanese, and California sardine, South African and Peruvian anchovy, Pacific salmon, Alaska pollock, and Chilean jack mackerel undergo decadal changes that are tightly correlated (correlation coefficients of 0.7–0.9) with the ACI and other indexes of climate variability. Similar conclusions were reached by Babcock Hollowed and colleagues (2001), who examined data sets for the North Pacific and Bering Sea. Strong associations between ENSO and PDO variability were revealed in time series data for catches of over 55 different fish stocks. The authors related these changes in fish stocks to changes in recruitment success driven by warmer or colder seas.

The relationship between climate variability and fish stocks is probably complex. In some cases, subtle changes may effect conditions at crucial stages in the life history of the fish species (e.g., estuaries; Attrill and Power 2002). Other effects may be broader in nature. The most widespread effects of climate occur on the primary and secondary production in marine ecosystems. Reid and colleagues (2001) revealed that the large increases in catches of the western stock of the horse mackerel (*Trachurus trachurus* L.) in 1987 were associated with an increase in phytoplankton and zooplankton stock over the same period, in turn due to warmer conditions in the North Sea in 1987.

Other studies have revealed declines in stocks associated with climate. Continuous plankton recorders (CPR), for example, have been deployed for more than 67 years in various oceans. The data generated by these devices tell an interesting story. Key planktonic species such as the copepod *Calanus* are highly correlated with climate indexes like the NOA, and the abundances of some key species have been declining since 1955, probably due to conditions that have altered the primary production of these areas of the global ocean.

Polar Marine Ecosystems

Most models of climate change indicate that the fastest rate of climate change will occur at the highest latitudes. Major changes in ice volume are already occurring as a result of increasing global temperature (de la Mare 1997; Kerr 1999), and temperatures in both polar regions have been the highest they have been for hundreds of years (Barbraud and Weimerskirch 2001). Consequently, it is important to consider the types of changes that are likely to occur in the earth's polar oceans. In both oceans, species have adapted

specifically to conditions associated with seasonally frozen sea scapes. As with coral reefs, there is abundant evidence that these types of organisms are changing as conditions warm.

Changes in sea temperature are also affecting plankton processes in polar as well as temperate regions. Reorganization of the North Atlantic copepod community structure includes warm-water species showing poleward range shifts, while cold-water species show range contraction (Beaugrand et al. 2002). These changes may have complex outcomes in terms of the organisms that live there. For example, one of the most spectacular marine phenomena involves mammals and birds attracted to krill associated with phytoplankton blooms during the Antarctic summer.

Krill are euphausid crustaceans that graze the rich seas of Antarctica during the late spring and summer. Densities of krill from the tip of the Antarctic peninsula from 1984–85 until 1995–96 were on average an order of magnitude less than those reported from periods prior to 1984 (Loeb et al. 1997). This decrease in population density was matched by the opposite trend by salps (pelagic tunicate *Salpa thompsoni*), which, as filter feeders, benefit from open waters. Krill (late larvae, juveniles, and adults), in contrast, feed on the microalgae that proliferate on the bottom of sea ice during the late winter. The net effect is that sustained warming, by reducing the extent of sea ice, is affecting a fundamental component of the Antarctic food webs. Evidence is growing that the decrease in krill abundance has implications for other members of this marine food chain. Numbers of Adele penguin (a key predator of krill) have fallen since the mid-1970s (Loeb et al. 1997). Since the mid-1970s fledgling survival has declined, and population size decreased by 70 percent since 1987. A similar relationship appears to hold for emperor penguins. Since the 1970s, emperor penguin numbers have declined by 50 percent, largely due to a major change in adult survivorship associated with warmer seas and lower sea ice cover (Barbraud and Weimerskirch 2001). Again, the reduced abundance of krill as seas warm is a compelling explanation for the large decrease in adult survival.

The extent of polar sea ice has implications for other aspects of the polar biology. Reduced sea ice has an impact on the reproductive biology of seals in the Artic (Kelly 2001). Ringed seals are vulnerable to earlier snowmelts due to the premature destruction of their subnivean (below the snow) lairs, exposing pups to extreme weather and increased predation. The exact position of the ice edge can have other effects. Pacific walruses, for example, need ice to support pups while adults feed on benthic invertebrates. For ice haul-outs to be suitably close to foraging grounds, water at the ice edge needs to be less than 100 m in depth. Recent sea ice retreat in the Beaufort and Chukchi seas has led to a decreased water depth and a reduction in available foraging areas, and has been cited as one of the reasons for the reduced reproductive success of walruses in recent decades (Kelly 2001).

PROJECTIONS OF FUTURE CHANGE: MARINE ECOSYSTEMS IN 2050?

Global temperatures and carbon dioxide concentrations are now higher than they have been for at least the past 400,000 years (IPCC 2001). Because these changes are likely to seriously affect the world's oceans, it is important to understand how biological systems will respond to these changes. This information is important to adaptive measures (if they exist) that may be taken by humans as they try to minimize the biological and socio-economic impacts resulting from these changes.

The Role of Acclimation, Adaptation, and the Rate of Change

Biological systems are not constant and have some ability to respond to environmental change. This ability can be expressed in various ways and has the potential to ameliorate the impact of a changing climate in some instances.

Coral reefs are instructive in this respect, since they are already experiencing the effects of climate change. Corals and their symbiotic dinoflagellates are able to acclimatize (change physiologically) to changes that may occur on a diurnal, weekly, or even yearly basis (e.g., Berkelmans and Willis 1999; Gates and Edmunds 1999). Symbiotic dinoflagellates also show strong abilities to acclimatize to higher light intensities during the day by modifying the concentration and redox state of biochemical constituents like xanthophylls (Brown et al. 1999). These changes can reduce the impact of high light conditions on the symbiotic dinoflagellates and are a prime example of physiological acclimation. Nonetheless, there are limits to both the extent and rate at which the physiological properties can change. If light levels are increased too much or are changed too rapidly, the symbiotic dinoflagellates of corals will experience chronic photoinhibition with resulting photo-bleaching and even mortality (Hoegh-Guldberg and Smith 1989; Baker 2001).

Acclimation by symbiotic dinoflagellates to sea temperatures appears to be no different from the process of acclimatizing to light. As discussed above, the symbiotic dinoflagellates of the coral *Pocillopora damicornis* have acclimated to the lower temperatures of seawater in winter (Berkelmans and Willis 1999). The corals and their symbiotic dinoflagellates in the same study did not have an infinite capacity for acclimation and died when temperatures exceeded their capacity for acclimation.

Some authors have suggested that coral populations might be able to adapt (evolve) to the warmer conditions of the future (Done 1999; Baker 2001; but see Hoegh-Guldberg et al. 2002). These studies point to the fact that corals from different parts of the world have different thermal optima, and those closer to the equator having higher thermal thresholds. However, corals are slow-growing and largely asexual organisms that are unlikely to change rapidly. While corals are adapted to different water temperatures across the globe, the time span over which this probably developed probably measures in the hundreds if not thousands of years (Hoegh-Guldberg 1999).

Range Shifts

There is ample evidence of a poleward movement of species ranges within marine ecosystems. This movement matches that seen for terrestrial organisms (Parmesan et al. 2000). As discussed above, intertidal organisms (Barry et al. 1995; Southward et al. 1995) and benthic fish (Holbrook et al. 1997) have moved to higher latitudes as a result of recent warming trends. These changes have management implications such as warm-water species replacing colder-water species and the invasion of pest organisms and the individual effect that species might have on the structure of the habitat (e.g., more or less grazing pressure might alter the standing stock of marine algae).

Not all organisms will experience range shifts. Reef-building corals, for example, have been proposed to move poleward as water temperatures at high latitudes warm, leading to coral reef development off coastlines that currently do not have significant coral reef development. However wonderful this scenario may sound (coral reefs off New York or Sydney), corals are ultimately limited by light levels and possibly carbonate alkalinity (which decreases in a poleward direction). These

factors are likely to limit coral reefs to small changes in their latitudinal range.

Phase Shifts

When the capacity for physiological acclimation and/or migration has been exceeded, mortality rates increase dramatically. If mortality differs among community members, it will cause a shift in the community dominance. For example, in many parts of the world, coral communities have shifted over to communities dominated by macroalgae under the influence of reduced grazing pressure (caused by overfishing), elevated nutrients, and other factors (e.g., Hughes 1994; Jackson et al. 2001).

Some coral reefs have already undergone phase shifts as a result of climate change exacerbated by other anthropogenic influences. In the 1998 global event, communities of corals were devastated by warmer than normal conditions (GCRMN 2000). In the Indian Ocean, some regions lost all their reef-building coral communities and the overall average loss for the Indian Ocean was 46 percent. In parts of the Caribbean, the total shift from branching Acroporid communities appears to have been unique in the past 3000 years (Aronson et al. 2002).

Projections of how changing conditions will influence the health and abundance of coral communities in the future can be derived from observations of how reefs have responded to conditions over the past 30 years. Combining these data with projections of future climates provides a snapshot of what might happen to coral reef communities under sustained warming (Fig. 16.4) (Hoegh-Guldberg 1999). In this regard, the size and length of thermal anomalies in sea surface temperature appear to be invaluable predictors of whether a bleaching event will occur or not and whether it will result in a mass mortality event (Hoegh-Guldberg 1999; Strong et al. 2000; Hoegh-Guldberg 2001).

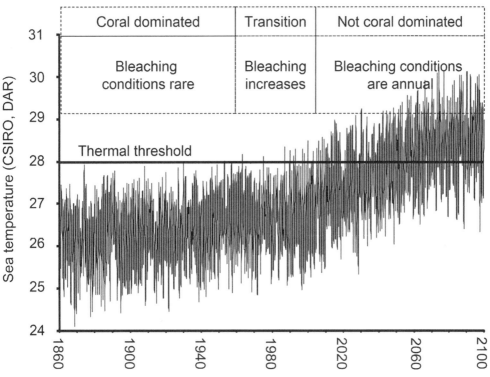

Figure 16.4. Projection of sea temperatures in the central Pacific (Rarotonga, Cook Islands) from 1860–2100 using the CSIRO DAR model (Hoegh-Guldberg 1999). The line indicates the thermal threshold for corals and symbiotic dinoflagellates from Rarotonga (3–4-week exposure). The projected sea temperatures that exceed the thermal threshold are used to project changes in the frequency of bleaching and mortality of reef-building corals. From these projections, some broad projections of how the abundance of reef-building corals will change over the next hundred years can be derived.

These types of analyses all have similar conclusions. Using the methodology of Strong and colleagues (2000; http://www.osdpd.noaa.gov / PSB / EPS / SST / dhw__retro.html), degree-heating-weeks (anomalies multiplied by time of exposure) can be calculated and compared to the outcome of recent events. These analyses reveal that thermal events that exceed 8 degree-heating weeks have almost always (99 percent of the time) resulted in coral bleaching. If conditions improve, bleached corals will recover their symbionts and

hence their brown color. If heating continues, however, and the degree-heating-weeks attains values of 13 or more, the event is likely to result in large-scale coral mortality.

Projections of how the frequency and intensity of thermal events will change over the next hundred years result in some rather worrying and inescapable conclusions. Even under a relatively mild scenario such as the IS92a (a doubling of carbon dioxide equivalents by 2100), bleaching is set to increase dramatically until it becomes an annual event by 2030–50 for most parts of the world (Hoegh-Guldberg 1999). If the degree-heating-weeks of these future events is calculated, the values rise to well over 40 for most tropical oceans by 2100 (Hoegh-Guldberg 2001). Given that major and almost complete mass mortality events occur for degree-heating-weeks values that exceed 13, it may be estimated that conditions such as those seen in the worst areas of the world's tropical oceans in 1998 will be annual events by 2030–50. Coral dominated ecosystems may become remnants of the past if sea temperatures continue to climb.

SUMMARY

The ramifications of climate change for the biodiversity of marine ecosystems are complicated by the variety of responses seen by ecosystems so far. This chapter has attempted to present an overview of the possible changes that are likely to be seen by marine ecosystems as greenhouse gas concentrations increase in the atmosphere. The types of organisms that will be living off our coastlines in 20 years' time will be substantially different from those found there today. They may be dramatically different in as few as 50 years.

Some but not all members of current ecosystems may move poleward, creating new ecological relationships and, indeed, new ecosystems. For example, fishes that are normally associated with coral reefs may migrate northward yet the corals that normally house them may not. Some coral dependents within these fish groups will not move. These new ecological relationships will produce new challenges for the managers of marine ecosystems, who will need to understand the dynamics of these new ecological relationships. Understanding the impact of climate change on marine ecosystems is in its infancy. Nonetheless, the projection that some structural elements such as corals will be largely removed (Hoegh-Guldberg 1999) has huge implications for the numerous other species that are dependent on coral reefs for their habitat. As many of the organisms and most of the interactions are still unknown to science, there is an imperative to document and understand this important element of the earth's biosphere before it is irrevocably changed.

REFERENCES

Aarset, A. V. 1982. Freezing tolerance in intertidal invertebrates (a review). *Comp. Biochem. Physiol.* 73A: 571–580.

Abele, L., Patton, W. K. 1976. The size of coral heads and the community biology of associated decapod crustaceans. *Journal of Biogeography* 3: 35–47.

Alheit, J., Hagen, E. 1997. Long-term climate forcing of European herring and sardine populations. *Fish. Oceanogr.* 6(2): 130–139.

Aronson, R. B., Precht, W. F., Macintyre, I. G., Murdoch, T. J. T. 2000. Ecosystems: Coral bleach-out in Belize. *Nature* 405: 36.

Aronson, R. B., I. G. Macintyre, W. F. Precht, T. J. T. Murdoch, and C. M. Wapnick. 2002. The expanding scale of species turnover events on coral reefs in Belize. *Ecological Monographs* 72: 233–249.

Attrill, M., Power, M. 2002. Climatic influence on a marine fish assemblage. *Nature* 417: 275–278.

Babcock Hollowed, A., Hare, S. R., Wooster, W. S. 2001. Pacific Basin climate variability and patterns of Northeast Pacific marine fish production. *Progress in Oceanography* 49: 257–282.

Baker, A. J. 2001. Reef corals bleach to survive change. *Nature* 411: 765–766.

Barbraud, C., Weimerskirch, H. 2001. Emperor penguins and climate change. *Nature* 411: 183–186.

Barnett, T. P., Pierce, D. W., Schnur, R. 2001. Detection of anthropogenic climate change in the worlds oceans. *Science* 292: 270–274.

Barry, J. P., Baxter, C. H., Sagarin, R. D. 1995. Climate related, long-term faunal changes in California rocky intertidal community. *Science* 267: 672–675.

Beaugrand, G., Reid, P. C., Ibanez, F., Lindley, J. A., Edwards, M. 2002. Reorganization of North Atlantic marine copepod biodiversity and climate. *Science* 296: 1692–1694.

Berkelmans, R., Oliver, J. K. 1999. Large-scale bleaching of corals on the Great Barrier Reef. *Coral Reefs* 18(1): 55–60.

Berkelmans, R., Willis, B. L. 1999. Seasonal and local spatial patterns in the upper thermal limits of corals on the inshore Central Great Barrier Reef. *Coral Reefs* 18(3): 219–228.

Black, R., Prince, J. 1983. Fauna associated with the coral Pocillopora damicornis at the southern limit of its distribution in western Australia. *Journal of Biogeography* 10: 135–152.

Brown, B. E. 1997. Coral bleaching: Causes and consequences. *Coral Reefs* 16, Suppl: S129–S138.

Brown, B. E., Ambarsari, I., Warner, M. E., Fitt, W. K., Dunne, R. P., Gibb, S. W., Cummings, D. G. 1999. Diurnal changes in photochemical efficiency and xanthophyll concentrations in shallow water reef corals: Evidence for photoinhibition and photoprotection. *Coral Reefs* 18: 99–105.

Bryant, D., Burke, L., McManus, J., Spalding, M. 1998. *Reefs at Risk: A Map-Based Indicator of Threats to the World's Coral Reefs*. Washington, D.C.: World Resources Institute.

Buddemeier, R. W., and Fautin, D. G. 1993. Coral bleaching as an adaptive mechanism: A testable hypothesis. *BioScience* 43(5): 320–326.

Church, J. A., Gregory, J. M., Huybrechts, P., Kuhn, M., Lambeck, K., Nhuan, M. T., Qin, D., and Woodworth, P. L. 2001. Changes in sea level. In *Climate Change (2001). The Scientific Basis. Contribution of Working Group 1 to the Third Assessment Report of the Intergovernmental Panel on Climate Change*, Houghton, J. T., Ding, Y., Griggs, D. J., Noguer, M., van der Linden, P., Dai, X., Maskell, K., and Johnson, C. I., eds., Cambridge: Cambridge University Press, 639–694.

Clark, P. U., Mitrovica, J. X., Milne, G. A., Tamisiea, M. E. 2002. Sea level fingerprinting as a direct test for the source of global meltwater pulse IA. *Science* 295: 2376–2377.

Coles, S. L., Jokiel, P. L., and Lewis, C. R. 1976. Thermal tolerance in tropical versus subtropical Pacific coral reefs. *Pac. Sci.* 30: 159–166.

Darwin, C. R. 1842. *The Structure and Distribution of Coral Reefs*. London: Smith Elder and Company.

de la Mare, W. 1997. Abrupt mid-twentieth-century decline in Antarctic sea-ice. *Nature* 389: 57–60.

Dickson, R., Yashayaev, I., Meincke, J., Turrell, W., Dye, S., Holfort, J. 2002. Rapid freshening of the deep North Atlantic Ocean over the past four decades. *Nature* 416: 832–837.

Done, T. J. 1999. Coral community adaptability to environmental change at the scales of regions, reefs, and reef zones. *American Zoologist* 39: 66–79.

Dove, S. G., Hoegh-Guldberg, O., Ranganathan, S. 2001. Major colour patterns of reef-building corals are due to a family of GFP-like proteins. *Coral Reefs* 19(3): 197–204.

Edmunds, P. J. 1994. Evidence that reef-wide patterns of coral bleaching may be the result of the distribution of bleaching-susceptible clones. *Mar. Biol.* 121: 127–142.

Feder, M. E., Hoffmann, G. E. 1999. Heat-shock proteins, molecular chaperones, and the stress response: Evolutionary and ecological physiology. *Annu. Rev. Physiol.* 61: 243–282.

Gagan, M. K., Ayliffe, L. K., Beck, J. W., Cole, J. E., Druffel, E. R. M., Dunbar, R. B., Schrag, D. P. 2000. New views of tropical paleoclimate from Corals. *Quaternary Science Reviews* 19: 167–182.

Gates, R. D. 1990. Seawater temperature and sublethal bleaching in Jamaica. *Coral Reefs* 8: 193–197.

Gates, R. D., and Edmunds, P. J. 1999. The physiological mechanisms of acclimatization in tropical reef corals. *American Zoologist* 39: 30–43.

Gattuso, J.-P., Frankignoulle, M., Bourge, I., Romaine, S., Buddemeier, R. W. 1998. Effect of calcium carbonate saturation of seawater on coral calcification. *Global Planetary Change* 18: 37–47.

GCRMN. 2000. *Status of Coral Reefs of the World*, ed. R. Wilkerson. Queensland: Australian Institute of Marine Science.

Gille, S. T. 2002. Warming of the Southern Ocean since the 1950s. *Science* 295: 1275–1277.

Hatcher, B. G. 1988. Coral reef primary productivity: A beggar's banquet. *Trends in Ecology and Evolution* 3: 106–11.

Hendy, E. J., Gagan, M. K., Alibert, C. A., McCulloch, M. T., Lough, J. M., Isdale, P. J. 2002. Abrupt decrease in tropical Pacific sea surface salinity at end of Little Ice Age. *Science* 295: 1511–1514.

Hochachka, P. W., Somero, G. N. 1984. *Biochemical Adaptations*. New York: Wiley.

Hodgson, G., and Liebeler, J. 2001. The global coral reef crisis: Trends and solutions 1997–2001. Reef Check Foundation Publications, USA. Available from http://www.reefcheck.org

Hoegh-Guldberg, O. 1999. Coral bleaching, climate change and the future of the world's coral reefs. *Marine and Freshwater Research Mar. Freshwater Res.* 50: 839–866.

Hoegh-Guldberg, O. 2001. The future of coral reefs: Integrating climate model projections and the recent behaviour of corals and their dinoflagellates. *Proceedings of the Ninth International Coral Reef Symposium*. October 23–27, 2000, Bali, Indonesia.

Hoegh-Guldberg, O., Jones, R. J., Ward, S., Loh, W. 2002. Is coral bleaching really adaptive? *Nature* 415: 601–602.

Hoegh-Guldberg, O., Salvat, B. 1995. Periodic mass-bleaching and elevated sea temperatures: Bleaching of outer reef slope communities in Moorea, French Polynesia. *Mar. Ecol. Prog. Ser.* 121: 181–190.

Hoegh-Guldberg, O., Smith, G. J. 1989. The effect of sudden changes in temperature, light and salinity on the population density and export of zooxanthallae from the reef corals *Stylophora pistillata* Esper and *Seriatopora hystrix* Dana. *J. Exp. Mar. Biol. Ecol.* 129: 279–303.

Holbrook, S. J., Schmitt, R. J., Stephens Jr., J. A. 1997. Changes in an assemblage of temperate reef fishes associated with a climate shift. *Ecological Applications* 7: 1299–1310.

Hughes, T. P. 1994. Catastrophes, phase-shifts and large-scale degradation of a Caribbean coral reef. *Science* 265: 1547–1551.

IPCC. 2001. Intergovernmental Panel on Climate Change. Summary for policymakers. A Report of Working Group I of the Intergovernmental Panel on Climate Change, pp. 1–20. R. T. Watson, ed. Cambridge: Cambridge University Press.

Jackson, J. B. C., Kirby, M. X., Berger, W. H., Bjorndal, K. A., Botsford, L. W., Bourque, B. J., Bradbury, R. H., Cooke, R., Erlandson, J., Estes, J. A., Hughes, T. P., Kidwell, S., Lange, C. B., Lenihan, H. S., Pandolfi, J. M., Peterson, C. H., Steneck, R. S., Tegner, 2001. Historical overfishing and the recent collapse of coastal ecosystems. *Science* 293: 629–638.

Kelly, B. P. 2001. Climate change and ice-breeding pinnipeds. In *Fingerprints of Climate Change*, ed. Walther, G.-R., Burga, C. A., and Edwards, P. J. New York: Kluwer Academic/Plenum, pp. 43–56.

Kepler, C. B., Kepler, A. K., Ellis, D. H. 1994. The natural history of Caroline Atoll, Southern Line Islands, 1. Part II. Seabirds, other terrestrial animals, and conservation. *Atoll-Research-Bulletin* 0 (398) I-III: 1–61.

Kerr, R. A. 1999. Climate change: Will the Arctic Ocean lose all its ice? *Science* 286: 1828–1829.

Kleypas J. A., Buddemeier R. W., Archer D., Gattuso J.-P., Langdon C., Opdyke B. N. 1999. Geochemical consequences of increased atmospheric carbon dioxide on coral reefs. *Science* 284: 118–120.

Klyashtorin, L. 1998. Long-term climate change and main commercial fish production in the Atlantic and Pacific. *Fisheries Research* 37: 115–125.

Kokita, T., Nakazono, A. 2001. Rapid response of an obligately corallivorous filefish *Oxymonacanthus longirostris* (Monacanthidae) to a mass coral bleaching event. *Coral Reefs* 20: 155–158.

Langdon, C., Takahashi, T., Marubini, F., Atkinson, M., Sweeney, C., Aceves, H., Barnett, H., Chipman, D., Goddard, J. 2000. Effect of calcium carbonate saturation state on the calcification rate of an experimental coral reef. *Global Biogeochemical Cycles* 14: 639–654.

Lindahl, U., Ohman, M. C., Schelten, C. K. 2001. The 1997/1998 mass mortality of corals: Effects on fish communities on a Tanzanian coral reef. *Marine-Pollution-Bulletin* 42: 127–131.

Linsley, B. K., Wellington, G. M., Schrag, D. P. 2000. Decadal sea surface temperature variability in the sub-tropical South Pacific from 1726 to 1997 A.D. *Science* 290: 1145–1148.

Loeb, V., Siegel, V., Holm-Hansen, O., Hewitt, R., Fraserk, W., Trivelpiecek, W., Trivelpiecek, S. 1997. Effects of sea-ice extent and krill or salp dominance on the Antarctic food web. *Nature* 387: 897–900.

Lough, J. M. 1999. Coastal climate of northwest Australia and comparisons with the Great Barrier Reef: 1960 to 1992. *Coral Reefs* 17: 351–367.

Maragos, J. E. 1994. Description of reefs and corals for the 1988 protected area survey of the northern Marshall Islands. *Atoll-Research-Bulletin* 419: 1–88.

Meyers, R. A., Worm, B. 2003. Rapid worldwide depletion of predatory fish communities. *Nature* 423: 280–283.

Muscatine, L. 1990. The role of symbiotic algae in carbon and energy flux in reef corals. *Coral Reefs* 25: 1–29.

Muscatine, L., and Porter, J. W. 1977. Reef corals: mutualistic symbioses adapted to nutrient-poor environments. *Bioscience* 27(7): 454–60.

Nichols, R., Hoozemans, F. M. J., Marchand, M. 1999. Increasing food risk and wetland losses due to global sea-level rise: Regional and global analyses. *Global Environmental Change* 9: S69–S87.

Odum, H. T., Odum, E. P. 1955. Trophic structure and productivity of a windward coral reef community on Eniwetok Atoll. *Ecological Monographs* 25: 291–320.

Overpeck, J., Hughen, K., Hardy, D., Bradley, R., Case, R., Douglas, M., Finney, B., Gajewski, K., Jacoby, G., Jennings, A., Lamoureux, S., Lasca, A., MacDonald, G., Moore, J., Retelle, M., Smith, S., Wolfe, A., Zielinski, G. 1997. Arctic environmental change of the last four centuries. *Science* 278: 1251–1256.

Parmesan, C., Root, T. R., Willig, M. R. 2000. Impacts of extreme weather and climate on terrestrial biota. *Bulletin of American Meteorological Society* 81: 443–450.

Parmesan, C., Yohe, G. 2003. A globally coherent fingerprint of climate change impacts across natural systems. *Nature* 421: 37–43.

Reid, P., de Fatima Borges, M., Svendsen, E. 2001. A regime shift in the North Sea circa 1988 linked to changes in the North Sea horse mackerel fishery. *Fisheries Research* 50: 163–171.

Sharp, V. A., Brown, B. E., Miller, D. 1997. Heat shock protein (hsp 70) expression in the tropical reef coral *Goniopora djiboutiensis*. *Journal of Thermal Biology* 22: 11–19.

Southward, A. J., Hawkins, S. J., Burrows, M. T. 1995. Seventy years' observations of changes in distribution and abundance of zooplankton and intertidal organisms in the western English channel in relation to rising sea temperature. *J. Therm. Biology* 20: 127–155.

Storey, K. B., Storey, J. M. 1988. Freeze tolerance in animals. *Physiol. Rev.* 68: 27–84.

Storey, K. B., Storey, J. M. 1992. Natural freeze tolerance in ectothermic vertebrates. *Ann. Rev. Physiol.* 54: 619–637.

Storey, K. B., Storey, J. M. 1996. Natural freezing survival in animals. *Annu. Rev. Ecol. Syst.* 27: 365–386.

Strong, A. E., Barrientos, C. S., Duda, C., Sapper, J. 1996. Improved satellite techniques for monitoring coral reef bleaching. *Proceedings of the Eighth International Coral Reef Society Symposium*. Panama.

Strong, A. E., K. K. Gjovig, E. Kearns, 2000. Sea surface temperature signals from satellites—An update. *Geophys. Res. Lett.* 27(11): 1667–1670.

Tomanek, L., and Somero, G. N. 2000. Time course and magnitude of synthesis of heat-shock proteins in congeneric marine snails (Genus *Tegula*) from different tidal heights. *Physiological and Biochemical Zoology* 73: 249–256.

Tomascik, T., Sander, F. 1987. Effects of eutrophication on reef-building corals. III. Reproduction of the reef-building coral porites. *Marine Biology* 94: 77–94.

Tudhope, A. W., Chilcott, C. P., McCulloch, M. T., Cook, E. R., Chappell, J., Ellam, R. M., Lea, D. W., Lough, J. M., Shimmield, G. B. 2001. Variability in the El Niño–Southern Oscillation through a glacial-interglacial cycle. *Science* 291: 1511–1517.

United Nations. 2002. World Population Prospects: The 2002 Revision. I. Comprehensive Tables.

Walther, G.-R., Post, E., Convey, P., Menzel, A., Parmesan, C., Beebee, T. J. C., Fromentin, J.-M., Hoegh-Guldberg, O., Bairlein, F. 2002. Ecological responses to recent climate change. *Nature* 416: 389–395.

CHAPTER SEVENTEEN

Climate Change and Freshwater Ecosystems

J. DAVID ALLAN, MARGARET PALMER,
AND N. LEROY POFF

To date, about 17,000 species exclusive of fish have been described in freshwater systems, and this is considered a gross underestimate because of the large number of undescribed species, particularly the small sediment fauna (Abell et al. 2000; Palmer et al. 2000). The most diverse groups (including estimated undescribed species) include the microbes and algae. Among the freshwater invertebrates, global species richness is highest in the nematodes and rotifers, insects, crustaceans, annelids, and mollusks, respectively (Palmer and Lake 2000). Many freshwater wetlands are extremely speciose, with over 2000 species at a site, with rivers and lakes typically harboring 80–1400 species and groundwater regions generally fewer than 150.

Some 41 percent of the 28,000 known fish species live in fresh water and about 1 percent of the remainder spend part of their lives in fresh water (Moyle and Cech 1996). East African rift valley lakes individually contain several hundred cichlid species and have attracted the attention of evolutionary biologists fascinated by their functional specialization and adaptive radiation. Lake Baikal, an ancient lake and the world's largest by volume, has many endemic species (65 percent of animals, 35 percent of plants; Burgis and Morris 1987). North America is rich in both vertebrate and invertebrate freshwater species (Allan and Flecker 1993), including nearly one-third of the world's described freshwater mussels and over 60 percent of the world's described freshwater crayfish (Master et al. 1998). Unfortunately, North America's freshwater environments have been identified as among those most threatened (Abell et al. 2000).

Extinction rates of freshwater fauna are extremely high. Ricciardi and Rasmussen (1999) documented current extinction rates of 0.4 percent per decade for freshwater fish, 0.1 percent for crayfish, and 0.8 percent for gastropods. Assuming extinction rates will continue to increase at the same rate as they have in the past century, Ricciardi and Rasmussen estimated future rates of 2.4 percent, 3.9 percent, and 2.6 percent for these same groups, respectively. These data compare to much lower current and future extinction rates for most terrestrial fauna (e.g., birds = 0.3 percent now, 0.7 percent future). The threats to freshwater fauna fall into several broad categories: nutrient enrichment, hydrological modifications, habitat loss and degradation, pollution, and the spread of invasive species. A changing climate and increasing levels of UV light pose additional risks that superimpose upon existing threats. The combination of rapid land-use change, habitat alteration, and a changing climate is viewed as a particularly serious challenge to aquatic ecosystems (Carpenter et al. 1992; Meyer et al. 1999; Lake et al. 2000). This is one example of the synergies that appear to be all too common between climate forcing and other stressors.

THE IMPORTANCE OF FRESH WATER AND FRESHWATER ECOSYSTEMS

Surface fresh waters are a small fraction of global water. Freshwater lakes constitute 0.009 percent of water in the biosphere, and rivers are one hundred–fold less by volume (Wetzel 2001). By contrast, polar ice and glaciers make up 2.08 percent of the global total. Because the freshwater supply is unevenly distributed over land surfaces and threatened by contamination in many places, the real supply is even less than these small numbers indicate. It has been estimated that humans appropriate roughly one quarter of the global renewable freshwater supply (Postel et al. 1996), and further population growth will cause this fraction to grow substantially (Vörösmarty et al. 2000), placing increasing pressure on freshwater ecosystems.

Healthy freshwater ecosystems provide vital ecosystem services to human societies (Naiman et al. 1995; Gleick 1998; Carpenter and Lunetta 2000), including the provision of clean water for drinking, for agriculture, for fisheries, and for recreation (Table 17.1). While the value of such ecosystem services is not easy to quantify in economic terms (Toman 1997), the impact of the loss of such services is often deeply felt. Many regions in the world have insufficient clean water to meet even the minimal demands for human survival (Postel 1997; Gleick et al. 2002), some countries are experiencing an increase in water-borne diseases (Sattenspiel 2000), while others are experiencing rapid declines in freshwater fishery yields (WRI 2000).

Maintenance of a diverse freshwater biota may be key to the retention of services provided by freshwater ecosystems (Covich et al. 1997; Palmer et al. 1997, 2000). Assemblages that are diverse may be able to utilize resources more efficiently, resulting in more productive systems; or they may offer "insurance" against ecosystem collapse in the face of disturbance (Loreau et al. 2001; Cardinale et al. 2002; Naeem 2002). Human actions that harm ecosystem health thus may also threaten human health and require costly replacement of damaged ecosystem services. For example, the city of New York has embarked on a watershed protection plan based on land acquisition at a cost of approximately $300 million, to prevent the need to construct a several-billion-dollar filtration system (Featherstone 1996), demonstrating the economic value of an undamaged freshwater resource.

The remainder of this chapter reviews the most likely and best anticipated impacts of future climate change on freshwater biodiversity, with an emphasis on

Table 17.1. Goods and Services Provided by Freshwater Ecosystems

Goods and Services	Streams and Rivers	Lakes and Ponds	Freshwater Wetlands	Coastal Wetlands
Water Supply				
Drinking, cooking, washing, and other household uses	X	X	—	—
Manufacturing, thermoelectric power generation, and other industrial uses	X	X	—	—
Irrigation of crops, parks, golf courses, etc.	X	X	X	—
Aquaculture	X	X	X	X
Supply of goods other than water				
Fish	X	X	X	X
Waterfowl	X	X	X	X
Clams, mussels, other shellfish, crayfish	X	X	X	X
Timber products	X	—	X	—
Nonextractive benefits				
Biodiversity	X	X	X	X
Flood control	X	X	X	—
Transportation	X	X	—	X
Recreational swimming, boating, etc.	X	X	X	X
Pollution dilution and water quality protection	X	X	X	X
Hydroelectric generation	X	—	—	—
Bird and wildlife habitat	X	X	X	X
Enhanced property values	X	X	X	X
Coastal shore protection	—	—	—	X

Source: Poff et al. (2002); modified from Postel and Carpenter (1997).

mid- and high-latitude regions, where effects are expected to be most severe.

CLIMATE CHANGE AND THE HYDROLOGIC CYCLE

Freshwater ecosystems will naturally be sensitive to changes in the hydrologic cycle, and these are difficult to predict. A warmer climate will result in greater evaporation from water surfaces and greater transpiration by plants, which will result in a more vigorous water cycle. However, whether rainfall will increase or decrease in a particular region is uncertain, and surface waters will decline even if precipitation increases, if evapotranspiration (ET) increases by a greater amount. General cir-

culation models (GCMs) are not yet able to reliably predict how precipitation and water levels will change at the local or regional level (Wigley 1999; NAST 2000), although downscaling and regional modeling efforts are in progress in many parts of the world in order to predict localized impacts of climate change (Neff et al. 2000; Polsky et al. 2000). There is still a great deal of uncertainty in climate change forecasts (Elzen and Schaeffer 2002; Forest et al. 2002; Heal and Kristrom 2002). For example, while earlier projections called for declines in levels of the Great Lakes of one meter or more (Mortsch and Quinn 1996), more recent models predict both increases and decreases, and of generally lesser magnitude (Lofgren et al. 2002). Frederick and Gleick (1999) examined

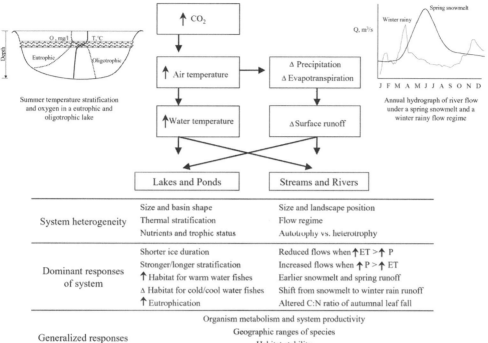

Figure 17.1. Linkages between atmospheric increases in CO_2 and environmental drivers of temperature and precipitation that regulate many physical and ecological processes in lakes and ponds (left) and rivers and streams (right). Studies of climate change impacts on lakes have emphasized responses to warming, which are affected by vertical temperature stratification. Studies of climate change impacts on rivers have emphasized responses to altered flow regime, including changes to magnitude, frequency, duration, and timing of discharge events. Some biological responses shown at the bottom of the figure are general.

runoff for 18 water resource regions of the United States using two contrasting GCMs and found that predictions often were in disagreement. The two models predicted the same direction of change in runoff in only 9 of the 18 regions, and where the direction was similar, often the magnitude was not. Perhaps the greatest single challenge in evaluating aquatic ecosystem response to future climate change is the considerable uncertainty regarding the local and regional responses of the hydrologic cycle.

IMPACTS ON FRESHWATER ECOSYSTEMS

In general, analyses of impacts on lakes have emphasized responses to changing temperature, while analyses of rivers and streams have emphasized changes to the amount and timing of flow (Fig. 17.1). The approach of this review is similar, although clearly some impacts of climate warming, such as higher metabolism and productivity, and poleward range shifts, are very general. In addition, forested streams are highly dependent upon inputs of terrestrial organic matter, especially leaf fall, for their energy supply, and so shifts in terrestrial vegetation and changes in leaf chemistry provide another, quite intricate set of pathways by which stream biota and ecosystems can be affected. Due to space limitations, wetlands are not included in this review. However, those wetlands dependent upon surface runoff (rather than groundwater) are especially sensitive to drying, and coastal wetlands to saltwater intrusion due to rising sea levels (IPCC

2001; Pearsall, this volume). Melting of permafrost, drying of peat, and resultant release of methane and CO_2 are of special concern not only because of the loss of many small boreal wetlands, but also because peat-accumulating wetlands are a carbon sink that could become a carbon source and positive feedback under future climate warming.

Lakes and Ponds

Freshwater lakes range in size from Lake Baikal, the deepest (1620 m maximum depth) and largest (23,000 km^3 by volume) lake, and the Laurentian Great Lakes, with a collective volume of 24,620 km^3, to literally millions of small lakes that dot glaciated landscapes and are usually less than 10 m in depth (Wetzel 2001). Subarctic lakes and prairie potholes are usually very small and shallow. Differences among lakes in surface area, depth, latitude and elevation, and water residence times are all factors that will influence their response to climate change. The more than 36,000 dams over 15 m high in operation worldwide have a combined maximum impoundment volume of approximately 8400 km^3, about seven times the volume of natural river water (Vörösmarty et al. 1997), and millions of farm ponds add substantially to the extent of standing fresh water.

Lakes can be classified according to their vertical temperature profile and the seasonality of that profile. During summer at mid-latitudes, lakes of sufficient depth typically develop thermal stratification due to the density difference between warmer and cooler water, resulting in a warm layer (the epilimnion), a cold and deep layer (the hypolimnion), and a zone of rapid temperature change in between known as the thermocline. The epilimnion is oxygenated and biologically productive, while the hypolimnion may experience oxygen limitation due to decomposition of organic matter raining down from

above. Climate interacts with lake basin shape and wind (which affects the depth of mixing of surface waters) to determine the strength and duration of thermal stratification.

PHYSICAL EFFECTS

The duration of ice cover for northern hemisphere lakes provides one of the strong signals of global climate change over the past 150 years (Magnuson et al. 2000; IPCC 2001). On average, from 1846 to 1995 the freeze date was 8.7 days later and the ice breakup 9.8 days earlier (Fig. 17.2). As well, the inter-annual variability in freeze dates, thaw dates, and ice duration has increased (Kratz et al. 2001). Climate-induced shortening of ice duration will affect evaporation rates and lake metabolism (IPCC 2001).

Future climate change will directly affect lake ecosystems through warmer temperatures and changes to the hydrologic cycle. However, the heterogeneity of lake types and locations will interact with

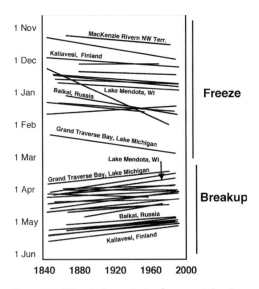

Figure 17.2. Historical trends in freeze and breakup dates of lakes and rivers in the Northern Hemisphere; 37 of the 39 trend slopes are in the direction of warming. Source: IPCC (2001); modified from Magnuson et al. (2000).

climate drivers in diverse ways. Lake size and depth, exposure to wind, and latitude or altitude will determine the existence, strength, and duration of thermal stratification, and therefore the seasonal amount of cold-, cool-, and warm-water habitat available. The water supply to lakes may be differently affected by future climate change depending on whether the water originates from glacier, snowmelt, rain-fed, or groundwater-fed sources. Water supplied by glaciers and snowmelt may increase initially but decrease in the long run, while small wetlands and prairie pothole lakes are examples of systems that may simply disappear depending on the balance between precipitation and evapotranspiration. Increases in evapotranspiration brought about by higher temperatures, longer growing seasons, and extended ice-free periods, unless offset by equal or greater increases in precipitation, are likely to result in reduced lake levels and river inputs. In cases where precipitation and evapotranspiration both increase, lake levels might change little but water residence time in lakes would be expected to be shortened.

IMPACTS ON BIOTA

The freshwater biota is dominated by cold-blooded organisms, and in general ectotherms increase their metabolism with increase in temperature until they approach their upper temperature tolerances. Rates generally increase by a factor of 2–4 with each 10°C increase in water temperature, up to about 30°C (Regier et al. 1990). In a review of approximately 1000 studies of macroinvertebrate production, Benke (1993) estimated a 3–30 percent increase in production for each 1°C increase in temperature. Thus while there may be complex and unpredictable changes in species composition, an overall increase in system productivity is likely to be a common response to climate warming.

A strong case can be made that future climate warming will alter the extent of habitat available for cold-, cool-, and warm-water organisms depending upon region, and result in range expansions and contractions (Fig. 17.3). Species at the

Figure 17.3. Ten-year average lake temperatures simulated using the Canadian Climate Center Atmosphere Ocean General circulation model as input data. Source: Poff et al. (2002); based on Hostetler and Small (1999).

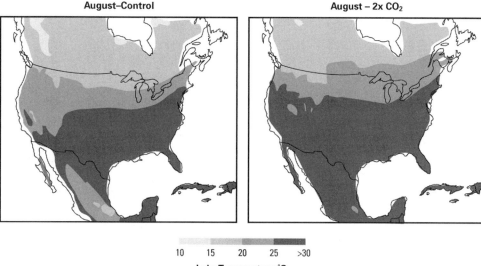

August–Control   **August – 2x CO$_2$**

10   15   20   25   >30
Lake Temperature °C

southern extent of their geographical distribution (in the north temperate zone) will shift northward and face local extirpation at their southern limit, while expanding at the northern limit of their range. It is estimated that a 4°C warming results in a 640-km northward latitudinal shift in thermal regimes for macroinvertebrates (Sweeney et al. 1992) and a roughly 500-km northward shift for smallmouth bass and yellow perch. While useful as a first-order approximation, such projections assume that water temperatures warm about as much as air temperatures (valid for shallow, well-mixed systems), that dispersal corridors are available, and that other system effects including biological interactions are unaltered. While many aquatic insects have aerial dispersal, fish and other organisms that require an aquatic dispersal corridor may lack the opportunity to disperse, due to the isolated nature of some lakes, dams that block connecting rivers, and land divides that separate basins such as the Laurentian Great Lakes from the Nelson and Mackenzie basins of Canada.

The effect of climate warming on lakes is further complicated by their thermal stratification. In large deep lakes such as Lakes Huron, Superior, and Michigan, a warming of 3.5°C is expected to increase suitable thermal habitat for warm-water fishes that occupy the epilimnion during summer. Cold- and cool-water fishes of the hypolimnion are also expected to benefit, because slight warming will promote metabolic activity yet remain well within thermal tolerances (Magnuson et al. 1997). In contrast, smaller and shallower lakes may experience sufficient loss of cool hyoplimnetic volume that fish requiring cold water (many top predators) may experience reduced habitat. In eutrophic (high nutrient input) lakes with a restricted area of hypolimnion, bottom waters may become depleted in oxygen due to the decomposition of algae and other organic matter that settle out of surface waters. As climate warming reduces the volume of the hypolimnion and increases the productivity of surface waters, oxygen depletion in the deep waters may become more severe.

A simulation study by Stefan and colleagues (2001) incorporated much of this complexity by exploring how 27 lake types (3 categories each of depth, area, and productivity) responded to a temperature change expected under a doubling of CO_2. Based on projected changes in temperature and hypolimnetic oxygen availability they forecast an overall 45 percent decrease in habitat for cold-water and 30 percent for cool-water fish, with greatest impact in lakes of shallow and medium depth. Warm-water fish benefited in all lake types, however.

While it is tempting to imagine these effects as simple adjustments, with some winners offsetting some losers, it is important to stress that the biological consequences of altered species assemblages are difficult to forecast and unlikely to be benign. Temperature also sets the northern range limit for harmful invasive species such as the zebra mussel (Strayer 1991), and so a northward range expansion seems highly probable. Given the well-established negative impacts of invasive species on freshwater ecosystems (Allan and Flecker 1993), native biodiversity may be adversely affected by such range shifts.

Eutrophication results from excessive nutrient inputs, promoting high biological production of algae and a preponderance of nuisance algae including the blue-green algae or Cyanobacteria (Carpenter et al. 1998). Water clarity and quality are reduced, deep and bottom waters may become anoxic from excessive decomposition, and some blue-green algae release toxins. Due to human activities, many more lakes today receive excessive nutrients from their catchments and from internal recycling, and warming is expected to increase lake productivity (IPCC 2001). However, because climate-influenced processes have interacting, seasonal, and often opposing effects (Magnuson et al. 1997; Schindler 1997), the relation between eutrophication and climate change is complex. A longer period of summer stratification will increase the likelihood of summer anoxia below the thermocline, while a shorter ice duration will reduce the likelihood of winter anoxia (Stefan and Fang 1993). More nutrients likely will be delivered to lakes by their catchments under wetter climates, and less under dryer climates, which in the latter case will result in longer water residence times (Schindler et al. 1996) and increase the importance of internal recycling. Light penetration will increase if less dissolved organic carbon (DOC) is exported from catchments into lakes, as is expected under drier conditions, and this could result in increased primary production at greater depths. These complex and offsetting interactions make it extremely difficult to predict how lake ecosystems will respond to alternative climate scenarios (Lathrop et al. 1999; IPCC 2001).

The three-way interactions among lake acidification, UV-B radiation, and climate warming, termed a "triple whammy" by Schindler (1998), are an instructive example of complex system responses. Lakes with low buffering capacity, including many north temperate lakes, experience lowered pH due to acid deposition, resulting in a number of biological and chemical changes including reduced DOC and greater water clarity. Climate warming potentially can limit the supply of buffering cations via reduced river inflow, which exacerbates the acidification process. DOC affects water clarity, especially in boreal lakes receiving DOC that is colored with dissolved humic material, and climate-induced reductions in streamflow can reduce DOC inputs to lakes. Thus acidification and climate warming interact to increase water clarity. This in turn influences the depth to which damaging UV-B radiation is able to penetrate. UV-B radiation, which has increased due to the reduction in stratospheric ozone, can directly damage the biota via molecular damage and oxidative stress (Häder et al. 2003), as well as furthering the depletion of DOC by accelerating the photolysis of organic macromolecules (Wetzel 2001) in yet another aspect of positive feedback. The ultimate effects of increased water clarity are difficult to state, because it potentially allows algal photosynthesis to occur at greater depth while permitting harmful UV-B radiation to reach greater depths as well.

Unexpected, synergistic effects are well illustrated by several of the previous examples. In one of the most thoroughly investigated areas of climate change and freshwater ecosystem response, future warming interacts with lake size and depth, position, and nutrient status to determine changes in temperature and oxygen in epilimnetic and hypolimnetic waters, thereby influencing habitat available to cold-, cool-, and warm-water fishes. Warming interacts with system connectedness and taxon dispersal capability to determine the opportunity for range shifts, and it will not be surprising if already successful invasives turn out to be best able to exploit opportunities and adversely affect native species. Warming may interact with human augmentation of nutrient supply to increase the extent of eutrophic waters, although complexities abound when we try to foresee the consequences. In some boreal lakes, acidification may interact with decreases in DOC due to reduced streamflows to increase water clarity, allowing the higher incidence of UV-B light to cause photolytic damage at greater depth than before. Clearly, shifting ranges and species substitutions are only the beginning of the anticipated impacts.

Rivers and Streams

Running waters contain only about 1 percent by volume of the fresh water occurring in lakes, but they are enormously important for transport of water and dissolved and suspended materials. Some 20 large rivers (less than 2000 km in length) contribute a major share of riverine export of water to oceans; the Amazon alone contributes 15 percent of the total (Allan 1995). As with lakes, the vast majority of running waters comprises small rivers, streams, and tributaries of larger systems. Stream order designates the smallest, permanent stream as first-order, the confluence of two stream of order n creates a stream of order n + 1, and so on. Of the approximately 5,200,000 km of rivers in the United States, nearly 50 percent are first-order, and the total for first- through third-order combined is just over 85 percent. Examples of large rivers include the Allegheny (seventh-order), the Columbia (ninth-order) and the Mississippi (tenth-order). Stream order, latitude and elevation, the relative contributions of surface water to groundwater, seasonal timing of precipitation and evapotranspiration, and additional variables of the drainage basin likely will influence response to climate change.

The ecology of rivers varies with their size (order) and landscape setting (Vannote et al. 1980). Lower order and headwater streams tend to derive much of their energy as organic matter inputs from the terrestrial ecosystem, and in forested locations they may have a closed canopy, low light levels, and stable flows. Larger and higher-order rivers tend to be more integrative of upstream influences, open to the sun, and periodically connected to their floodplains, at least in low-lying areas and where human engineering is minimal. Dams alter flow regimes and upstream–downstream connectivity, and many populated rivers receive a wide variety of contaminants along their lengths.

PHYSICAL EFFECTS

Streams and rivers respond rapidly to changes in air temperature because they are relatively shallow and well-mixed. Thus, future warming can be expected to directly increase the seasonal water temperatures of most running water ecosystems, with greater effects at more northerly (poleward) latitudes. Warm-season river temperatures usually closely approximate air temperatures, typically with a time lag of weeks or less, although streams and smaller rivers with a large component of groundwater or meltwater may be considerably cooler than summer air temperatures, corresponding to mean annual air temperature for the region. While small streams receiving glacier, snowmelt, or groundwater can experience substantial day–night temperature fluctuations, larger rivers are thermally stable over the 24-hour cycle due to the volume of water, and show little or no temperature profile, due to vertical mixing. Unlike lakes, the distribution of heat in river water is very uniform vertically.

River flow or discharge (m^3/s) typically is variable on intra-annual, inter-annual, and very long time scales, and although some groundwater-fed systems are highly stable, they are the exception. Variability in the magnitude, duration, frequency, timing, and rate of change of river flows collectively characterizes the flow regime of a river (Poff et al. 1997) and varies with region due to the influence of climate, vegetation, and geology on river flow (Poff and Ward 1989). The flow regimes of rivers already are altered by a number of human actions, most notably dams, which often regulate and stabilize river flows, and land-use change, which often causes more rapid runoff from land to receiving stream systems and makes flow more responsive to rainfall extremes.

Climate change is expected to significantly alter flow regimes as well as the total volume of river runoff by changing precipitation, evapotranspiration, and their relative magnitude. Precipitation falling on watersheds is translated into stream runoff by direct overland flow and by ground-

Present

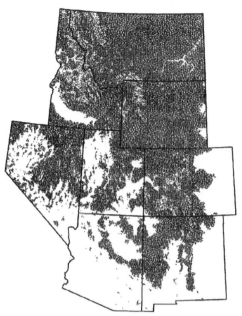

+3°C

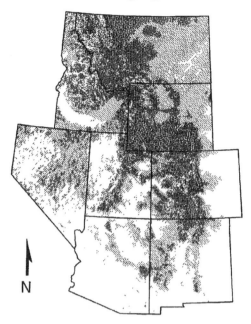

N

water flow. In humic, vegetated regions the majority of runoff follows subsurface pathways, and the majority of precipitation returns to the atmosphere as evapotranspiration. On average in the United States, about one-third of precipitation becomes stream runoff. Thus, how climate change alters terrestrial vegetation and its transpiration rates will influence runoff, even where seasonal shifts in precipitation do not occur. For example, in well-vegetated humid regions, summer flows are low even though precipitation may equal that in winter (Benke et al. 2000). In the winter rainfall area of southeastern Australia, rains have shifted to spring, and the associated increase in plant transpiration is causing a "green drought" in which streamflows are reduced while the surrounding terrestrial vegetation flourishes (P. S. Lake, personal communication). Whether due to climate change or other fluctuations, this example demonstrates how seasonal shifts in the timing of rainfall can influence the amount of water in streams.

Expected changes in regional flow

Figure 17.4. Present and future potential distribution of stream segments that could support cold-water trout (dark shading) and habitat loss (light shading) in the Rocky Mountains given a 3°C warming in July air temperatures. Source: Poff et al. (2002); based on Keheler and Rahel (1996).

regime change are uncertain due to inability to predict changes in the balance between precipitation and evapotranspiration; nevertheless, some outcomes are expected with reasonably high certainty. Glacier-fed streams are likely to experience an increase in discharge for years to decades, followed by declines. Warmer temperatures will cause a shift from winter snow to rain and snow, or primarily rain, depending upon latitude. As a result, streamflows will reflect earlier spring snowmelt or the transition to a variable winter flow regime in response to rain. In regions that currently experience a mix of rain and snow during winter, floods often result from "rain on snow" events, when snow and frozen ground cause much of the land to behave as an impervious surface, and so rainfall produces immediate and extreme runoff. In some areas of

Canada a shift in dominant mode of pre-cipitation from snow to rain has already been observed (Frederick and Gleick 1999). The location of a watershed relative to the ocean may mediate its response to climate change, however. For example, high-latitude maritime watersheds cur-rently receive less precipitation as snow compared to inland watersheds, which are therefore more vulnerable to earlier runoff as snow shifts to rain (Loukas and Quick 1999).

The magnitude and timing of runoff is a critical factor influencing the aquatic biota and ecosystem processes (Poff et al. 1997). Variation in flow regime among water-sheds is believed to maintain high regional diversity, because different combinations of frequency, magnitude, duration, and timing of flow influence the variety of habitat conditions afforded. Thus, changes in flow regimes in response to climate change can have profound effects on aquatic ecosystems.

BIOLOGICAL IMPACTS

Rapid climate change has many nega-tive implications for the biodiversity of rivers and streams. On a global scale, the vulnerability of stream systems appears to increase as one moves from the tropics to the poles (Poff et al. 2001), where dispro-portionately greater warming will occur, along with associated dislocations of the hydrologic cycle.

In general, streams are coolest in the headwaters, and a warming will tend to push species upstream to find thermally optimal habitats. However, small streams are effectively like the tops of mountains, in that once these cool-water refuges are warmed there is no escape route for indi-viduals or populations that are trapped there. Thus, stream networks in low-gradi-ent, lowland areas are more vulnerable to climate warming than are those in areas with high topographic relief. Aquatic or-ganisms in mountainous areas have the potential to move upstream to higher-alti-tude refuges that may remain thermally suitable during the course of climate change; however, the overall extents of the species' ranges will contract. This is seen for trout in the Rocky Mountain region (Fig. 17.4).

Climate change may cause extinction at several taxonomic levels. At the species level, those species that are highly re-stricted in their geographic distribution or that are very specialized ecologically (and thus occupy narrow habitat types) are vul-nerable to global extinction. This is true for fish (Angermeier 1995; Poff et al. 2001), where there are regional differences in the proportional occurrence of specialized species (Poff et al. 2001). It is also proba-bly true for invertebrates such as mussels, whose biology is closely tied to host fish species and whose habitat is greatly di-minished by river regulation (Duncan and Lockwood 2001).

Particular geographic locations are es-pecially vulnerable. For example, fish in the southern Great Plains and the desert Southwest of the United States cannot move northward because those streams and rivers tend to run west and east. With just a few degrees of warming, up to 20 native fish species in these regions are at risk of extinction (Matthews and Zimmer-man 1990; Covich et al. 1997).

If related species with particular ecolog-ical traits are more sensitive to climate change, then higher levels of taxonomic diversity may also be threatened. For ex-ample, because darter species in genera of the family Percidae are uniformly small, are ecologically specialized, and generally have narrow geographic ranges, they ap-pear vulnerable as a group to extinction (Poff et al. 2001).

Even if species do not go extinct, reduc-tion in regional abundances can result in biodiversity loss at the population genetic level. Streams and rivers are naturally rela-tively isolated due to watershed divides, and within-watershed genetic differentiation has occurred for many groups. Severe isolation has led to speciation, as with the galaxiid fishes of New Zealand (e.g., Allibone and Townsend 1997). Many salmonid fish species that migrate to the sea have developed a high affinity for particular watersheds, where environmental conditions vary, leading to local adaptation, meaning genetic differentiation at the subpopulation level. A change in climate that causes the elimination of a subpopulation results in some loss of genetic variation from the larger species population.

The production of salmon and other high-value fishes that spend part of their lives in fresh water may be vulnerable to climate change. Beamish and colleagues (1999) document a synchrony between climate indexes and both Pacific salmon and Pacific sardines that may have occurred for centuries, and is well established for the past 100 years. This raises the possibility that global warming will result in large and abrupt changes in abundance, termed regime shifts, in commercially important species. Within the river ecosystem, used for spawning and rearing of juveniles, bioenergetic models predict that warmer spring-summer temperatures will result in higher predation rates by important piscivores of juvenile salmon (Petersen and Kitchell 2001). Increased mortality of juvenile coho salmon in the Columbia River basin may result from warmer temperatures and reduced streamflows, and thermal barriers may impede the migration of adult salmon (Mote et al. 1999). On the other hand, populations of sockeye salmon in Alaska appear to benefit from a warming of surface sea temperatures, observed as bouts of natural climatic variation over the past 300 years, apparently due to increased ocean productivity (Finney et al. 2000). Anadromous fishes potentially will be affected by climate change impacts to both marine and freshwater environments.

Streams draining forested landscapes and with forested riparian zones derive much of their energy as organic matter inputs from the terrestrial environment (Allan 1995). Leaves and leaf fragments, termed coarse particulate organic matter (CPOM), are quickly colonized by fungi and bacteria, and subsequently consumed by a wide variety of invertebrates. Climate change is likely to affect the processing of detritus and functioning of the microbial-shredder food web linkage in complex ways. Altered carbon-to-nitrogen ratios of the leaves likely will reduce palatability, temperature changes will affect leaf processing rates, and floods may export leaf matter before it can be processed (Rier and Tuchman 2002; Tuchman et al. 2002, 2003). Figure 17.5 illustrates that these interactions are complex and potentially offsetting, making the overall impact of climate on this important energy supply difficult to predict.

In their natural state, river networks typically are connected systems in which biota can move during periods of rapid environmental change. Indeed, in previous bouts of climate change, such movements have been critical to sustained species survival for fishes (e.g., Briggs 1986) and aquatic invertebrates such as stoneflies (Zwick 1981). In the contemporary landscape, rivers are fragmented by a variety of human activities (Allan and Flecker 1993; Dynesius and Nilsson 1994) and species and populations are less capable of moving easily along river corridors. This isolation poses one of the largest threats to aquatic biodiversity during climate change (Poff et al. 2001, 2002).

SUMMARY

Human demands for freshwater quantity and quality now pose severe threats to freshwater biodiversity, and human demands are projected to increase dramatically (Postel et al. 1996; Vörösmarty et al. 2000). The multiple human stressors of

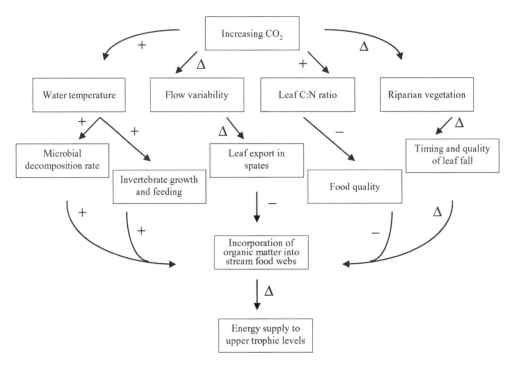

Figure 17.5. A rise in atmospheric CO_2 is expected to have complex effects on organic matter inputs, especially leaf fall, which is a principal energy supply to small streams. A warmer climate will increase most rate processes, increased hydrologic variability may adversely affect the retention and thus processing of organic matter, a higher carbon to nitrogen ratio of leaves is expected to reduce palatability, and climate change may cause shifts in the composition of riparian species. Source: Based on studies by Rier and Tuchman (2002) and Tuchman et al. (2002, 2003).

freshwater ecosystems (Allan and Flecker 1993; Palmer et al. 2000) will interact with future climate change (Schindler 2001; Poff et al. 2002) to further compromise the biodiversity and function of freshwater ecosystems. Physical impacts due to the direct effects of warming and altered hydrology are understood in broad outline but are manifested differently across lakes, ponds, rivers, and streams owing to system heterogeneity including size, depth, landscape position, latitude, and many additional factors (see Fig. 17.1). Higher biological rates of metabolism, growth and production are expected to be widespread in temperate and boreal regions, because the majority of organ-

isms of freshwater ecosystems are ectotherms. Biological responses often are unpredictable due to the complexity of the system, including nonlinear responses, sometimes offsetting interactions, and the importance of exceptional, stochastic events. The potential for negative synergies is of special concern. The impacts due to warming of lakes depend on depth, which determines the extent of the hypolimnion, and lake productivity, which influences oxygen depletion in deep waters.

Poleward range shifts are expected to be a very general response in river and stream biota, although both natural and manufactured migration barriers are widespread and north–south dispersal corridors may be absent. Invasive species also will undergo range shifts, posing a risk to native biodiversity in regions they have not yet colonized. The combined effects of climate change on the food webs of forested streams, which are highly dependent upon organic matter inputs of terrestrial origin, are difficult to predict due to the offsetting nature of multiple pathways of impact (see Fig. 17.5).

A number of studies have evaluated changes in available habitat for fishes due to climate warming. This is more straightforward in streams, which are well mixed and track air temperature closely, than in lakes, where thermal stratification and depth affect the relative change in habitat for cold-, cool-, and warm-water fishes. Rivers and streams will experience altered flow regimes due to changes in both precipitation and evapotranspiration. Because water budgets depend on the balance of these terms and both are difficult to predict at the catchment or regional scale, it usually is uncertain whether streamflows will increase or decrease, and in what season. However, spring runoff is likely to occur earlier and snowmelt runoff regimes change to winter rainy regimes, a trend already observed in some locations.

Mitigation and adaptation measures need far more investigation before they can be described with confidence. Many mitigation measures likely will require addressing interacting stressors of human origin. Invasive species pose potent threats to native biodiversity and continue to be facilitated by fish and wildlife agencies and habitat degradation. Removal of riparian vegetation in farmland warms streams by as much as $3-5°C$ (Abell and Allan 2002), and riparian tree plantings likely would lower stream temperatures by a similar amount. Dams and inhospitable habitat create barriers to dispersal by aquatic organisms that will need dispersal corridors to cope with changing temperatures. Acid deposition is expected to interact with climate-induced changes in delivery of dissolved organic carbon to boreal lakes to enhance water clarity and permit damaging UV-B rays to penetrate to greater depth. It may be easier to deal with the already existing sources of negative synergies, than with climate change itself. Freshwater ecosystems will adapt to climate change as they have adapted to land-use change, acid rain, habitat degradation, and multiple forms of pollution. Unfortunately, that adaptation is likely to entail a diminishment of native biodiversity.

REFERENCES

Abell, R. A., et al. 2000. *Freshwater Ecoregions of North America.* Washington, D.C.: Island Press.

Abell, R. A., and J. D. Allan. 2002. Riparian shade and stream temperatures in an agricultural catchment, Michigan, USA. *Verhandlungen der Internationalen Vereinigung für Theoretische und Angewandte Limnologie* 28:1–6.

Allan, J. D. 1995. *Stream Ecology: Structure and Function of Running Waters.* The Netherlands: Kluwer Academic.

Allan, J. D., and A. S. Flecker. 1993. Biodiversity conservation in running waters. *BioScience* 43:32–43.

Allibone, R. M., and C. R. Townsend. 1997. Distribution of four recently discovered galaxiid species in the Taieri River, New Zealand: The role of macrohabitat. *Journal of Fish Biology* 51:1235–1246.

Angermeier, P. L. 1995. Ecological attributes of extinction-prone species—loss of freshwater fishes of Virginia. *Conservation Biology* 9:143–158.

Beamish, R. J., D. J. Noakes, G. A. McFarlane, L. Klyashtorin, V. V. Ivanov, and V. Kurashov. 1999. The regime concept and natural trends in the production of Pacific salmon. *Canadian Journal of Fisheries and Aquatic Sciences* 56:516–526.

Benke, A. C. 1993. Concepts and patterns of invertebrate production in running waters. *Verhandlungen der Internationalen Vereinigung für Theoretische und Angewandte Limnologie* 25:15–38.

Benke A. C., I. Chaubey, G. M. Ward, and E. L. Dunn. 2000. Flood pulse dynamics of an unregulated river floodplain in the southeastern US coastal plain. *Ecology* 81:2730–2741.

Briggs, J. C. 1986. Introduction to the zoogeography of North American fishes, pp. 1–16. In *The Zoogeography of North American Freshwater Fishes*, C. H. Hocutt and E. O. Wiley, eds. New York: John Wiley and Sons.

Burgis, M. J., and P. Morris. 1987. *The Natural History of Lakes.* Cambridge, U.K.: Cambridge University Press.

Cardinale, B. J., M. A. Palmer, and S. L. Collins. 2002. Species diversity enhances ecosystem functioning through interspecific facilitation. *Nature* 415:426–429.

Carpenter, D. E., and R. S. Lunetta. 2000. Challenges in forecasting the long-term impacts of multiple stressors on a mid-Atlantic region, USA. *Environmental Toxicology and Chemistry* 19:1076–1081.

Carpenter, S. R., N. F. Caraco, D. L. Correll, R. W. Howarth, A. N. Sharpley, and V. H. Smith. 1998. Nonpoint pollution of surface waters with phosphorus and nitrogen. *Ecological Applications* 8:559–568.

Carpenter, S. R., S. G. Fisher, N. B. Grimm, and J. F. Kitchell. 1992. Global change and freshwater ecosystems. *Annual Reviews of Ecology and Systematics* 23:119–139.

Covich, A. P., S. C. Fritz, P. J. Lamb, R. D. Marzolf, W. J. Matthews, K. A. Poiani, E. E. Prepas, M. B. Richman, and T. C. Winter. 1997. Potential effects of climate change on aquatic ecosystems of the Great Plains of North America. *Hydrological Processes* 11:993–1021.

Duncan, J. R., and J. L. Lockwood. 2001. Spatial homogenization of the aquatic fauna of Tennessee: Extinction and invasion following land use change and habitat alteration, pp. 245–257. In *Biotic Homogenization*, J. L. Lockwood and M. L. McKinney, eds. New York: Kluwer Academic/Plenum.

Dynesius, M., and C. Nilsson. 1994. Fragmentation and flow regulation of river systems in the northern third of the world. *Science* 266:753–762.

Elzen, D. M., and M. Schaeffer. 2002. Responsibility for past and future global warming: Uncertainties in attributing anthropogenic climate change. *Climatic Change* 54 (1–2):29–73.

Featherstone, J. 1996. Conservation in the Delaware River Basin. *Journal of the American Water Works Association* 88:42.

Finney, B. P., I. Gregory-Eaves, J. Sweetman, M. S. V. Douglas, and J. P. Smol. 2000. Impacts of climate change and fishing on Pacific salmon abundance over the past 300 years. *Science* 290:795–799.

Forest, C. E., P. H. Stone, A. P. Sokolov, M. R. Allen, and M. D. Webster. 2002. Quantifying uncertainties in climate system properties with the use of recent climate observations. *Science* 295:113–117.

Freckman, D. W., J. Brussaard, P. Snelgrove, and M. A. Palmer. 1997. Biodiversity and ecosystem functioning of soils and sediments. *Ambio* 26:556–562.

Frederick, K. D., and P. H. Gleick. 1999. *Water and Global Climate Change: Potential Impacts on U.S. Water Resources*. Arlington, Va.: Pew Center on Global Climate Change.

Gleick, P. H. 1998. *The World's Waters: The Biennial Report on Fresh Water Resources*. Washington, D.C.: Island Press.

Gleick, P., W. Burns, E. Chalecki, M. Cohen, K. Cushing, A. Mann, R. Reyes, G. Wolff, and A. Wong. 2002. *The World's Water: The Biennial Report of Freshwater Resources 2002–2003*. Washington, D.C.: Island Press.

Häder, D. P., H. D. Kumar, R. C. Smith, and R. C. Worrest. 2003. Aquatic ecosystems: Effects of solar ultraviolet radiation and interactions with other climate change factors. *Photochemical and Photobiological Science* 2:39–50.

Heal, G., and B. Kristrom. 2002. Uncertainty and climate change. *Environmental and Resource Economics* 22(1–2):3–39.

Hostetler, S. W., and E. E. Small. 1999. Response of North American freshwater lakes to simulated future climates. *Journal of the American Water Resources Association* 35:1625–1637.

IPCC. 2001. Third report of the working group of the intergovernmental panel on climate change. *International Panel on Climate Change*. www.ipcc.ch

Keleher, C. J., and F. J. Rahel. 1996. Thermal limits to salmonid distribution in the Rocky Mountain region and potential habitat loss due to global warming: A geographic information system (GIS) approach. *Transactions of the American Fisheries Society* 125:1–13.

Kratz, T. K., B. P. Hayden, B. J. Benson, and W. Y. B. Chang. 2001. Patterns in the interannual variability of lake freeze and thaw dates. *Verhandlungen der Internationalen Vereinigung für Theoretische und Angewandte Limnologie* 27:2796–2799.

Lake, P. S., M. A. Palmer, P. Biro, J. Cole, A. P. Covich, C. Dahm, J. Gilbert, W. Goedkoop, K. Martens, and J. Verhoeven. 2000. Global change and the biodiversity of freshwater ecosystems: Impacts on linkages between above-sediment and sediment biota. *BioScience* 50:1099–1107.

Lathrop, R. C., S. R. Carpenter, and D. M. Robertson. 1999. Summer water clarity responses to phosphorus, Daphnia grazing, and internal mixing in Lake Mendota. *Limnology and Oceanography* 44:137–146.

Lofgren, B. M., F. H. Quinn, A. H. Clites, R. A. Assel, A. J. Eberhardt, and C. L. Luukkonen. 2002. Evaluation of potential impacts on Great Lakes water resources based on climate scenarios of two GCMs. *Journal of Great Lakes Research* 28 (4):537–554.

Loreau, M., S. Naeem, P. Inchausti, J. Bengtsson, J. P. Grime, A. Hector, D. U. Hooper, M. A. Huston, D. Raffaelli, B. Schmid, D. Tilman, and D. A. Wardle. 2001. Biodiversity and ecosystem functioning: Current knowledge and challenges. *Science* 294:806–808.

Loukas, A., and M. C. Quick. 1999. The effect of climate change on floods in British Columbia. *Nordic Hydrology* 30:231–256.

Magnuson, J. J., D. M. Robinson, B. J. Benson, R. H. Wynne, D. M. Livingstone, T. Arai, R. A. Assel, R. G. Barry, V. Card, E. Kuusisto, N. G. Granin, T. D. Prowse, K. M. Stewart, and V. S. Vuglinski. 2000. Historical trends in lake and river cover in the Northern Hemisphere. *Science* 289:1743–1746.

Magnuson, J. J., K. E. Webster, R. A. Assel, C. J. Bowser, P. J. Dillon, J. G. Eaton, H. E. Evans, E. J. Fee, R. I. Hall, L. R. Mortsch, D. W. Schindler, and F. H. Quinn. 1997. Potential effects of climate change on aquatic systems: Laurentian Great Lakes and Precambrian Shield area. *Hydrological Processes* 11:825–871.

Master, L. L., S. R. Flack, and B. A. Stein, eds. 1998. *Rivers of Life: Critical Watersheds for Protecting Freshwater Biodiversity*. Arlington, Va.: The Nature Conservancy.

Matthews, W. J., and E. G. Zimmerman. 1990. Potential effects of global warming on native fishes of the southern Great Plains and the Southwest. *Fisheries* 15:26−32.

Meyer, J. L. 1997. Stream health: Incorporating the human dimension to advance stream ecology. *Journal of the North American Benthological Society* 16:439−447.

Meyer, J. L., M. J. Sale, P. J. Mulholland, and N. L. Poff. 1999. Impacts of climate change on aquatic ecosystem functioning and health. *Journal of the American Water Resources Association* 35:1373−1386.

Mortsch, L. D., and F. H. Quinn. 1996. Climate change scenarios for Great Lakes ecosystem studies. *Limnology and Oceanography* 41:903−911.

Mote, P., D. Canning, D. Fluharty, R. Francis, J. Franklin, A. Hamlet, M. Hershman, M. Holmberg, K. G. Ideker, W. Keeton, D. Lettenmaier, R. Leung, N. Manuta, E. Miles, B. Noble, H. Parandvash, D. W. Peterson, A. Snover, and S. Willard. 1999. *Impacts of Climate Variability and Change in the Pacific Northwest*. Seattle: University of Washington.

Moyle, P. B., and J. J. Cech, Jr. 1996. *Fishes: An Introduction to Ichthyology*. Prentice-Hall.

Naeem, S. 2002. Ecosystem consequences of biodiversity loss: The evolution of a paradigm. *Ecology* 83:1537−1552.

Naiman, R. J., J. J. Magnuson, D. M. McKnight, and J. A. Stanford. 1995. *The Freshwater Imperative*. Washington, D.C.: Island Press.

(NAST) National Assessment Synthesis Team. 2000. *Climate Change Impacts on the United States: The Potential Consequences of Climate Variability and Change*. Washington, D.C.: U.S. Global Change Research Program.

Neff, R., H. Chang, C. G. Knight, R. G. Najjar, B. Yarnal, and H. A. Walker. 2000. Impact of climate variation and change on Mid-Atlantic Region hydrology and water resources. *Climate Research* 14:207−218.

Palmer, M. A., A. P. Covich, B. Finlay, J. Gibert, K. D. Hyde, R. K. Johnson, T. Kairesalo, P. S. Lake, C. R. Lovell, R. J. Naiman, C. Ricci, F. F. Sabater, and D. L. Strayer. 1997. Biodiversity and ecosystem function in freshwater sediments. *Ambio* 26:571−577.

Palmer, M. A., A. P. Covich, S. Lake, P. Biro, J. J. Brooks. J. Cole, C. Dahm, J. Gibert, W. Goedkoop, K. Martens, J. Verhoeven, and W J. van de Bund. 2000. Linkages between freshwater benthic ecosystems and above sediment life: Drivers of biodiversity and ecological processes? *BioScience* 50:1062−1068.

Palmer, M. A., and P. S. Lake. 2000. Invertebrate biodiversity in freshwaters, pp. 531−542. In *Encyclopedia of Biodiversity*, Volume 3, Simon Levin, ed. London: Academic Press.

Petersen, J. H., and J. F. Kitchell. 2001. Climate regimes and water temperature changes in the Columbia River: Bioenergetic implications for predators of juvenile salmon. *Canadian Journal of Fisheries and Aquatic Sciences* 58:1831−1841.

Poff, N. L., J. D. Allan, M. B. Bain, J. R. Karr, K. L. Prestegaard, B. D. Richter, R. E. Sparks, and J. C. Stromberg. 1997. The natural flow regime: A paradigm for river conservation and restoration. *BioScience* 47:769−784.

Poff, N. L., P. A. Angermeier, S. D. Cooper, P. S. Lake, K. D. Fausch, K. O. Winemiller, L. A. K. Mertes, M. W. Oswood, J. Reynolds, and F. J. Rahel. 2001. Fish diversity in streams and rivers, pp. 315−350. In *Global Biodiversity in a Changing Environment: Scenarios for the 21st Century*, F. S. Chapin III, O. E. Sala and E. Huber-Sannwald, eds. New York: Springer-Verlag.

Poff, N. L., M. M. Brinson, and J. W. Day, Jr. 2002. *Aquatic Ecosystems and Global Climate Change*. Arlington, Va.: Pew Center on Global Climate Change.

Poff, N. L., and J. V. Ward. 1989. Implications of streamflow variability and predictability for lotic community structure: A regional analysis of streamflow patterns. *Canadian Journal of Fisheries and Aquatic Sciences* 46:1805−1818.

Polsky, C., J. Allard, N. Currit, R. Crane, and B. Yarnal. 2000. The Mid-Atlantic region and its climate: Past, present, and future. *Climate Research* 14:161−173.

Postel, S. 1997. *Last Oasis: Facing Water Scarcity*. New York: W.W. Norton.

Postel, S. L., and S. R. Carpenter. 1997. Freshwater ecosystem services, pp. 195−214. In *Natures Services: Societal Dependence on Natural Systems*, G. D. Daily, ed. Washington D.C.: Island Press.

Postel, S., G. C. Dailey, and P. R. Ehrlich. 1996. Human appropriation of renewable freshwater. *Science* 271:785−788.

Regier, H. A., J. A. Holmes, and D. Pauly. 1990. Influence of temperature change on aquatic ecosystems: An interpretation of empirical data. *Transactions of the American Fisheries Society* 119:374−389.

Ricciardi, A., and J. B. Rasmussen. 1999. Extinction rates of North American freshwater fauna. *Conservation Biology* 13:1220−1222.

Rier, S. T., and N. C. Tuchman. 2002. Elevated-CO_2-induced changes in the chemistry of quaking aspen (*Populus tremuloides* Michaux) leaf litter: Subsequent mass loss and microbial response in a stream ecosystem. *Journal of the North American Benthological Society* 21:16−27.

Rogers, D. J., and S. E. Randolph. 2000. The global spread of malaria in a future, warmer world. *Science* 289:1763.

Sattenspiel L. 2000. Tropical environments, human activities, and the transmission of infectious diseases. *Yearbook of Physical Anthropology* 43:3−31.

Schindler, D. W. 1997. Widespread effects of climatic warming on freshwater ecosystems. *Hydrologic Processes* 11:1043−1067.

Schindler, D. W. 1998. A dim future for boreal waters and landscapes. *BioScience* 48:157−164.

Schindler, D. W. 2001. The cumulative effects of climate warming and other human stresses on

Canadian freshwaters in the new millennium. *Canadian Journal of Fisheries and Aquatic Sciences* 58:18–29.

Schindler, D. W., S. E. Bayley, B. R. Parker, K. G. Beatty, D. R. Cruikshank, E. J. Fee, E. U. Schindler, and M. P. Stainton. 1996. The effects of climatic warming on the properties of boreal lakes and streams at the Experimental Lakes Area, northwestern Ontario. *Limnology and Oceanography* 41:1004–1017.

Stefan, H. G., and X. Fang. 1993. Model simulations of dissolved oxygen characteristics of Minnesota lakes: Past and future. *Environmental Management* 18:73–92.

Stefan, H. G., X. Fang, and J. G. Eaton. 2001. Simulated fish habitat changes in North American lakes in response to projected climate warming. *Transactions of the American Fisheries Society* 130:459–477.

Strayer, D. L. 1991. Projected distribution of the zebra mussel, *Dreissena polymorpha*, in North America. *Canadian Journal of Fisheries and Aquatic Sciences* 48:1389–1395.

Sweeney, B. W., J. K. Jackson, J. D. Newbold, and D. H. Funk. 1992. Climate change and the life histories and biogeography of aquatic insects in eastern North America, pp. 143–176. In *Global Climate Change and Freshwater Ecosystems*, P. Firth and S. G. Fisher, eds. New York: Springer-Verlag.

Toman, M. A. 1997. Ecosystem valuation: An overview of issues and uncertainties, pp. 25–44. In *Ecosystem Function and Human Activities*, R. D. Simpson and N. L. Christensen, eds. New York: Chapman and Hall.

Tuchman, N. C., K. A. Wahtera, S. T. Rier, R. G. Wetzel, and J. A. Teeri. 2002. Elevated atmospheric CO_2 alters leaf litter nutritional quality for stream ecosystems: An *in situ* leaf decomposition study. *Hydrobiologia* 495:203–211.

Tuchman, N. C., K. A. Wahtera, R. G. Wetzel, N. M. Russo, G. M. Kilbane, L. M. Sasso, and J. A. Teeri. 2003. Nutritional quality of leaf detritus altered by elevated atmospheric CO2: Effects on development of mosquito larvae. *Freshwater Biology* 48:1432–1439.

Vannote, R. L., G. W. Minshall, K. W. Cummins, J. R. Sedell, and C. E. Cushing. 1980. The river continuum concept. *Canadian Journal of Fisheries and Aquatic Sciences* 37:130–137.

Vörösmarty, C. J., P. Green, J. Salisbury, and R. B. Lammers. 2000. Global water resources: Vulnerability from climate change and population growth. *Science* 289:284–288.

Vörösmarty, C. J., K. P. Sharma, B. M. Fekete, A. H. Copeland, J. Holden, J. Marble, and J. A. Lough. 1997. The storage and aging of continental runoff in large reservoir systems of the world. *Ambio* 26:210–219.

Wetzel, R. G. 2001. *Limnology*, third edition. San Diego, Calif.: Academic Press.

Wigley, T. M. L. 1999. *The Science of Climate Change: Global and U.S. Perspectives*. Arlington, Va.: Pew Center on Global Climate Change.

WRI. 2000. *World Resources 2001–2002—People and Resources: The fraying web of life*. Washington, D.C.: World Resources Institute. www.wri.org/wr2000

Zwick, P. 1981. Das Mittelmeergebiet als galziales Refugium für Plecoptera. *Acta Entomological Jugoslavica* 17:107–111.

Climate Change Impacts on Soil Biodiversity in a Grassland Ecosystem

Diana H. Wall

Understanding the breadth and wealth of the soil biota and their roles in ecosystem processes is essential to addressing how global change alters soil communities and to determining the effects on the ecosystem services provided by soil organisms. Soil biodiversity comprises an amazing wealth of forms, sizes, densities, life histories, and complex interactions. Examining the increasing size of the organisms—from the minute actinomycetes, bacteria, and fungi to the world of micoarthropods (mites, Collembola), unsegmented roundworms (Nematoda), and the larger, visible isopods, centipedes, millipedes, spiders, ants, Enchytraeidae, earthworms, termites, and vertebrates such as gophers, prairie dogs, and lizards—it becomes apparent why each group requires several taxonomic specialists to determine species richness. The number of described species is most well known for the larger organisms, but it is the lesser-known organisms, for example, the mites and nematodes, whose species numbers could substantially increase estimates of global biodiversity (May 1988). The densities of organisms in a gram of soil, (10^8 actinomycetes, 10^{5-6} fungi, 10^{3-6} micro-algae, 10^{3-5} protozoa, 10^{1-2} nematodes, 10^{3-5} other invertebrates), the magnitude of species diversity, the person-hours and taxonomic expertise and resources required, make learning about the fascinating life in soil a challenge equivalent to exploring life on Mars. This rich community will play an important role in determining the fate of biodiversity during climate change (see Box 17.1).

The shortgrass steppe (SGS) region of northern Colorado (latitude 40°49'N, longitude 104°46'W) borders the central and northern Great Plains ecosystems and is a semi-arid mosaic of native grassland and agroecosystems. Climate change models predict that soil moisture will decrease over large areas of semi-arid grassland in North America as global warming progresses (Manabe and Wetherald 1987; National Assessment Synthesis Team (NAST) 2001). The decline in precipitation and the warming over the central region of the Great Plains of North America result in a marked increase of the drought indexes and water demand by grasslands (Ojima et al. 1999). Additionally, both land use change and invasive species are contributing to modifications in the soil habitat at the SGS.

The SGS evolved with many herbivores, aboveground (bison, antelope, deer, rodents, insects) and belowground (arthropods, nematodes). More recently, livestock has become a major grazer in the system.

Estimates indicate about 5–10 percent of net primary productivity (NPP) is consumed equally by livestock, arthropods, and soil nematodes (Stanton 1983; Ingham and Detling 1984; Milchunas and Lauenroth 1991).

Three native bunch grasses dominate the region: blue grama (*Bouteloua gracilis* (H.B.K.) Lag. Ex.), western wheatgrass (*Agropyron smithii*), and needle and thread grass (*Stipa comata*). Their roots are the major source of carbon inputs to the soil food web, which is speciose and composed of microflora, protozoa, arthropods (mites, Collembola), nematodes, annelids, and small vertebrates (Moore et al. 2004). Nematodes occupy many trophic levels in the soil food web, and changes in the abundance and biomass of functional groups and/or species composition reflect even small changes occurring in the soil habitat (chemical, physical properties) and their food source (bacteria, fungi, microfauna,

Box 17.1. **The Belowground Connection**

David A. Perry

Mycorrhizal fungi and some soil bacteria fix nitrogen, converting it from a form that plants can't use to a form in which it can be used in plant metabolism and tissue production. This usable nitrogen is absorbed by plants from nodules of the fungi or bacteria that surround plant rootlets. Based on these relationships, Perry and colleagues (1990) discussed two hypothetical situations related to species migrations during climate change: (1) belowground communities supported by outgoing plants meet the needs of immigrants, and there is sufficient overlap between the two to stabilize the belowground, resulting in a smooth transition; (2) overlap between compatible plants is insufficient to stabilize the belowground, weeds capture the site, and soils lose the ability to support productive plant communities. The latter case implies reduced carbon sinks, hence positive feedback to warming.

Mycorrhizal fungi seem likely to play a key role in community transitions, and are by far the most studied. Diversity of mycorrhizal fungi promotes plant diversity in old field communities (Van der Heijden et al. 1998) and, in forests, differing species of trees and shrubs form guilds based on common mycorrhizal symbionts (Janos 1987; Borchers and Perry 1990; Amaranthus and Perry 1994; Smith et al. 1995; Hagerman et al. 2001). Moreover, studies in both forests and grasslands show that mycorrhizal fungi hyphae link different plant species into belowground networks through which carbon and nutrients move (Read 1994; Simard et al. 1997), though the effects of such movements on plant community composition are unclear. Studies in quite different systems have shown that successful plant establishment can be strongly influenced by the presence or absence of the "right" kinds of mycorrhizal fungi (Jumpponen et al. 1998; Horton et al. 1999; Klironomos 2002; Rojas et al. 2002). It follows that migration routes may be strongly influenced by mycorrhizal fungi (Perry et al. 1990), and in fact Horton and colleagues (1999) found that ectomycorrhizal Douglas fir successfully invaded chaparral dominated by ectomycorrhizal *Arctostaphylous* sp., but not adjacent areas dominated by arbuscular mycorrhizal *Ceanothus* sp., an effect resulting from mycorrhizal associations and not other environmental differences between the vegetation types. While most or all woody plants are obligately mycorrhizal, many weeds are facultatively mycorrhizal, raising questions about the fate of mycorrhizal communities on sites captured by weeds. Even if weeds do maintain mycorrhizal fungi communities, they are unlikely to be the types required by many woody species, including conifers, oaks, birches, beeches, alders, and ericaceous shrubs, leading to questions about the persistence of mycorrhizal fungi in the absence of hosts.

Details of plant migration with climate change will vary, but it seems clear the structure of the plant community and the structure of the belowground community are intimately tied to each other, a relationship that is both a source of strength and an Achilles heel. The composition of soil microbial communities has recently been shown to vary across landscapes with changes in plant community composition (Myers et al. 2001), raising the possibility that migrating plants will encounter a strange and perhaps discordant soil biota. The implications for successful establishment on new sites, and how quickly the soil community would adapt to new plant species are open questions.

roots) (Bongers and Ferris 1999). *Bouteloua gracilis* has 188 nematode species associated with it at the SGS (Wall, unpublished).

A four-year experiment at the SGS, using native grassland as a control and open-topped chambers with ambient (360 ppm) and elevated (700 ppm) CO_2 and temperature, significantly increased fungivorous and predaceous mites (Cryptostigmata and Mesostigmata) in the elevated CO_2 treatments (Moore et al. 2004). While these indirect effects on soil biota are results of short-term study, evidence from this and other studies clearly indicates a response of different soil organisms to experiments with increased CO_2 and temperature (Wall et al. 2001a; Cole et al. 2002).

Plant production and soil microbial processes are limited by soil water content in the SGS. Increased soil moisture under elevated CO_2 due to improved plant water use efficiency is expected to improve soil conditions for plant and microbial growth. As a result, microbial populations that mediate processes such as soil organic matter decomposition, methane oxidation (conversion of atmospheric CH_4 to CO_2), and nitrification (the oxidation of ammonium to nitrite) might be expected to increase under elevated CO_2. The end result would be increased CO_2, NO_x, and N_2O emissions from soil under elevated CO_2. The SGS study suggests that competition between plant roots and microbes for scarce mineral nitrogen supply limits nitrifier activity, as observed from greater N uptake by plants and lower NO_x and N_2O emissions (Mosier et al. 2002). Soil CO_2 emissions were higher occasionally when the difference in soil water content between elevated and ambient CO_2 sites was greatest, while CH_4 consumption increased under the improved soil water status at these times.

In the SGS, CO_2-induced increases in aboveground production are accompanied by reductions in shoot N concentration.

These N reductions may be the result of more efficient plant resource use or simply of increased soil nutrient demand driven by faster-growing plants. Root symbionts like mycorrhizae may be especially important in meeting increased nutrient demands under elevated CO_2, and increases in infection under elevated CO_2 may reflect that importance. Mycorrhizae may also contribute to the improved plant–water relations that occur under CO_2-enriched atmospheres, but they may be less important in that function as water becomes more available. Too few data have been collected on the SGS to allow definitive interpretations of mycorrhizal responses to global changes, but the information at hand does suggest the potential for significant responses to CO_2, temperature, and precipitation. More work will be required to determine the significance of variable plant species and associated mycorrhizal responses to global changes, and the adaptive significance of these responses for individual plants and the plant communities (see Box 17.1).

Climate change is affecting soil diversity, soil food webs, and community composition, but the responses of the biota differ with the taxonomic group, making it impossible to generalize the responses of soil biodiversity across all ecosystems. Whether the loss of soil biodiversity has consequences for ecosystem function has recently been examined by a model developed for the food web at the SGS. Results indicate that soil fauna as a whole are very important for plant production, but if functional groups are lost, there will be compensation by the remaining groups (Hunt and Wall 2002). However, in this model and in most elevated CO_2 experiments, there are few studies of soil biodiversity at the species level. Thus, actual species-level responses and species losses affecting ecosystem processes such as hydrology or plant productivity or rates of decomposition may not be noted. Nevertheless, the soil biota of some ecosystems appears to be more vulnerable to global change, particularly in areas, such as these grasslands, where the soil food web is critical for mediating herbivores and plant allocation of nutrients. Future scenarios of biodiversity change for soils (Wall et al. 2001a) are more likely to be positive where regional management conserves the integrity of the plant–soil relationship.

REFERENCES

Amaranthus, M. P., and D. A. Perry. 1994. The functioning of ectomycorrhizae in the field: linkages in space and time. *Plant and Soil* 159:133–140.

Bongers, T., and H. Ferris. 1999. Nematode community structure as a bioindicator in environmental monitoring. *Trends in Ecology & Evolution* 14:224–228.

Borchers, S. L., and D. A. Perry. 1990. Growth and ectomycorrhiza formation of Douglas-fir seedlings grown in soils collected at different distances from pioneering hardwoods in SW Oregon. *Can. Jour. For. Res.* 20:712–721.

Cole, L., R. D. Bardgett, P. Ineson, and J. K. Adamson. 2002. Relationships between enchytraeid worms (Oligochaeta), climate change, and the release of dissolved organic carbon from blanket peat in northern England. *Soil Biology & Biochemistry* 34:599–607.

Elliott, E. T., K. Horton, J. C. Moore, D. C. Coleman, and C. V. Cloe. 1984. Mineralization dynamics in fallow drylands wheat plots, Colorado. *Plant and Soil* 76:149–155.

Freckman, D. W., and C. H. Ettema. 1993. Assessing nematode communities in agroecosystems of varying human intervention. *Agriculture, Ecosystems, and Environment* 45:239–261.

Hagerman, S. M., S. M. Sakakibara, and D. M. Durall. 2001. The potential for woody understory plants to provide refuge for ectomycorrhizal inoculum at an interior Douglas-fir forest after clear-cut logging. *Can. J. For. Res.* 31:711–721.

Hendrix, P. F., ed. 1995. *Earthworm Ecology and Biogeography in North America.* Boca Raton, Fla.: Lewis Publishers.

Horton, T. R., T. D. Bruns, and V. T. Parker. 1999. Ectomycorrhizal fungi associated with *Arctostaphylous* contribute to *Pseudotsuga menziesii* establishment. *Can. Jour. Bot.* 77:93–102.

Hunt, H. W., and D. H. Wall. 2002. Modelling the effects of loss of soil biodiversity on ecosystem function. *Global Change Biology* 8:33–50.

Ingham, R. E., and J. K. Detling. 1984. Plant-herbivore interactions in a North American mixed-grass prairie: 3. Soil nematode populations and root biomass on Cynomys ludovicianus colonies and adjacent uncolonized areas. *Oecologia* (Berlin) 63:307–313.

Jackson, R. B., J. L. Banner, E. G. Jobbagy, W. T. Pockman, and D. H. Wall. In press. Ecosystem carbon loss with woody plant invasion of grasslands. *Nature.*

Janos, D. P. 1987. VA mycorrhizas in humid tropical systems, pp. 107–134. In G. R. Safir, ed. *Ecophysiology of VA Mycorrhizal Plants.* Boca Raton, FL: CRC Press.

Jumpponen, A., K. Mattson, J. M. Trappe, and R. Ohtonen. 1998. Effects of established willows on primary succession on Lyman Glacier forefront, North Cascade Range, Washington, USA: Evidence for simultaneous canopy inhibition and soil facilitation. *Arctic and Alpine Research* 30:31–39.

Klironomos, John. 2002. Feedback with soil biota contributes to plant rarity and invasiveness in communities. *Nature* 417:67–70.

Lubchenco, J., A. M. Olson, L. B. Brubaker, S. R. Carpenter, M. M. Holland, S. P. Hubbell, S. A. Levin, J. A. MacMahon, P. A. Matson, J. M. Melillo, H. A. Mooney, C. H. Peterson, H. R. Pulliam, L. A. Real, P. J. Regal, and P. G. Risser. 1991. The sustainable biosphere initiative: An ecological research agenda. *Ecology* 72:371–412.

Manabe, S., and R. T. Wetherald. 1987. Large-scale changes of soil wetness induced by an increase in atmospheric carbon dioxide. *Journal of the Atmospheric Sciences* 44:1211–1235.

Matson, P. A., W. J. Parton, A. G. Power, and M. J. Swift. 1997. Agricultural intensification and ecosystem properties. *Science* 277:504–509.

May, R. M. 1988. How many species are there on Earth? *Science* 241:1441–1449.

Milchunas, D. G., and W. K. Lauenroth. 1991. A quantitative global assessment of the effects of grazing by large herbivores on vegetation and soils. *Bulletin of the Ecological Society of America* 72:195.

Moore, J. C. 1986. Micro-mesofauna dynamics and functions in dryland wheat-fallow agroecosystems. Ph.D. Dissertation. Colorado State University, Fort Collins.

Moore, J. C., J. Sipes, A. A. Whittemore-Olson, H. W. Hunt, D. H. Wall, P. C. d. Ruiter, and D. C. Coleman. 2004. Trophic structure and nutrient dynamics of the belowground food web within the rhizosphere of the shortgrass steppe. In W. K. Lauenroth and I. C. Burke, eds., *Ecology of the Shortgrass Steppe: Perspectives from Long-term Research.* Cambridge, U.K.: Oxford University Press.

Mosier, A. R., J. A. Morgan, J. Y. King, D. LeCain, and D. G. Milchunas. 2002. Soil-atmosphere exchange of CH_4, CO_2, NO_x and N_2O in the Colorado shortgrass steppe under elevated CO_2. *Plant and Soil* 240:201–211.

Myers, Rachel T., Donald R. Zak, David C. White, and Aaron Peacock. 2001. Landscape-level patterns of microbial community composition and substrate use in upland forest ecosystems. *Soil Sci. Soc. Am. J.* 65:359–367.

National Assessment Synthesis Team (NAST). 2001. *Climate Change Impacts on the United States: The Potential Consequences of Climate Variability and Change.* Cambridge, U.K.: Cambridge University Press.

Ojima, D., L. Garcia, E. Elgaali, K. Miller, T. G. F. Kittel, and J. Lackett. 1999. Potential climate change impacts on water resources in the Great Plains. *Journal of the American Water Resources Association* 35:1443–1454.

Perry, D. A., J. G. Borchers, S. L. Borchers, and M. P. Amaranthus. 1990. Species migrations and ecosystem stability during climate change: The belowground connection. *Cons. Biol.* 4:266–274.

Porazinska, D. L., R. D. Bardgett, M. B. Blaauw, H. W. Hunt, A. N. Parsons, T. R. Seastedt, and D. H. Wall. In press. Relationships at the aboveground-belowground interface: Plants, soil microflora and microfauna, and soil processes. *Ecology.*

Read, D. J. 1994. Plant-microbe mutualisms and community structure. pp. 181–209. In E.-D. Schulze and H. A. Mooney, *Biodiversity and Ecosystem Functions.* Berlin: Springer.

Rojas, N. S., D. A. Perry, C. Y. Li, and L. M. Ganio. 2002. Interactions among soil biology, nutrition, and performance of actinorhizal plant species in the H. J. Andrews Experimental; Forest of Oregon. *Applied Soil Ecology* 19:13–26.

Schlesinger, W. H., J. F. Reynolds, G. L. Cunningham, L. F. Huenneke, W. M. Jarrell, R. A. Virginia, and W. G. Whitford. 1990. Biological feedbacks in global desertification. *Science* 247:1043–1048.

Simard, S. W., D. A. Perry, M. D. Jones, D. D. Myrold, D. M. Durall, and R. Molina. 1997. Net transfer of carbon between ectomycorrhizal tree species in the field. *Nature* 388:579–582.

Smith, J. E., R. Molina, and D. A. Perry. 1995. Occurrence of ectomycorrhizas on ericaceous and coniferous seedlings grown in soils from the Oregon coast range. *New Phytol.* 129:73–81.

Stanton, N. L. 1983. The effect of clipping and phytophagous nematodes on net primary production of blue grama, *Bouteloua gracilis. Oikos* 40:249–257.

Swift, M. J., J. Vandermeer, P. S. Ramakrishnan, J. M. Anderson, C. K. Ong, and B. A. Hawkins. 1996. Biodiversity and ecosystem function, pp. 261–298. In H. A. Mooney, J. H. Cushman, E. Medina, O. E. Sala, and E. D. Schulze, eds., *Functional Roles of Biodiversity: A Global Perspective.* New York: John Wiley and Sons.

Van Der Heijden, M. G., J. N. Klironomos, M. Ursic, P. Moutogliss, R. Streitwold-Engel, T. Boller, A. Wiemkin, and I. R. Sanders. 1998. Mycorrhizal fungal diversity determines plant biodiversity, ecosystem variability and productivity. *Nature* 396:69–72.

Wall, D. H., G. A. Adams, and A. N. Parsons. 2001a. Soil biodiversity, pp. 47–82. In F. S. Chapin, III, O. E. Sala, and E. Huber-Sannwald, eds., *Global Biodiversity in a Changing Environment: Scenarios for the 21st Century.* New York: Springer-Verlag.

Wall, D. H., P. V. R. Snelgrove, and A. P. Covich. 2001b. Conservation priorities for soil and sediment invertebrates, pp. 99–124. In M. E. Soule and G. H. Orians, eds., *Conservation Biology: Research Priorities for the Next Decade.* Washington, D.C.: Island Press.

Wall, D. H., and R. A. Virginia. 2000. The world beneath our feet: Soil biodiversity and ecosystem functioning, pp. 225–241. In P. R. Raven and T. Williams, eds., *Nature and Human Society: The Quest for a Sustainable World.* Washington, D.C.: National Academy of Sciences and National Research Council.

Wall-Freckman, D., and S. P. Huang. 1998. Response of the soil nematode community in a shortgrass steppe to long-term and short-term grazing. *Applied Soil Ecology* 9:39–44.

CHAPTER EIGHTEEN

Synergistic Effects

BERT G. DRAKE, LESLEY HUGHES,
E. A. JOHNSON, BRAD A. SEIBEL,
MARK A. COCHRANE, VICTORIA J. FABRY,
DANIEL RASSE, AND LEE HANNAH

Climate change affects species in concert with rising atmospheric CO_2, and its direct impacts on species may be coupled with indirect impacts through effects on disturbance regimes, pathogens, and other factors. Emerging understanding of these multiple effects suggests that they are synergistic—the combined impact on species is often greater than the sum of the individual effects.

The greatest synergy from a conservation viewpoint will be between climate change effects and habitat loss. Existing stresses to biodiversity will interact with dynamics induced by climate change to produce some of the most damaging effects on biodiversity. The synergies between climate change, habitat loss, and other nonclimate stressors are discussed in Chapter 19. A second set of synergies, the subject of this chapter, involve climate change effects in interaction with one another, including the direct effect of the buildup of atmospheric CO_2 on both terrestrial and marine systems.

ATMOSPHERIC CO_2 AND TERRESTRIAL ECOSYSTEMS

The consequences of the responses of photosynthesis and transpiration to rising atmospheric CO_2 for most plants are stimulation of growth; reduction of the numbers of stomata, increased tolerance of water stress and drought; alteration of the allocation of biomass to leaves, stems, and roots; a change in the stoichiometry of mineral constituents; reduced rates of specific respiration; and other effects (reviews by Poorter 1993; Lloyd and Farquhar

Box 18.1. **C_3, C_4, CAM, and CO_2**

Three major forms of photosynthesis are found in terrestrial plants: C_3, C_4, and CAM. C_3 is ancestral, and utilizes the enzyme Rubisco in primary carboxylation. The CAM and C_4 forms of photosynthesis are thought to have evolved to overcome inefficiencies in this process, and utilize PEP-carboxylase in primary carboxylation, allowing nighttime carbon assimilation (and higher water use efficiency) in CAM plants and a higher affinity for CO_2 and higher water and nutrient use efficiencies in C_4 plants. Most CAM plants are succulents of desert and semi-arid ecosystems and thus only a small contributor to global NPP. C_4 photosynthesis is found in many tropical grass species, and forms the basis for herbaceous productivity in tropical and subtropical grassland and savanna ecosystems. Increasing atmospheric CO_2 reduces the relative advantage of C_4 over C_3 photosynthesis and thus may have important implications for processes which currently determine the balance between these two functional types in mixed ecosystems.

1996; Curtis and Wang 1998). Elevated CO_2 reduces stomatal conductance in both C_3 and C_4 plants (see Box 18.1) (Wand et al. 1999; Lodge et al. 2001), which reduces transpiration (Jarvis et al. 1999; Li et al., 2003). When combined with a stimulation of photosynthesis, water use efficiency in most plants is increased. (Box 18.1) (Drake et al. 1997).

Given that elevated CO_2 alters the performance of C_3 and C_4 plants, how do these leaf-level responses translate to ecosystem-level effects? At issue is the extent to which ecosystem processes are altered by increasing availability of carbohydrates to plants, lowered protein content in leaves, and enhanced soil water availability, and whether these effects are sustained over time and in combination with altered temperature and water balance. Early views saw temperature increases as driving increased respiration, perhaps leading to a "runaway" greenhouse effect. Later research showed that photosynthesis is stimulated by elevated CO_2, balancing the respiration effect of temperature. The next question became—where is all the carbon going? This question arose because CO_2 stimulus of growth observed under experimental conditions was suppressed in whole-ecosystem settings, due to photosynthetic down-regulation. Part of the answer is that underground processes were being strongly affected, even when aboveground growth impacts were slight. Processes play out over longer time periods than short-term measurement of growth, however; therefore the long-term consequences of elevated CO_2 in combination with changed climate are still unclear.

Experimental plant communities grown under elevated CO_2 have shown differential responses between C_3 species, C_4 species, and N-fixers (He et al. 2002), and between grasses, nonlegume dicots, and legumes (Lüscher et al. 1998; Teyssonneyre et al. 2002), for example. Nitrogen fixers appear to be generally highly responsive (Poorter 1993), possibly because increased carbohydrate availability increases resource allocation to N-fixing symbionts. Weeds appear to outperform crop species in terms of photosynthetic and growth stimulation when assessed individually (Ziska and Bunce 1997), but when grown in mixture the competitive outcome will also be influenced by other environmental conditions such as temperature (Alberto et al. 1996). Similarly, invasive plant species generally possess traits which allow them to respond strongly to elevated CO_2, thus creating the potential for enhanced dominance and range expansion in native ecosystems (Dukes and Mooney 1999; Smith et al. 2000).

In ecosystems characterized by frequent disturbance (e.g., fire), enhanced regrowth responses in elevated CO_2 by woody resprouters as compared to herbaceous species and reseeders could fundamentally alter community structure over the long term (Dijkstra et al. 2002). Regeneration of woody resprouters following fire is significantly enhanced by CO_2 enrichment, particularly in the presence of high nutrient availability (Hoffmann et al. 2000). This effect may already be contributing to tree invasions and thickening in grass-dominated ecosystems, including savannas (Bond and Midgley 2000).

Long-term Free-air CO_2 Enrichment (FACE) studies in C_3 woody and mixed C_3/C_4 herbaceous ecosystems during the past 15 years have shown that elevated CO_2 had impacts throughout the ecosystem and that the effects have been sustained for as long as studies have been maintained: in the Chesapeake Bay C_3/C_4 herbaceous wetland, 17 years, and in Florida scrub oak, 6 years. In these field studies, long-term exposure to elevated CO_2 caused higher rates of photosynthesis (Fig. 18.1); (Drake and Leadley 1991; Jacob et al. 1995; Li et al. 1999; Rasse et al. 2002; Hymus et al. 2002; Ainsworth et al. 2002), stimulation of growth of shoots (Dijkstra et al. 2002; Rasse and Drake, 2004) and roots (Day et al. 1996), reduced stomatal conductance (Lodge et al. 2001), reduced transpiration (Li et al. 2003), and increased soil water (Hungate et al. 2002). In the wetland study, much of the additional carbon assimilated by canopy photosynthesis is respired by soil microbes and exported from the ecosystem in soil water (Marsh et al. 2004), although some carbon also escapes as methane (Dacey et al. 1994). The increase in soil carbon in the wetland ecosystem stimulated increased numbers of fora-

minifera and nematodes, increased nitrogen fixation, and increased soil respiration. Reduction in tissue N caused increased C:N in plant tissues, increased N use efficiency, and reduced insect grazing in the wetland study (Thompson and Drake 1994), and reduced insect herbivory in the scrub oak ecosystem (Stiling et al. 1999). In both of these ecosystems, the most important effects were a consequence of stimulation of CO_2 assimilation, acclimation of photosynthesis, and reduced transpiration.

An interesting consequence of the combined effects of rising CO_2 and temperature is that photosynthesis is expected to be stimulated in low light (Long and Drake 1991; Long and Drake 1992). Elevated CO_2 increases quantum yield and reduces the light compensation point. At 25°C, doubling CO_2 caused the light compensation point to decline by about 40 percent. This effect alone will stimulate ecosystem photosynthesis through an increase in the time during the photoperiod when photosynthesis is positive: for species growing in the deep shade of tropical forests, this could be a substantial increase in carbon accumulation. There is no known mechanism by which the plant can regulate this stimulation of photosynthesis at low light by increasing atmospheric CO_2, in contrast to growth of plants in high light at high CO_2 in which acclimation of the photosynthetic apparatus can be accompanied by a reduction in the amount and the activity of Rubisco (see Box 18.1), thus reducing photosynthesis and growth.

Models based on realistic parameters derived from long-term field studies are essential to evaluate the effects of rising CO_2. Research has demonstrated that mechanistic models produce realistic estimates of photosynthesis, ecosystem gas exchange and plant productivity under ambient and elevated CO_2 confirming that when our best understanding of plant physiological processes is translated into

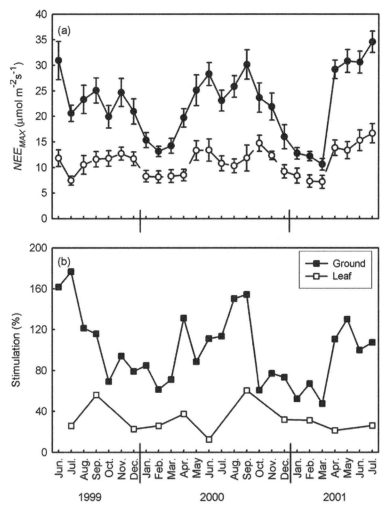

Figure 18.1. (a) Maximum net ecosystem gas exchange in the scrub oak ecosystem. Closed symbols, elevated CO_2; open symbols, ambient. (b) The stimulation of net ecosystem gas exchange (%) per unit ground area (closed squares) and per unit leaf area (open squares).

mathematical equations, the predicted response to elevated CO_2 closely matches the measured one. In addition, such mechanistic models are capable of accurately simulating the long-term growth of perennial species, such as temperate forest trees, over the entire twentieth century.

Such a model, ASPECTS, was applied to predict the response of European beech forests to twenty-first-century environmental conditions. Simulations conducted by coupling a rising CO_2 scenario (IS92a

from the IPCC) to a downscaled GCM scenario from the Canadian Global Coupled Model indicated that plant productivity would be increasingly enhanced by the increasing CO_2 during the first 30 years of the twenty-first century and subsequently reach a stimulation plateau averaging + 33 percent until the end of the century (Fig. 18.2). This predicted stimulation plateau is the consequence of the positive effect of rising CO_2 on photosynthesis being counterbalanced by increased respiratory losses due to higher biomass and higher temperatures. Other vegetation models predict a greater increase, with plant productivity reaching about 50 percent greater than present conditions by the end of the

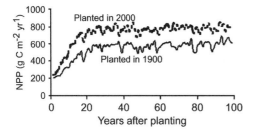

Figure 18.2. Simulated net primary productivity (NPP) for Belgian beech trees planted in 1900 as compared to trees planted in 2000. Simulations were conducted with the IS92a CO_2-rise scenario and a downscaled version of the Canadian Global Coupled Model GCM. Realistic forest management strategies were implemented. Simulations were conducted with water and nitrogen cycles coupled to the carbon cycle.

twenty-first century (Grant & Nalder 2000; Cramer et al. 2001). The results presented in Figure 18.2 include the effects of modified water stress under changing CO_2 and climate conditions. The model predicted that water stress did not increase for temperate European forests in the course of the twenty-first century because transpiration was more affected by the effect of stomatal closure under high CO_2 than by the water-spending effect of a higher vapor pressure deficit associated with a warmer climate (Misson et al. 2002). These results concur with those of a range of global vegetation models that predict an increase in water use efficiency in the course of the twenty-first century (Cramer et al. 2001). Model simulations suggest that the positive effect of rising CO_2 on plant productivity will be larger than the potentially adverse effects of climate changes as currently predicted by most GCM scenarios.

In conclusion, research during the past 20 years has shown that rising CO_2 will improve the efficiency of key physiological processes including photosynthesis, respiration, and nitrogen and water use. Rising CO_2 will stimulate increased yield in crops and will stimulate primary production in most native ecosystems. Single-species experiments show differential response to CO_2 in C_3 and C_4 plants, but differences in growth are seldom sustained in whole-ecosystem CO_2 enrichment experiments. However, below-ground changes and changes in process are pronounced in whole-ecosystem studies. Disturbance regime changes, such as drought, currently trigger compositional shifts between C_3 and C_4 grasses, changes that may be altered or intensified by altered CO_2 and climate. While fundamental information on each of the key physiological processes is lacking, and it is not possible to predict which species will eventually dominate and what this change in species mix may mean for ecosystem processes, long-term field experiments show that there is strong reason to expect significant carbon accumulation in native ecosystems.

MARINE BIOTIC RESPONSE TO ELEVATED CARBON DIOXIDE

The carbon dioxide–carbonate system is arguably the most important chemical equilibrium in the ocean. It influences nearly every aspect of marine science, including the ecology and, ultimately, the biodiversity of the oceans. It is largely responsible for controlling the pH of seawater and thus affects directly many other chemical equilibria as well. By the middle of this century atmospheric carbon dioxide (CO_2) is expected to reach double preindustrial levels (Houghton et al. 2001). This expected increase in atmospheric CO_2 will give rise via passive diffusion to a twofold increase in surface ocean CO_2 concentrations and cause a drop in surface water pH of about 0.4 units because of the CO_2-carbonate buffer system (Fig. 18.3).

A pH reduction of approximately 0.1 unit in surface waters has occurred already due to anthropogenic CO_2 input. Moreover, the atmospheric CO_2 signal is apparent to depths of more than 1000 m in some regions of the oceans (Goyet et al. 1999). These pH changes represent a sig-

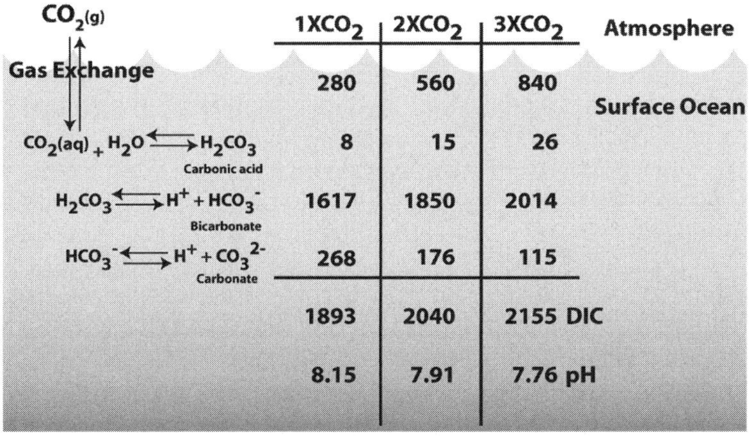

	1XCO$_2$	2XCO$_2$	3XCO$_2$	Atmosphere
CO$_{2(g)}$				
Gas Exchange	280	560	840	
				Surface Ocean
CO$_{2(aq)}$ + H$_2$O $\rightleftharpoons$ H$_2$CO$_3$ Carbonic acid	8	15	26	
H$_2$CO$_3$ $\rightleftharpoons$ H$^+$ + HCO$_3^-$ Bicarbonate	1617	1850	2014	
HCO$_3^-$ $\rightleftharpoons$ H$^+$ + CO$_3^{2-}$ Carbonate	268	176	115	
	1893	2040	2155 DIC	
	8.15	7.91	7.76 pH	

Figure 18.3. A schematic representation of the changes in the CO$_2$-carbonate system with elevated atmospheric CO$_2$. From the pre-industrial concentration (1 × CO$_2$) to a doubling (2 × CO$_2$) and tripling (3 × CO$_2$) of atmospheric CO$_2$, the projected decrease in carbonate ion concentration in surface water is 35% and 58%, respectively. Different model scenarios predict that atmospheric CO$_2$ will double before the year 2100. Source: Adapted from Feely et al. (2001) and Houghton et al. (2001); courtesy of R. A. Feely.

nificant fraction of the natural variation of seawater pH throughout the world's oceans (7.6 to 8.3), especially considering that pH is a log scale of proton concentration. Within the past few years fossil fuel CO$_2$, superimposed on the natural background levels, became a "major ion" of seawater (Box 18.2).

The impact of elevated CO$_2$ on climate, and the effect of climate alterations on marine ecosystems, has been given considerable attention (Hoegh-Guldberg, this volume). Significant research effort has also been focused on elucidating the ocean's role in buffering elevated changes in atmospheric CO$_2$. The ocean plays a central role in the global carbon cycle as a vast reservoir that takes up a substantial portion of anthropogenic carbon from the atmosphere. Because of its enormous volume, purposeful disposal of CO$_2$ in the ocean is being seriously considered as a means of mitigating greenhouse warming (Box 18.3) (Brewer et al., 1999; Brewer 2000; Seibel and Walsh 2001). However,

the effects of anthropogenic CO$_2$—whether passively or actively disposed in the ocean—on marine biota have received relatively little attention to date.

Photosynthesis and Primary Productivity

Most species of marine phytoplankton are able to use HCO$_3^-$ in photosynthesis and, owing to the large pool of HCO$_3^-$ in seawater, their growth is not limited by carbon (Raven 1997; Tortell et al. 1997). Thus, unlike terrestrial systems in which elevated atmospheric CO$_2$ may have a fertilization effect, global net primary productivity of marine phytoplankton is not expected to be stimulated by increased dissolved [CO$_2$] (Raven 1994). Increased temperature in the upper ocean will increase density stratification, however, leading to reduced upwelling of nutrients, which decreases primary productivity (Falkowski et al. 1998; Cox et al. 2000). Shifts in species composition are likely to occur as a consequence of differential growth rates at different nutrient concentrations and values of ambient seawater pH (c.f. Hinga 2002).

In contrast to marine algae, seagrasses utilize HCO$_3^-$ inefficiently and their light-saturated photosynthetic rates are limited by dissolved [CO$_{2 (aq)}$] (Zimmerman et al. 1997; Short & Neckles 1999; Invers et al. 2001). Increases in growth rates

---

Box 18.2.   **Carbon Dioxide and the Marine Carbonate System**

Dissolved carbon dioxide readily reacts with water to form carbonic acid. The acid then dissociates in two steps to HCO_3^- and CO_3^{2-} resulting in an increase in the concentration of protons (hydrogen ions, H^+), some of which combine with CO_3^{2-} to form HCO_3^-. Thus, total dissolved inorganic carbon (DIC) exists in seawater in three major forms: $[CO_{2(aq)}]$, $[HCO_3^-]$, and $[CO_3^{2-}]$. At a pre-industrial atmospheric concentration of 280 ppm and a surface seawater pH of 8.15, the relative concentrations of the major forms of carbon are $[CO_{2(aq)}] = 0.5\%$, $[HCO_3^-] =$ 85%, and $[CO_3^{2-}] = 14\%$. Within the next 50 years, the concentration of atmospheric CO_2 is expected to increase to two times the pre-industrial concentration. This addition of anthropogenic CO_2 will decrease the concentration of CO_3^{2-} as well as seawater pH. Carbon dioxide in seawater also interacts with calcium carbonate ($CaCO_3$) such that addition of CO_2 enhances $CaCO_3$ dissolution. Calcium carbonate saturation state is determined largely by CO_3^{2-} and so will decrease, with dramatic consequences for calcareous organisms, upon further addition of fossil fuel CO_2.

---

and biomass with elevated dissolved inorganic carbon (DIC) concentrations have been reported in many seagrass species (c.f. Short and Neckles 1999). While these results suggest that additions to atmospheric CO_2 will lead to higher primary productivity in seagrass communities, the long-term response will depend on interactions with other potentially limiting factors such as light and nutrients.

Acid-Base Regulation in Animals

Carbon dioxide is a waste product of routine animal metabolism whereby food molecules are broken down and converted to energy in the form of ATP. Marine species with low metabolic rates have low rates of CO_2 production and are therefore expected to tolerate only low concentrations of CO_2 and protons, all else being equal (Seibel and Walsh 2001). Low rates of metabolism typically correlate with lower concentrations of ion transport proteins such as Na^+/K^+ and H^+-ATPases (Gibbs and Somero 1990), suggesting reduced capacities of acid–base balance. If compensation of acid–base imbalance is not achieved, reduced intra- and extracellular pH and elevated CO_2 may depress

metabolism in some species (Guppy and Withers 1999; Hand 1991; Pörtner and Reipschläger 1996). Pörtner and Reipschläger (1996) point out that, in active squids, a change in arterial pH of as much as 0.15 unit could result from the anticipated increase in seawater PCO_2 over the next century and that this is sufficient to impair oxygen transport and limit the scope for activity in such species.

Biogenic Calcification and Carbonate Saturation State

The major sources of oceanic calcium carbonate ($CaCO_3$) are the tests or shells precipitated by planktonic foraminifera, coccolithophorids, and thecosomatous pteropods. In addition, coral reef ecosystems contain many benthic organisms that produce $CaCO_3$ including scleractinian corals and calcareous algae. Planktonic foraminifera and coccolithophorids precipitate $CaCO_3$ in the form of calcite, whereas pteropod molluscs, calcareous green algae, and corals produce carbonate skeletons of aragonite, a metastable form of $CaCO_3$ that is about 50 percent more soluble in seawater than is calcite (Mucci 1983). Other organisms, including calcareous red algae, echinoderms, bryozoans, and benthic foraminifera, are commonly associated with reef habitats and precipitate $CaCO_3$ in the form of magnesian calcite (less than 5 mole % $MgCO_3$). The solubility of magnesian calcite is slightly greater than that of aragonite (Walter and Morse 1984; Bischoff et al. 1987). Thus, under corrosive conditions, aragonite and magnesian calcite will dissolve faster than calcite in a mixed assemblage of carbonate particles. As the saturation of each mineral form of $CaCO_3$ is largely determined by the seawater $[CO_3{}^{-2}]$, oceanic uptake of anthropogenic CO_2 decreases the saturation state of $CaCO_3$ in seawater.

Field and laboratory studies with a variety of calcifying organisms demonstrate that the degree or extent of carbonate supersaturation has a direct and profound effect on individual species and community calcification rates. This effect now has been well-documented for corals and coral reef communities. For example, a positive correlation between $CaCO_3$ production and degree of supersaturation has been reported for coralline algae (c.f. Gattuso et al. 1999), reef-building coral species (Gattuso et al. 1998; Leclercq et al. 2000), and coral reef ecosystems (Langdon et al. 2000; Leclercq et al. 2002). Empirical data and model projections predict that a doubling of atmospheric CO_2 will result in a decrease in coral reef calcification of 14 to 40 percent (Gattuso et al. 1998; Kleypas et al. 1999; Langdon et al. 2000; Leclercq et al. 2000; Leclercq et al. 2002). Reduced calcification of coral reef–building organisms will likely result in weaker skeletons and increased susceptibility to erosion, and may lead to shifts in species composition and community structure (Done 1999; Kleypas et al. 1999).

Organisms that may be especially sensitive to elevated CO_2 are listed in Table 18.1. Animals with low metabolic rates, such as gelatinous zooplankton, appear to lack mechanisms for compensation of acid-base imbalance and may also be particularly sensitive, but no data exist at present. Organisms reliant on supersaturation of $CaCO_3$ for deposition of shells and tests will be most directly affected. Surface ocean uptake of CO_2 will reduce aragonite saturation state, for example, by 30 percent (Kleypas et al. 1999). Carbonate saturation is also temperature-dependent such that elevated CO_2 will push calcareous organisms toward the equator. For example, the aragonite saturation isoline of 3.5 will shift toward the equator by as much as 15 degrees latitude (Kleypas et al., 1999). In contrast, organismal ranges will shift poleward due to an indirect effect of carbon dioxide on climate, creating conflict between temperature and carbonate saturation effects (see Hoegh-Guldberg, this volume). For instance, corals affected by coral bleaching in the tropics may not be able to establish farther north or south, because saturation levels decrease with latitude, an effect that will become more pronounced as atmospheric CO_2 levels continue to rise. Range reduction may occur within the water column, as well as with latitude. Vertical migration patterns of species such as thecosomatous pteropods (molluscs) may be disrupted as anthropogenic CO_2 penetrates to depths where carbonate saturation is already low. Thus, direct and indirect CO_2 effects will result synergistically in a narrowing and shoaling of inhabitable water.

DISTURBANCE REGIME INTERACTIONS

Climate change affects the abundance and distribution of organisms, and hence biodiversity, by affecting ecosystem processes. Disturbances, for example, have a significant impact on population, energy, and nutrient cycles. Only in the past several decades have ecologists begun to understand the integral part that disturbances

Table 18.1. Examples of Marine Organisms Sensitive to Elevated CO_2 and Reduced pH

Species	Description	Seawater pH/CO_2	Sensitivity	Reference
Bacteria				
Aerobic heterotrophic bacteria		7.5	Reduced growth	Reviewed in Knutzen 1981
Phytoplankton				
Cricosphaera elongata		7.4	Reduced growth rate	Reviewed by Hinga 2002
Emiliania huxleyi		7.9	Calcification reduced 16%	Riebesell et al. 2000
		7.6	Growth rate only 10% of control pH 8.1	Reviewed in Hinga 2002
		7.9	Calcification reduced 45%	
Gephyrocapsa oceanica		7.6–7.8	Mortality relative to control pH 8.2	Reviewed in Hinga 2002
Gonyaulax polyedra		7.6–7.8	Irregular growth relative to pH 8.2	
Thoracosphaera hemii				
Cnidaria				
Corals		$2 \times CO_2$ pre-industrial	40% drop in calcification	Kleypas et al. 1999; Langdon et al. 2000
Mollusca				
Clio pyramidata	Pteropod	7.56–7.39	Shell dissolution	Fabry, unpublished data
Mytilus edulis	Sea mussel	7.1/6.5 mm Hg	Shell dissolution	Lindinger et al. 1984
Pinctata fuscata	Japanese pearl oyster	7.7	Shell dissolution, reduced growth	Reviewed in Knutzen 1981
		>7.4	Increasing mortality	
Illex illecebrosus	Epipelagic squid	7.5/1.5 mm Hg	Impaired oxygen transport/reduced scope for activity due to highly sensitive respiratory protein	Pörtner and Reipschläger 1996
Arthropoda				
Euphausia pacifica	Krill	7.54, LC_0	Mortality observed below LC_0; in all cases, survival declined with increasing exposure time, decreasing pH	Yamada and Ikeda 1999
Paraeuchaeta elongata	Mesopelagic copepod	7.41, LC_0		
Conchoecia sp.	Ostracod	7.79, LC_0		
Chaetognatha				
Sagitta elegans	Chaetognath	7.76, LC_0	See above	Yamada and Ikeda 1999
Echinodermata				
Strongylocentrotus purpuratus	Urchin		High sensitivity inferred from lack of pH regulation and passive buffering via test dissolution during emersion.	cf.: Burnett et al. 2002; Spicer 1995
Cystechinus sp.	Deep-sea urchin	7.8	80% mortality under simulated CO_2 sequestration.	Barry et al. 2002
Vertebrata				
Scyliorhinus canicula	Dogfish	7.7/1 mm Hg	Increased ventilation	Reviewed in Truchot 1987

Box 18.3.   **Ocean Sequestration of Carbon Dioxide**

The growing concern over the potential for global warming has spurred the development of technologies for capture, storage, and disposal of carbon dioxide. In addition to a variety of terrestrial alternatives, the ocean is being evaluated as a reservoir for storage of anthropogenic carbon. Two strategies are being seriously considered: enhancement of photosynthetic uptake of carbon dioxide by marine algae via iron fertilization of the surface; and direct injection of carbon dioxide gas or liquid to great depths in the ocean. Either approach will compound the disruptions to biological systems that will be caused by passive invasion of anthropogic CO_2 (Seibel and Walsh 2001). Marine iron fertilization will cause major changes to food webs due to the massive algal blooms it produces. Deep ocean disposal will kill all marine life over perhaps hundreds of cubic kilometers, due to changes in pH and direct toxicity of CO_2. Further, the practical benefits of both approaches are limited. Experimental iron fertilization in the southern oceans has shown an order of magnitude lower response than predicted by theory, raising serious doubts about its feasibility. Much CO_2 disposed in deep oceans as liquid eventually enters the oceanic carbonate cycle and escapes, so deep ocean "disposal" delays but does not fully avoid atmospheric effects.

---

play in the operation of ecosystems. This understanding has led to a shift away from an equilibrium view of ecosystem dynamics to a view that ecosystems are flexible organizations that are used to a constantly changing environment (Sprugel 1991; Wu and Loucks 1995).

Disturbance regimes of possible importance in interaction with climate change include fire, drought, and windfall. Drought and length of dry season are particularly important to high-biodiversity tropical forests, where they help regulate flowering and reproduction (Bush, this volume). Tree fall in high winds can be an important source of canopy openings, and hence diversity, in forest, and large storms may help create disturbances important to maintaining site-specific diversity in coral reefs (Connell 1978).

Fire plays a critical role, and one likely to be modified by climate change, in both fire-adapted and non-fire-adapted ecosystems. Effects in systems that are adapted to fire will be very different from those which have not historically experienced repeated fires. The variety of effect will de- pend heavily on the specific changes in climate that take place and how they interact with the biology of the site. Two very different systems—boreal and tropical forests—illustrate the possible contrasting impacts of climate change on disturbance regimes and biodiversity.

Tropical Forests

Fire is particularly important in tropical forests, where it can have devastating impacts in systems not adapted to burn. Once among the world's last great wildernesses, tropical forests have come under increasing pressure from growing human populations in recent decades. Rates of deforestation and forest degradation have greatly increased the world over. The main human-mediated disturbances include deforestation, forest fragmentation, logging, and forest fire. Natural disturbances include drought and windfall. Formerly rare, fire is becoming common in many forests and is frequently connected with deforestation, fragmentation, and logging. Climate change influences in tropical forests may be most pronounced in fire and drought regimes.

The effects of drought on tropical forests are difficult to quantify. In a detailed study of 205 tree and shrub species, 70 percent were shown to have increased mortality rates during periods of extensive drought. Mortality rates differed by species, guild (e.g., colonist versus generalist), and size, but exceptions existed for all observed trends signifying a complex response of forest communities to drought (Condit et al. 1995). Droughts of increased frequency may also have cascading effects on animal populations. For example, in Panama, El Niño droughts lead to cycles of high and then low levels of fruit production, which in turn lead to substantial famine events for frugivorous and granivorous mammals (Wright et al. 1999).

Since tropical forests are often organized along environmental gradients—principally precipitation (Pyke et al. 2001)—frequent droughts may change large-scale forest structure and distribution. Specific effects may include a greater prevalence of deciduous species or behavior (Condit et al. 2000) and greater selection for deeper rooting species (Kleidon and Lorenz 2001).

Droughts also indirectly lead to more frequent and devastating forest fires. Although forest fires occur every year, multitemporal studies have shown that the largest fires occur in El Niño years, accounting for 90 percent of all burning (Cochrane et al. 1999). The severe El Niño of 1997–98 resulted in the burning of approximately 8 million hectares of forest in Southeast Asia (Barber and Schweithelm 2000) and over 9.2 million hectares throughout Latin America (UNEP 2002).

If increasing atmospheric CO_2 levels and deforestation lead to hotter, dryer weather as predicted (Costa and Foley 2000; Zhang et al. 2001; DeFries et al. 2002) then forests will burn more frequently. Fire in tropical forests, especially evergreen rain forests, is a severe disturbance. There is evidence of previous fire occurrence in these forests (Sanford et al. 1985; Saldarriaga and West 1986; Hammond and ter Steege 1998; Turcq et al. 1998) but it has generally been very infrequent, on the order of hundreds or thousands of years between events (Mueller-Dombois 1981; Hammond and ter Steege 1998). Consequently, the forest had been ill prepared by evolution for such frequent disturbance (Uhl and Kauffman 1990). Some species of trees have significant resprouting capacity and other fortuitous traits that may assist in post-fire regeneration, but only in the absence of future fires. Faunal response to fire may be both guild- and species-specific (Barlow et al. 2002; Fredericksen and Fredericksen 2002) but will ultimately depend upon the severity of fire impact to the forest. Landscape fragmentation and land cover change interact synergistically to expose more of the forest to fire and consequently raise the risk of unintended fires occurring across the entire landscape (Cochrane 2001).

The new fire dynamic for many tropical forests is one of frequent fire incursions and increasing fire severity. The first fire in an intact closed canopy forest is unimpressive. Except for tree-fall gaps and other areas of unusual fuel structure, the fire will spread as a thin, slowly creeping ribbon of flames a few tens of centimeters in height (Cochrane et al. 1999). Over much of the burned area, the fire will consume little besides leaf litter. Such a fire may not seem worrisome but, in fact, its impacts are very severe. In terms of the energy released, the fire-line intensity is low, being similar in intensity to prescribed fires (50 kW/m) in temperate forests. However, the slow advance of tropical fires makes them deadly due to the long residence time flames have at the base of fire-contacted trees. In evergreen tropical forests, most of the trees have very thin bark and are therefore highly susceptible to damage by fire (Uhl and Kauffman 1990). A typical fire as described would kill nearly 40 percent of the trees (greater than or equal to 10 cm diameter) but reduce living biomass by as little as 10 percent, since few large trees, which compose the majority of the biomass, would be killed (Cochrane and Schulze 1999).

If the forest reburns within a few years of the initial fire, the fires will be much worse. In recurrent fires, flame lengths, flame depths, spread rates, residence times, and fire-line intensities are all significantly higher. A second fire may kill another 40 percent of the remaining trees, this time corresponding to 40 percent of the living biomass. In forests that reburn, large trees have no survival advantage over smaller trees because the changes in fire behavior overwhelm even the defenses of the largest, thickest-barked trees (Cochrane et al. 1999). Weedy vines and grasses, some of which are quite flammable even when green, quickly invade twice-burned forests. The process is clear; burning of these closed canopy evergreen forests creates a positive feedback in both fire susceptibility and fire severity. This means that the fires become not only more frequent but also much worse each time. The end result can be the complete destruction of the affected forests.

Boreal Forests

A contrasting picture of the role of climate change in disturbance is presented by the boreal forest. Lightning-caused wildfires are the predominant disturbance in the North American boreal forest (Johnson 1992; Payette 1992). There may be important interactions of boreal fire with other disturbance regimes. Spruce budworm (*Choristoneura flumiferana*) is the most widespread disturbance along with fire in the eastern boreal forest. Drought has allowed budworm (and other insects) to become epidemic in regions in which they are usually endemic. Modeling shows that future climate change may greatly increase

the area affected by outbreaks of these insects. There is little rigorous evidence that insect-attacked forests are more susceptible to fire but some evidence to the contrary. One serious biodiversity concern, however, is that the increased extent and frequency of insect epidemics—particularly spruce budworm, jack pine budworm (*Choristoneura flumiferana*), and forest tent caterpillar (*Malacosoma disstria*)—will hasten the conversion of ecosystems in changing climates.

The weather that precedes a crown fire in boreal forests is rather specific. In order to dry fuels, a period of persistent hot-dry weather must occur. Normal weather in the boreal forest is an alternating sequence of surface high- and low-pressure systems moving from west to east, and thus a sequence of generally dry and wet periods. However, on occasion a high-pressure system persists, providing a longer period of warmer and drier weather. These "blocking" high-pressure systems deflect the normal sequence of highs and lows north and south, producing meridional flow. Two patterns of these anomalous highs have been recognized over the boreal forest. The Pacific North American Pattern (PNA) has a high over western boreal North America, and the Hudson Bay high has a high over central and eastern boreal North America (Wallace and Gutzler 1981; Knox and Lawford 1990; Johnson and Wowchuk 1993; Bonsal and Lawford 1999; Skinner et al. 1999).

The causes of these patterns are as yet not well understood, but appear to be related to anomalies in surface water temperature in the North Pacific (Walsh and Richman 1981). The spring and summer occurrences of the "blocking" high-pressure systems are known for only a few decades (Vega et al. 1995). Fortunately, a record of fire intervals for the past three centuries does exist in time-since-fire distributions (for methods see Johnson and Van Wagner 1985; Johnson and Gutsell 1994; Reed 2000). The time-since-fire distribution is derived from a map of the polygons which give the date of the last crown fire. The cumulative proportion of the study area in each time-since-fire age class gives the time-since-fire distribution. This distribution is an estimate of the survivorship of units of the landscape. If one assumes a connection between the frequency of "blocking" high-pressure systems and the fire intervals, one can tentatively suggest how climate change and fire occurrence interact.

The main conclusion to be drawn from these time-since-fire distribution studies (Flannigan et al. 1998) is that there have been at least three changes in the average interval between fires in the past 300 years. Two of these changes seem to mark the beginning and the end of the last cold period of the Little Ice Age (mid-1700s and late 1800s), and the third perhaps (the record is not yet conclusive) started in the 1940s. Although the Little Ice Age was a period of cooler and moister climate, it had shorter intervals between fires, which suggests that the frequency of "blocking" high pressures was higher. From approximately the 1890s to the 1940s, the climate was warmer and drier and the fire intervals were longer, suggesting that the frequency of "blocking" high pressures was lower. After the 1940s, the interval appears to be even longer, but the variation in intervals is large; that is, standard error is large. Thus counterintuitively, the cooler-moister climate appears to be connected with shorter fire intervals and perhaps more frequent blocking systems, while warmer-drier climates appear to have longer fire intervals and perhaps less frequent blocking systems. Early studies of climate change (e.g., IPCC 1996) and climate change and fire weather suggested that the increase in CO_2 would lead to warmer-drier conditions over the boreal forest and hence an increase in fire occurrence. However, these conclusions appear to be at variance with past fire occurrences and climate changes. A recent study by

Flannigan and colleagues (1998) using daily, instead of monthly, data in a GCM found fire weather more consistent with the time-since-fire distributions in the past.

From our present understanding of forest dynamics, fire interval would have to be shorter than the age of first reproduction or longer than the life-span of different plant species in order to affect regeneration and species diversity. At least for trees, this would mean that the fire interval would have to be less than 20–30 years (the age at which trees must begin to produce large numbers of seeds) or longer than 300–400 years (when most reach maximum age in competitive environments). Herbaceous and shrub species probably also fit into this range, since most reproduce vegetatively after fire and can survive by clonal growth for hundreds of years (Whitford 1949). Consequently, species richness may have changed due to fire interval, but given the short periods between these climate changes (often 100 years or less), these changes would probably be overshadowed by other fire effects in the boreal system.

ECOLOGICAL INTERACTIONS

Soils and Plant-Microbial Interactions

Global change impacts on plant symbioses are arguably the most important of all to understand because these interactions directly affect carbon sequestration and nutrient cycling, therefore having the potential to influence feedbacks to further atmospheric change.

Most (more than 90 percent) plant species form mycorrhizal associations (Smith and Read 1997), and about two-thirds of these plants are symbiotic with arbuscular mycorrhizal (AM) fungi (Fitter et al. 2000). Estimates of the amount of carbon allocated to mycorrhizal fungi vary from 4–20 percent of the plant's total carbon budget (Graham 2000). As obligate symbionts, these fungi depend on the plant as a carbon source and thus form an additional carbon sink significant enough to enhance carbon fixation under elevated atmospheric CO_2 and to alleviate photosynthetic acclimation (Staddon et al. 1999).

Growth of both endo- and ectomycorrhizal plant species is generally more stimulated by elevated CO_2 than that of nonmycorrhizal species because the fungal symbionts respond positively to increased root carbon and exudates (e.g., O'Neill 1994; Olesniewicz and Thomas 1999). It has generally been hypothesized that as a result, the services provided by the fungus such as phosphate uptake might be enhanced in the future (Fitter et al. 2000). Mycorrhizal plants may thus be relieved of a potential nutrient deficiency and therefore be capable of a greater response to elevated CO_2, until constrained by some other limitation such as water.

Vector-Borne Diseases

Warmer temperatures directly increase both vector and parasite reproduction and development, frequency of blood feeding, and the rate at which pathogens are acquired, despite shortening mean daily survivorship (Patz et al. 2000; Liang et al. 2002). Related changes in rainfall, seasonality, humidity, and large-scale meteorological phenomena such as ENSO may also alter the quality and quantity of vector-breeding sites (Liang et al. 2002). Transmission of many parasitic diseases is confined to the rainy season (Patz et al. 2000), and small changes in seasonality may be very important because transmission rates tend to increase exponentially rather than linearly through the season (Kovats et al. 2001). Some arthropod vectors may undergo more generations per year and shorter or milder winters will increase parasite survival, especially for subtropical species (Sutherst 2001).

As the geographic distributions of vector-borne pathogens change in both time and space, host populations will be exposed to longer transmission seasons and immunologically naive populations will be exposed to newly introduced pathogens (Kovats et al. 2001). Tropical vectors such as mosquitoes will most likely expand their ranges poleward and upward in elevation (see LaPointe et al., this volume), although in the case of diseases such as human malaria, the extent of the possible range expansion has been hotly debated (e.g., Rogers and Randolph 2000). Temperate species are likely to experience a range contraction in the tropics and subtropics (Sutherst 2001). Modeling of tick-borne encephalitis (TBEv), for example, suggests that it may not survive along the southern edge of its present range (Slovenia, Croatia, and Hungary) but may have new foci in Scandinavia (Randolph 2001).

The greatest effect of warming on disease transmission will most likely be observed at the extremes of the range of temperatures at which transmission occurs. For many vector-borne diseases these lie in the range 14–18°C at the lower end and 35–40°C at the upper end. (Githeko et al. 2000). Warming in the lower range has a significant and nonlinear impact on the extrinsic incubation period of parasites and consequently on disease transmission, while at the upper end, transmission would cease. The faster rise in minimum compared to maximum temperatures over the past few decades (Easterling et al. 1997), thus, has particular significance for vector-borne pathogens (Kovats et al. 2001).

Hosts and Parasitoids

Any increase in temporal asynchrony between the susceptible stages of the host and the foraging of adult parasitoids because of changes in emergence patterns may have significant effects on the persistence of the interaction, and hence on the stabilizing effect of the natural enemy on the host (Godfray et al. 1994; Cannon 1998). In addition, parasitoid development time, longevity, and attack rate are influenced by temperature (Kingsolver 1989), and warming may differentially affect rates of host and parasitoid development and their activity. Parasitoid infection of both the peach potato aphid, *Myzus persicae* (Bezemer et al. 1998), and larvae of the moth *Epirrita autumnata* (Virtanen and Neuvonen 1999), for example, has been shown to increase at elevated temperatures (Bezemer et al. 1998). Parasitoids may also be affected via the effects of elevated CO_2 on their hosts. The lower nutritional quality of leaves grown at high CO_2 has been shown to create herbivores of lower nutritional quality in some studies (Roth and Lindroth 1995). Increased development time of leaf-mining insects at elevated CO_2 has also been found to result in greater mortality from parasitoids while in the mine (Stiling et al. 1999, 2002).

Plants and Pathogens

Changes in temperature, precipitation, soil moisture, and relative humidity will all influence the sporulation and colonization success of plant pathogens and the degree of synchrony between pest occurrence and susceptible stages in the host plants (Ayres and Lombandero 2000; Yamamura and Yokozawa 2002). Many of the direct impacts of elevated CO_2 on plant morphology and physiology will also affect both host resistance and the probability of pathogen infection (Chakraborty et al. 2000). Significant increases in the rate of net photosynthesis allow increased mobilization of resources into host resistance at elevated CO_2 (Hibberd et al. 1996). Other changes—including production of papillae, accumulation of silicon at sites of appressorial penetration, greater accumulation of carbohydrate in leaves, more waxes, extra layers of epidermal cells, increased fiber content, lowered nutrient concentrations, and greater number of

mesophyll cells—can all influence host resistance (Hibberd et al. 1996; Chakraborty et al. 2000). Reduced pathogen penetration may also result from a reduction in stomatal density and stomatal conductance at high CO_2 (Hibberd et al. 1996). Few generalizations can currently be made as to the direction and magnitude of disease incidence and severity under elevated CO_2. While some diseases have been shown to cause more severe growth reductions under elevated CO_2, others are mitigated, and the nutrient and water status of the plants often interact significantly with CO_2 treatment (Thompson et al. 1994, Chakraborty et al. 2000). The potential for accelerated pathogen evolution to threaten host resistance is an interesting possibility. Despite initial delays in host penetration, established colonies have been shown to grow faster in host tissue at elevated CO_2 (Chakraborty et al. 2000).

FEEDBACK FROM IMPACTS ON HUMAN SYSTEMS

Climate change also affects human systems in ways that result in feedbacks to biodiversity. Human systems dependent on the relatively stable climate typical of the past 11,000 years may be disrupted by climate change. Where the relationship between human systems and biodiversity is negative, such destabilization may make it more so. Even neutral or positive relationships may become negative when destabilized, although in a smaller number of cases, the biodiversity feedback may be positive. These feedback effects may play out in different ways in tropical, temperate, and freshwater systems.

Climate change is highly likely to change agricultural conditions in ways that will depress yields, at least temporarily, in many areas (Ramankutty et al. 2002b). Decline of agricultural productivity may lead to farming over more extensive areas—at the expense of existing natural habitat. Traditional ("subsistence") agriculture in the tropics often relies on surrounding forest as a "hardship buffer." When crop yields are down, forest products, forest-derived food, or logging can be used to supplement incomes. As farmers exploit forests or other natural areas in the struggle to maintain incomes, the damage to biodiversity may be severe. If forest is cleared for additional extensive agriculture, the damage may be permanent, even if the need is transitory.

Sea level rise is another effect of climate change that will impinge on human systems. A rise in sea level in areas such as Bangladesh and Taiwan may dislocate millions of coastal plain dwellers. Because other lowlands in these regions are already densely populated, natural areas in uplands will likely feel the direct or indirect pressures of this human migration.

Climate change will raise frost lines in the tropics, in turn raising the elevation of cultivable land in tropics at the expense of montane forest. In the Andes, for example, the altiplano is heavily populated, and frost-limited crops such as coffee are increasingly replacing mid-elevation forests. As frost lines rise with climate change, agriculture will press farther up the flank of the Andes. Forest species, however, will not be able to move up with rising temperatures, because habitat is no longer available at the altiplano, and the net effect will be a squeeze on mid-montane forests—some of the highest biodiversity forests on the planet.

Another feedback effect will be felt in ocean ecosystems. The anticipated decline in agricultural productivity is likely to increase the use—and runoff—of fertilizers. Over 50 dead zones, associated with intense agricultural runoff, have been identified worldwide and are usually offshore from industrialized countries (Malakoff 1998). In industrialized countries, where farmers are more likely to try to maintain yields by increasing their use of fertilizer rather than the area under cultivation, ocean "dead zones" downstream from these applications may become more extensive.

Not all climate change–human feedbacks will have negative impacts on biodiversity. For example, rising sea level may make human occupation of low marine islands impossible, thus reducing fishing and other pressures on reefs. Of course, reefs face multiple negative effects of climate change, so this benefit to marine ecosystems may be of limited importance.

The system closest to its limit and therefore most sensitive to human feedbacks from climate change may be freshwater ecosystems. Water demand already accounts for 60 percent of all available freshwater flows. Population growth coupled with a warming climate may push demand for freshwater beyond available supplies (Allan and Poff, this volume).

Whatever the regional dynamics, the global picture is one of total human appropriation of freshwater. Warming will increase evaporative loss, likely reducing freshwater habitats. It will encourage human water use, causing more withdrawals and increasing pressure on dwindling supplies. The fate of freshwater biodiversity in a future, more populous world is tenuous, a vulnerability that will be deepened by climate change and deepened again by human feedbacks from climate change.

REFERENCES

Ainsworth E. A., P. A. Davey, G. J. Hymus, B. G. Drake, and S. P. Long. 2002. Long-term response of photosynthesis to elevated carbon dioxide in a Florida scrub-oak ecosystem. *Ecological Applications* 12: 1267–1275.

Alberto, A. M. P., L. H. Ziska, C. R. Cervancia, and P. A. Manalo. 1996. The influence of increasing carbon dioxide and temperature on competitive interactions between a C_3 crop, rice (*Oryza sativa*) and a C_4 weed (*Echinochloa glabrescens*). *Australian Journal of Plant Physiology* 23: 795–802.

Ayres, M. P., and M. J. Lombardero. 2000. Assessing the consequences of global change for forest disturbance from herbivores and pathogens. *The Science of the Total Environment* 262:262–286.

Barber, C. V., and J. Schweithelm. 2000. *Trial by Fire: Forest Fire and Forestry Policy in Indonesia's Era of Crisis and Reform.* Washington, D. C.: World Resources Institute.

Barlow, J., T. Haugaasen, and C. A. Peres. 2002. Effects of ground fires on understory bird assemblages in Amazonian forests. *Biological Conservation* 105: 157–169.

Barry, J., B. A. Seibel, J. Drazen, M. Tamburri, C. Lovera, and P. Brewer. 2002. Field experiments on direct ocean CO_2 sequestration: The response of deep-sea faunal assemblages to CO_2 injection at 3200 m off Central California. *EOS Transactions* [Amer. Geophysical Union] 83: OS51P-02.

Bezemer, T. M., T. H. Jones, and K. J. Knight. 1998. Long-term effects of elevated CO_2 and temperature on populations of the peach potato aphid *Myzus persicae* and its parasitoid *Aphidius matricariae.* *Oecologia* 116: 128–135.

Bischoff, W. W., F. T. Mackensie, and F. C. Bishop. 1987. Stabilities of synthetic magnesian calcites in aqueous solution: Comparison with biogenic materials. *Geochimica et Cosmochimica Acta* 51: 1413–1423.

Bond, W.J., and G. F. Midgley. 2000. A proposed CO_2-controlled mechanism of woody plant invasion in grasslands and savannas. *Global Change Biology* 6: 865–869.

Bonsal, B. R., and R. G. Lawford. 1999. Teleconnections between El Niño and La Niña events and summer: Extended dry spells on the Canadian prairies. *International Journal of Climatology* 19: 1445–1458.

Brewer, P. G. 2000. Contemplating action: Storing carbon dioxide in the ocean. *Oceanography* 13: 84–92.

Brewer, P. G., G. Friederich, E. T. Peltzer, and F. M. J. Orr. 1999. Direct experiments on the ocean disposal of fossil fuel CO_2. *Science* 284: 943–945.

Burkhardt, S., G. Amoroso, U. Riebesell, and D. Sültemeyer. 2001. CO_2 and HCO_3^- uptake in marine diatoms acclimated to different CO_2 concentrations. *Limnology and Oceanography* 46: 1378–1391.

Burnett, L., N. Terwilliger, A. Carroll, D. Jorgensen, and D. Scholnick. 2002. Respiratory and acid-base physiology of the purple sea urchin. *Strongylocentratus purpuratus*, during air exposure: Presence and function of a facultative lung. *The Biological Bulletin* 203: 42–50.

Bush, M. 2002. Distributional change and conservation on the Andean flank: A palaeoecological perspective. *Global Ecology & Biogeography* 11: 475–484.

Cannon, R. J. C. 1998. The implications of predicted climate change for insect pests in the UK, with emphasis on non-indigenous species. *Global Change Biology* 4: 785–796.

Charkraborty, S., A. V. Tiedemann, and P. S. Teng. 2000. Climate change: Potential impact on plant diseases. *Environmental Pollution* 108: 317–326.

Cochrane, M. A. 2001. Synergistic interactions between habitat fragmentation and fire in evergreen tropical forests. *Conservation Biology* 15(6): 1515–1521.

Cochrane, M. A., A. Alencar, M. D. Schulze, C. M. Souza Jr., D. C. Nepstad, P. Lefebvre, and E. Davidson. 1999. Positive feedbacks in the fire dynamic of closed canopy tropical forests. *Science* 284: 1832–1835.

Cochrane, M. A., and M. D. Schulze. 1999. Fire as a recurrent event in tropical forests of the eastern Amazon: Effects on forest structure, biomass, and species composition. *Biotropica* 31(1): 2–16.

Condit, R., S. P. Hubbell, and R. B. Foster. 1995. Mortality rate of 205 Neotropical tree and shrub species and the impact of a severe drought. *Ecological Monographs* 65: 419–439.

Condit, R., K. Watts, S. A. Bohlman, R. Perez, R. B. Foster and S. P. Hubbell. 2000. Quantifying the deciduousness of tropical forest canopies under varying climates. *Journal of Vegetation Science* 11: 649–658.

Connell, J. 1978. Diversity in tropical rain forests and coral reefs. *Science* 199: 1302–1310.

Costa, M. H., and J. A. Foley. 2000. Combined effects of deforestation and doubled atmospheric CO2 concentrations on the climate of Amazonia. *Journal of Climate* 13: 18–34.

Cox, P. M., R. A. Betts, C. D. Jones, S. A. Spall, and I. J. Totterdell. 2000. Acceleration of global warming due to carbon-cycle feedbacks in a coupled climate model. *Nature* 408: 184–187.

Cramer, W., A. Bondeau, F. I. Woodward, I. C. Prentice, R. A. Betts, V. Brovkin, P. M. Cox, V. Fisher, J. A. Foley, A. D. Friend, C. Kucharik, M. R. Lomas, N. Ramankutty, S. Sitch, B. Smith, A. White, and C. Young-Molling. 2001. Global response of terrestrial ecosystem structure and function to CO_2 and climate change: Results from six dynamic global vegetation models. *Global Change Biology* 7: 357–373.

Curtis, P.S., and X.Wang. 1998. A meta-analysis of elevated CO_2 effects on woody plant mass, form, and physiology. *Oecologia* 113: 299–313.

Dacey, J. W. H., B. G. Drake, and M. J. King. 1994. Stimulation of methane emission by carbon dioxide enrichment of marsh vegetation. *Nature* 370: 47–49.

Day, F. P., E. P. Weber, C. R. Hinkle, and B. G. Drake. 1996. Effects of elevated CO_2 on fine root length and distribution in an oak-palmetto scrub ecosystem in central Florida. *Global Change Biology* 2: 101–106.

DeFries, R. S., L. Bounoua, and G. J. Collatz. 2002. Human modification of the landscape and surface climate in the next fifty years. *Global Change Biology* 8: 438–458.

Dijkstra, P., G. J. Hymus, D. Colavito, D. Vieglais, C. Cundari, D. P. Johnson, B. A. Hungate, C. R. Hinkle, and B. G. Drake. 2002. Elevated atmospheric CO_2 stimulates aboveground biomass in a fire-regenerated scrub-oak ecosystem. *Global Change Biology* 8: 90–103.

Done, T. J. 1999. Coral community adaptability to environmental change at the scales of regions, reefs and reef zones. *Amer. Zool.* 39: 66–79.

Drake, B. G., M. A. Gonzalez-Meler, and S. P. Long. 1997. More efficient plants: A consequence of rising atmospheric CO_2? *Annual Reviews of Plant Physiology and Plant Molecular Biology* 48: 607–637.

Drake, B. G., and P. W. Leadley. 1991. Canopy photosynthesis of crops and native plant communities exposed to long-term elevated CO_2 treatment. *Plant, Cell and Environment* 14 (8): 853–860.

Dukes, J. S., and H. A. Mooney. 1999. Does global change increase the success of biological invaders? *Trends in Ecology and Evolution* 14: 135–139.

Easterling, D.R., B. Horton, P. D. Jones, T. C. Peterson, T. R. Karl, D. E. Parker, M. J. Salinger, V. Razuvayev, N. Plummer, P. Jamason, and C. K. Folland. 1997. Maximum and minimum temperature trends for the globe. *Nature* 277: 364–367.

Falkowski, P. G., R. T. Barber, and V. Smetacek. 1998. Biogeochemical controls and feedbacks on ocean primary production. *Science* 281: 200–206.

Feely, R., C. Sabine, T. Takahashi, and R. Wanninkhof. 2001. Uptake and storage of carbon dioxide in the oceans. *Oceanography* 14:3–32.

Fitter, A. H., A. Heinemeyer, and P. L. Staddon. 2000. The impact of elevated CO_2 and global climate change on arbuscular mycorrhizas: A mycocentric approach. *New Phytologist* 147: 179–187.

Flannigan, M. D., Y. Bergeron, O. Engelmark, and B. M. Wotton. 1998. Future wildfire in circumboreal forest in relation to global warming. *Journal of Vegetation Science* 9: 469–476.

Flannigan, M. D., T. J. Lynham, and P. C. Ward. 1989. An extensive blowdown occurrence in Northwestern Ontario. Presented at the Tenth Conference on Fire and Forest Meteorology, April 17–21, 1989, Ottawa, Canada.

Fredericksen, N. J., and T. S. Fredericksen. 2002. Terrestrial responses to logging and fire in a Bolivian tropical humid forest. *Biodiversity and Conservation* 11: 27–38.

Gattuso, J.-P., D. Allemand, and M. Frankignoulle. 1999. Photosynthesis and calcification at cellular, organismal and community levels in coral reefs: A review on interactions and control by carbonate chemistry. *Amer. Zool.* 39: 160–183.

Gattuso, J.-P., M. Frankignoulle, I. Bourge, S. Romaine, and R. W. Buddemeier. 1998. Effect of calcium carbonate saturation of seawater on coral calcification. *Global Planet. Change* 18: 37–46.

Gibbs, A. H., and G. N. Somero, 1990. Na^+-K^+ adenosine triphosphatase activities in gills of marine teleost fishes, changes with depth, size and locomotory activity level. *Mar. Biol.* 106: 315–321.

Githeko, A. K., S. W. Lindsay, U. E. Confalonieri, and J. A. Patz. 2000. Climate change and vector-borne diseases: A regional analysis. *Bulletin of the World Health Organization* 78: 1136–1147.

Godfray, H. C. J., M. P. Hassell, and R. D. Holt. 1994. The population dynamic consequences of phenological asynchrony between parasitoids and their hosts. *Journal of Animal Ecology* 63: 1–10.

Goyet, C., C. Coatanoan, G. Eischeid, T. Amaoka, K. Okuda, R. Healy, and S. Tsunogai. 1999. Spatial variation of total CO_2 and total alkalinity in the northern Indian Ocean: A novel approach for the quantification of anthropogenic CO_2 in seawater. *J. Mar. Res.* 57: 135–163.

Graham, J.H. 2000. Assessing costs of arbuscular mycorrhizal symbiosis in agroecosystems, pp. 127–140. In G. K. Podila and D. D. Douds, eds., *Current Advances in Mycorrhizal Research*. St Paul, Minn.: APS Press.

Grant, R. F., and I. A. Nalder. 2000. Climate change effects on net carbon exchange of a boreal aspen-hazelnut forest: Estimates from the ecosystem model ecosys. *Global Change Biology*. 6: 183–200.

Guppy, M. and P. Withers. 1999. Metabolic depression in animals: Physiological perspectives and biochemical generalizations. *Biol. Rev.* 74: 1–40.

Hammond, D. S., and H. ter Steege. 1998. Propensity for fire in Guianan rainforests. *Conservation Biology* 12: 944–947.

Hand, S. C. 1991. Metabolic dormancy in aquatic invertebrates. In *Advances in Comparative and Environmental Physiology*, vol. 8, pp. 1–50. Heidelberg: Springer-Verlag.

He, J. S., F. A. Bazzaz, and Schmid, B. 2002. Interactive effects of diversity, nutrients and elevated CO_2 on experimental plant communities. *Oikos* 97: 337–348.

Hibberd, J. M., R. Whitbread, and J. F. Farrar. 1996. Effect of elevated CO_2 concentration on infection of barley by *Erysiphe graminis*. *Physiological and Molecular Plant Pathology* 48: 37–53.

Hinga, K. R. 2002. Effects of pH on coastal marine phytoplankton. *Mar. Ecol. Prog. Ser.* 238: 281–300.

Hoffmann, W. A., F. A. Bazzaz, N. J. Chatterton, P. A. Harrison, and R. B. Jackson. 2000. Elevated CO_2 enhances resprouting of a tropical savanna tree. *Oecologia* 123: 312–317.

Houghton, J. T., Y. Ding, D. J. Griggs, M. Noguer, P. J. van der Linden, and D. Xiaosu. 2001. Climate change 2001: The scientific basis. In IPCC *Third Assessment Report: Climate Change 2001*, p. 944. Cambridge: Cambridge University Press.

Hungate, B. A., M. Reichstein, P. Dijkstra, D. Johnson, G. Hymus, J. D. Tenhunen, and B. G. Drake. 2002. Evapotranspiration and soil water content in a scrub-oak woodland under carbon dioxide enrichment. *Global Change Biology* 8: 289–298.

Hymus G. J., D. P. Johnson, S. Dore, P. Dijkstra, H. P. Anderson, C. R. Hinkle, and B. G. Drake. 2003.

Effects of elevated atmospheric CO_2 on net ecosystem CO_2 exchange of a scrub-oak ecosystem. *Global Change Biology* 9: 1802–1812.

Hymus G. J., T. Snead, D. Johnson, B. Hungate, and B. G. Drake. 2002. Acclimation of photosynthesis and respiration to elevated CO_2 in two scrub-oak species. *Global Change Biology* 8: 317–328.

Invers, O., R. C. Zimmerman, R. S. Alberte, M. Perez, and J. Romero. 2001. Inorganic carbon sources for seagrass photosynthesis: An experimental evaluation of bicarbonate use in species inhabiting temperate waters. *J. Exp. Mar. Biol. Ecol.* 265: 203–217.

IPCC. 1996. *Climate Change 1995 Impacts, Adaptation and Mitigation of Climate Change: Scientific-Technical Analyses.* Intergovernmental Panel on Climate Change. Cambridge: Cambridge University Press.

Jacob, J., C. Greitner, and B. G. Drake. 1995. Acclimation of photosynthesis in relation to Rubisco and non-structural carbohydrate contents and *in situ* carboxylase activity in *Scirpus olneyi* grown at elevated CO_2 in the field. *Plant, Cell and Environment* 18: 875–884.

Jarvis, A. J., T. A. Mansfield, and W. J. Davies. 1999. Stomatal behaviour, photosynthesis and transpiration under rising CO_2. *Plant, Cell and Environment* 22: 639–648.

Johnson, E. A. 1992. *Fire and Vegetation Dynamics.* Cambridge: Cambridge University Press.

Johnson, E. A., and S. L. Gutsell. 1994. Fire frequency models, methods and interpretations. *Advances in Ecological Research* 25: 239–287.

Johnson, E. A., and C. E. Van Wagner. 1985. The theory and use of two fire history models. *Canadian Journal of Forest Research* 15: 214–220.

Johnson, E. A., and D. R. Wowchuk. 1993. Wildfires in the southern Canadian Rocky Mountains and their relationship to mid-tropospheric anomalies. *Canadian Journal of Forest Research* 23: 1213–1222.

Kingsolver, J. G. 1989. Weather and the population dynamics of insects: Integrating physiological and population ecology. *Physiological Zoology* 62: 314–334.

Kleidon, A., and S. Lorenz. 2001. Deep roots sustain Amazonian rainforest in climate model simulations of the last ice age. *Geophysical Research Letters* 28: 2425–2428.

Kleypas, J. A., J. W. McManus, L. A. B. Menez. 1999. Environmental limits to coral reef development: Where do we draw the line? *Amer. Zool.* 39: 146–159.

Knox, J. L., and R. G. Lawford. 1990. The relationship between Canadian prairie dry and wet months and circulation anomalies in the mid-troposphere. *Atmosphere-Ocean* 28(2): 189–215.

Knutzen, J. 1981. Effects of decreased pH on marine organisms. *Marine Pollution Bulletin* 12: 25–29.

Kovats, R. S., D. H. Campbell-Lendrum, A. J. McMichael, A. Woodward, and J. S. H. Cox. 2001. Early effects of climate change: Do they include changes in vector-borne disease? *Proceedings of the Royal Society of London Series B* 356: 1057–1068.

Langdon, C., R. Takahashi, C. Sweeney, D. Chipman, J. Goddard, F. Marubini, H. Aceves, H. Barnett, and M. J. Atkinson. 2000. Effect of calcium carbonate saturation state on the calcification rate of an experimental coral reef. *Global Biogeochemical Cycles* 14: 639–654.

Leclercq, N., and J.-P. Gattuso. 2002. Primary production, respiration, and calcification of a coral reef mesocosm under increased CO_2 partial pressure. *Limnol. Oceanogr.* 47: 558–564.

Leclercq, N., J.-P. Gattuso, and J. Jaubert. 2000. CO_2 partial pressure controls the calcification rate of a coral community. *Global Change Biol.* 6: 329–334.

Li, J.-H., P. Dijkstra, C. R. Hinkle, R. M. Wheeler, and B. G. Drake. 1999. Photosynthetic acclimation to elevated atmospheric CO2 concentration in the Florida scrub-oak species Quercus geminata and Quercus myrtifolia growing in their native environment. *Tree Physiology* 19: 229–234.

Li, J.-H., W. A. Dugas, G. J. Hymus, D. P. Johnson, C. R. Hinkle, B. G. Drake, and B. A. Hungate. 2003. Direct and indirect effects of elevated CO_2 on transpiration from Quercus myrtifolia in a scrub oak ecosystem. *Global Change Biology* 9: 96–105.

Liang, S. Y., K. J. Linthicum, and J. C. Gaydos. 2002. Climate change and the monitoring of vector-borne disease. *Journal of the American Medical Association* 287: 2286.

Lindinger, M. I., D. J. Lauren, and D. G. McDonald. 1984. Acid-base balance in the sea mussel, *Myrilus edulis*, III: Effects of environmental hypercapnia on intra- and extracellular acid-base balance. *Marine Biology Letters* 5: 371–381.

Lloyd, J., and G. D. Farquhar. 1996. The CO_2 dependence of photosynthesis, plant growth responses to elevated atmospheric CO_2 concentrations and their interaction with soil nutrient status. I. General principles and forest ecosystems. *Functional Ecology* 10: 4–32.

Lodge R. J., P. Dijkstra, B. G. Drake, J. I. L. Morison. 2001. Stomatal acclimation to increased CO_2 concentration in a Florida scrub oak species Quercus myrtifolia Willd. *Plant Cell and Environment* 24: 77–88.

Long, S. P. and B. G. Drake. 1991. The effect of the long-term CO_2 fertilization in the field on the quantum yield of photosynthesis in the C_3 sedge *Scirpus olneyi*. *Plant Physiology* 96: 221–226.

Long, S. P., and B. G. Drake. 1992. Photosynthetic CO_2 assimilation and rising atmospheric CO_2 concentrations, pp. 69–103. In N. R. Baker and H. Thomas, eds., *Crop Photosynthesis: Spatial and Temporal Determinants.* Amsterdam: Elsevier Science.

Lüscher, A., G. R. Hendrey, and J. Nösberger. 1998. Long-term responsiveness to free air CO_2 enrichment of functional types, species and genotypes of plants from fertile permanent grassland. *Oecologia* 113: 37–45.

Malakoff, D. 1998. Death by suffocation in the Gulf of Mexico. *Science* 281:190–193.

Marsh, A., P. Megonigal, and B. G. Drake. 2004. Export of carbon from a Chesapeake Bay wetland stimulated by elevated atmospheric CO_2. *Global Change Biology*.

Misson, L., D. P. Rasse, C. Vincke., M. Aubinet, and L. M. François. 2002. Predicting transpiration from forest stands in Belgium for the 21st century. *Agriculture and Forest Meteorology* 11: 265–282.

Mucci, A. 1983. The solubility of calcite and aragonite in seawater at various salinities, temperatures and 1 atmosphere total pressure. *American Journal of Science* 238: 780–799.

Mueller-Dombois, M. 1981. Fire in tropical ecosystems. *Proceeding of the Conference—Fire Regimes and Ecosystem Properties*. GTR WO-26, 137–176.

Olesniewicz, K. S. and R. B. Thomas. 1999. Effects of mycorrhizal colonization on biomass production and nitrogen fixation of black locust (*Robinia pseudoacacia*) seedlings grown under elevated atmospheric carbon dioxide. *New Phytologist* 142: 133–140.

O'Neill, E. G. 1994. Responses of soil biota to elevated atmospheric carbon dioxide. *Plant and Soil* 165: 55–65.

Patz, J. A., T. K. Graczyk, N. Geller, and A. Y. Vittor. 2000. Effects of environmental change on emerging parasitic diseases. *International Journal for Parasitology* 30: 1395–1405.

Payette, S. 1992. Fire as a controlling process in the North American boreal forest, pp. 144–169. In H. H. Shugart, R. Leemans, and G. B. Bonan, eds., *A Systems Analysis of the Global Boreal Forest*. Cambridge: Cambridge University Press.

Poorter, H. 1993. Interspecific variation in the growth response of plants to an elevated ambient CO_2 concentration. *Vegetatio* 104/105: 77–97.

Pörtner, H. O., and A. Reipschläger. 1996. Ocean disposal of anthropogenic CO_2: Physiological effects on tolerant and intolerant animals, pp. 57–81. In B. Ormerod and M. V. Angel, eds. *Ocean Storage of Carbon Dioxide. Workshop 2-Environmental Impact*. Cheltenham, U.K.: IEA Greenhouse Gas R&D Program.

Pyke, C. R., R. Condit, S. Anguilar, and S. Lao. 2001. Floristic composition across a climate gradient in a neotropical lowland forest. *Journal of Vegetation Science* 12: 553–566.

Ramankutty, N., J. A. Foley, J. Norman, and K. Mc-Sweeney. 2002. The global distribution of cultivable lands: Current patterns and sensitivity to possible climate change. *Global Ecology & Biogeography* 11: 377–392.

Randolph, S. E. 2001. The shifting landscape of tick-borne zoonoses: Tick-borne encephalitis and Lyme borreliosis in Europe. *Proceedings of the Royal Society of London Series B* 356: 1045–1056.

Rasse, D. P., and B. G. Drake. 2004. Seventeen-year sedge response to elevated CO_2 is cumulative and salt-induced. *Global Change Biology*.

Rasse, D. P., S. Stolaki, G. Peresta, and B. G. Drake. 2002. Patterns of canopy-air CO_2 concentration in a brackish wetland: Analysis of a decade of measurements and the simulated effects on the vegetation. *Agriculture and Forest Meteorology* 144: 59–73.

Raven, J. A. 1994. Carbon fixation and carbon availability in marine phytoplankton. *Photosynthesis Res.* 39: 259–273.

Raven, J. A. 1997. Inorganic carbon acquisition by marine autotrophs. *Advances in Botanical Research* 27: 85–209.

Reed, W. J. 2000. Reconstructing the history of forest fire frequency—Identifying hazard rate change points using the Bayes' Information Criterion. *Canadian Journal of Statistics* 28: 353–365.

Riebesell, U., I. Zondervan, B. Rost, P. D. Tortell, R. E. Zeebe, and F. M. M. Morel. 2000. Reduced calcification of marine plankton in response to increased atmospheric CO_2. *Nature* 407: 364–367.

Rogers, D. J., and S. E. Randolph. 2000. The global spread of malaria in a future, warmer world. *Science* 289: 1763–1766.

Roth, S. K., and R. L. Lindroth. 1995. Elevated atmospheric CO_2: Effects on phytochemistry, insect performance and insect-parasitoid interactions. *Global Change Biology* 1: 173–182.

Roumet, C., and J. Roy. 1996. Prediction of the growth response to elevated CO_2: A search for physiological criteria in closely related grass species. *New Phytologist* 134: 615–621.

Saldarriaga, J. G., and D. C. West. 1986. Holocene fires in the northern Amazon Basin. *Quaternary Research* 26: 358–366.

Sanford, R. L., J. Saldarriaga, K. Clark, C. Uhl, and R. Herrera. 1985. Amazon rainforest fires. *Science* 227: 53–55.

Seibel, B. A. and P. J. Walsh. 2001. Potential impacts of CO_2 injection on deep-sea biota. *Science* 294: 319–320.

Short, F. T. and H. A. Neckles. 1999. The effects of global climate change on seagrasses. *Aquatic Bot.* 63: 169–196.

Skinner, W. R., B. J. Stocks, D. L. Martell, B. Bonsal, and A. Shabbar. 1999. The association between circulation anomalies in the mid-troposphere and area burned by wildland fire in Canada. *Theoretical and Applied Climatology* 63: 105.

Smith, S. D., T. E. Huxman, S. F. Zitzer, T. N. Charlet, D. C. Housman, J. S. Coleman, L. K. Fenstermaker, J. R. Seemann, and R. S. Nowak. 2000. Elevated

CO$_2$ increases productivity and invasive species success in an arid ecosystem. *Nature* 408(6808): 79–82.

Smith, S. E., and D. J. Read. 1997. *Mycorrhizal Symbiosis*. London: Academic Press.

Spicer, J. J. 1995. Oxygen and acid-base status of the sea urchin *Psammechinus miliaris* during environmental hypoxia. *Marine Biology* 124: 71–76.

Sprugel, D. G. 1991. Disturbance, equilibrium and environmental variability: What is natural vegetation in a changing environment? *Biological Conservation* 58: 1–18.

Staddon, P. L., A. H. Fitter, and D. Robinson. 1999. Effects of mycorrhizal colonization and elevated carbon dioxide on carbon fixation and below ground carbon partitioning in *Plantago lanceolata*. *Journal of Experimental Botany* 50: 853–860.

Stiling, P., M. Cattell, D. C. Moon, A. M. Rossi, B. A. Hungate, G. Hymus, and B. G. Drake. 2002. Elevated atmospheric CO$_2$ lowers herbivore abundance, but increases leaf abscission rates. *Global Change Biology* 8: 658–667.

Stiling, P., A. M. Rossi, B. Hungate, P. Dijkstra, C. R. Hinkle, W. M. Knott III, and B. G. Drake. 1999. Decreased leaf-miner abundance in elevated CO$_2$: Reduced leaf quality and increased parasitoid attack. *Ecological Applications* 9: 240–244.

Sutherst, R. W. 2001. The vulnerability of animal and human health to parasites under global change. *International Journal for Parasitology* 31: 933–948.

Teyssonneyre, F., C. Picon-Cochard, R. Falcimagne, and J. F. Soussana. 2002. Effects of elevated CO$_2$ and cutting frequency on plant community structure in a temperate grassland. *Global Change Biology* 8: 1034–1046.

Thompson, G., and B. G. Drake. 1994. Insects and fungi on a C$_3$ sedge and a C$_4$ grass exposed to elevated atmospheric CO$_2$ concentrations in open top chambers in the field. *Plant, Cell and Environment* 17: 1161–1167.

Tortell, P. D., J. R. Reinfelder, and F. M. M. Morel. 1997. Active uptake of bicarbonate by diatoms. *Nature* 390: 243–244.

Truchot, J. P. 1987. *Comparative Aspects of Extracellular Acid-Base Balance*. Berlin: Springer-Verlag.

Turcq, B., A. Sifeddine, L. Martin, M. L. Absy, F. Soubies, K. Suguio, and C. Volkmer-Ribeiro. 1998. Amazonia rainforest fires: A lucustrine record of 7000 years. *Ambio* 27: 139–142.

Uhl, C. and J. B. Kauffman. 1990. Deforestation, fire susceptibility, and potential tree responses to fire in the eastern Amazon. *Ecology* 71: 437–449.

UNEP. Cochrane, M.A. 2002. *Spreading Like Wildfire—Tropical Forest Fires in Latin America and the Caribbean: Prevention, Assessment and Early Warning*. United Nations Environment Program, Regional Office for Latin America and the Caribbean. http://www.Rolac.UNexp.MX/dewalac/eng/five_ingles.ddf

Vega, A. J., K. G. Henderson, and R. V. Rohli. 1995. Comparison of monthly and intramonthly indices for the Pacific/North American Teleconnection Pattern. *Journal of Climate* 8: 2097–2103.

Virtanen, T., and S. Neuvonen. 1999 Performance of moth larvae on birch in relation to altitude, climate, host quality and parasitoids. *Oecologia* 120: 92–101.

Wallace, J. M., and D. S. Gutzler. 1981. Teleconnections in the geopotential height field during the northern hemisphere winter. *Monthly Weather Review* 109: 784–811.

Walsh, J. E., and F. M. B. Richman. 1981. Seasonality in the associations between surface temperature over the United States and the north Pacific Ocean. *Monthly Weather Review* 10: 767–782.

Walter, L. M., and J. W. Morse. 1984. Magnesian calcite stabilities: A reevaluation. *Geochimica et Cosmochimica Acta* 48: 1059–1069.

Wand, S. J. E., G. F. Midgley, M. H. Jones, and P. S. Curtis. 1999. Responses of wild C$_4$ and C$_3$ grass (Poaceae) species to elevated atmospheric CO$_2$ concentration: A meta-analytic test of current theories and perceptions. *Global Change Biology* 5: 723–741.

Whitford, P. B. 1949. Distribution of woodland plots in relation to succession and clonal growth. *Ecology* 30: 199–208.

Wright, S. J., C. Carrasco, O. Calderon, and S. Paton. 1999. The El Niño–Southern Oscillation variable fruit production and famine in a tropical forest. *Ecology* 80: 1632–1647.

Wu, J., and O. L. Loucks. 1995. From balance of nature to hierarchical patch dynamics: A paradigm shift in ecology. *Quarterly Review of Biology* 70: 439–466.

Yamada, Y., and T. Ikeda. 1999. Acute toxicity of lowered pH to some oceanic zooplankton. *Plankton Biology and Ecology* 46: 62–67.

Yamamura, K., and M. Yokozawa. 2002. Prediction of a geographic shift in the prevalence of rice stripe virus disease transmitted by the small brown leafhopper, *Laodelphax striatellus* (Fallen) (Hemiptera: Delphacidae), under global warming. *Applied Entomology and Zoology* 37: 181–190.

Zhang, H., A. Henderson-Sellers, and K. McGuffie. 2001. The compounding effects of tropical deforestation and greenhouse warming on climate. *Climate Change* 49: 309–338.

Zimmerman, R. C., D. G. Kohrs, D. L. Steller, and R. S. Alberte. 1997. Impacts of CO$_2$ enrichment on productivity and light requirements of eelgrass. *Plant Physiol.* 115: 599–607.

Ziska, L. H., and J. A. Bunce. 1997. Influence of increasing carbon dioxide concentration on the photosynthetic and growth stimulation of selected C$_4$ crops and weeds. *Photosynthesis Research* 54: 199–208.

Avian Malaria, Climate Change, and Native Birds of Hawaii

Dennis LaPointe, Tracy L. Benning, and Carter Atkinson

As with many other insular environments, the Hawaiian Islands have experienced extreme rates of extinction among their native flora and fauna. Approximately 90 species of endemic birds have become extinct since the arrival of humans between 1500–2000 years B.P. (Steadman 1995). At least 40 percent of the known Hawaiian passerines have become extinct in the past century (Scott and Sincock 1985). Hawaiian forest birds, particularly the endemic honeycreepers (Drepanidinae), have been hardest hit. This assemblage of 50 known species originated from a single finch ancestor and represents an extraordinary example of adaptive radiation that rivals Darwin's finches. Unfortunately, nearly half of these species have become extinct, 9 more species are currently listed as endangered, and most remaining species populations are in decline (James and Olson 1991; Jacobi and Atkinson 1995).

Limiting factors responsible for extinctions and declines in Hawaiian forest bird populations are numerous and dynamic. Loss of habitat by commercial logging, cattle ranching, and sugar cane cultivation, as well as habitat modification by feral ungulates, continues to have an impact on forest bird populations (Berger 1988). More troubling are distributional anomalies such as the absence of many species from intact low elevation habitat that were observed as early as the turn of the twentieth century (Henshaw 1902). Today native forest bird communities attain their highest density and diversity in forests located above 1500 meters on the larger islands of Hawaii (4205 m) and Maui (3055 m), while species diversity and density have been most greatly reduced on the low, smaller islands of Oahu (1231 m) and Molokai (1515 m) (Scott et al. 1986).

While introduced mammalian predators, such as the black rat *Rattus rattus* (Atkinson 1977), and avian competitors (Mountainspring and Scott 1985) may have contributed to species extinction and decline in these low-elevation forests, there is a growing consensus that Hawaiian forest bird populations are most threatened by introduced mosquito-borne avian disease, particularly avian malaria (Warner 1968; van Riper et al. 1986; Atkinson et al. 1995).

The intercellular parasitic protozoa *Plasmodium relictum* is the causative organism of avian malaria in the Hawaiian Islands (Laird and van Piper 1981). As with other avian malaria, *P. relictum* requires a culicine mosquito to complete its life cycle and be transmitted to the next host. Since mosquitoes were not native to the Hawaiian Islands, native forest birds evolved in the absence of this cosmopolitan continental parasite. The southern house mosquito *Culex quinquefasciatus*, a highly competent vector, was accidentally introduced in the early 1800s and quickly became established throughout the island chain. *Culex quinquefasciatus* is found throughout the tropical and subtropical regions of the world where it is well adapted to the domestic environs of humans. In the Hawaiian Islands *C. quinquefasciatus* has also become established in native forest bird habitat through use of an aquatic habitat created by the feeding of feral pigs on native tree ferns, *Cibotium spp.*

While the route and date of *P. relictum's* arrival in the islands are unknown, it is believed that many of the extinctions and population declines occurring since the 1920s were driven by avian malaria (van

317

Riper et al. 1986). For decades the altitudinal distribution of native Hawaiian forest birds was interpreted as a negative correlation with the altitudinal distribution of the vector mosquito. As native birds became restricted to increasingly higher-elevation forests (Scott et al. 1986), the great fear among resource managers was that the ultimate altitudinal range of C. quinquefasciatus would completely overlap the range of the remaining forest birds—effectively driving many species to extinction. But forest bird populations at high-elevation forest appear to be stable, even with some presence of vector mosquitoes, suggesting that other factors may be limiting transmission.

Like most vector-borne diseases, transmission of avian malaria may be profoundly affected by ambient temperature. The onset, duration, and completion of the parasite's development to the infective stage in the vector are determined by temperature. For most Plasmodium species development can occur in the range of 16–30°C (with optimal temperatures around 28–30°C). Temperatures above 30°C are often lethal, while temperatures lower than 16°C greatly inhibit parasite development. A minimum threshold temperature for development can be derived from modeling the developmental rate of the parasite at various constant temperatures. For the Hawaiian variant of P. relictum, development is greatly prolonged at temperatures below 17°C and effectively ceases at 13°C (LaPointe 2000). In environments where the mean ambient temperature is 13°C or cooler, transmission of avian malaria is precluded. Between 13°C and 17°C limited transmission of avian malaria might be expected.

The elevation coinciding with the 13°C isotherm on high Hawaiian Islands is 1800 m—the elevation at which the highest densities of endangered and common native birds occur. These thermal constraints to parasite development may adequately explain the relictual populations

of Hawaiian forest birds on the high islands of Maui and Hawaii, and the broad forested extent of the Alakai Swamp on Kauai. The rapid and almost complete extinction of endemic forest avifauna on the low islands of Molokai and Oahu would also support this model.

But does this explanation of high-altitude refugia create a false sense of security? Could global climatic change, in synergy with introduced avian malaria, lead to the extinction of susceptible native forest bird species? A 2°C increase in mean ambient temperature due to global warming could greatly reduce the extent of disease-free or limited forest bird habitat in the Hawaiian Islands. Landscape analysis of three of the largest remaining protected tracts of Hawaiian forest bird habitat clearly demonstrates the loss of disease-limited or disease-free habitat with increasing ambient temperature (Fig. 18.4).

It could be argued that the impact of an upward advance of avian malaria would be countered by the altitudinal expansion of native forest into a disease-free temperature zone. But here the effects of adjacent land use can greatly confound the expected outcome of climate change. Three distinct scenarios emerge in the analysis of Hanawi Refuge, the Hakalau National Wildlife Refuge, and the Alakai Swamp—the largest remaining tracts of protected native forest in the Hawaiian Islands (Benning et al. 2002).

On the island of Maui a 2°C warming would reduce the disease-free area of the Hanawi Refuge by half, due to the shape of the reserve and the inverse relationship between area and elevation on mountains (Fig. 18.4a). Rare, susceptible species may be lost following such a drastic reduction in disease-free habitat, but ultimately forest might expand into the higher elevations maintaining a disease-free refugia for more common species. The situation would be quite different, however, at Hakalau National Wildlife Refuge on the island of Hawaii (Fig. 18.4b). A 2°C warm-a)

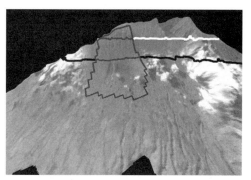

b)

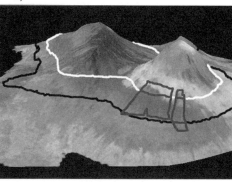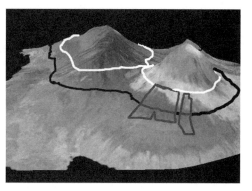

c)

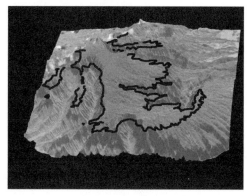

Figure 18.4. Present positions (left panels) and projected future positions (right panels) of 17°C (black) and 13°C (white) isotherms relative to important conservation areas in the Hawaiian Islands. Changes are shown for: (a) Hanawi Refuge (black boundary) on the island of Maui; (b) Hakalau Refuge (black boundary) on Hawaii; and (c) the Alakai Swamp region on the island of Kauai. The right-hand panels indicate loss of disease-free habitat above the 13°C isotherm under a 2°C warming scenario. Images were created using a 1995 SPOT false-color composite image draped over 30-m digital elevation models for each island. *Source:* Courtesy of Tracy L. Benning.

ing at Hakalau would effectively eliminate all disease-free habitat. This catastrophic loss of refugia would be compounded by the near total loss of native forests upslope from the refuge that has resulted from nearly two centuries of cattle ranching. The increasingly rare akiapolaau *Hemignathus munroi* and the cavity nesting Hawaii akepa *Loxops coccineus* might easily succumb under this scenario. Restoration of upslope forests at Hakalau National Wildlife Refuge is critical for the maintenance of disease-free refugia and ultimately, the long-term survival of these species.

The honeycreepers of Kauai may be the most vulnerable to the disease-related impacts of climate change. The high central plateau of the Alakai Swamp rises to a maximum elevation of just under 1600 m and covers a vast area where mean ambient temperatures should result in limited disease transmission. A 2°C warming, however, would reduce this area by 85 percent (Fig. 18.4c). With no upward expansion of the forest to create refugia, extinctions of rare honeycreepers and general population declines would be expected. The endemic honeycreepers of Kauai, which remained largely intact after two centuries of human encroachment, might be driven to extinction in less than 100 years. Development of an effective disease management program is the only real option for long-term conservation of Kauai honeycreepers.

While the severe susceptibility of Hawaiian passerines to *Plasmodium relictum*, combined with habitat constraints, appears to be a unique threat to regional avifauna, there is ample evidence that climate change and disease will have a profound global effect on marine and terrestrial biota (Harvell et al. 2002). Temperature influences all life history traits of arthropod vectors and the pathogens they transmit. In most cases increased environmental temperature results in increased transmission locally or expansion in the geographical range or seasonal occurrence

of disease. Continental avifauna restricted to high elevation or high latitudes where vectors or pathogens do not currently occur may be particularly vulnerable to avian malaria or arbovirus, like West Nile, as the environment becomes more favorable to disease.

REFERENCES

Atkinson, C. T., K. L. Woods, R. J. Dusek, L. S. Sileo, and W. M. Iko. 1995. Wildlife disease and conservation in Hawai'i: Pathogenicity of avian malaria (*Plasmodium relictum*) in experimentally infected Iiwi (*Vestiaria coccinea*). *Parasitology*. 111: S59-S69.

Atkinson, I. A. E. 1977. A reassessment of factors, particularly *Rattus rattus* L., that influence the decline of endemic forest birds in the Hawaiian Islands. *Pac. Sci*. 31: 109-133.

Benning, T. L, D. A. LaPointe, C. T. Atkinson, and P. M. Vitousek. 2002. Interactions of climate change with biological invasions and land use in the Hawaiian Islands: Modeling the fate of endemic birds using geographic information system. *PNAS*. 99(22): 14246-14249.

Berger, A. J. 1988. *Hawaiian Birdlife*, 2nd edition. Honolulu: University Press of Hawaii.

Harvell, C. D., C. E. Mitchell, J. R. Ward, S. Altizer, A. P. Dobson, R. S. Ostfeld, and M. D. Samuel. 2002. Climate warming and disease risks for terrestrial and marine biota. *Science*. 292: 2158-2162.

Henshaw, H. W. 1902. *Birds of the Hawaiian Islands*. Honolulu: Thos. G. Thrum.

Jacobi, J. D., and C. T. Atkinson. 1995. Hawaii's endemic birds, pp. 376-381. In E. T. LaRoe, G. S Farris, C. E. Puckett, P. D. Doran, and M. J. Mac, eds., *Our Living Resources: A Report to the Nation on the Distribution, Abundance, and Health of U.S. Plants, Animals, and Ecosystems*. Washington, D.C.: U.S. Government Printing Office.

James, H. F., and S. L. Olson. 1991. Descriptions of thirty-two new species of birds from the Hawaiian Islands: Part II. Passeriformes. *Ornithol. Monog.* No. 46. 1-88.

Laird, M., and C. van Riper III. 1981. Questionable reports of *Plasmodium* from birds in Hawaii, with recognition of *P. relictum* ssp. *capistranoae* (Russell, 1932) as the avian malaria parasite there, pp. 159-165. In E. V. Canning, ed., *Parasitological Topics*. Lawrence, Kans: Allen Press.

LaPointe, D. A. 2000. Avian malaria in Hawaii : The distribution, ecology and vector potential of forest-dwelling mosquitoes. Ph.D thesis, University of Hawaii, Honolulu.

Mountainspring, S., and J. M. Scott. 1985. Interspecific competition among Hawaiian forest birds. *Ecol. Monog.* 55: 219–239.

Scott, J. M., S. Mountainspring, F. I. Ramsey, and C. B. Kepler. 1986. Forest bird communities of the Hawaiian Islands: Their dynamics, ecology, and conservation. *Studies in Avian Biology*, No. 9.

Scott, J. M., and J. L. Sincock. 1985. Hawaiian birds, pp. 548–562. In R. L. DiSilvestro, ed., *Audubon Wildlife Report*. New York: National Audubon Society.

Steadman, D. W. 1995. Prehistoric extinctions of Pacific island birds: Biodiversity meets zooarchaeology. *Science*. 267: 1123–1131.

van Riper, C. III, S. G. van Riper, M. L. Goff, and M. Laird. 1986. Epizootiology and ecological significance of malaria in Hawaiian land birds. *Ecol. Monog.* 56: 327–344.

Warner, R. E. 1968. The role of introduced diseases in the extinction of the endemic Hawaiian avifauna. *Condor*. 70: 101–120.

PART V

Conservation Responses

It is impossible to say whether conservation will overcome the two great challenges it faces this century—winning the battle against nonclimate stressors, and coping on top of that with the dynamics introduced by climate change. It is certain, however, that without increasingly dynamic conservation strategies, much biodiversity will be lost due to climate change.

Conservation is visibly losing the battle against habitat loss in areas such as insular Southeast Asia, and losing against other nonclimate stressors such as the bushmeat trade in many regions. Climate change will not play a major role in determining the outcome of these struggles, since they will be decided in the next decade or less. In many areas, even the fundamental resources of staff and funding required to prevent erosion of protected areas are missing. In these areas, climate change may play a decisive role, because systems that cannot function well in a relatively static climate will suffer as dynamics increase.

A key to success will be the resources to move conservation to the landscape scale. New staff and funding will be required to complete the current site-scale (protected areas) system, and additional new resources to expand systems to be effective in landscape conservation. This implies a level of funding that may be an order of magnitude larger than is currently available for protected areas worldwide, and sustained on a long-term basis. Some nations, many with high levels of biodiversity, will not have the budget for this increase. New funding mechanisms may be required to move funding to these countries on a long-term basis and at magnitudes far exceeding current conservation transfers. Political will is required for budget allocation in those countries that do have the resources.

Conservationists must be able to pinpoint the strategies, mechanisms, and resources required to meet the climate change challenge. The financing needed to close the funding gap in existing protected areas is relatively well known, from several global analyses (James and Green 1999; Bruner et al. 2003). Identifying resources to create the dynamic system to deal with climate change is in a much earlier phase. The expanded strategy must clearly be described, so that costing can follow. Rob Peters began this job in the 1980s, and it is increasingly imperative that his revolutionary work be completed in detail. This is the task taken on by the authors of Part V.

This part opens with an examination of the challenge—how climate change interacts with major nonclimate stressors (threats) such as habitat loss and invasive species, what biodiversity concerns must drive climate change–integrated conservation strategies, and some indication of the foundations that currently exist for such a response. The subsequent chapter describes the design of dynamic conservation strategies, followed by two chapters on the implementation of these strategies in protected areas and the matrix. The matrix is all unprotected land, and it forms the framework in which protected areas

and connectivity must exist. Matrix management has the most new elements and is perhaps the greatest challenge for conservation. The work required to adapt protected areas to climate change is addressed in the final chapter. Examples in all chapters draw on marine and terrestrial systems in turn, as marine systems are in many ways at the forefront of climate change conservation challenge.

Together these chapters offer a more complete view of the technical steps required to introduce climate dynamics into conservation strategies. They should also make readers aware that even greatly improved conservation strategies face limits in the amount of change that can be accommodated. To be successful, these new strategies on the ground will need to be coupled with strategies to limit change, by stabilizing greenhouse gases in the atmosphere—a challenge that will be examined in the final part of this book.

REFERENCES

Bruner, A., L. Hannah, J. Hanks, and A. Balmford. 2003. *The Long-Term Cost of a Complete Global Protected Areas System*. Gland, Switzerland: IUCN.

James, A. N., and M. J. B. Green. 1999. *Global Review of Protected Area Budgets and Staff*. 1–35. Cambridge, UK: World Conservation Monitoring Center.

CHAPTER NINETEEN

Conservation with a Changing Climate

THOMAS E. LOVEJOY

The pioneering work of Ruth Patrick (Patrick 1948, 1949, 1961) demonstrated that the biological diversity of a region or ecosystem reflects not only its characteristic physical environment and natural history but also all the effects of the full array of human-driven stressors. The changes to biodiversity essentially integrate all of them. This was first established for freshwater systems, and rivers in particular, but applies equally well to terrestrial and marine ecosystems.

Little recognized as the fundamental principle underlying environmental science and management, this Patrick Principle (Lovejoy 2000) has to be the basis for all conservation strategies including those introduced by a changing climate.

IMPACTS

By implication one of the major differences biodiversity is encountering with anthropogenic climate change is the synergies with other concurrent stresses from human activities. They range across the entire spectrum from pollution (from the soup of manufactured chemicals) to habitat destruction. Chemical pollution impacts range from particular chemicals and groups of organisms (e.g., ibuprofen and various vulture species in India to regional effects such as acid rain and anoxic zones in the ocean. Habitat destruction includes habitat fragmentation, a particularly problematic factor under climate change. And the problem of alien and invasive species, so favored by nonnatural disturbance, is only greater when climate change is added.

Habitat loss and invasive species are perhaps the two greatest current threats to biodiversity, and both will experience major interaction with climate change. Estimates of habitat loss indicate that most of the habitable area of the earth is now dominated by human uses (Hannah et al. 1994; Sanderson et al. 2002). Other estimates of high-biodiversity areas indicate loss of 70 percent or higher (Myers et al. 2000). Unfortunately, a positive correlation has been noted between biodiversity and habitat loss (Brooks et al. 2001), perhaps because climatic factors are favorable to both human and non-human life in certain areas.

The impact of climate change in this heavily fragmented world may be immense. In many instances, species will no longer be able to adjust their ranges to track changing climatic conditions. Implications of this reduced response may include genetic impoverishment or extinction, as previous authors in this volume have indicated.

For alien invasives, climate change may mean a world of new opportunities. Fragmentation means that when species ranges retract due to climate change, new species may not be able to disperse to fill the void. Weeds will be the exception—these rapidly dispersed pioneers may find many vacated niches, and exploit them. These and other changes have serious implications for biodiversity that will require new conservation strategies.

RESPONSE

In one sense all the traditional tools of conservation are still valid. There is a need to have priorities such as species that are most immediately vulnerable (endangered species and endemics with very restricted range). There needs to be a thorough analysis of all threats. Protected areas, corridors, landscape conservation, ecosystem management, adaptive management, monitoring, and ex-situ conservation all have critical roles to play.

Nonetheless, conservation that traditionally has operated in a relatively static world must now attempt to succeed in one of considerable flux. Ecosystems will not pick up and move like Birnam Wood with all constituent species in concert; rather, species will respond individualistically. There will be considerably more change in phenology, populations, and genetics. Natural succession will not reliably lead to the community composition it would have previously. Further, there is the very distinct possibility of abrupt climate change, with scenarios currently quite difficult to envision.

All of this will require strategies not only to provide great flexibility for species to respond successfully and avoid extinction but also in approaches to ongoing proactive management. There will inevitably be a lot of learning in the process because conservation will be a less predictable applied science.

INSTITUTIONAL ASPECTS

Institutional coordination, always vital, will be required as never before. The management of landscapes and seascapes will require integration of the human or development agenda with the conservation agenda to a degree rarely seen before. This kind of management will be needed at all scales from the local to the regional, national, and international. It will be needed between agencies and all levels of government.

PROTECTED AREAS IN A WORLD OF CHANGING CLIMATE

The current protected area system in the world is insufficient, as it stands for a climate-stable world. For those reasons alone it needs to be augmented. But in a climate changing world, the current protected area system, which will undoubtedly retain great value, needs reanalysis. That insight will help considerably in an overall increase and redesign to create a system resilient in the face of changing climate.

With climate change, even the best-designed protected area system cannot aspire to conserve biological diversity if it consists mostly of isolated units. While it has been true that all landscapes need to be biologically functional, this becomes an even greater imperative. Restoration of connections in landscapes between protected areas, always important for various migratory species, is central to conservation under climate change (Chapter 20). The closer society can approach the spatial model of human society existing within nature, the greater will be the resilience to climate change.

Indeed, the matrix in which protected areas exist is fundamental to successful conservation of biological diversity under the dynamics of climate change. The matrix, just as the protected areas themselves, will require active management to an extent never envisioned previously (Chapters 21 and 22). That management must also take into account climate-induced changes in human pressures on landscapes and seascapes, such as changes in agricultural area.

SCIENTIFIC AND OTHER
RESEARCH NEEDS

All the above considerations require a major investment in research and monitoring. Improved climate modeling at scales relevant to conservation will favor better conservation planning. Improved understanding of dispersal biology will improve landscape planning and matrix management. Monitoring at a level of detail and on a massive scale is essential not only to know what is actually going on, and to measure success, but also to uncover new

dimensions of climate change as unraveling ecosystems and assembling novel ones reveal new aspects which must be added to the conservation challenge.

Clearly a central tenet of conservation with changing climate is to reduce the negative synergistic impacts caused by nonclimatic stresses. Coral reefs, for example, are less vulnerable to coral bleaching from rising sea temperatures if stresses like sedimentation are reduced or eliminated—or at least they are up to a point. This principle holds for all ecosystems. While that consists of doing what makes conservation sense anyway, the added conservation value given climate change can help in efficient allocation of conservation funds.

There are important implications for planning in order for a great deal of biological diversity to make it through the climate change gauntlet. Planning has to include much longer time frames as well as the current short ones, 50 and 100 years as well as 5 and 10. The planning has to include scales relevant to processes: continental in some cases, down to the local (e.g., Scott, this volume; Midgely and Millar, this volume). It has to plan in a world of greater flux, in which current human uses may change, and in which landscape strategies may have to change as the picture of climate change and responses by humans and the remainder of life on earth become more apparent.

Phenological change will require monitoring at a level of species and organisms to reveal the pattern and thus a required adaptive management response. Genetic change is perhaps the hardest to address in the sense that even monitoring it for a major segment of the living world seems an impossibly huge task. Yet, there are some sensible and practical things that should be possible. Populations can be given priority that maximizes genetic diversity when possible. Subspecies and variants can be used as rough indicators of a degree of genetic diversity and thus accorded some conservation priority. Perhaps it will be possible to identify populations with genetic resilience to climate change (such as insects with genes for wings suited for long distance dispersal).

In addition to reducing nonclimatic stressors, ecosystem change can be addressed in part by habitat restoration. As contrasted to trying to maintain the status quo—this could play an important role in providing dispersal opportunities. So can management of disturbance regimes. Ecotones and gradients will play important roles because they already are subject and resilient to change in any event.

The total challenge for conservation of biodiversity clearly is enormous. That speaks all the more strongly for the need to curtail climate change beyond the level of "dangerous interference" with ecosystems (Chapters 23 and 24).

SUMMARY

Conservation planning must now be seen in the context of human influence on ecosystems that is pervasive. Climate change is the next great set of human changes that may alter biodiversity and its conservation.

Nonclimate stressors are likely to exhibit synergies with climate change. Habitat loss and alien invasive species are particularly likely to interact in negative ways with climate change. Since these are often considered the two greatest threats to biodiversity, conservation responses will be challenged by these interactions.

New conservation strategies are necessary that are dynamic and well-coordinated. The following chapters outline needs in collaboration, limiting nonclimate stressors, management, and monitoring to respond to the challenges mounted by climate change. Early testing and refinements of these methods will be critical to mounting an appropriate conservation response to the challenge posed by climate change.

REFERENCES

Brooks, T., Balmford, A., Burgess, N., Fjeldsa, J., Hansen, L. A., Moore, J., Rahbek, C., & Williams, P. 2001. Toward a blueprint for conservation in Africa. *BioScience*, 51, 613−624.

Hannah, L., Lohse, D., Hutchinson, C., Carr, J. L., & Lankerani, A. 1994. A preliminary inventory of human disturbance of world ecosystems. *Ambio*, 23, 246.

Lovejoy, T. 2000. Biodiversity, pp. 22−35. In Patten C., et al., eds., *Respect for the Earth* (The 2000 Reith Lectures). London: Profile Books, i−xiii, 1−173.

Myers, N., Mittermeier, R. A., Mittermeier, C. G., Da Fonseca, G. A. B., & Kent, J. 2000. Biodiversity hotspots for conservation priorities. *Nature*, 403, 853−858.

Oaks, J. L., Gilbert, M., Virani, M. Z., Watson, R. T., Meteyen, C. U., Rideout, B. A., Shivaprasad, H. L., Ahmed, S., Chandry, M. J. I., Arshad, M., Mahmood, S. M., Ali, A., and Khan, A. A. 2004. Diclofenac residues as the cause of vulture population decline in Pakistan. *Nature*, 427, 630−632.

Patrick, R. 1948. Factors affecting the distribution of diatoms. *Bot. Review* 14, 473−524.

Patrick, R. 1949. A proposed biological measure of stream conditions based on a survey of Conestoga Basin, Lancaster County, Pennsylvania. *Proc. Acad. Nat. Sci. Phila.*, 101, 277−341.

Patrick, R. 1961. A study of the numbers and kinds of species found in rivers in eastern United States. *Proc. Acad. Nat. Sci. Phila.*, 113, 215−258.

Sanderson, E., Jaiteh, M., Levy, M. A., Redford, K. A., Wannebo, A. & Woolmer, G. 2002. Human footprint and the last of the wild: A quantitative evaluation of human influence on the land's surface and its implications for conservation. *BioScience*, 333−356.

CHAPTER TWENTY

Designing Landscapes and Seascapes for Change

LEE HANNAH AND LARA HANSEN

Future climate change will drive dynamics in biodiversity that have in the past played out across entirely natural landscapes. Current landscapes are highly fragmented, meaning that natural responses such as range shifts may now require careful management to succeed. Central to this effort is the planning of landscape elements, both static and dynamic, to form an integrated web of land uses that can support biotic change.

The marine setting is considerably different, but connectivity remains a central issue. Physical habitat destruction for benthic species and alterations to the food web due to overfishing for pelagic species are critical impediments to a natural response to climate change (Roy and Pandolfi, this volume; Hoegh-Guldberg, this volume). Coral bleaching due to warming water temperatures is already strongly affecting coral reefs, making connectivity important for recolonization and recovery. Many marine species have pelagic juvenile stages, so connectivity may be facilitated by networks of protected areas linked by prevailing currents.

This chapter describes principles that may be applied to the design of landscapes and seascapes for conservation of dynamic biodiversity. It begins with an overview of present planning tools, focusing on advances in bioinformatics and spatial analysis that will facilitate effective responses to climate change. The definition of planning areas and conservation targets is emphasized. The chapter then examines the specific steps of designing landscapes and seascapes for climate change, focusing on protected areas and the matrix. The matrix is all unprotected land, across which land uses have major effects on protected areas and connectivity (Gascon et al. 2000). The chapter concludes with an overview of the elements of an integrated plan.

PLANNING TOOLS

Designing landscapes and seascapes for biodiversity conservation is a rapidly emerging field, thanks to advances in bioinformatics, spatial analysis, phylogenetics, and other fields. It is rooted in the concepts of landscape ecology and related attempts to understand spatial relationships among biological, physical, and human systems (Forman 1995). Practical applications of these ideas have emphasized the habitat and movement needs of large carnivores as an organizing principle (Terborgh et al. 1999). More recently, landscape connectivity has been recognized as important in accommodating the biological dynamics of climate change (Hannah et al. 2002).

The goals of landscape-level biodiversity conservation plans have become more complex as the science has matured and as data availability has improved. A complete plan would consider all aspects of biodiversity—from genetic diversity, to species diversity, to ecosystem diversity and processes (Noss and Harris 1986). Factors that have been suggested for inclusion include gene flow (Dobson et al. 1999), accommodation of large-scale disturbance regimes, the conservation of evolutionary and other processes (Noss and Harris 1986; Smith et al. 1997), weighting for phylogenetic diversity (Vane-Wright et al. 1991), and threat dynamics (Channell and Lomolino 2001).

GIS and Reserve Selection Algorithms

Spatial analysis tools are rapidly increasing in sophistication. GIS systems now commonly available allow complex spatial analysis to be conducted on a personal computer. Various land uses, economic values, and biodiversity attributes may be overlain and compared. Modified GIS programs permit optimization of biodiversity and social variables at the landscape scale (e.g., TAMARIN; see Chapter 21). They may also be used to allow stakeholders to examine trade-offs, for instance, between biodiversity and economic production.

Tools for selection of protected areas, known as "reserve selection algorithms," are automated computer routines that apply optimization principles to large biodiversity data sets. C-Plan, Worldmap, and SITES are among the conservation software packages that incorporate reserve selection algorithms (Pressey et al. 1997). Protected areas systems may be designed using complementarity (addition of areas which contribute the most new biodiversity), irreplaceability (addition of areas with unique species or attributes), and other rules with these systems (Pressey et al. 1997). The value of the reserve selection algorithms depends heavily on information on biodiversity distribution (ranges of species) available as input, so advances in bioinformatics that allow rapid access to species' distributional data improve the utility of these tools (Cabeza and Moilanen 2001). Industrial optimization programs may also be applied to biodiversity problems, but this has been less commonly applied in conservation planning.

Tools that facilitate access to the data required for spatial analysis are constantly improving. Advances in bioinformatics allow specimens scattered in museums across multiple continents to provide data for a conservation plan in the region in which they were collected (Sugden and Pennisi 2000). Online databases allow the exchange of taxonomic, genetic, and other information over the Internet, often in a matter of seconds (Bisby 2000). Coupled with increasing availability of social, demographic, land use, and economic data in digital format, this increase in availability of biological information makes it possible to plan and compare trade-offs in multiple biological and social variables across a landscape (Groves et al. 2002).

Climate change is the next challenge. Alterations in climate will change species' distributions, fitness, and physiology, thus altering patterns of diversity in a landscape (Walther et al. 2002). These changes must be accounted for in designing landscape or seascape connectivity. Not all connectivity is created equal—connectivity designed with climate change as an explicit consideration will have a much greater effect than connectivity that is ad hoc with respect to climate change (Hannah et al. 2002). Additionally, traditional protected areas and connectivity alone will likely not ensure enduring landscapes or seascapes. We will need to learn how to limit climate stresses even within protected areas, as well as identifying natural climate refugia and tolerant populations.

Setting Targets

Siting and structuring conservation elements within a landscape requires the definition of targets—specific biological elements or properties to be conserved, and desired outcomes adequate for their conservation (Groves et al. 2002). The biological composition of the planning area will define the range and number of possible targets. There is increasing consensus that planning units should be defined in biological terms, the most widely accepted of which is the "ecoregion" (Groves et al. 2002). In the following discussion, ecoregions will be assumed to be the planning unit.

Targets may be selected that are specific to each of the major elements of biodiversity—genetic diversity, species diversity, and ecosystem diversity (Groves et al. 2002). Species diversity targets are the most straightforward, since the unit of selection (species) is relatively easily defined. Genetic targets may include sub-

species, races, or priorities based on phylogenetic diversity. Ecosystem targets may include processes such as migrations, or locations important for maintaining processes such as evolutionary diversification (e.g., ecotones).

Climate change affects selection of targets in two fundamental ways. First, it will alter ecoregion boundaries. For instance, climate change will result in different biome boundaries in the future, which may change the size or configuration of an ecoregion. Changes in ecoregions can be assessed in several ways. GCM projections have been used to create maps of future vegetation zones, which may be equated to ecoregions (Kittel et al. 1995), or ecoregions as defined by abiotic factors may be mapped directly using future climate projections (Hargrove and Hoffman 1999). Such ecoregion transformations may be considered when setting planning unit boundaries, and target selection revised accordingly.

The second way in which climate change can affect target selection is by exacerbating existing threats. When climate change makes a species more vulnerable to extinction (e.g., Thomas et al. 2004), it is more likely to be identified as threatened and therefore more likely to become a target for conservation. Similar considerations hold for genetic- or ecosystem-level targets. Therefore, climate change is likely to increase the number of conservation targets, through both changing the planning area and by increasing the number of species at threat.

PLANNING FOR CLIMATE CHANGE

The data and tools necessary to incorporate climate change into landscape or seascape conservation plans are increasingly available. Present and projected climate data are now widely available for many areas at resolutions suitable for conservation planning (e.g., the Hadley Centre RCM, "PRECIS").

Models for projecting species range shifts with such data are available online from several sources (http:www.ento.csiro.au// climex; www.DIVA-GIS.org; www.life-mapper.org/desktopgarp/). The remainder of this chapter outlines means of incorporating climate change into landscape and seascape conservation plans at various stages—assessing change in targets, selecting protected areas, assessing connectivity, and creating an integrated plan.

Assessing Change in Targets

Assessing future changes in targets and threats is an essential part of any conservation plan. Climate change, habitat loss, invasive species, and other changes will influence the spatial distribution of species through time. These changes may act synergistically, resulting in even more dramatic changes than when each is acting alone. Ignoring these changes may lead to conservation plans that are flawed in the long term (Hannah et al. 2002). Sound planning therefore requires identification of factors that may change, placing bounds on possible change, modeling change in spatially explicit ways, and balancing the effects of different changes. Climate change is a major consideration in this analysis.

Assessing the effect of climate change on conservation targets requires information on present and possible future climates, the distribution of species (or other target), and the relationship between those distributional characteristics and climate. Models or general observed patterns of movement may be used to infer possible responses of species to future climate change (Parmesan, this volume; Peterson et al., this volume). Where species' distributional data are sufficiently detailed and understanding of the relationship between climate and distribution sufficiently well-understood, spatially explicit modeling is possible. Spatially explicit and qualitative approaches may be mixed in a single

ecoregion plan, depending on the state of knowledge associated with individual targets.

Projections of future climate for use in such modeling must be at a scale suitable for landscape conservation planning. In general, GCM results are too coarse-scale to be directly applicable (Raper and Giorgi, this volume). Statistical downscaling or RCMs should therefore be used to assess change in targets.

Modeling of changes in species distributions can be conducted using a variety of methods, including CLIMEX, GARP, genetic algorithms, artificial neural networks, and others (Peterson et al., this volume). The models are relatively quick and simple to run once the climatology has been developed, so it is feasible to run them for all target species with sufficient distributional data (more than ~20 records). Advances in bioinformatics provide rapid access to distributional information so these models may be run for large numbers of species (Peterson et al. 2002).

Projected range shifts may be large, and uncertainty high, in long-range projections (2080–2100), making a focus on near- and intermediate-term projections (2030–2050) preferable. A medium-term time frame reduces uncertainty yet provides adequate lead time for adjustments such as additional protected areas or connectivity as climate change progresses. It is, therefore, useful to use near- and medium-term projections as the primary planning tool in range shift modeling, coupled with long-range projections for risk assessment (e.g., to ensure that options for possible long-distance movement are not foreclosed).

Changes in process or genetic targets are more difficult to model. Where genetic targets involve taxonomically distinct entities such as subspecies, range shift models may be applicable. Process targets and nontaxonomic genetic targets may require other modeling techniques. For instance, population dynamics may be simulated using metapopulation models.

Change assessment of all species and other targets provides quantitative (spatially explicit) or qualitative input into the planning of a landscape or seascape for biodiversity conservation. This information is central to the three principal design steps—defining fixed conservation elements (protected areas), planning connectivity, and limiting external pressures that will act synergistically with climate change (e.g., extractive uses, pollution).

Protected Areas Selection

Protected areas provide the fixed elements in a dynamic conservation plan. Geographically fixed does not imply static—flexible management in concert with the surrounding landscape ensures that protected areas are a dynamic part of the landscape / seascape conservation strategy. Fixed elements serve to protect unchanging features, such as the proportion of a species range not projected to move, as well as "nodes" or junctions that are important in many overlapping movements.

Species range shift models are particularly useful in protected areas selection. Their spatially explicit projections of pos-

sible range movements are suitable for use in reserve selection algorithms. For this purpose, modeled range shifts may be divided into three groups based on spatial scale of movement: "stay-at-home" species have substantial overlap in current and future ranges; "neighborhood movers" have range shifts that may be accommodated within a single large protected area; and "cross-country movers" experience range shifts on a large scale that requires land between protected areas.

An example of a "stay-at-home" species from modeling of the proteas of South Africa is given in Figure 20.1. Where species have substantial overlap in present and future range, the area of overlap may be used as their effective range for protected areas planning. This allows both their present and future ranges to be simultaneously considered in reserve selec-

Figure 20.1. A protea species with substantial overlap between present range and future range as projected by general additive modeling (a "stay-at-home" species—see text). Current range lost in 2050 (gray), projected future range (dark gray), and overlap (black) are shown. In these range models for *Serruria glomerata*, all future range overlaps with present range. Future climatic conditions were derived from a HadCM2 doubled CO_2 2050 projection with downscaling for South Africa.

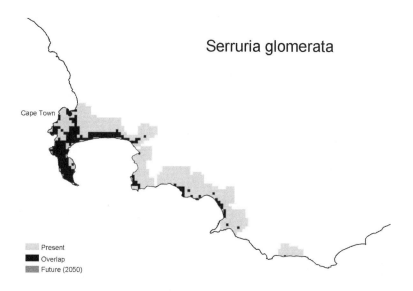

Serruria glomerata

Cape Town

▨ Present
■ Overlap
▨ Future (2050)

Figure 20.2. A protea species with adjacent present and future (2050) range (a "neighborhood mover" species—see text). Current range lost in 2050 (gray), projected future range (dark gray), and overlap (black) are shown. Modeling specifications as for Figure 20.1.

tion. The area of overlap may be used as input data for a reserve selection algorithm such as Worldmap or SITES. Principles of cost-effective protected area selection, including complementarity and irreplaceability (Pressey et al. 1994), may be directly applied to data in this form, facilitating an optimized protected area system design based on both present and future ranges (Pressey et al. 1993).

"Neighborhood movers" have little or no overlap in present and medium-term future ranges, but they have future and present ranges in proximity at a scale that could be captured in a single, large protected area. Management within a single protected area can facilitate range shifts in these species. Figure 20.2 illustrates a "neighborhood mover" from the South Africa protea study. Such meso-scale movements can define needed additions to existing protected areas which would incorporate the scale of movement, or priority sites for new protected areas. Repeating patterns in neighborhood range shifts may be found, as illustrated in Figure 20.3. Areas in which multiple neighborhood

movements overlap can be prioritized for protected area expansion or addition.

Longer-distance movements require connectivity at a larger scale, one exceeding the likely limits of any individual protected area. Facilitating these range shifts requires management coordination between multiple protected areas and across the intervening land use matrix. Figure 20.4 illustrates a "cross-country mover." These species help define priority areas for connectivity. They are also high priority for monitoring, to determine whether model predictions are accurate and investments in connectivity warranted.

Similar analysis may be possible for genetic, process, and other targets. While spatially explicit modeling may not be possible for these nonspecies targets, landscape elements necessary for their conservation may nonetheless emerge. For instance, "rule of thumb" analysis of poleward and upward shifts might make the northernmost mountain range in an ecoregion (or southernmost, in the Southern Hemisphere) a priority for a protected area, even in the absence of spatial modeling. Balancing uncertainty in spatially explicit modeling with greater certainty but lower spatial specificity in qualitative analyses is an ongoing and inherent process in climate change analysis. It is best overcome by applying a broad range of analytical tools and expertise, at multiple scales and in iterative designs (Root and Schneider 1995).

Definition of fixed conservation elements in the landscape begins to help specify areas in which connectivity is needed. Such analysis will be more valuable where it includes consideration of land transformation and species' dispersal capabilities, as in Figures 20.3 and 20.4 (see figure captions for details). Ideally, range shift models would be combined with models of future land transformation and models of other threat dynamics. With fixed elements defined, connectivity may be assessed.

Assessing Connectivity

Landscape or seascape connectivity involves multiple attributes across multiple spatial scales (Noss and Harris 1986; Forman 1995). Assessment of connectivity requirements depends on the selection of conservation targets (Groves et al. 2002) to bring the problem within reach of the limited resources of conservation managers and planners.

The same modeling tools used to select protected areas are useful in defining connectivity. Species or other targets not captured in the fixed landscape elements (protected areas) become the focus of connectivity analysis. For example, species identified by modeling as "cross-country movers" (see above) may be used to define connectivity needs associated with range shifts. Figure 20.4 illustrates a long-distance ("cross-country mover") range shift in one of the protea species (*Serruria linearis*) modeled for South Africa. To maintain even 10 percent of its present range, this species needs connectivity on at least three fronts. Analysis of land use and habitat quality would suggest preferred patterns of connectivity. In this medium-term (2050) projection with an assumption of limited dispersal, the edge-to-edge distances of connectivity required are on the order of 10 km. Using this approach, individual connectivity plans may be constructed for each target species. Most ecoregions will have no more than several hundred target species (e.g., endemic and threatened taxa), of which only a fraction will need long-distance connectivity in the medium term. Thus, designing connectivity where targets are clearly identified is probably achievable in most instances. However, the South Africa data suggest that, after 2050, increasing long-distance range shifts will make connectivity more and more difficult.

The types of connectivity needed include continuous natural habitat through which range shifts can occur in multiple species, as well as degraded, secondary, or human land uses with key resources sufficient to permit range shifts in one or a few species. The use of target species allows identification of resource needs and timing at a level of precision suitable to drive management action. Distributed resources adequate to provide food or habitat values that fluctuate with changes in phenology are also important. These needs may vary over time, so one type of connectivity may grade into another, both spatially and temporally.

The uncertainty associated with future connectivity needs requires flexible land uses to maintain options for future restoration to natural habitat. Low-intensity land uses that offer a mix of habitat values that may be changed over time to respond to shifting biological needs provide an important complement to undisturbed natural habitat. For instance, low-intensity agroforestry may preserve soil properties that maintain the possibility of future reversion to natural vegetation (Hannah 2004). Such systems may also include mixes of cover trees and crop trees that may be manipulated for their food or habitat values to birds (Hannah 2004). Innovative landowner agreements and incentive schemes could allow these lands to

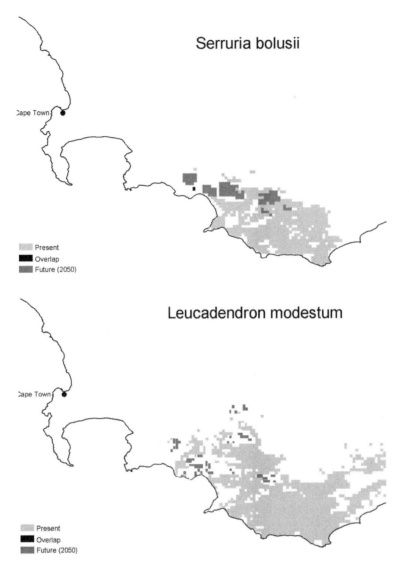

Figure 20.3. Examples of a repeating pattern in protea range shifts. All 4 panels represent species with adjacent present and future (2050) range. Current range lost in 2050 (gray), projected future range (dark gray), and overlap (black) are shown. Range shifts poleward in this region of South Africa are constrained by limited land area. As a result, range shifts are predominantly upslope, and northward from the lowland coastal plain. The area in which all species show suitable future range is a series of low mountains, in which several small protected areas currently exist.

be managed to revert to forest, or could allow for future crop mixes to be shifted as species' spatial needs change due to changes in range or phenology. Encourag-ing these types of land uses can improve flexibility in system connectivity. For example, landholder agreements might be developed in several possible locations to accommodate a projected species range shift, but only activated in one area, based on monitoring results indicating the actual primary axis of the shift.

In the case of seascapes, connectivity can use tools such as marine protected area networks. A system of distinct protected areas designed to allow for migration between sites will facilitate range shifts along thermal and latitudinal gradients. Conve-

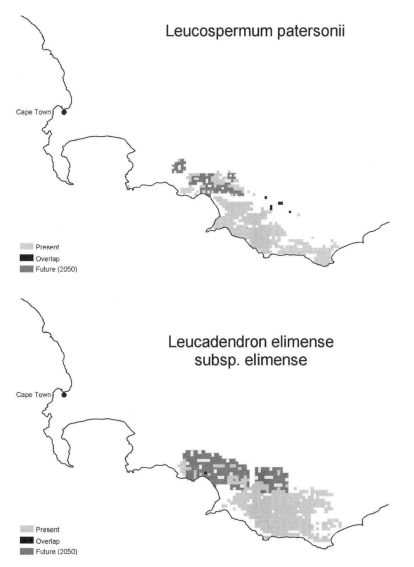

Figure 20.3. Continued.

niently, the network protected area approach is already being championed for protection of marine biodiversity from conventional threats, such as overfishing (Dayton et al. 2000). Connectivity of marine protected areas may assist in natural adjustments to coral bleaching, which is a dramatic impact of climate change. Coral species are sensitive to changes in temperature of as little as 1°C over annual average maximum sea temperatures. One possible natural refuge for coral will be found in latitudinal range shifts. The Eastern Africa Marine Ecoregion (Olson and Dinerstein 1998) is an example of a possible network of marine protected areas (MPAs) along a latitudinal, and concurrent thermal, gradient. It contains 4600 km of migratory range crossing the Equator, along the coasts of five countries (Somalia, Kenya, Tanzania, Mozambique, and South Africa), and already contains 28 marine protected areas. While more protected areas are needed even without climate change, such linked networks are a good initial step.

Assessing connectivity in the face of cli-

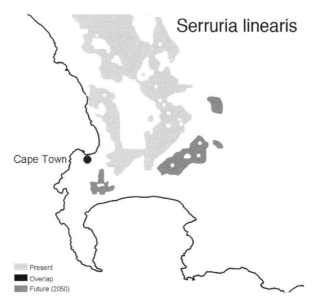

Figure 20.4. A protea species for which present and future ranges are largely disjunct (a "cross-country mover"—see text). Current range lost in 2050 (gray), projected future range (dark gray), and overlap (black) are shown. Although some areas of adjacent present and future range are evident, maintaining a viable population would require longer-distance movements on three or more fronts.

mate change therefore involves consideration of several factors, including: possible climate-related changes in targets; patterns of future landscape connectivity with respect to these changes; and degree of management control possible within different time frames. Where future target needs and present landscape connectivity are compatible, a plan may be developed that seeks to preserve this relationship. Where expected future connectivity does not match biological needs or is unknown, directed or flexible management should be provided.

Space alone will not ensure the success of a conservation strategy under the additional pressure of climate change, because while the process of range shifts is occurring, many other landscape/seascape modifications will be taking place. Precipitation will change and affect not only freshwater availability, but also hydrology and dilution factors. Changes will be oc-

curring from broad-scale ecosystem function to individual physiological responses. To increase the tolerance of systems to climate change at all these scales, it will be necessary to limit overall nonclimate stress as much as possible. A large literature of regional impact studies on various sectors (e.g., agriculture, human health) has been initiated by the IPCC (IPCC 2001). These studies provide a valuable resource for identifying climate-related changes in threats.

CREATING A DYNAMIC LANDSCAPE CONSERVATION PLAN

A dynamic landscape conservation plan is a combination of fixed and dynamic spatial elements, coupled with management guidelines for all target species, genetic resources, and ecosystems in the planning area. A completed plan has adequate precision for all conservation actors to understand their respective roles in managing, monitoring, and adapting to dynamic change. It is sufficiently open to permit individual actors to design dynamic responses specific to their site or management mandate.

Protected areas provide the fixed, fully

natural landscape elements of the plan. They may be selected by using reserve selection algorithms or other systematic reserve planning tools, based on both species' current and projected future ranges. Management of specific targets may evolve over time in protected areas, but the overall land use is fully natural.

Dynamic landscape elements, in contrast, may range from fully natural to disturbed, and their land use and management may change considerably over time (see Chapter 21). The desired condition of each planning unit in the matrix should be defined for both the present and the future, as well as intermediate steps necessary to bridge the two. For example, a range shift in a target species might require land that is currently in forest plantation to be fully natural habitat by 2050—present land use would need to transition over several decades to fully restored habitat.

Management guidelines suggest a means of achieving desired land use or resource availability in both the present and the future. For example, land acquisition might be specified as a management action where natural habitat needs to be preserved, or management of farmlands to leave waste grain on the ground might be specified where this is an important food source for migrating birds. Each management guideline would be associated with an implementing agency. A government conservation department might be identified to acquire land in natural habitat, while an NGO might be specified in work with farmers to maintain practices friendly to migratory birds.

The completed plan is therefore both a spatial plan and a road map for management. Where target species' needs evolve with climate change, the complexity of the plan multiplies, since spatial and management design must be specified for various points in the future, as well as the present. Given these complexities, perhaps the most important part of the plan is its provisions for monitoring and adaptation.

These elements give the plan flexibility in dealing with uncertainties, coupled with the ability to mount a management response to unforeseen developments as climate changes.

SUMMARY

The creation of a dynamic landscape plan for the conservation of biodiversity, once an almost impossibly complex task, is now a surmountable problem. Planning begins with the selection of targets. Data on these species or other elements are gathered, and, where possible, future changes are modeled. Until recently, this effort alone would have required hundreds or thousands of hours to create a georeferenced database from museum records, but recent advances in bioinformatics increasingly make such data available online. RCMs are now available for use on personal computers, with boundary conditions available for multiple GCMs. Systems for modeling species range shifts are available for desktop computers. Information on spatial and functional habitat needs of species, as far as they are known, can be gathered through traditional literature searches. The time required for these data preparation and modeling steps is increasingly within reach of a planning effort that spans 2–3 years.

The most complex task in creating a dynamic plan is the assignment of spatial elements and management tasks across the landscape. Modeled species' range shifts can be used to define protected areas as fixed elements in the landscape. Reserve selection algorithms exist as tools for identifying fixed landscape elements. They may be adapted for climate change with several modifications. Connectivity needs can be grouped by spatial scale, but this analysis remains complex. No computer tools comparable to reserve selection algorithms now exist for this part of the process, so it is time-consuming and largely dependent on the judgment of individual planners. Particularly difficult is the balance of spatial and management elements necessary to optimize multiple outcomes.

A completed plan divides the landscape into spatial, fixed or connective, and non-spatial elements, each with management and temporal attributes specific to each target. Protected areas management plans will specify management actions, monitoring variables and triggers of management change specific to each target species as conditions change. Plans for managing the connectivity matrix between protected areas will have similar management, monitoring, and management change features, but applied in a much different context and with different implementation mechanisms. Efforts to manage nonclimate change stresses are requisite to the long-term function of protected areas and connective efforts. Management in both protected areas and the matrix is, therefore, critical to the implementation and revision of the plan over time, and is the subject of the following two chapters.

REFERENCES

Bisby, F. A. 2000. The quiet revolution: Biodiversity informatics and the Internet. *Science* 289:2309–2312.

Cabeza, M., and A. Moilanen. 2001. Design of reserve networks and the persistence of biodiversity. TRENDS in Ecology & Evolution 16:242–248.

Channell, R., and M. V. Lomolino. 2001. Dynamic biogeography and conservation of endangered species. *Nature* 403:84–86.

Dayton, P. K., E. Sala, M. J. Tegner, and S. Thrush. 2000. Marine reserves: Parks, baselines, and fishery enhancement. *Bulletin of Marine Science* 66(3):617–634.

Dobson, A., K. Ralls, M. Foster, M. E. Soule, D. J. Simberloff, D. Doak, J. A. Estes, L. S. Mills, D. Mattson, R. Dirzo, H. Arita, S. Ryan, E. A. Norse, R. F. Noss, and D. Johns. 1999. Corridors: Reconnecting fragmented landscapes. In M. E. Soule and J. Terborgh, eds., *Continental Conservation: Scientific Foundations of Regional Reserve Networks*, pp. 129–170. Washington D.C.: Island Press.

Forman, R.T.T. 1995. Land mosaics: The ecology of landscapes and regions. Cambridge: Cambridge University Press.

Gascon, C., G. B. Williamson, and G. A. B. da Fonseca. 2000. Receding forest edges and vanishing reserves. *Science* 288(5470):1356–1358.

Groves, C. R., D. B. Jensen, L. L. Valutis, K. H. Redford, M. L. Shaffer, J. M. Scott, J. V. Baumgartner, J. V. Higgins, M. W. Beck, and M. G. Anderson. 2002. Planning for biodiversity conservation: Putting conservation science into practice. *BioScience* 52:499–512.

Hannah, L. 2004. Agroforestry and climate change-integrated conservation strategies. In S. Schroth, ed., *Agroforestry and Biodiversity Conservation*. Amsterdam: Kluwer.

Hannah, L., G. F. Midgley, T. E. Lovejoy, W. J. Bond, M. Bush, J. C. Lovett, D. Scott, and F. I. Woodward. 2002. Conservation of biodiversity in a changing climate. *Conservation Biology* 16:264–268.

Hargrove, W. W., and F. M. Hoffman. 1999. Using multivariate clustering to characterize ecoregion borders. *Computing in Science & Engineering* 1:18–25.

IPCC. 2001. Climate Change 2001: Impacts, Vulnerability and Adaptation; contribution of Working Group II to the Third Assessment Report of the Intergovernmental Panel on Climate Change. IPCC Working Group 2. 2001. Port Chester, N.Y.: Cambridge University Press. 1919.

Kittel, T.G.F., N. A. Rosenbloom, T. H. Painter, D. S. Schimel, and V. M. Participants. 1995. The VEMAP integrated database for modeling United States ecosystem/vegetation sensitivity to climate change. *Journal of Biogeography* 22:857–862.

Noss, R. F., and L. D. Harris. 1986. Nodes, networks, and MUMs: Preserving diversity at all scales. *Environmental Management* 10:299–309.

Olson, D. M., and E. Dinerstein. 1998. The global 200: A representation approach to conserving the earth's most biologically valuable ecoregions. *Conservation Biology* 12(3):502–515.

Peterson, A. T., M. A. Ortega-Huerta, J. Bartley, V. Sanchez-Cordero, J. Soberon, R. H. Buddemeier, and D. R. Stockwell. 2002. Future projections for Mexican faunas under global climate change scenarios. *Nature* 416:626–629.

Pressey, R. L., C. J. Humphries, C. R. Margules, R. I. Vanewright, and P. H. Williams. 1993. Beyond opportunism—Key principles for systematic reserve selection. *Trends in Ecology & Evolution* 8:124–128.

Pressey, R. L., I. R. Johnson, and P. D. Wilson. 1994. Shades of irreplaceability—Towards a measure of the contribution of sites to a reservation goal. *Biodiversity & Conservation* 3:242–262.

Pressey, R. L., H. P. Possingham, and J. R. Day. 1997. Effectiveness of alternative heuristic algorithms for identifying indicative minimum requirements for conservation reserves. *Biological Conservation* 80:207–219.

Root, T. L., and S. H. Schneider. 1995. Ecology and climate: Research strategies and implications. *Science* 269:334−341.

Smith, T. B., R. K. Wayne, D. J. Girman, and M. W. Bruford. 1997. A role for ecotones in generating rainforest biodiversity. *Science* 276:1855−1857.

Sugden, A., and E. Pennisi. 2000. Diversity digitized. *Science* 289:2305.

Terborgh, J., J. A. Estes, P. Paquet, K. Ralls, D. Boyd-Heger, B. J. Miller, and R. F. Noss. 1999. The role of top carnivores in regulating terrestrial ecosystems. In M. E. Soule and J. Terborgh, eds., *Continental Conservation: Scientific Foundations of Regional Reserve Networks*, pp. 39−64. Washington D.C.: Island Press.

Thomas, C. D., A. Cameron, R. E. Green, M. Bakkenes, L. J. Beaumont, Y. C. Collingham, B. F. N. Erasmus, M. Ferreira de Siqueira, A. Grainger, L. Hannah, L. Hughes, B. Huntley, A. S. Van Jaarsveld, G. E. Midgley, L. Miles, M. A. Ortega-Huerta, A. T. Peterson, O. L. Phillips, and S. E. Williams. 2004. Extinction risk from climate change. *Nature* 427:145−148.

Vane-Wright, R. I., C. J. Humphries, and P. Williams. 1991. What to Protect?—Systematics and the Agony of Choice. *Biological Conservation* 55:235−254.

Walther, G., E. Post, P. Convey, A. Menzel, C. Parmesan, T.J.C. Beebee, J. Fromentin, O. Hoegh-Guldberg, and F. Bairlein. 2002. Ecological responses to recent climate change. *Nature* 416:389−395.

Integrating Climate Change
into Canada's National Park System

Daniel Scott

Canada is the world's second largest nation and contains approximately 20 percent of the world's remaining wilderness areas (Parks Canada 2000). Many of these areas are at the high northern latitudes projected to be particularly strongly affected by climate change. This poses a strong challenge to the nation's parks policy and planning frameworks.

The Canadian national park system consists of 39 terrestrial parks that occupy over 244,000 km². Given the large area and the wilderness status of many of these parks, they are of global significance for the conservation of temperate and arctic ecosystems. Parks Canada is the federal agency responsible for establishing and managing a national park system representative of the ecosystems found across Canada. To accomplish this mandate, a national park system plan was developed in 1971 to provide a long-term framework for park selection, with the goal of protecting 39 "natural regions" (defined by vegetation classification and physiological features), deemed representative of landscapes across Canada. Although only 24 of the natural regions are currently represented in the national park system, the government of Canada has repeatedly reaffirmed its commitment to completing the system.

Amendments to the National Parks Act in 1988 require that the maintenance of ecological integrity, defined as "the conditions of an ecosystem where the structure and function of the ecosystem are unimpaired by stresses induced by human activity, and the ecosystem's biological diversity and supporting processes are likely to persist," be the primary objective of park management (Parks Canada 1998: 23). Parks Canada goes on to state, "Ecosystems are inherently dynamic and change does not necessarily mean a loss of integrity. A system with integrity may exist in several states, but the change occurs within *acceptable limits*" (1998: 24, their emphasis). The magnitude and pace of climate change and associated ecosystem responses are expected to accelerate ecosystem change beyond what science has observed to be the natural range in some regions (Overpeck et al., this volume). Consequently, it remains uncertain as to how ecosystem change "within acceptable limits" is to be interpreted by park managers within the context of climate change.

The core policies of completing the national park system plan and maintaining ecological integrity are illustrative of the sensitivities of existing conservation policy and planning frameworks to climate change. Like protected area planning around the world, Canada's national park system plan was developed with the assumptions of climatic and biogeographic stability. A large body of research now indicates that these assumptions are no longer tenable (Lovejoy, this volume). To assess the impact climate change might have on the Canadian system, Scott and colleagues (2002) used modeling results from equilibrium global vegetation models forced with multiple climate change scenarios to examine possible changes in biome representation (Fig. 20.5). Their results revealed that regardless of the vegetation scenario used, substantial changes in biome representation occurred. A new biome type appeared in over half the national parks. The proportional representation of biomes in the national park system also changed, with diminishing representation of northern biomes (tundra, tundra/taiga, and boreal forest) and additional representation of more southerly biomes (temperate evergreen and temperate mixed forests in particular). Although equilibrium vegetation modeling results can be considered only indicative of the probable trajectory of vegetation change and not predictive of the future distribution and composition of biomes (Peterson et al., this volume), these and other vegetation modeling results in Canada (Rizzo and Wiken 1992; Hogg and Hurdle 1995; Price and Apps 1996) indicate that the twenty-first century is likely to be characterized by ecological communities in transition.

The policy of completing the current national park system plan should be reassessed in light of these results. Short-term climate change adaptations might mean new park establishment should focus on biomes projected to lose representation (tundra, taiga/tundra, and boreal forest), on protecting areas expected to remain refugia for current ecological communities, and on maximizing the potential of species to respond to climate change (e.g., increasing connectivity or protecting outlier populations). To do so would require rethinking current park selection criteria and perhaps the designation of new types of parks.

Perhaps the most critical question regarding the role of Parks Canada in an era of climate change is whether the national parks are to continue to protect current ecological communities or facilitate ecosystem adaptation to climate change. An examination of existing policy and planning frameworks suggests either interpretation is possible. These two distinct management paradigms have direct implications for park management decisions. For instance, the translocation of populations at risk from climate change to areas more suitable, which is commonly proposed in the literature, could be interpreted as inconsistent with maintaining ecological integrity if the species in question was not native to the destination region and could have an adverse impact on species in existing communities.

A review of management plans from several national parks revealed additional climate change sensitivities at the individual park level. These included park objective statements, wildfire management strategies, individual species management plans and contingencies for species at risk, nonnative species management programs, and species reintroduction programs. For example, the stated purpose of Prince Albert National Park is to "Protect for all time the ecological integrity of a natural area of Canadian significance representative of the southern boreal plains and plateaux. . . ." Yet all six vegetation change scenarios examined by Scott and colleagues (2002) projected the eventual loss of boreal forest in this park (see Fig. 20.5 for two scenarios), suggesting that the park's mandate would be untenable in the long term. Furthermore, the decision to reintroduce natural fire regimes in ecotonal parks like Prince Albert National Park, where vegetation models project a shift from boreal forest to grasslands, would hasten the transition to grassland communities and therefore be in conflict with the current park purpose.

The policy and planning sensitivities identified above demonstrate the need for a Parks Canada policy on climate change. Such a policy would enable a coordinated response strategy that would provide a framework for planning decisions at the park level. Parks Canada cannot unilaterally develop comprehensive contingency plans for climate change, however, without modification of the agency's mandate, policy framework, and resource allocations. This would require ministerial approval and likely legislative changes to the National Parks Act.

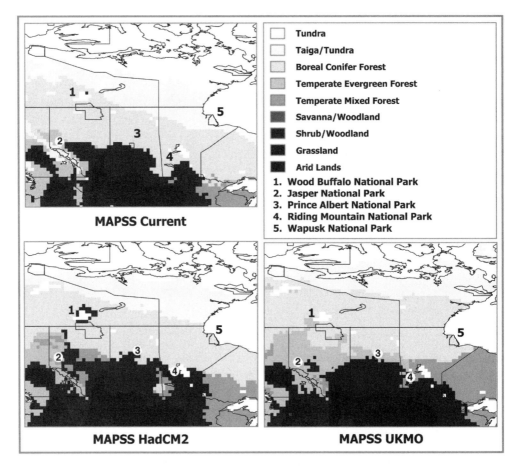

Figure 20.5. Modeled biome distribution change in national parks of central Canada. See Scott et al. (2002) for national park analysis and methods and Neilson (1998) for additional information on the MAPSS vegetation model and climate change scenarios.

Climate change represents a fundamentally different threat to Canada's national park system. Never before has there been an ecological stressor that has raised questions about the adequacy of the system to protect representative samples of Canadian ecosystems. Over a decade ago, Lopoukhine (1990) suggested that "climate change is poised to alter the rate of evolution in (Parks Canada) policies and Act." Parks Canada has taken early initiatives to examine the implications of climate change for its parks and mandate. Since 2000, Parks Canada has commissioned a wide-ranging report on climate change impacts, developed climate change scenarios for each park, and conducted professional training workshops on climate change.

The integration of climate change into existing institutional frameworks of government conservation agencies will be both complex and challenging. Governments and conservation stakeholders cannot shy away from this challenge and must work collaboratively so that as the scientific community continues to advance understanding of the impacts of climate change on physical and biological systems, there exists the institutional capacity to develop and implement pragmatic conservation responses.

REFERENCES

Boer, G., Flato, G., & Ramsden, D. 2000. A transient climate change simulation with greenhouse gas and aerosol forcing: Projected climate to the twenty-first century, *Climate Dynamics*, 16, 427–450.

Halpin, P. 1997. Global climate change and natural-area protection: Management responses and research directions, *Ecological Applications*, 7, 828–843.

Hill, M., Downing, T., Berry, P. Coppins, B., Hammond, P., Marquiss, M., Roy, D., Tefler, M., and Welch, D. 1999. Climate changes and Scotland's natural heritage: An environmental audit. *Scottish Natural Heritage Research Survey and Monitoring Report*, No. 132.

Hogg, E., and Hurdle, P. 1995. The Aspen Parkland in Western Canada: A dry-climate analogue for the future boreal forest? *Water, Air and Soil Pollution*, 82, 391–400.

Lopoukhine, N. 1990. National parks, ecological integrity and climatic change. In: *Climatic Change: Implications for Water and Ecological Resources*, ed. Wall, G., and Sanderson, M., pp. 317–328. Department of Geography Publication Series, Occasional Paper No. 11, University of Waterloo, Waterloo, Ontario, Canada.

Neilson, R. 1998. Simulated changes in vegetation distribution under global warming. The regional impacts of climate change: An assessment of vulnerability, ed. by Watson, R., Zinyowera, M., & Moss, R., pp. 441–456. A Special Report of IPCC Working Group 2, Cambridge University Press, Cambridge, UK.

Parks Canada. 1998. State of the parks 1997 Report, Parks Canada, Ottawa, Canada.

Parks Canada. 2000. Unimpaired for future generations? Protecting ecological integrity with Canada's national parks—Volume 2—Setting a new direction for Canada's national parks, Report of the Panel on the Ecological Integrity of Canada's National Parks, Ottawa, Canada.

Price, D., and Apps, M. 1996. Boreal forest responses to climate-change scenarios along an ecoclimatic transect in central Canada, *Climatic Change*, 34, 179–190.

Rizzo, B., and Wiken, E. 1992. Assessing the sensitivity of Canada's ecosystems to climatic change, *Climatic Change*, 21, 37–55.

Scott, D., Malcolm, J., and Lemieux, C. 2002. Climate change and biome representation in Canada's national park system: Implications for system planning and park mandates, *Global Ecology and Biogeography*, 11, 475–484.

Scott, D., and Suffling, R. 2000. Climate change and Canada's National Parks, Environment Canada, Toronto.

Villers-Ruiz, L., & Trego-Vazquez, I. 1998. Climate change on Mexican forests and natural protected areas, *Global Environmental Change*, 8, 141–157.

CHAPTER TWENTY-ONE

Managing the Matrix

GUSTAVO A. B. DA FONSECA, WES
SECHREST, AND JUDY OGLETHORPE

Human-induced climate change will profoundly affect how managers plan, design,
and manage landscapes for biodiversity
conservation (Peters 1991). Existing natural and semi-natural areas in the human-
altered matrix surrounding protected areas constitute a vast majority of the
planet's land surface—a full 83 percent of
land globally is estimated to have been directly affected by humans (Sanderson et al.
2002). With a little over 10 percent of the
surface of the planet under some form of
protection, management of the intervening matrix plays a vital role in the future of
biodiversity conservation.

In the past few decades, there has been
an increased understanding of ecological
responses to climate change (McCarty
2001; Walther et al. 2002). Clearly, the
current practice of segregating human and
natural landscapes into independently
managed units is not appropriate given the
climate change dynamics. How can current human practices and land use be altered and improved to stem the negative
biological effects of climate change?

The composition of matrix areas is
highly variable, but almost always less than
ideal for the conservation of most "natural" biodiversity features. This must be
reversed if matrix regions are to become
transitional pathways for range shifts, crucial habitats in themselves, and potential
future protected areas. One key way to
achieve this reversal is to reduce the
"harshness" of the matrix, broadly defined
as its effects on surrounding natural habitats, which can especially affect small fragments (Gascon et al. 2000). For many protected areas, managing and reducing the
effects of matrix degradation will become

increasingly important for medium- to long-term conservation of biodiversity. Chapter 20 has outlined methods for planning a landscape for climate change. This chapter examines challenges and mechanisms associated with landscapes already affected by human activity.

PLANNING TOOLS

The utilization of species as measures of biodiversity remains the most reliable and efficient basis for developing strategies to reduce climate change impacts on biodiversity. Nonetheless, assessing biodiversity at all levels from ecosystems to species to genetic diversity is crucial in the face of climate change. Many of these other aspects of biodiversity are less readily quantified, but they can be used to supplement species information such as range and population estimates.

The initial step of setting targets includes using threatened and restricted range species, as well as others that are identified as being a high conservation priority, such as montane species. Conservation should focus not only on protecting sustainable populations and areas but also on maintaining and restoring species to as close to their natural roles as possible, while acknowledging that this is not a static target.

A risk inherent in setting simple targets is that complexity remains unaccounted for. Particularly in the matrix, species that are not specifically targeted in conservation plans are likely to be lost. It is therefore important that a robust set of targets is used for protected area selection, so that many nontarget species are protected in the fully natural elements of the landscape. In the matrix, the most important species to target will be those that require territories too large to be captured in any single protected area, species that have foraging or other resource requirements outside of reserves, and species that will undergo large-scale range shifts due to climate change.

Planning the matrix for conservation of these species requires knowledge of their range and status, and scenarios of future change. Currently, efforts are under way from several organizations to provide sound scientific information on species' status. An important first step includes the conservation assessments of IUCN—the World Conservation Union—which include full assessments of species groups, including all mammals, birds, amphibians, and reptiles within the next few years through the IUCN Red List Programme (http://www.redlist.org), as well as other groups including fishes, plants and invertebrates. These data will be readily accessible by conservation managers through the IUCN Species Information Service (http://www.iucn.org/themes/ssc/programs/sisindex.htm). Several other large initiatives are under way to compile data on biodiversity, especially species information, including the Global Biodiversity Information Facility (http://www.gbif.org). These and other efforts by governments and organizations are needed for evaluating biodiversity.

The traditional planning methods used by conservation biologists to counteract human impacts on biodiversity must be expanded to take into account climate change. The incorporation of climate change into target setting should include the use of new technologies and methods, especially those used to predict the extent of change to species and ecological processes. For species, this can be accomplished using already available methods to model potential distributions (Hannah and Hansen, Chapter 14, this volume).

Planning for conservation of key species in the matrix requires spatial analysis tools. Within the context of a dynamic landscape conservation plan, matrix management will have to be defined in detail, often involving choices between many possible options for achieving a given set of conservation goals. Making these choices will involve balancing conservation goals and information with land use and economic potentials of the land.

Geographic information systems (GIS) provide one important and widely available set of tools for this analysis. Use of a GIS platform permits the comparison and weighting of conservation values such as area, edge, and fragmentation in the matrix, side by side with consideration of current land use, access to markets, and economic potential. The data needed for these analyses are increasingly available through remote sensing (e.g., intact habitat, land cover, market data), government agencies (e.g., land tenure, population, income levels), and research (e.g., biodiversity value, economic potential, market trends).

More sophisticated and interactive planning is possible with tools derived from GIS. For example, TAMARIN is a GIS-based spatial decision support system developed through a partnership among local nongovernment organizations (NGOs), the World Bank, and governments to assist biodiversity conservation planning for the southern Bahian region of Brazil (Fig. 21.1). TAMARIN integrates principles from conservation biology with economic theory in both the design and evaluation of alternative scenarios and with respect to trade-offs between environmental and other social goals (http://www .biogeog.ucsb.edu/projects/wb/wb.html provides information about obtaining the TAMARIN model and associated GIS data). Based on conservation planning principles, TAMARIN seeks to achieve representation, resilience, and redundancy objectives at the lowest economic opportunity cost. It permits the assessment, based on criteria of environmental benefits and economic costs, of any proposed configuration of land use. In a workshop application with local NGOs and government agencies, simulations were made using TAMARIN to allocate a hypothetical biodiversity fund.

Spatial analysis tools may be used in interactive planning with stakeholders in an adaptive management process. This can help illuminate issues and produce consensus among stakeholders, allowing the plan to move more immediately toward implementation. Adaptive management involving stakeholders has been applied to a variety of problems, and may offer one of the best hopes for stakeholder buy-in and conservation implementation across a wide variety of land use, tenure, and ownership regimes in the matrix.

COORDINATED MANAGEMENT

Strong spatial and sectoral coordination will be crucial to secure space in the matrix for conservation and to limit stresses to biodiversity. This will require action at various levels: local, national, and regional. It will involve coordination both within the conservation sector and with many other sectors that have impact on biodiversity.

Within the conservation sector, coordination and communication among researchers, policy makers, and practitioners on the ground will be essential to ensure that scientific understanding of climate change implications is applied in the best way possible, and as soon as it is available. Information must flow back to researchers so that research programs and models can be refined based on current needs and developments in field programs.

Coordination will also be essential between conservationists and other sectors—for example, agriculture, water, energy, health, education, infrastructure, and transport. All these sectors will experience climate-induced changes of their own, which, in turn, will affect biodiversity in the matrix. Many impacts on different sectors will be interrelated—for example,

Figure 21.1. The implementation area of TAMARIN in southeastern Brazil. Areas prioritized for connectivity are referred to as "corridors" in this example. Two corridors within the ecoregion: the Discovery Corridor and the Serra do Mar Corridor.

changes in agricultural crop ranges may result in abandonment of some agricultural land, and pressure to open up new natural areas. Similarly, there will be movement of people from areas abandoned due to coastal flooding and declining agricultural potential or water supplies. Responses will need to be well coordinated and multi-sectoral to limit additional stresses on biodiversity.

Spatially coordinated management is important because climate change–mediated effects on species will span regional, local, and national boundaries. At the regional level, multi-sectoral land-use planning at larger spatial scales is key, which means that conservation and sound natural resource management should be mainstreamed in the planning process. Conservation champions, researchers, policy makers, and practitioners must participate in planning processes to negotiate and help reconcile biodiversity needs with changing human requirements. Adaptation for conservation can obtain greatest leverage at this level.

At the local level, a wide range of individual actions by many different stakeholders can directly influence resistance and resilience of biological systems to climate change. Stakeholders include landowners, farmers, natural resource users, local governments, private companies, NGOs, and local communities, and coordination will be key to promoting awareness of consequences and reducing impacts. Local government will play a key role in coordinating adaptation responses, especially in developing countries. NGOs can facilitate collaboration between government and communities, and can also help to build community capacity to adapt natural resource management and farming practices in response to climate change.

The national level also has a strong policy role, particularly in supporting regional assessments and responses. This may involve setting aside national differences to participate in regional assessments and policies. In both regional and national assessments, it will be important for biologists to advocate the consideration of biodiversity alongside human responses to climate change. Regional climate scenarios developed for these assessments can serve both human and biological analyses.

Collaboration across boundaries is essential, because many matrix landscapes cross jurisdictional borders, including those between neighboring communities, land owners, local and provincial/state governments, and nations. Among the goals of collaboration across these boundaries are maintaining corridors, ensuring space for climate-induced range shifts, and ensuring that natural resource harvesting is adapted sustainably to new productivity levels. Policies and land-use planning, as well as conservation imple-

mentation, need to be harmonized across boundaries.

Building trust and shared vision among transboundary stakeholders takes time, particularly at larger scales and higher levels where it tends to be much more formal (e.g., between countries) (van der Linde et al. 2001). Good political will is essential for any successful transborder collaboration, and this often requires long-term fostering and formal legal mechanisms for collaboration (e.g., memoranda of understanding). In order to prepare for climate change adaptation, these processes need to be initiated as soon as possible, and the additional capacity and resources required should not be underestimated.

The need for coordination is not only horizontal (e.g., between local stakeholders or between countries) but often also vertical. For example, strong coordination is needed between national and local governments to ensure common or compatible policies. National conservation strategies should be harmonized within regions, so that all nations in a connected region can adopt harmonized and complementary adaptation strategies.

Stakeholders should be involved from an early stage to ensure participation and buy-in to the planning process, and to ensure sound implementation. Mutually beneficial situations must be sought in order to provide incentives, and this requires effective communication of scientific predictions of climate change impacts and response scenarios. It will be important to use existing coordination, collaboration, and communication mechanisms and to create new ones when needed.

REDUCING NONCLIMATE STRESSORS

Climate change effects on ecosystems and biodiversity may be reduced by effectively protecting and repairing already affected landscapes. In the matrix outside of protected areas, the goal is twofold—to slow or halt land degradation and to improve land that is currently heavily affected by human activity. This goal, if it incorporates rigorous science, will improve the social and economic benefits of local regions as well as their overall environmental health and status.

The most important principle for matrix management is that the closer a system can be returned to a natural state, the less active human management will be needed to prevent negative responses to climate change (Trexler and Haugen 1995). In economic terms, this will be the most efficient method, in that natural processes are allowed to operate more freely. For those landscapes that remain matrix habitat, management should be focused on how to best protect biodiversity and ecosystems in light of the current changing climate. Vulnerable terrestrial habitats include freshwater, montane, island, and coastal areas, each of which presents different conservation challenges. Minimizing the effects of climate change on these areas will be important to maintaining many ecosystem processes as well as biodiversity more generally. The unpredictability of many biological responses will require highly flexible management strategies and tactics. Biological responses will differ depending on the taxa and region in question, so there is no formula or single optimal plan for every situation. Efforts designed specifically for stressors known to be critical for target species are therefore the most effective.

Where knowledge of the biology of species is lacking, generic solutions may substitute for responses tailored to individual situations. For instance, the removal of barriers to dispersal is a tactic that will often help mitigate climate change effects on biodiversity. Roads can be designed for lower impact—wildlife corridors over and under roads have been shown to help dispersal, though requirements for different species vary. For instance, the width, length, vegetation coverage, and hydrol-

ogy dictate whether species such as amphibians and small mammals will move through corridors under roads. Corridor establishment has been shown to be beneficial for plant and animal species by facilitating movement and population persistence (Tewksbury et al. 2002). This benefit must be extended to include species responses to climate change, such as improving matrix habitat to allow for species movement.

Incompatible land-use practices remain the largest threat to biodiversity, and their impacts could increase, especially as agricultural systems adapt to climate change (Tilman et al. 2001). The majority of the matrix is composed of agricultural areas, including rangeland, farmland, and forest plantations, in addition to previously affected areas and urban areas. Management of matrix habitat should include minimizing the need for additional land for human uses when possible, and maximizing the efficiency of current land used. Resource consumption, including use of fresh water, must be slowed and reversed.

Biodiversity will not be adapting in a fixed land-use pattern—land uses are going to change drastically with climate change—agricultural patterns will shift as crop ranges shift, and it is very likely that there will be pressure to open up new areas as their agricultural potential increases, or as old areas become unfarmable. There are likely to be large resettlement movements, away from low-lying coastal areas. People may use natural resources differently as their livelihood security changes, particularly in the developing world where people do not have the means or knowledge to change to alternative crops or production systems quickly.

Some other threats to biodiversity will likely be exacerbated in the near future. Invasive species are already causing many extinctions, a threat that could increase with climate change (Peterson et al., Chapter 14, this volume). Reducing these stressors now will greatly reduce the likelihood of follow-on impacts due to synergy with climate change. Matrix habitat will remain a key component of minimizing the spread of invasive species, which will most likely be able to persist and spread in human-altered habitats.

PLANNING THE MATRIX

Assignment of conservation values to different portions of the landscape (or seascape) was addressed in Chapter 20, and will be only briefly reviewed here. A well-planned landscape design includes multiple possible options for achieving connectivity goals, to allow for flexibility and adaptation during implementation. Planning for the matrix is a more detailed continuation of the planning process begun at the landscape level. It forms the basis for implementation, and adjustments to management through monitoring, in an iterative process.

Among the best-developed tools for planning the matrix are those for assessing the habitat and economic values of land units. For instance, the TAMARIN software developed by the University of California allows planners to examine trade-offs between habitat values and economic values of land units and connectivity configurations (Sanderson et al. 2003). Habitat values may be denominated in generic descriptors such as forest cover, or in more sophisticated terms based on the known habitat needs of target species. Economic values may be defined as purchase costs or as easement values for purchasing limited conservation restrictions on land. Stakeholders may manipulate habitat and social values and examine outcomes in what is essentially a consensus-building process.

Assessing conservation needs is integral to developing habitat values for use in such applications. As described in Chapter 20, this involves the setting of conservation targets—usually species, populations, and processes—and determining the specific habitat needs for each. Large carnivores requiring extensive areas are especially important in this assessment, since they structure ecosystems and require large, connected areas of habitat. Maintaining or restoring top predators requires large areas of habitat and connectivity, and therefore can dictate much of the spatial structure of matrix connectivity. In most areas, there are few options for large connected habitats, and those necessary for large predators will be, by default, those few large natural or semi-natural areas that remain connected. These large areas will also capture many of the other biodiversity targets of the region, so that completing habitat needs for other target species is essentially a gap analysis.

Once existing habitat needs have been determined, analysis of the impact of climate change allows estimation of likely spatial changes in the distribution of habitat needs at different points in the future. An integrated plan can then be constructed (Fig. 21.2), which will necessarily have to balance habitat needs of various species, present versus future habitat needs, and habitat needs versus changing social values and economic costs. Planning should take into account the increased magnitude and frequency of climatic events such as storms. Planning should be done also at larger spatial scales to incorporate as many ecological processes as possible.

The final step in planning the matrix is bridging landscape potentials with habitat needs. In some cases, existing natural habitat in the landscape may be adequate for all target habitat needs. In other cases, semi-natural areas may have to be targeted for management to maintain specific habitat values (for instance, maintaining overstory trees in shade-grown coffee as migratory corridors for a bird species), or restoration of habitat in geographically critical areas may be required. Since bridging landscape potentials and habitat needs involves economic and social trade-offs,

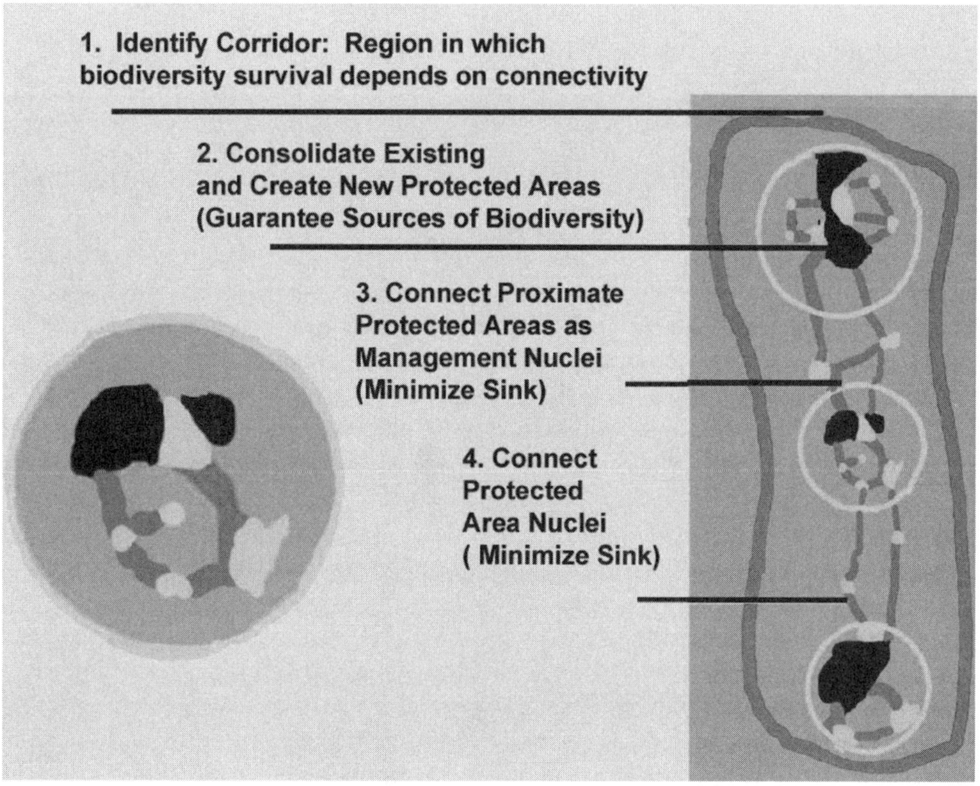

1. **Identify Corridor: Region in which biodiversity survival depends on connectivity**

2. **Consolidate Existing and Create New Protected Areas (Guarantee Sources of Biodiversity)**

3. **Connect Proximate Protected Areas as Management Nuclei (Minimize Sink)**

4. **Connect Protected Area Nuclei (Minimize Sink)**

Figure 21.2. Elements of a hypothetical landscape conservation plan. Matrix elements such as easements on private land and biodiversity-friendly agriculture provide connectivity between new and existing protected areas.

this process will best be carried out with stakeholders rather than as a technical desk study, and as part of an integrated multiple land-use planning process. In any event, the plan developed will have to be modified as implementation experience builds, in an adaptive process with management and monitoring (see below).

MANAGING THE MATRIX

Management of the matrix relies on combinations of regulatory and incentive-based approaches (Fig. 21.3). Private landholders, major players in matrix conservation, seldom have conservation as their primary objective. Therefore, ele-

ments of regulatory coercion, such as land-use zoning and endangered species laws, play a role. At the same time, excessive regulation brings political resistance, so that incentive-based approaches are critical to maintaining landholder support and generating voluntary cooperation.

Regulatory approaches will likely provide a large base for conserving or minimizing other human factors such as agriculture to further affect the most critical areas identified. The IUCN database of protected areas includes a system of categorization, in addition to other data. Although it is impossible to fully categorize and differentiate among the many types and degrees of protection, the IUCN categories allow separation of protection from fully protected land to land reserved primarily for sustainable use of natural resources. Category I and II protected areas—Nature Reserves and National Parks—are discussed in more detail in

Figure 21.3. Regulatory and incentive-based approaches in landscape connectivity. Statutory protection such as parks and nature reserves are linked with incentive-based modifications of land use such as easements, or agroforestry.

Chapter 22. Categories III through VI offer less explicit protection for biodiversity and include Natural Monument, Habitat/Species Management Area, Protected Landscape/Seascape, and Managed Resource Protected Area. Expansion of managed landscapes and resource management areas offers an important regulatory tool for matrix management. These are essentially forms of land-use zoning that protect important biological features, while allowing land to be used primarily for human production. Where extensive, low-intensity management for biodiversity is required, these tools may be very effective.

The use of incentive-based approaches is an alternative that may be more effective where landholder acceptance is highly variable. Incentive-based approaches allow a market to develop, in which interested landholders may participate. One such method, conservation concessions, is modeled after logging concessions. In return for a direct payment or establishment of a trust fund, land is placed under conservation. This approach has been used successfully in an increasing number of critical areas, and can be used in varying degrees to offer protection to areas of the matrix that are not adequately managed at present. Another method is to purchase and then lease land, which allows for better control of land management decisions.

Other methods include landholder agreements, such as easements and contractual conservation, which can establish land management on both public and private lands. A specific method that can be increasingly effective for dealing with climate change is conservation futures. These futures are similar to concessions or contractual agreements, with the added flexibility of conservation action in the future depending on variable climate change. These strategies have been used successfully in some current static conservation plans, but their use in dynamic plans that incorporate rapid changes should implement improved science and monitoring, and focus on prevention.

The focus on prevention and flexibility should help maintain options for the future. A landscape-based approach should be used not only to identify areas that are critical now, but also to manage other areas which could become critical. Land- and seascapes must be regarded as whole systems that have currently suffered losses and reductions in many of their components but which ideally should be maintained and restored to best conserve nature in the long term. While the reality of the situation dictates that priorities be established, many areas that are not priorities are key components of long-term persistence of biodiversity. Thus, land uses should be promoted that minimize negative effects and maximize the benefits and future possibilities for restoration of natural habitats.

Restoration is an important option, particularly where the course of climate change dynamics is uncertain. Conservation plans may want to hold options open to restore lands to natural habitat, for instance as it becomes apparent that they are essential to species' range shifts. It may therefore be appropriate to use regulatory or incentive-based approaches to encourage present land uses that are compatible with restoration. For instance, forest plantations and agroforestry, which may have little current value for biodiversity conservation, may retain soil structure and be more suitable for future restoration than for other land uses (e.g., urban development).

Oportunistic restoration may also offer valuable opportunities for biodiversity. Degraded areas, such as abandoned agriculture fields and settlements, after being stripped of their available resources, are often rendered useless by human activity. Slash and burn agriculture has devastated landscapes, leaving ecosystems that are a shell of their past in terms of biodiversity and functional capability, as well as largely useless to humans. The process of ecological restoration of such areas is most often successfully accomplished by natural dispersal and colonization. The large-scale fragmentation and destruction of habitat does not often lend itself to rapid natural succession. These areas are often unable to be recolonized by native flora and fauna, for reasons ranging from nutrient loss, hydrological changes, failure of species to disperse, and establishment of nonnative species. Whereas temperate forests in the eastern United States have increased in the past century due to abandonment of farms, tropical areas in Madagascar have become a vast wasteland containing mostly introduced grasses. One main difference lies in the soil nutrients, which are plentiful in the glacial soils of North America but mostly tied into the biomass and thin soil layers that have eroded throughout Madagascar.

The addition of climatic change adds pressure to the need to rapidly turn the tide of natural ecosystem losses. Landscapes that are not beyond recovery have shown regeneration of native vegetation and recolonization of degraded areas surrounding fragments (Laurance et al. 2002). Often relationships between habitat and species are key to maintaining natural ecosystems, as evidenced by the crucial need for lemurs in the restoration of dry deciduous forests in Madagascar (Ganzhorn et al. 1999). Improved connectivity of fragments lies in restoring the matrix to conditions closer to their original natural state. Any type of vegetation cover is an improvement over severely degraded land such as livestock pastures. Nonforest areas surrounding the matrix can support biodiversity, as demonstrated by small vertebrate species within the matrix landscapes in central Amazonia (Gascon et al. 1999). Restoration of forests can help increase the rate of carbon dioxide removed from the atmosphere, and thus help in the quest for carbon sequestration. It has been increasingly evident that some areas require human intervention to facilitate natural processes of succession. Indeed, successful attempts at ecological restoration have occurred in many different terrestrial ecosystems.

MONITORING AND ADAPTATION

There are currently few attempts to monitor biodiversity in the matrix. As climate change necessitates a greater role for the matrix in biodiversity management, this situation must change. Monitoring of conservation target species in the matrix will need to be complementary to monitoring in protected areas. The overall monitoring systems must be less intensive and cover much larger areas. They will entail an important part of the increased cost of managing biodiversity in a changing climate.

Monitoring of climate-induced changes in biodiversity will be essential to design flexible responses and to evaluate which approaches work best to promote system resistance and resilience in particular circumstances. Climate change impacts are uncertain, so monitoring provides crucial information on evolving impacts. Monitoring can confirm expected impacts, identify impacts that have not been foreseen, or suggest refinements to change projections. Each of these functions provides improved information for management, which can reduce management costs by permitting early response or by curtailing inappropriate management.

Adaptation of land and natural resource management regimes in light of monitoring results will be essential in order to respond most effectively to changes. The only certainty in the coming years is that biodiversity will be affected on increasingly shorter temporal scales and larger spatial scales; plans must have the flexibility to change responsively and rapidly once new information arises.

Monitoring is needed at local, national, and regional levels. Biological monitoring includes gathering information on species distribution, abundance, persistence, and dispersal. The fact remains that the scientific research that underlies management of human-dominated landscapes is still developing, and any management strategy must remain flexible to include new research and unexpected results. The ability of management to respond by using the most current science will inevitably lead to better overall results.

Conservation targets are primary focal points for monitoring in the matrix. Sensitive species constitute a second group of high priority for monitoring efforts. However, a sample of nontarget and nonsensitive species should also be monitored to help identify unanticipated impacts.

Biodiversity surveys are important in determining where biodiversity is and how best to protect it. These must rely on adequate sampling, often of a selected group of species, that can help direct local management decisions. Many protected areas include surveys of areas in and around their borders. Surveys are often done once to establish baseline data for areas—this baseline must be further extended with additional surveys and monitoring of temporal changes. These data can serve to establish medium- to long-term trends in species distribution and abundance, which will aid in determining the effects of climate change on local biodiversity.

Biological monitoring will help to assess the direct impacts of climate change on biodiversity. However, it is important to monitor trends in nonclimate stressors too, especially as they are influenced by climate change. Therefore, economic, political, and social indicators must also be integrated into monitoring plans. In some cases they may give results earlier than biological indicators. Monitoring impacts of other key activities such as agriculture, transport, the energy sector, and human settlement is also required. Information on these pending and actual changes in land and resource use must be used to adjust and adapt conservation interventions.

SUMMARY

Human land use, especially in the tropics, in particular via deforestation, is the largest factor contributing to the current loss of species (Steadman 1991). The addition of global climate change increases threats to global biodiversity. The most important regional goals to achieve in order to address this challenge are to establish extensive protected area networks, with landscape corridors providing connectivity, and to improve management of the intervening matrix (Hannah et al. 2001). The current IUCN database of protected areas and their status allows for a solid basis for management on local and regional scales. This can provide a framework for identifying and improving other areas to conserve biodiversity and ecological processes. Multiple-use matrix landscapes could become a key component in primary habitat conservation following climatic changes. Sound science and the monitoring of such resources as species distributions, regional climate models, and tools and methods for efficient conservation are crucial. The integration of landscapes must coincide with coordination of regional political systems, including both subnational and international cooperation. Implementation of regulatory and incentive-based mechanisms for conservation, including the use of new or unconventional tools, is necessary for long-term conservation in the face of climate change.

In many terrestrial regions, the matrix surrounding remnant forests and other natural areas is the dominant landscape. Species persistence in fragments differs, though the results in most cases indicate a dramatic shift in distribution and abundances of many species. The planet has been and is currently intensely threatened by human activity, with climate change emerging as a daunting threat. The ways that governments and societies manage the matrix will be of key importance to the future of global biodiversity.

REFERENCES

Didham, R. K., and J. H. Lawton. 1999. Edge structure determines the magnitude of changes in microclimate and vegetation structure in tropical forest fragments. Biotropica 31:17–30.

Ganzhorn, J. U., J. Fietz, E. Rakotovao, D. Schwab, and D. Zinner. 1999. Lemurs and the regeneration of dry deciduous forest in Madagascar. Conservation Biology 13:794–804.

Gascon, C., T. E. Lovejoy, R. O. Bierregaard, J. R. Malcolm, P. C. Stouffer, H. Vasconcelos, W. F. Laurance, B. Zimmerman, M. Tocher, and S. Borges. 1999. Matrix habitat and species persistence in tropical forest remnants. Biological Conservation 91:223–229.

Gascon, C., G. B. Williamson, and G. A. B. Fonseca. 2000. Receding edges and vanishing reserves. Science 288:1356–1358.

Hannah, L., G. F. Midgley, T. Lovejoy, W. J. Bond, M. Bush, J. C. Lovett, D. Scott, and F. I. Woodward. 2001. Conservation of biodiversity in a changing climate. Conservation Biology 16:264–268.

Harvell, C. D., C. E. Mitchell, J. R. Ward, S. Altizer, A. P. Dobson, R. S. Ostfeld, and M. D. Samuel. 2002. Climate warming and disease risks for terrestrial and marine biota. Science 296:2158–2162.

Hilton-Taylor, C. 2002. Red List of Threatened Species. Gland, Switzerland: IUCN.

Houghton, J. T., Y. Ding, D. J. Griggs, M. Noguer, P. J. van der Linden, and D. Xiaosu, eds. 2001. Climate Change 2001: The Scientific Basis. Contribution of Working Group I to the Third Assessment Report of the Intergovernmental Panel on Climate Change (IPCC). Cambridge: Cambridge University Press.

Hughes, J. B., G. C. Daily, and P. R. Ehrlich. 1997. Population diversity: Its extent and extinction. Science 280:689–692.

Laurance, W. F., S. G. Laurance, and P. Delamonica. 1998. Tropical forest fragmentation and greenhouse gas emissions. *Forest Ecology and Management* 110:173–180.

Laurance, W. F., T. E. Lovejoy, H. L. Vasconcelos, E. M. Bruna, R. K. Didham, P. C. Stouffer, C. Gascon, R. O. Bierregaard, S. G. Laurance, and E. Sampaio. 2002. Ecosystem decay of Amazonian forest fragments: A 22-year investigation. *Conservation Biology* 16:605–618.

Lester, R. T., and J. P. Myers. 1991. Double jeopardy for migrating wildlife. In R. L. Wyman, ed., *Global Climate Change and Life on Earth.* New York: Chapman and Hall.

Lima, M. G. de, and C. Gascon. 1999. The conservation value of linear forest remnants in central Amazonia. *Biological Conservation* 91:241–247.

Margules, C. R., and R. L. Pressey. 2000. Systematic conservation planning. *Nature* 405:243–253.

McCarty, J. P. 2001. Ecological consequences of recent climate change. *Conservation Biology* 15:2:320–331.

Mesquita, R., P. Delamonica, and W. F. Laurance. 1999. Effects of surrounding vegetation on edge-related tree mortality in Amazonian tree fragments. *Biological Conservation* 91:129–134.

Myers, N. 1998. Lifting the veil on perverse subsidies. *Nature* 392:327–328.

Peters, R. L. 1991. Consequences of global warming for biological diversity. In R. L. Wyman, ed., *Global Climate Change and Life on Earth.* New York: Chapman and Hall.

Redford, K., and G. A. B. da Fonseca. 1986. The role of gallery forests in the zoogeography of the cerrado's non-volant mammalian fauna. *Biotropica* 18:125–135.

Sanderson, E. W., M. Jaiteh, M. A. Levy, K. H. Redford, A. V. Wannebo, and G. Woolmer. 2002. The human footprint and the last of the wild. *Bioscience* 52:891–904.

Sanderson, J., G. Fonseca, C. Galindo-Leal, K. Alger, V.H. Inchausty, K. Morrison. 2003. *Biodiversity Corridors: Consideration for Planning, Implementation, and Monitoring of Sustainable Landscapes.* Washington, D.C.: Center for Applied Biodiversity Science, Conservation International.

Steadman, D. W. 1991. Extinction of species: Past, present, and future. In R. L. Wyman, ed., *Global Climate Change and Life on Earth.* New York: Chapman Hall.

Stenseth, N. C., A. Mysterud, G. Ottersen, J. W. Hurrell, K.-S. Chan, and M. Lima. 2002. Ecological effects of climate fluctuations. *Science* 23:1292–1296.

Tewksbury, J. J., D. J. Levey, N. M. Haddad, S. Sargent, J. L. Orrock, A. Weldon, B. J. Danielson, J. Brinkerhoff, E. I. Damschen, and P. Townsend. 2002. Corridors affect plants, animals, and their interactions in fragmented landscapes. *Proceedings of the National Academy of Sciences* 99:12923–12926.

Tilman, D. 1999. Global environmental impacts of agricultural expansion: The need for sustainable and efficient practices. *Proceedings of the National Academy of Sciences* 96:5995–6000.

Tilman, D., J. Fargione, B. Wolff, C. D'Antonio, A. Dobson, R. Howarth, D. Schindler, W. H. Schlesinger, D. Simberloff, and D. Swackhamer. 2001. Forecasting agriculturally driven global environmental change. *Science* 292:281–284.

Trexler, M. C., and C. Haugen. 1995. *Keeping It Green: Tropical Forestry Opportunities for Mitigating Climate Change.* Washington, D.C.: World Resources Institute.

van der Linde, H., J. Oglethorpe, T. Sandwith, D. Snelson and Y. Tessema. 2001. Beyond boundaries: Transboundary natural resource management in Sub-Saharan Africa. Washington D.C.: Biodiversity Support Program.

Vane-Wright, R. I., C. J. Humphries, and P. H. Williams. 1991. What to protect?—Systematics and the agony of choice. *Biological Conservation* 55, 235–254.

Walther, G.-R., E. Post, P. Convey, A. Menzel, C. Parmesan, T. J. C. Beebee, J.-M. Fromentin, O. Hoegh-Guldberg, and F. Bairlein. 2002. Ecological responses to recent climate change. *Nature* 416:389–395.

Managing for Future Change on the Albemarle Sound

Sam H. Pearsall, III

The sea is rising. In northeastern North Carolina, it currently is rising at the rate of 4–5 mm per year (Permanent Service for Mean Sea Level 2002). The rate of sea level rise is accelerating as the result of global warming (IPCC 2002). Coastal ecosystems, especially estuarine beaches and mudflats, peat-lands, tupelo and cypress swamps, and salt and freshwater marshes, will likely be eliminated. A rising sea requires these ecosystems to migrate inland to escape local extirpation; but for that to happen, lands that might support new tidal ecosystems must be held in natural vegetation until those tidal systems can be established. In other words, along the coast where sea level is rising, humans must prepare the way for freshwater ecosystems to escape and transform while new brackish and saline systems become established in their places (Noss 2001).

Rising sea level results in a great many interactive and synergistic stresses in coastal ecosystems. As waters rise, shallow estuary bottoms, near shore slopes, and salt marshes well adapted to saltwater environments are overwhelmed by the increasing depth of the water and the loss of light, nutrient flows, and by the loss of oxygen in the case of organisms that require access to air. Coastal plain soils typically are unconsolidated sediments and peat that erode rapidly when they are exposed to waves and currents (Rogers and McCarty 2000).

Rising seawater can impinge on organisms not well adapted to salt (Williams et al. 1999; Rogers and McCarty 2000). Full-strength seawater is toxic to brackish water species, and brackish water is toxic to most freshwater species. Permanent inundation is not necessary for salt poisoning to occur. Storm surges transfer toxic levels of salt to new locations well inland of mean high tide. Salt spray (aerosols of seawater and airborne salt crystals) can kill vegetation several miles inland and downwind of a beach with strong wave action.

In much of northeastern North Carolina, a protective buffer of salt marshes is growing on very dense root mats; behind are soils composed mainly of peat with occasional clay lenses or surface veneers of alluvial silts and sands. The peat layer can

be many meters deep, but it is extraordinarily vulnerable to the effects of sea level rise. When seawater reaches peat soils, a group of sulfate-metabolizing anaerobic bacteria digest peat at rates far exceeding those of the normal methane-producing anaerobes active in brackish water. These bacteria release sulfides in solution which kill surface vegetation, including salt-tolerant marsh species, and thus remove protection from the surface and water's edge. Brief and episodic exposure to salt water is sufficient to activate the sulfate bacteria, but is generally not sufficient to kill the methane producers. Full immersion results in the replacement of methane producers with sulfate bacteria, and as a consequence, peat virtually melts away in salt water (Hackney and Yelverton 1990; C. Hackney, personal communication, University of North Carolina, Wilmington). In addition, channels dredged through the salt marshes to allow boat access to the sounds or in attempts to drain the land encourage the intrusion of salt water.

The vulnerability of peat soils to climate change–induced sea level rise contributes to the very rapid release of sequestered carbon into the atmosphere as CO_2 and

CH$_4$. These greenhouse gases contribute further to global warming, giving the system a pronounced positive feedback loop. In addition, where people have cleared drained peat lands for agriculture, the peat oxidizes very rapidly, releasing more carbon, and the absence of surface vegetation increases vulnerability to both wind and water erosion.

In response to this problem, The Nature Conservancy has established a project in northeastern North Carolina to develop and test strategies for accommodating sea level rise through ecosystem restoration and management. The project includes about 200,000 ha of existing and proposed conservation lands at risk from rising seas in the Mid-Atlantic Coastal Plain around Albemarle, Croatan, Currituck, Roanoke, and northern Pamlico Sounds (Fig. 21.4).

The project invests in land and water conservation to test the efficacy of various conservation strategies in response to rising seas. Implementing this approach will take many years. Land included in the program will be protected from conversion, and many converted areas will be restored. Sites within the program will be less vulnerable to sea level rise as a result.

Conservation measures vary according to land type. Five measures are used to classify all the land in the project below 3.5 m into over 100 land types. These measures are: (1) condition of the shore (native beach or salt marsh, artificial dune, nonnative and/or planted vegetation, hard armor [e.g., seawalls, rip rap, bulkheads, dikes—excluded from the study]); (2) presence or absence of adjacent oyster beds or reefs; (3) condition of native ecosystems inland of and contiguous with the shore (relatively undisturbed, disturbed, replaced); (4) presence of paved surfaces that isolate potential inland habitat from the shore (width equals or exceeds 10 m excluded); and (5) ditched and successfully drained, ditched but not successfully drained, not ditched. There

are at least two replicates for each appropriate treatment on each land type. One treatment (the control) always includes no action beyond prevention of further degradation.

Treatments comprise a variety of measures, including establishing or restoring near-shore oyster beds or reefs, establishing or restoring dunes, establishing or restoring native vegetation (submerged aquatic vegetation, salt and brackish marsh species, shrubs, and trees), establishing noninvasive, nonnative vegetation, scarifying or removing small roads, plugging ditches. Each land type is subject to appropriate combinations of these treatments.

The program will be evaluated at multiple levels, with a graduated timeline. Evaluation and validation of management strategies address questions of the efficacy of different management strategies in the short run (e.g., does plugging ditches with local borrowed materials effectively exclude salt from previously ditched lands?). Local threat abatement will take longer to assess (up to 25 years) and concerns questions about the efficacy of management strategies as expressed in local ecological responses (e.g., does excluding salt from previously ditched land result in slower peat decomposition and reduced surface vegetation mortality?). Local and regional ecological responses will unfold over decades. Evaluation for these factors will include issues such as whether slowing peat decomposition and reducing surface mortality results in successful colonization by salt-tolerant vegetation and whether the local colonization by salt-tolerant vegetation ultimately results in the inland and upland migration of coastal vegetation types. Finally, in the long term, the project's impacts on biodiversity will be evaluated across the entire project area and, by extrapolation, the Mid-Atlantic Coastal Plain Ecoregion.

At every level, the program design includes strategies and schedules for data collection, standards for data evaluation,

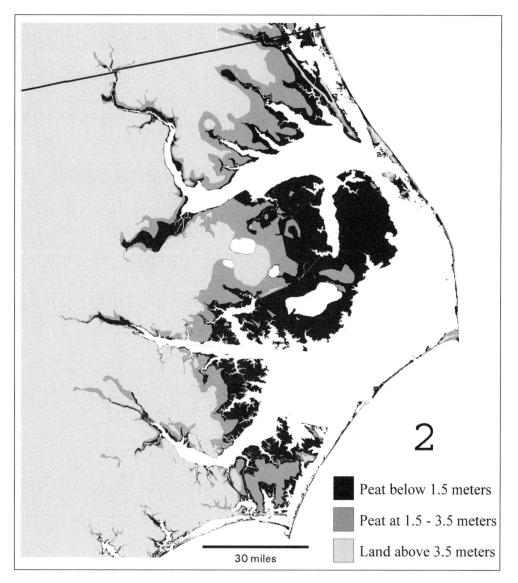

Figure 21.4. Map of peat soils of Albemarle Sound.

the duration of the decision cycle, and the conditions that must prevail for changing the management strategy. By the twenty-fifth year, based on feedback from the data for each land type, all the replicates should be subject to the same strategy. By the hundredth year there should be good indication of whether the adaptive management strategies will prevent or minimize loss of biodiversity due to rising seas. Validation at this scale should continue as long as the land is managed for biodiversity and the sea continues to rise.

REFERENCES

Hackney, C. T., and G. F. Yelverton. 1990. Effects of human activities and sea level rise on wetland ecosystems in the Cape Fear River estuary, North Carolina, USA, pp. 55–61. In D. F. Whigham et al., eds., *Wetland Ecology and Management: Case Studies.* Netherlands: Kluwer Academic.

IPCC (Intergovernmental Panel on Climate Change). 2002. *Climate Change 2001: Synthesis Report,*

Watson, R. T., et al., eds. Cambridge, U.K.: Cambridge University Press.

Moorhead, K. K., and M. M. Brinson. 1995. Response of wetlands to rising sea level in the lower coastal plain of North Carolina. *Ecological Applications* 5:261–271.

Noss, R. 2001. Beyond Kyoto: Forest management in a time of rapid climate change. *Conservation Biology* 15:578–590.

Permanent Service for Mean Sea Level. 2002. *http://www.pol.ac.uk/psmsl/*

Rogers, C., and J. P. McCarty. 2000. Climate change and ecosystems of the mid-Atlantic region. *Climate Research* 14:235–244.

Williams, K., K. C. Ewel, R. P. Stumpf, F. E. Putz, and T. W. Workman. 1999. Sea-level rise and coastal forest retreat on the west coast of Florida, USA. *Ecology* 80:2045–2059.

CHAPTER TWENTY-TWO

Protected Areas Management in a Changing Climate

LEE HANNAH AND ROD SALM

The role of protected areas will change and expand as climate changes (Peters and Darling 1985; Peters and Lovejoy 1992). New areas will be required to protect species in both their present and future ranges (Lovejoy, this volume). The management of both existing and future protected areas will need to be adapted to account for climate change–related impacts (Peters and Darling 1985; Halpin 1997). And, as protected areas assume a role in a broader plan for conservation across a landscape, the goals and methods of their management must change (Fig. 22.1).

Protected areas are the most important and most effective component of current conservation strategies (Bruner et al. 2001). There is strong reason to believe that they will continue to be central in conservation strategies designed for climate change (Hannah et al. 2002b). Area under protection is expanding, while remaining undisturbed habitat is declining (Sanderson et al. 2002), so that by the time climate change impacts are pronounced, protected areas may represent most of the remaining natural areas of the planet. Protected areas provide the least disturbed natural habitat, and therefore the best hope for natural response (e.g., range shifts) to changing climate. Consequently, protected areas will play a dominant role in efforts to conserve biodiversity in the future, as they do today.

Management in a landscape designed for conservation is considerably different from management of isolated units, however. As landscape dynamics become an important design consideration, management must serve to integrate protected and unprotected (matrix) elements as well as

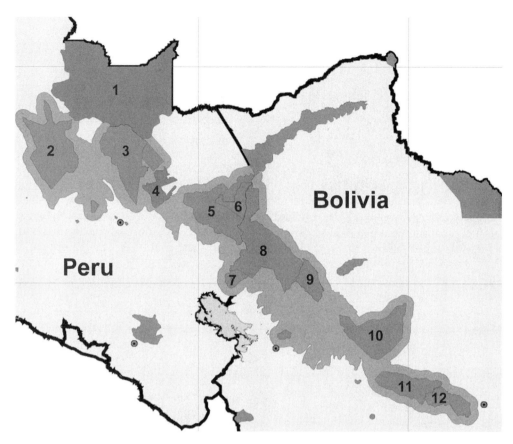

Figure 22.1. Protected areas in a landscape connectivity plan for a portion of the tropical Andes in Peru and Bolivia. In this area, a large number of protected areas allow a connectivity plan to be designed with protected areas as its dominant land use. Coordination of management among these areas will be one of the primary means of responding to landscape-level alterations in biodiversity brought about by climate change. Key to protected areas: 1-Alto Purus Reserved Zone; 2-Vilcabamba Reserved Zone; 3-Manu Reserved Zone; 4-Amarakaeri Reserved Zone; 5-Tambopata-Candamo Reserved Zone; 6-Bahuaja-Sonene National Park; 7-Ulla Ulla National Fauna Reserve; 8-Madidi National Park and Integrated Management Area; 9-Pilon Lajas Biosphere Reserve; Isiboro Secure National Park; 11-Carrasco National Park; 12-Amboro National Park.

coordinate at multiple spatial and temporal scales (Noss and Harris 1986). For example, a corridor to allow a species range to expand from one reserve to another can be successful only if one reserve manages the species to allow dispersal and the other reserve manages to allow habitat changes favorable to the species.

New conservation elements have been described in the preceding two chapters that allow changes in biodiversity in response to climate alterations. These elements cannot be fully effective unless they are integrated, both spatially and functionally, into a system of protected natural areas, carefully designed, and effectively managed. This chapter outlines the essential changes in protected areas management necessary to facilitate dynamic landscape conservation, using primarily examples from the marine realm.

PLANNING TOOLS

Protected areas have been classified according to the degree of protection by the

World Conservation Union (IUCN). In this chapter, the term "protected area" refers to Category I and II areas, the equivalent of national parks or nature reserves. However, the other protected areas categories can be strong regulatory planning tools, as described in Chapter 21. For Category I and II areas, planning takes place within the general framework described by the IUCN.

For each individual protected area, a management plan is the fundamental tool for planning (Groves et al. 2002). Traditionally, management plans have been prepared on a 3–5-year cycle, with a planning horizon seldom exceeding 10 years. For climate change, expanded time horizons of 50 and 100 years are appropriate (Root and Schneider 1993). GCM scenarios are commonly available for these horizons. Planning should examine contingencies suggested by general circulation model (GCM) or regional scenarios, balance uncertainties, and apply the precautionary principle where feasible.

Balancing uncertainties implies weighting of action toward present patterns of species' occurrence, since these are known with relative certainty. Medium-term, 30–50-year climate changes can be taken into account and preliminary plans and investments made to accommodate responses such as range shifts (Hannah et al. 2002b). Longer-term (80–100-year) climate projections carry greater uncertainty but should be scanned for extreme consequences to plan for small, preventative investments. In cases where actions to reduce extreme consequences can be taken with little cost to present objectives, small investments may be warranted even where uncertainty is substantial (the precautionary principle) (IPCC 2001). Where investment requirements are substantial, the effect (e.g., range shift) involved can be targeted for monitoring, so that investment in responses can be undertaken promptly if the certainty of occurrence increases.

SPATIALLY COORDINATED MANAGEMENT

Spatially coordinated management is needed to capture biological dynamics among multiple protected areas and across the landscape. Many nations maintain national or regional planning frameworks to ensure consistency of methods and quality control of protected areas planning, but none of these currently considers future biological dynamics in harmonizing plans at multiple sites (Scott et al. 2002). Canada is perhaps the most advanced in this respect, but even there the coordination process has not progressed much beyond examination of possible change scenarios (Scott, this volume).

A regional protected areas plan for climate change must set spatially coordinated management objectives for protected areas in the region as well as for the matrix (Fig. 22.2). This implies adjusting management practices in one protected area to help meet objectives in another protected area or across the region as a whole. For example, a management objective of allowing a transition from forest to native grassland in response to climate warming at an individual site is meaningless without a source of native grass dispersal nearby.

Regional coordination must bridge national boundaries as well, so that inclusion of biological dynamics in national plans is only a step in complete regional coordination. A more complete response would encompass an entire ecoregion, involving all relevant authorities at multiple levels— site, state or provincial, national, international—and across agency jurisdictions (including protected areas managers as well as managers responsible for managing and regulating land use in the matrix).

REDUCING NONCLIMATE STRESSORS

Climate change introduces a new urgency to reducing nonclimate stressors in pro-

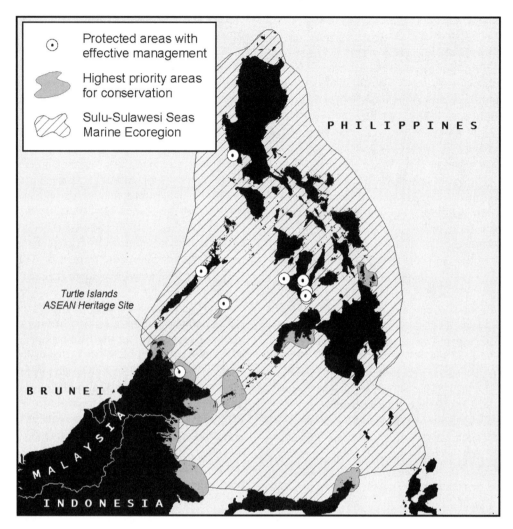

Figure 22.2. An ecoregional plan for marine protected areas. In this example from the Philippines, protected areas and high-priority areas for conservation in the matrix have been identified, often with long-distance connectivity as a design factor. *Source:* Courtesy of Lara Hansen, WWF-US.

tected areas. One of the best safeguards against climate change impacts is to reduce stress on biodiversity from nonclimate sources. This will increase resilience to possible climatic impacts and simplify management of dynamics arising from climate change. For example, coral reefs stressed by pollution or sediment may be more susceptible to coral bleaching due to climate change (Done 2001). Manage-

ment to reduce pollution and sediment loads can therefore delay the effects of climate change and lessen those effects when they do arrive.

Reduction of nonclimate stresses is the current major focus of most protected areas management, so further reduction may not be possible without new resources, new information, or new stakeholder acceptance. Information about climate change can help protected areas managers attain these new means, however. Budgetary justifications can draw on the added and imminent threat of climate change, information about possible synergies with climate change can suggest priorities for management of nonclimate stressors, and the new optic of climate change and biological dynamics can help influence stakeholder opinion in favor of new action. As evidence of climate change and its impacts grows, these arguments will carry increasing credibility. It is part of the job of protected areas managers to make the case credibly before impacts arrive, when it is essentially too late.

PLANNING PROTECTED AREAS FOR CLIMATE CHANGE

A landscape design for climate change is the critical prerequisite to site planning. Essential elements of the protected area role in a landscape plan include focusing on conservation targets, planning for dynamics in concert with the surrounding matrix and other protected areas, and assigning staff time and resources for coordination outside of the reserve. These plan elements initially will be modest but may be expected to grow in scope and resource requirements as climate change progresses.

The ecoregional plan will identify species and other targets for inclusion in site planning (Groves et al. 2002). The role of the protected area plan is to outline in detail the occurrences of the species in the protected area, how they will be managed, and when coordination of management with other protected areas or the matrix may be required. For example, an ecoregional plan may identify a wetland bird as a conservation target and a certain protected area as containing important habitat for the species. The protected area plan would identify which specific wetlands within the reserve contain the target species, the pressures on those wetlands, and how they should be managed. If the species forages outside the reserve during key dry spells, the protected area plan would specify actions to be taken with conservation managers in the matrix to protect those external resources, both for current use and for potential increased use as climate changes.

Consideration of dynamics both inside and outside the protected area is essential to planning for climate change. For example, in marine protected areas containing coral reefs, planning steps to reduce coral bleaching have been identified (Salm et al. 2001). Areas known to be prone to bleaching can be protected from heavy use. Efforts to reduce other stressors, such as pollution and sedimentation, can be targeted on sensitive areas. Important source areas for recolonization of damaged reefs can be identified and prioritized for management.

The connectivity of marine systems provides an example of the importance of planning for dynamics that extends outside the reserve boundary. A coral reef on the periphery of a reserve might be an important source of recolonization for neighboring reefs, even if it is not an important source for the reserve itself (Glynn 2001). Similarly, working with local fishermen to reduce impacts on neighboring reefs might play a critical role in allowing recolonization of damaged areas within the reserve.

Zoning may also be influenced by climate change considerations. Tourism or other uses might be restricted in sensitive areas. For instance, zoning to protect areas resistant to coral bleaching has been proposed for marine protected areas, since resistant areas are crucial to recovery and reestablishment of damaged areas following a bleaching event (Salm et al. 2001).

Increased frequency of extreme events may be an important consideration in zoning plans. Coral reefs damaged by tropical storms are a source of diversity, but they are fragile during recovery. Human use may prevent normal recovery by increasing local extinction rates or by reducing the establishment of immigrant species. Coral bleaching is expected to exacerbate these effects, making zoning to reduce or prevent recreational use in recovering areas an important option for protecting reef diversity.

Planning that anticipates increased staff and resource burden can help provide for mounting management needs. Sensitive systems, such as coral reefs, will require more resources early on. Eventually, all protected areas will require additional management resources, so it is important that system plans, policy makers, and national governments recognize this need in advance.

Working outside of reserve boundaries and managing species more intensively will require new staff skills. Park staff accustomed to focusing within the reserve and letting natural processes dominate management strategies will find an increasing need to work outside the reserve and actively manage species and processes. This may require new skills, both in dealing with matrix land users outside the protected area and in more intrusive and technical management of species and processes. Sound plans will anticipate these needs and will identify training needs and requirements for new staff.

Where ecoregional or landscape plans do not yet exist, protected areas managers can initiate this process by focusing on species endemic to their region that may be sensitive to climate change. Gradually, as more and more reserves plan for climate change and coordination outside of their own boundaries, a regional landscape plan will begin to emerge.

MANAGING PROTECTED AREAS FOR CLIMATE CHANGE

Managing for climate change is like learning to hit a moving target—it means that more nimble and innovative responses are required (Hannah et al. 2002a). While management of protected areas is more straightforward (limited area, primarily regulatory) than management within the matrix (large areas, regulatory- and incentive-based trade-offs), significant adjustments in management will be needed. More aggressive and interventionist management is likely to be required at most sites, balanced with a complementary ability to back off or reverse interventions if conditions change. Climate change will dictate greater emphasis on management outside the protected areas boundary and, therefore, greater use of incentive systems relative to past protected areas practice.

Climate change will require pro-active management responses. For example, area closures can be used to protect and promote regeneration in areas where coral bleaching has occurred (Salm et al. 2001). Managers may close an area to tourism when bleaching is observed or tie management to known causes of bleaching—permitting dive tourism in La Niña years but restricting use in El Niño years.

Management of populations is likely to require higher levels of intervention and greater innovation. For example, in the Monteverde cloud forest, one of the first sites in which population-level climate change effects have been recorded (Pounds et al. 1999; Pounds et al., this volume), only unusual management options seem likely to maintain target species populations. It is likely that maintaining amphibian populations at Monteverde would require introduction of artificial moisture, perhaps in the form of "drip irrigation" for habitats. Upper-elevation birds, such as the quetzal, are suffering increasing nest predation from toucans that have moved upslope in response to changed climatic conditions (Pounds et al., this volume)—maintaining their populations might require artificial nest enclosures conceptually similar to those used to protect snowy plovers from ground predators in western North America.

Management responses that are incremental and reversible will be useful, since it may be difficult to distinguish normal climate variability from long-term trends.

In the short-term window of observation available to protected areas managers, many long-term trends may be initially indistinguishable from climate variability. A directional trend in temperature, for instance, may initially be manifested as a series of unusually warm summers. Management thresholds may be exceeded by these variations (for instance, the population of a target species requiring cool summers may decline), triggering management action. This is appropriate, as inaction may result in the need for more and more expensive management if the variation is indeed a trend. However, if the changes prove to be part of natural climate variability (i.e., values return to near-historic levels in the long term), it will be desirable to reverse the management action. Actions that are staged in increments can be implemented earlier and at less cost, and those with greater degrees of reversibility will minimize unnecessary cost and management input. It is essential to begin some management before a trend is confirmed, since much experience to date indicates that population declines may be difficult (coral bleaching) or impossible (Monteverde amphibian declines) to reverse once the trend is obvious.

MONITORING AND ADAPTIVE MANAGEMENT

Monitoring linked to management adaptation is perhaps the most important part of planning and managing for climate change (Hannah et al. 2002b). Few of the myriad effects of climate change on biodiversity in a park will be known in advance. The vast majority of change will take place unheralded, and will be detected only if carefully planned and effective monitoring systems are in place.

Emphasis on conservation targets is essential to a cost-effective system. Since many or most species may ultimately be affected by climate change, it is impossible

to design a monitoring system that can detect all relevant change. Instead, a system must target those species, processes, and genetic resources that are the focus of management action. Hence, design of a sound monitoring scheme requires an ecoregional or landscape plan.

Since climate change can threaten species and change their conservation status, it is important to monitor some non-target species as well. Species believed to be sensitive to climate change, such as mountaintop species or those with limited ranges, are good candidates for monitoring (Hannah et al. 2002b). In addition, a sample of all other species may be targeted for monitoring to help in the detection of unanticipated effects.

Once targets are clearly set, a monitoring program can be laid out to detect changes that are keyed to management response thresholds. For example, it is likely that a landscape management plan for the Monteverde region would target the endemic amphibians within Monteverde itself. Rapid decline in these populations would be a trigger for management action, so that monitoring results showing decline, such as have already been observed, would trigger management response.

The monitored variables should relate to management targets but do not necessarily need to involve them. For example, areas known to have suffered coral bleaching in the past could be used to detect the onset of bleaching events, triggering a suite of more intensive management actions such as zoning restrictions and additional monitoring of target reefs. By limiting intensive monitoring to the time of expected impact, the resources needed for monitoring could be greatly reduced by such a system.

Survey work will take on increasing importance in protected areas management (Hannah et al. 2002b). Surveys to increase understanding of current species distributions within a protected area will be critical to detecting species range shifts. As environmental factors that confer resistance or sensitivity to climate change are identified, survey work can help document the extent of these factors within a protected area. For example, surveys of marine protected areas can help identify environmental factors that contribute to bleaching resistance, such as cold-water upwelling, physical shading from land features, and rapid currents. Once identified, sensitive areas can be used to guide improved reserve zoning. Survey work can also help track the progress of responses to climate change, such as bleaching events, and will be critical to this and other monitoring efforts.

SUMMARY

Managing for climate change is like learning to hit a moving target. Managers must set their actions within the context of a regional plan, so that movement in one reserve is coordinated with similar or contradictory species dynamics in other reserves and the matrix. The IUCN classification of protected areas provides the basic definition of what is primarily a regulatory management framework. Nonetheless, climate change will require reserve managers to work increasingly in coordination with the surrounding matrix of land uses, and to draw on incentive-based efforts in doing so.

Management planning requires the consideration of the regional context and expanded time horizons. Most protected areas plans currently have time horizons of a decade or less. Adding 30–50-year and 80–100-year horizons to correspond to commonly available GCM projections can help protected areas managers integrate climate change into their planning processes.

Management strategies are likely to have to be more innovative and more interventionist. Incremental management responses allow response to changes in cli-

mate before it is clear whether they signal long-term change or are simply part of normal variability. Readily reversible management actions will minimize the costs associated with "false positives"—apparent trends that trigger management action but later turn out to be only part of normal climate variability.

Additional staff, resources, and training will be required to respond to climate change. Initially, national or regional systems can redirect resources to parks most vulnerable to the initial impacts of climate change. In the long term, however, all parks will be affected, and overall increases in budget, personnel, and training will be required.

Monitoring in synchrony with management will be critical to detecting unanticipated responses. This does not imply a need for large, unfocused monitoring programs, however. Strategic focus on target species and processes can ensure that monitoring is cost-effective and relevant to management action.

REFERENCES

Bruner, A. G., R. E. Gullison, R. E. Rice, and G. A. da Fonseca. 2001. Effectiveness of parks in protecting tropical biodiversity. *Science* 291:125–128.

Done, T. 2001. Scientific principles for establishing MPAs to alleviate coral bleaching and promote recovery. In R. V. Salm and S. L. Coles, eds., *Coral Bleaching and Marine Protected Areas*, pp. 53–59. Honolulu: The Nature Conservancy.

Glynn, P. W. 2001. History of significant coral bleaching events and insights regarding amelioration. In R. V. Salm and S. L. Coles, eds., *Coral Bleaching and Marine Protected Areas*, pp. 36–39. Honolulu: The Nature Conservancy.

Groves, C. R., D. B. Jensen, L. L. Valutis, K. H. Redford, M. L. Shaffer, J. M. Scott, J. V. Baumgartner, J. V. Higgins, M. W. Beck, and M. G. Anderson. 2002. Planning for biodiversity conservation: Putting conservation science into practice. *BioScience* 52:499–512.

Halpin, P. N. 1997. Global climate change and natural-area protection: Management responses and research directions. *Ecological Applications* 7:828–843.

Hannah, L., G. F. Midgley, T. Lovejoy, W. J. Bond,

M.L.J.C. Bush, D. Scott, and F. I. Woodward. 2002a. Conservation of biodiversity in a changing climate. *Conservation Biology* 16:11–15.

Hannah, L., G. F. Midgley, and D. Millar. 2002b. Climate change-integrated conservation strategies. *Global Ecology & Biogeography* 11:485–495.

IPCC. 2001. Climate Change 2001: Impacts, Vulnerability and Adaptation; contribution of Working Group II to the Third Assessment Report of the Intergovernmental Panel on Climate Change. IPCC Working Group 2. Port Chester, N.Y.: Cambridge University Press. 1919.

Noss, R. F., and L. D. Harris. 1986. Nodes, networks, and MUMs: Preserving diversity at all scales. *Environmental Management* 10:299–309.

Peters, R. L., and J. D. Darling. 1985. The greenhouse effect and nature reserves. *BioScience* 35:707.

Peters, R. L., and T. E. Lovejoy. 1992. *Global Warming and Biological Diversity.* London: Yale University Press.

Pounds, J. A., M.P.L. Fogden, and J. H. Campbell. 1999. Biological response to climate change on a tropical mountain. *Nature* 398:611–615.

Root, T. L., and S. H. Schneider. 1993. Can large-scale climatic models be linked with multiscale ecological studies? *Conservation Biology* 7:256–270.

Salm, R. V., S. L. Coles, J. M. West, G. Llewellyn, T. Done, B. D. Causey, P. W. Glynn, W. Heyman, P. Jokiel, D. Obura, and J. Oliver. 2001. *Coral Bleaching and Marine Protected Areas.* Honolulu: The Nature Conservancy.

Sanderson, E., M. Jaiteh, M. A. Levy, K. H. Redford, A. Wannebo, and G. Woolmer. 2002. The human footprint and the last of the wild. *Bioscience* 52:891–904.

Scott, D., J. R. Malcolm, and C. Lemieux. 2002. Climate change and modeled biome representation in Canada's national park system: Implications for system planning and park mandates. *Global Ecology & Biogeography* 11:475–484.

PART VI

Policy Responses

The ultimate challenge for biodiversity conservationists may lie in the atmosphere. Biologists, like most people, have taken the atmosphere for granted, ignoring the large amounts of human-generated gases it has been forced to absorb. It is now virtually certain that these almost unnoticed acts of pollution will now come back to damage biodiversity. As truly global climate change manifests itself, no area will be free from its effects. Within decades, climatic conditions may exist that are without past analogue. This will take management of biodiversity into uncharted waters and make it impossible to say exactly what constitutes a "natural" or "baseline" ecosystem condition. Avoiding these problems is apparently now impossible, but its extent can be limited, and this has substantial implications for biodiversity.

The previous chapters of this book have made a clear case for serious biodiversity impacts in as few as 50 years and with about a doubling of pre-industrial CO_2 levels. A conservative risk management stance would dictate that biologists advocate stabilizing at these levels and avoiding more pronounced effects. However, this is a mammoth task, as will be described in the pages that follow.

The two chapters in this part address, in turn, the social policy issues surrounding greenhouse gas stabilization and the policy advocacy choices facing biologists in this debate. Stabilization of greenhouse gases is not even on the table in current international negotiations, which focus on mechanisms of cooperation, verification and emissions reduction. This places biologists in the position of advocating a revolutionary position in a debate in which they have been largely silent. Can biologists influence such a contentious high-stakes debate? The long-term future of biodiversity may hinge on the answer to this question.

CHAPTER TWENTY-THREE

Emissions Reductions and Alternative Futures

ROBERT T. WATSON

Human-induced climate change is one of the most important environmental issues facing society world-wide. The overwhelming majority of scientific experts and governments recognize that while scientific uncertainties exist, there is strong scientific evidence demonstrating that human activities are changing the Earth's climate and that further human-induced climate change is inevitable. Changes in the Earth's climate have and will continue to adversely effect ecological systems and their biodiversity. Limiting these ecological changes will require stabilization of global greenhouse gas concentrations. Previous authors in this book have reviewed negative impacts on biodiversity with scenarios based on about a doubling of pre-industrial CO_2 levels. Limiting greenhouse gas–induced warming to these levels would require major changes in energy supply and social structure. It is therefore important for those concerned with biodiversity conservation to be aware of the drivers of climate change, the options available for curtailing those drivers, and the social choices implied in stabilizing greenhouse gases.

DRIVERS OF CLIMATE CHANGE

The main indirect drivers of climate change are demographic, economic, socio-political, and the rate and direction of technological change (IPCC 2000). Individual behavioral choices are also important. These driving forces determine the future demand for energy and changes in land use, which in turn affect emissions of greenhouse gases and aerosol precursors, which in turn change the radiative balance of the atmosphere resulting in changes in the Earth's climate.

The IPCC *Special Report on Emissions Scenarios* explored the implications of four equally plausible, but very different, worlds on greenhouse gas emissions, each assuming no globally coordinated effort to address human-induced climate change (IPCC 2000). The scenarios ranged from a world of very rapid economic growth, a global population that peaks in 2050 and declines thereafter, the rapid introduction of new and more efficient technologies and where the emphasis is on global "sustainable and equitable solutions," to a world that is very heterogeneous, where the underlying theme is self-reliance and local identity, where economic growth is regionally oriented and per capita economic growth and technological change are more fragmented and relatively slow, and where fertility patterns across regions converge slowly, resulting in a continuously increasing population.

• The demographic variables that have implications for greenhouse gas emissions include population size and rate of change over time (births and deaths), age and gender structure of the population, household distribution by size and composition, and spatial distribution (urban versus rural). The IPCC (2000) used three different population trajectories. The lowest trajectory assumed a population of 8.7 billion people in 2050, decreasing to 7 billion people by 2100. The middle trajectory assumed a population of 10.4 billion people by 2100, and the highest trajectory assumed a population of 15 billion people by 2100.

• The economic variables that have implications for greenhouse emissions include global domestic product, purchasing power parity, and per capita income ratio between developed countries and economies in transition to developing countries. All scenarios (IPCC 2000) assumed a more affluent world than today, with the gross world product rising to 10 times today's value of US $33 trillion by 2100 in the lowest trajectory to 26 times today's value by 2100 in the highest trajectory. A narrowing of income differences among world regions is assumed in many of the scenarios. The per capita income ratio between developed countries and economies in transition to developing countries, which averaged 16.1 in 1990, was assumed to decrease to between 1.5 and 4.2 in 2100.

• The rate and direction of technological change is at least as important a driving force as demographic change and economic development (IPCC 2000). Although technological change is related to economic development, very divergent paths for development of energy systems and land-use patterns can develop. The IPCC explored the implications of highly divergent paths ranging from continued dependence on fossil fuels, especially coal, to a major transition to renewable energy technologies and significant improvements in energy efficiency.

IMPLICATIONS OF THE IPCC SCENARIOS

These IPCC scenarios resulted in a broad range of projected greenhouse gas and aerosol precursor emissions and concentrations. For example, the atmospheric concentration of carbon dioxide, the major anthropogenic greenhouse gas, was projected to increase from the current level of about 370 ppm to between 540 and 970 ppm by 2100, without taking into account possible climate-induced additional releases from the biosphere in a warmer world (IPCC 2001b). These changes in the atmospheric concentrations of greenhouse gases and aerosols are projected to result in increases in global mean surface temperatures between 1990 and 2100 of 1.4 to 5.8°C, with land areas warming more than the oceans; globally averaged precipitation is projected to increase, but with increases and decreases in particular regions, accompanied by more intense precipitation events over most regions of the world; and global sea level is projected to rise by about 4 to 35 inches between 1990 and 2100. The incidence of extreme weather events is projected to increase—for example, hot days, floods, and droughts (IPCC 2001b).

Realizing the lowest projected changes in greenhouse gas emissions, which still resulted in a projected increase in global mean surface temperature of 1.4°C, could be accomplished without concerted global action to reduce greenhouse gas emissions, but only if the global population peaks in 2050 and declines thereafter, if that economic growth is accompanied by the rapid introduction of less carbon-intense and more efficient technologies, and if there is an emphasis on global "sustainable and equitable solutions" (IPCC 2001b; IPCC 2000). This world will not materialize with a "business-as-usual" attitude; it will require governments and the private sector world-wide to form a common vision of an equitable and sustainable world, new and innovative public–private partnerships, the development of less carbon-intensive technologies, and an appropriate policy environment. While the four IPCC scenarios were "nonclimate intervention" scenarios, the lowest scenarios contained many of the features required to limit human-induced climate change, that is, a significant transition to non-fossil-fuel technologies and energy-efficient technologies (IPCC 2000).

STABILIZATION SCENARIOS

While the near-term challenge for most industrialized countries is to achieve their Kyoto Protocol targets, the longer-term challenge is to meet the objectives of Article 2 of the United Nations Framework Convention on Climate Change, that is, stabilization of greenhouse gas concentrations in the atmosphere at a level that would prevent dangerous anthropogenic interference with the climate system, with specific attention being paid to food security, ecological systems, and sustainable economic development. Article 2 specifically states that such a level should be achieved within a timeframe sufficient to allow ecosystems to adapt naturally. To stabilize the atmospheric concentrations of carbon dioxide will require that emissions eventually will have to be reduced to only a small fraction of current emissions, that is, 5–10 percent of current emissions. To stabilize carbon dioxide at 550 ppm, emissions globally will have to peak between 2020 and 2030 and then be reduced below current emissions between 2030 and 2100 (Table 23.1) (IPCC 2001b).

The range of projected global mean annual temperature changes shown in Table 23.1 for each stabilization level is due to the different climate sensitivity factors of the models (the climate sensitivity factor is the projected change in temperature at equilibrium when the atmospheric concentration of carbon dioxide is doubled—it ranged from 1.7–4.2°C in Table 23.1). Consequently, the projected changes in temperature are very sensitive to the assumed value of the climate sensitivity factor, for example, the projected change in temperature for a stabilization level of 450 ppm and a high temperature sensitivity factor is comparable to stabilization at 1000 ppm with a low climate sensitivity factor. If we assume that the central value of the climate sensitivity factor is most likely, then the range of projected changes in temperature in 2100 and at equilibrium is 1.8–2.8°C and 2.7–5.8°C, respectively. These projected changes in mean temperature have taken only changes in carbon dioxide into account. If changes in the other greenhouse gases are taken into account, it would be approximately equivalent to assuming an additional 100 ppm of carbon dioxide. Even the lowest stabilization levels of carbon dioxide are projected to lead to significant changes in the magnitude and rate of change of temperature, thus threatening biodiversity.

The lowest IPCC SRES scenario (IPCC 2000) would allow stabilization of carbon dioxide at about 550 ppm, but none of the SRES scenarios would allow stabilization of carbon dioxide at 450 ppm. Thus stabilization below 550 ppm would require a radical change in the way energy is produced and consumed. One possibility is to increase efficient use of energy combined with producing a significant amount of the energy from modern biofuels whereby the carbon dioxide is captured and stored in the deep ocean or in depleted oil or gas wells, thus producing energy with "negative" emissions of carbon dioxide. A significant amount of effort is currently being expended on finding cost-effective ways of capturing and storing carbon dioxide from the effluents of fossil fuel power plants.

THE TECHNOLOGICAL CHALLENGE

Greenhouse gas emissions are highly dependent upon the development pathway, and approaches to mitigate climate change will be both affected by and have impacts on broader socio-economic policies and trends—those relating to development, sustainability, and equity (IPCC 2001d). Stabilization will require emissions reductions in all regions; that is, Annex I countries cannot achieve stabilization alone because of the large projected increases in emissions in developing countries (IPCC 2000). Lower emissions will require different patterns of energy resource development and utilization (trend toward decarbonization) and increases in end-use efficiency (IPCC 2001d).

The IPCC concluded that significant reductions in net greenhouse gas emissions are technically feasible due to an extensive array of technologies in the energy supply, energy demand, and agricultural and forestry sectors, many at little or no cost to society (IPCC 2001d). Indeed, the good news is that significant technical progress has been made in the past five years and at a faster rate than expected (wind turbines, elimination of industrial by-products, hybrid engine cars, fuel cell technology, underground carbon dioxide storage) (IPCC 2001d). However, realizing these emissions reductions involves the development and implementation of supporting policies to overcome barriers to the diffusion of these technologies into the marketplace, increased public and private sector funding for research and development, and effective technology transfer (N-S and S-S) (IPCC 2001d).

Reductions in greenhouse gas emissions from the energy sector can be accomplished through fuel switching (coal to oil to gas), increased power plant efficiency (30 percent to approximately 60 percent), increased use of renewable energy technologies (e.g., biomass, solar, wind, hydro) and nuclear power, and carbon dioxide capture and storage from power plants (IPCC 2001d). On the demand side, there are many opportunities in transportation, industry, and commercial and residential buildings to improve efficiency (IPCC 2001d). In addition, afforestation, reforestation, improved forest, cropland, and rangeland management, and agroforestry provide a wide range of opportunities to increase carbon uptake, and slowing deforestation an opportunity to reduce emissions (IPCC 2000; IPCC 2001d).

Realizing the potential of these technologies to reduce greenhouse gas emis-

Table 23.1.

Stabilization Level (ppm)	Date for Global Emissions to Peak	Date for Global Emissions to Fall Below Current Levels	Temperature Change by 2100 (K)	Equilibrium Temperature Change
450	2005–2015	before 2040	1.2–2.3 (1.8)	1.5–3.9 (2.7)
550	2020–2030	2030–2100	1.7–2.3 (2.3)	2.0–5.1 (3.4)
650	2030–2045	2055–2145	1.8–3.2 (2.5)	2.4–6.1 (4.1)
750	2050–2060	2080–2180	1.9–3.4 (2.7)	2.8–7.0 (4.6)
1000	2065–2090	2135–2270	2.0–3.5 (2.8)	3.5–8.7 (5.8)

sions will require supporting policies and programs including: energy pricing strategies and taxes, removal of subsidies that increase GHG emissions, domestic and international tradable emissions permits, voluntary programs, regulatory programs including energy-efficiency standards, and incentives for use of new technologies during market buildup; and education and training such as product advisories and labels (IPCC 2001d).

Key questions are: What is the rate at which a transition to a less carbon-intensive energy sector can be accomplished? and How does this compare to what has been accomplished in the past? Figure 23.1 shows the projected rates of change in energy intensity (energy per unit of gross domestic product (GDP)) and carbon intensity (carbon emissions per unit of energy) that would be required for the different SRES scenarios (IPCC 2001d). The figure shows that the historical rates of changes in energy intensity (1–1.5 percent per year) are consistent with those needed for stabilization of carbon dioxide concentrations at 650 and 750 ppm, and in some cases for 450 and 550 ppm, but the historical rates of changes in carbon intensity (less than 0.5 percent per year) are far slower than most of those needed for any stabilization level of carbon dioxide concentrations between 450–750 ppm. Thus business-as-usual changes in technology will not achieve the desired goals of a less carbon-intensive energy system. Changes in energy intensity can arise

from technological changes as well as through structural changes in the economy—for example, a move from heavy industry to a service economy—whereas changes in carbon intensity will require decarbonizing the energy sector at a rate much faster than any historical changes.

ECONOMIC IMPLICATIONS OF STABILIZATION

There is a wide range of estimates of the costs of mitigating climate change (IPCC 2001d). The IPCC estimated that half the projected increase in global emissions between now and 2020 could be reduced with direct benefits (negative costs), the other half at less than $100 per ton of carbon (IPCC 2001d). Reductions in emissions can be obtained at no or negative costs by exploiting no-regrets opportunities—that is, by reducing market or institutional imperfections (e.g., subsidies), by taking into account ancillary benefits (e.g., local and regional air quality improvements), and by using revenues from taxes or auctioned permits to reduce existing distortionary taxes through revenue recycling (IPCC 2001d).

In the absence of international carbon trading, the costs of complying with the Kyoto Protocol for industrialized countries range from 0.2–2 percent, whereas with full trading among industrialized countries the costs are halved to 0.1–1 percent. The equivalent marginal costs range from

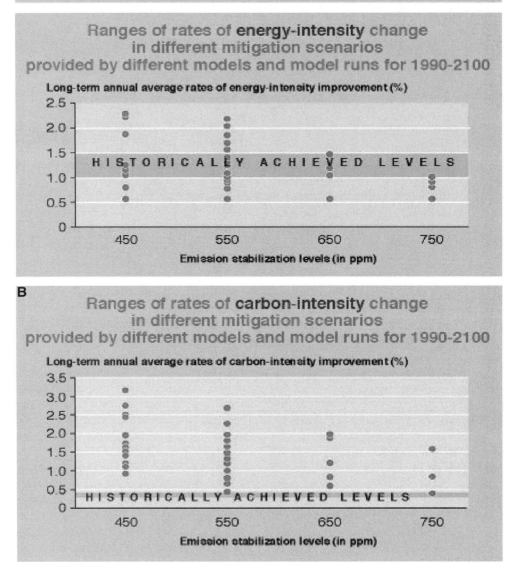

Figure 23.1. (a):The required rate of decrease in energy intensity (energy per unit GDP) in order to meet given carbon dioxide concentration stabilization targets between 450 and 750 ppm compared to historically achieved rates, and (b) the required rate of in carbon intensity (carbon emissions per unit of energy) to stabilize at levels between 450 and 750 ppm compared to historically achieved rates. Source: IPCC (2001d), reprinted with permission.

$76–322 in the United States without trading and $14–135 with trading (IPCC 2001d). These costs could be further reduced with use of sinks (carbon sequestration using reforestation, afforestation, decreased deforestation, and improved forest, cropland, and grassland management), project-based trading between industrialized countries and developing countries through the Clean Development

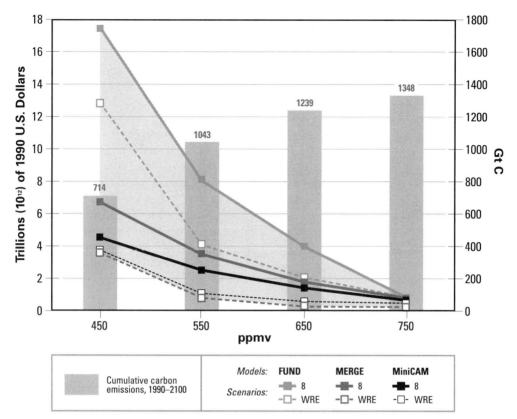

Figure 23.2. The mitigation costs (1990 U.S. dollars, present value discounted at 5% per year for the period 1990–2100) of stabilizing CO_2 concentrations at 450–750 ppm are calculated using three global models, based on different model-dependent baselines. *Source:* IPCC 2001d.

Mechanism (CDM),[1] and reducing the emissions of other greenhouse gases, for example, methane.

Known technological options could achieve stabilization of carbon dioxide at levels of 450–550 ppm over the next 100 years. The costs of stabilization are estimated to increase moderately from 750 ppm to 550 ppm, but significantly going from 550 ppm to 450 ppm (Fig. 23.2) (IPCC 2001d). However, it should be recognized that the pathway to stabilization as well as the stabilization level itself are key determinants of mitigation costs.

The costs of stabilization are relatively small—for example, stabilization at 550

ppm is estimated to cost between US $0.4 and 8 trillion over the next century—and will have only a minor impact on the rate of economic growth. The reduction in projected gross domestic product (GDP) averaged across all IPCC story lines and stabilization levels is lowest in 2020 (1 percent), reaches a maximum in 2050 (1.5 percent) and declines by 2100 to 1.3 percent. The annual 1990–2100 GDP growth rates over the twentieth century across all stabilization scenarios were reduced on average by only 0.003 percent per year, with a maximum reaching 0.06 percent per year (IPCC 2001d).

THE KYOTO PROTOCOL— THE POLITICAL SITUATION

All key industrialized countries except the United States and Australia have ratified the Kyoto Protocol, which contains a number of core elements: these are the flexibility mechanisms (i.e., carbon trading[2]); land use, land-use change and forestry (LU-LUCF) activities; funding for developing countries, and the compliance regime. The United States and Australia have publicly stated that they will not ratify. Russian ratification allowed the Kyoto Protocol to enter into force on February 16, 2005. The carbon trading mechanism now in place under the protocol provides a powerful market mechanism for efficient emissions reduction in Europe.

The United States has stated that the Kyoto Protocol is flawed policy because:

- *There are still significant scientific uncertainties*: While it is possible that the influence of human activities on the Earth's climate system has been overestimated, it is equally possible that the impact of human activities on the Earth's climate system has been underestimated. In any case, scientific uncertainty is no excuse for inaction (the precautionary principle).

- *High compliance costs would hurt the U.S. economy*: This is in contrast to the analysis of the IPCC. As shown earlier, the IPCC estimated that the costs of U.S. compliance would be between $14 and $135 per ton of carbon avoided with international emissions trading—to put this in perspective—a 5 cents per gallon gasoline tax would be equivalent to U.S. $20 per ton of carbon. The IPCC also noted that these costs could be further reduced by using carbon sinks and by carbon trading with developing countries (i.e., the CDM).

- *It is not fair*: The United States argues that the Kyoto Protocol is not fair because large developing countries such as India and China are not obligated to reduce their emissions. Fairness is an equity issue—the parties to the Kyoto Protocol agreed that industrialized countries had an obligation to take the first steps to reduce their greenhouse gas emissions, recognizing that about 80 percent of the total anthropogenic emissions of greenhouse gases have been emitted from industrialized countries (the United States currently produces about 25 percent of the global emissions—36 percent of Annex I emissions); per capita emissions in industrialized countries far exceed those from developing countries; developing countries do not have the financial, technological, and institutional capability of industrialized countries to address the issue; and increased use of energy is essential for alleviating poverty and for long-term economic growth.

- *It is not effective*: The United States argues that the Kyoto Protocol will not be effective because developing countries are not obligated to reduce their emissions. It is true that long-term stabilization of the atmospheric concentration of greenhouse gases cannot be achieved without global reductions, especially recognizing that most of the projected growth in greenhouse gas emissions over the next 100 years are from developing countries. Hence the issue is who should do what in the short-term, recognizing the long-term challenge.

On the positive side, a number of multinational companies have committed to significant cuts in greenhouse gas emissions. More than 40 multinational companies have voluntarily agreed to reduce their emissions of greenhouse gases and improve the energy efficiency of their products. Several of these companies have already met or exceeded their targets and have saved money in doing so.

THE IMPLICATIONS OF CLIMATE MITIGATION PROJECTS ON BIODIVERSITY

As noted earlier, reductions in greenhouse gas emissions can be accomplished by using renewable energy technologies (crop waste, solar energy, and wind power), which may have positive or negative effects on biodiversity depending upon site selection and management practices (IPCC 2002; CBD 2003). Substitution of fuelwood by crop waste, the use of more efficient wood stoves and solar energy, and improved techniques to produce charcoal can also take pressure from forests, woodlots, and hedgerows. Most studies have demonstrated low rates of bird collision with windmills, but the mortality may be significant for rare species. Proper site selection and a case-by-case evaluation of the implications of windmills on wildlife and ecosystem goods and services can avoid or minimize negative impacts.

Hydropower, which has been promoted as a technology with significant potential to mitigate climate change by reducing the greenhouse gas intensity of energy production, has potential adverse effects on biodiversity. The ecosystem impacts of specific hydropower projects vary widely and may be minimized depending on factors including type and condition of pre-dam ecosystems; type and operation of the dam (e.g., water flow management); and the depth, area, and length of the reservoir. Run of the river hydropower and small dams have generally less impact on biodiversity than do large dams, but the cumulative effects of many small units should be taken into account (IPCC 2002; CBD 2003).

Bio-energy plantations provide the potential to substitute fossil fuel energy with biomass fuels but may have adverse impacts on biodiversity if they replace ecosystems with higher biodiversity. However, bio-energy plantations on degraded lands or abandoned agricultural sites could benefit biodiversity (IPCC 2002; CBD 2003).

Land use, land-use change, and forestry activities can play an important role in reducing net greenhouse gas emissions to the atmosphere. Afforestation and reforestation can have positive, neutral, or negative impacts on biodiversity depending on the ecosystem being replaced, management options applied, and spatial and temporal scales. The reforestation of degraded lands will often produce the greatest benefits to biodiversity but can also provide the greatest challenges to forest management. Afforestation and reforestation activities that pay attention to species selection and site location can promote the return, survival, and expansion of native plant and animal populations. In contrast, clearing native forests and replacing them with a monoculture forest of exotics would clearly have a negative effect on biodiversity. Plantations of native tree species will usually support more biodiversity than will exotic species, and plantations of mixed tree species will usually support more biodiversity than will monocultures; but plantations of exotic species can contribute to biodiversity conservation when appropriately situated in the landscape (IPCC 2002; CBD 2003).

Slowing deforestation and forest degradation can provide substantial biodiversity benefits in addition to mitigating greenhouse gas emissions and preserving ecological services. Since the remaining primary tropical forests are estimated to contain 50–70 percent of all terrestrial plant and animal species, they are of great importance in the conservation of biodiversity. Tropical deforestation and degradation of all types of forests remain major causes of global biodiversity loss. Any project that slows deforestation or forest degradation will help to conserve biodiversity. Projects in threatened/vulnerable forests that are unusually species-rich, globally rare, or unique to that region can provide the greatest immediate biodiversity benefits. Projects that protect forests from land conversion or degradation in key watersheds have potential to substantially slow soil erosion, protect water resources, and conserve biodiversity (IPCC 2002; CBD 2003).

Agroforestry systems have substantial potential to sequester carbon and can reduce soil erosion, moderate climate extremes on crops, improve water quality, and provide goods and services to local people. In addition, there are a large number of agricultural management activities (e.g., conservation tillage, erosion control practices, and irrigation) that will sequester carbon in soils and which may have positive or negative effects on biodiversity, depending on the practice and the context in which they are applied. Improved management of grasslands (e.g., grazing management, protected grasslands and areas set-aside, grassland productivity improvements, and fire management) can enhance carbon storage in soils and vegetation while conserving biodiversity. Revegetation activities that increase plant cover on eroded, severely degraded, or otherwise disturbed lands have a high potential to increase carbon sequestration and enhance biodiversity (IPCC 2002; CBD 2003).

ADAPTATION STRATEGIES FOR ECOLOGICAL SYSTEMS

Conservation of ecosystem structure and function is an important climate change adaptation strategy because species- and genetically rich ecosystems have a greater potential to adapt to climate change. Adaptation activities can have negative or positive impacts on biodiversity, but positive effects may generally be achieved through: maintaining and restoring native ecosystems; protecting and enhancing ecosystem services; actively preventing and controlling invasive alien species; managing habitats for rare, threatened, and endangered species; developing agroforestry systems at transition zones; paying attention to traditional knowledge; and monitoring results and changing management regimes accordingly. Adaptation activities that can be beneficial to biodiversity include the establishment of a mosaic of interconnected terrestrial, freshwater, and marine multiple-use reserve protected areas designed to take into account projected changes in climate, and integrated land and water management activities that reduce nonclimate pressure on biodiversity and hence make the system less vulnerable to changes in climate. Adaptation activities can also threaten biodiversity either directly—through the destruction of habitats (e.g., building seawalls, thus affecting coastal ecosystems)—or indirectly—through the introduction of new species or changed management practices (e.g., mariculture or aquaculture) (IPCC 2002; CBD 2003).

SUMMARY

Biodiversity and the goods and services provided by ecological systems are essential for life on Earth and are the very foundation of sustainable development.

Given that biodiversity is sensitive to even small changes in the Earth's climate, applying the precautionary principle would argue that stabilization of the atmospheric concentrations of greenhouse gases at close to current levels is appropriate. Scientific uncertainties are no excuse for inaction, especially recognizing that the loss of biodiversity is irreversible, and that there are cost-effective and equitable ways to reduce greenhouse gas emissions in the energy supply, energy demand, and agricultural and forestry sectors, many at little or no cost to society.

Progress requires political will and moral leadership. The actions of today's generation will profoundly affect the Earth inherited by our children and future generations. Given that all major industrialized countries, except for the United States, have ratified the Kyoto Protocol, and given that the United States emits about 25 percent of global greenhouse gas emissions, U.S. policymakers should be urged to recognize that there is no dichotomy between economic growth and environmental protection, that addressing climate change provides economic opportunities to restructure and make a more efficient energy system, and that as the world's only super-power the United States has a moral obligation to demonstrate political leadership. We must also recognize that the Kyoto Protocol is only the first step on a very long journey to ultimately reduce greenhouse gas emission significantly below current emissions.

Unless we act now to limit human-induced climate change, history will judge us as having been complacent in the face of compelling scientific evidence that humans are changing the Earth's climate with predominantly adverse effects on human health, ecological systems, and socio-economic sectors. Do we really want our heritage to be that of sacrificing the Earth's biodiversity for cheap fossil fuel energy, ignoring the needs of future generations, and failing to the meet the challenge of providing energy in an environmentally and socially sustainable manner when so many choices were available? Leaders from government and industry must stand shoulder to shoulder to ensure that the future of the Earth is not needlessly sacrificed.

NOTES

1. The Clean Development Mechanism allows an investor in an industrialized country (industry or government) to invest in an eligible carbon mitigation project in a developing country and be credited with Certified Emission Reduction Units that can be used by investors to meet their obligation to reduce greenhouse gas emissions under the Kyoto Protocol.

2. The Kyoto Protocol allows: (i) industrialized countries to trade their allocation of carbon emissions among themselves (Article 17); (ii) industrialized governments or companies from industrialized countries to implement carbon mitigation projects jointly (Article 6) and share the Emissions Reductions Units that can be used to meet obligations to reduce greenhouse gas emissions under the Kyoto Protocol; and (iii) allows an investor in an industrialized country (industry or government) to invest in an eligible carbon mitigation project in a developing country (Article 12—the CDM—see footnote 1).

REFERENCES

CBD (Roster of Experts for the Convention on Biodiversity). 2003. *A Report on Climate Change and Biodiversity.* Quebec, Canada: United Nations Environment Program.

IPCC (Intergovernmental Panel on Climate Change). 2000. *A Special Report on Emissions Scenarios,* N. Nekicenovic and R. Swart, eds. New York: Cambridge University Press.

IPCC (Intergovernmental Panel on Climate Change). 2001a. *A Special Report on Land Use, Land-Use Change and Forestry,* R. T. Watson, I. R. Noble, B. Bolin, N. H. Ravindranath, D. J. Verado, and D. J. Dokken, eds. Cambridge, U.K.: Cambridge University Press.

IPCC (Intergovernmental Panel on Climate Change). 2001b. *Climate Change 2001, The Scientific Basis,* J. T. Houghton, Y. Ding, D. J. Griggs, M. Noguer, P. J. van der Linden, X. Dai, K. Maskell, and C. A. Johnson, eds. Contribution of Working Group I to the Third Assessment Report of the IPCC. Cambridge, U.K.: Cambridge University Press.

IPCC (Intergovernmental Panel on Climate Change). 2001c. *Climate Change 2001, Impacts, Adaptation, and Vulnerability,* J. J. McCarthy, O. F. Canziani, N. A. Leary, D. J. Dokken, and K. S. White, eds. Contribution of Working Group II to the Third Assessment Report of the IPCC. Cambridge, U.K.: Cambridge University Press.

IPCC (Intergovernmental Panel on Climate Change). 2001d. *Climate Change 2001, Mitigation*, B. Metz, O. Davidson, R. Swart, and J. Pan, eds. Contribution of Working Group III to the Third Assessment Report of the IPCC. Cambridge, U.K.: Cambridge University Press.

IPCC (Intergovernmental Panel on Climate Change). 2002. *A Technical Paper on Climate Change and Biodiversity*, H. Gitay, A. Suarez, R. T. Watson, and D. J. Dokken, eds. IPCC Technical Paper V. Cambridge, U.K.: Cambridge University Press.

CHAPTER TWENTY-FOUR

Global Greenhouse Gas Levels and the Future of Biodiversity

THOMAS E. LOVEJOY AND LEE HANNAH

Over a decade has passed since the publication of *Global Warming and Biological Diversity* (Peters and Lovejoy 1992). In that time, the term "climate change" has replaced "global warming" in recognition that human greenhouse gases are causing much more than just temperature shifts, while "biological diversity" today is known by its more familiar contraction, "biodiversity." Profound changes in science and policy have unfolded in this time as well. Bounds have been placed on likely global mean temperature change, and those bounds have risen steadily higher (IPCC 2001). Already there is clear evidence of species responding in nature to the climate change that has taken place to date. Public awareness and concern have grown across the globe, driven by new information.

The possibility of rapid, large climate change superimposed on long-term trends burst on the scene during the 1990s. Examination of ice core records from Greenland and Antarctica showed surprising variability in the climate of the past 200,000 years (Overpeck et al., this volume). Abrupt changes punctuated the record so often that rapid change appeared more the norm than the past 11,000 years of warm, stable climate. But even within the relative stability of the Holocene, significant variability is clear.

With the record of rapid change has come the realization that the climate system is both fragile and highly interconnected. A breakdown in the thermohaline circulation can result in huge and extremely rapid climate change in the North Atlantic, causing major temperature changes that are global in extent. Changes in El Niño are similarly important to patterns of temperature and rainfall in many parts of the tropics. These behaviors are consistent with a climate system that has multiple semi-stable states and is capable of shifting rapidly between them.

The vulnerability of the climate system to human interference therefore has come to be of much greater concern. In addition to the impacts of steady directional change comes the possibility that the buildup of greenhouse gases may unexpectedly push some part of the climate system over a major threshold. The ice core record gives the knowledge that thresholds exist, but understanding of their mechanisms is a much more complex, and therefore still incomplete, process. Pinpointing levels of human interference permissible to ensure that thresholds are not crossed remains difficult.

All these developments hold important implications for biodiversity. More rapid temperature rise means that conditions will more quickly exceed those of past interglacials that may have been warmer than the present. Rapid change challenges current concepts of long-distance dispersal capabilities. Thresholds and multiple stable states imply changes that conservation systems are not yet designed to withstand.

BIOLOGICAL LESSONS

Climate change biology has advanced in tandem with these advances in climate science. Among the insights that have come since the publication of *Global Warming and Biological Diversity* are new views of evolutionary potential, dispersal modes, and responses to change.

Rapid evolution of some traits, such as photoperiod, is now understood to be possible. Mismatches in photoperiod as organisms move poleward at unprecedented rates now seem unlikely to be problematic. Several experimental systems have shown rapid evolutionary adjustments in photoperiod in both animals and plants. Other traits may be subject to rapid genetic selection. Long-winged morphs of insects may be the result of long-term selection for recessive traits that facilitate response to climate change. Such traits may lie dormant for long periods, masked by dominant alleles, until climate change exerts strong selective pressure for rapid dispersal.

Range shifts are solidly supported as the dominant response to climate change, despite examples of genetic adaptation. Most studies of past and present changes reveal range shifts, rather than evolution or extinction. The broad expectation of poleward and upslope range shifts with warming has been confirmed in many paleoecological studies. Research reveals these same patterns dominating present species' responses.

Individualism in species responses to climate change is now strongly established. Biome models and movement of montane "zones" still retain some broad relevance, but in most instances species are expected to move in their own characteristic ways in response to climate change. This implies that vegetation communities may be torn apart and reassembled in novel ways. Predator–prey and species competitive interactions may vary greatly as a result. Conservation systems will have to be adaptable to adjust to these changes.

Yet biodiversity has survived these past rapid changes largely intact. There is no reliable record of mass extinctions in the Pleistocene, so plants and animals have been able to survive huge regional changes by modern standards, but the mechanisms for this are still under debate.

Northern and southern perspectives on rapid climate change are divergent. Substantial evidence has emerged from the Northern Hemisphere supporting major southerly (equatorward) refugia during glacial periods, implying long-distance dispersal as a major feature of rapid postglacial recolonization of the landscape. In the Southern Hemisphere, a different picture has emerged. In southern areas, massive ice sheets were not the rule, and the dominant view is that small remnant patches of vegetation ("micropockets") were the main sources of dispersal during climate reversals.

These northern and southern perspectives may now be converging. Evidence is mounting that relictual populations of plants and animals persisted surprisingly close to continental ice sheets in both North America and Europe. One reason may be that the height of the massive ice sheets actually blocked southward movement of cold polar air, creating some relatively equable microclimate near the ice. The genetic fingerprint of dispersal from such micropockets would be similar to that from long-distance colonization by a few pioneers from southern refugia (Hewitt and Nichols, this volume). Thus, it may be possible that micropockets have played an important role in rapid response to climate change in the biotas of both hemispheres (McGlone and Clark, this volume).

IMPLICATIONS FOR CONSERVATION

Rapid responses, dispersal modes, and community reorganization all carry major implications for conservation. Systems to conserve biodiversity in the face of such dynamics must engage entire landscapes. Chapters in this book have described how to design static and dynamic elements in a landscape, and suggested priorities for their management. These systems will be able to better respond to climate change than past, "parks only" strategies. However, there are clear limits to the change they can accommodate.

Rapid responses provide a precedent on which to base strategies for conservation during rapid future climate change. However, the potential for rapid response is greatly constrained by habitat loss. Both rapid genetic responses and range shifts are jeopardized by existing levels of habitat loss.

Rapid genetic response is compromised where high levels of habitat fragmentation result in loss of genetic diversity. In human-dominated landscapes, natural habitat may remain as small fragments, implying small populations of resident species which lose genetic diversity. Since the recessive traits necessary for rapid response to climate change are frequently less competitive in current climates, they may be lost in small, fragmented populations. This will reduce the pool of individuals capable of rapid response to climate change or eliminate the genetic variants for rapid response altogether.

Rapid range shifts, by whatever mechanism, will be limited by transformed landscapes. Long-distance dispersal is dependent on population size, destination area, and intermediate suitable habitats, all of which are greatly reduced in human-dominated landscapes. Similarly, micropockets may be eliminated by human land uses, and where they do survive they cannot serve as centers of expansion if surrounded by transformed land.

Limitations on response are compounded when communities reorganize in disturbed landscapes. As species' individualistic responses to climate change unfold, community composition will change. Constituents may leave and arrive randomly, sometimes leaving resources available for new species. Where this process unfolds surrounded by human landscapes harboring many weedy species, composition may shift toward human comensals. There will be no barrier to species leaving, since this only involves the death of individuals at a site. However, species arriving will have to traverse fragments of natural habitat, whereas weeds will be able to enter from abundant surrounding human-converted lands. The net effect may be one of simplification and dominance by weedy species.

For all these reasons, natural response to climate change will be constrained. Emerging tools in conservation can nonetheless allow this capacity to be tapped, but there are strong limits on what can be accomplished. Range shift modeling can help identify where suitable new climate space may appear, but it is difficult to estimate species' ability to reach newly suitable sites. Reserve selection algorithms can identify efficient configurations for sites where species' present and future ranges overlap, but species that must move substantial distances are excluded. Connectivity can be tailored for individual target species, but it will break down when too many species move long distances or where changes increase without limits.

Therefore, improved strategies will be meaningless unless change is kept within limits. Critical limits will be specific to each species, each site, and each region. Some species at some sites, such as the golden toad at Monteverde, may already have seen change in excess of their critical limits. Other limits will be reached soon, affecting mounting numbers of species the further into the future change is allowed to continue. The question then becomes, "How much climate change is too much?" The remainder of this chapter addresses this question.

GLOBAL GREENHOUSE GAS LEVELS AND THE FUTURE OF BIODIVERSITY

The foregoing conservation strategies notwithstanding, the future of biodiversity is highly dependent on dramatic reduction of greenhouse gases. Unless anthropogenic climate change can be first slowed and then stopped, it will result in large temperature increases, changes in precipitation, and alterations in teleconnections, extreme events, and other phenomena that are highly likely to affect biodiversity in ways that will eventually become impossible to manage. Avoiding unmanageable long-term outcomes requires stabilizing atmospheric greenhouse gas levels.

Stabilizing greenhouse gas levels in turn requires transition to an economy based on carbon-neutral sources of energy. Since this entails replacing essentially all current fossil fuel–based transportation and electricity production, biologists advocating stabilization need to understand the costs and environmental consequences of possible alternatives.

The means to achieve stabilization fall into three main categories: increased energy efficiency, alternate sources of energy, and carbon sequestration (see Chapter 23). Energy efficiency can make a significant contribution, especially if incentives are created to increase the rate of improvement in efficiency above that driven by current economic circumstances (that include widespread subsidies favoring inexpensive fossil fuels). This is not by itself, however, sufficient to provide the necessary reduction in greenhouse gases.

Alternate energies also can make significant contributions but often have their own particular environmental impacts. Nonetheless the contributions that can be made by solar power and wind can contribute in important ways. Hydroelectric power is widely perceived as clean from a greenhouse gas perspective, but can have significant impacts on freshwater biodiversity and in many instances produce greenhouse gas (methane release from rotting flooded vegetation). Biofuels have a special appeal in the sense that the carbon released by their combustion can be offset by the concurrent growth of the replacement biofuels. Nonetheless, biofuel production on a sufficient scale to be meaningful in the global energy budget is likely to require enormous land area, up to twice the current area used for agriculture (Hoffert et al. 2002), which would have a major impact on remaining natural habitat and its biodiversity. Nuclear power is occasionally mentioned as a possible alternate energy but remains expensive, with problems of waste disposal and worrisome nuclear proliferation potential in an unsettled world. Fusion would avoid many of these concerns, but fusion as a practical energy source remains technically elusive.

The third category is carbon sequestration. This includes geophysical sequestration in mineral reactions or disposal in geologic formations and biological sequestration, such as reducing deforestation, reforestation, and schemes to stimulate marine organisms to take up more carbon. This latter has significant environmental impact (Seibel and Fabry 2003) and consequently would seem an undesirable approach. Sequestration can also include engineering solutions, with the carbon injected in subsurface locations (suitably, in some instances, into locations from which oil and gas have been extracted). Such locations may ultimately be limited, so engineering solutions with inert products like magnesium carbonate have very significant potential.

It is clear in any case, whatever the final set of approaches, that none by itself is sufficient to the scale and the urgency of the task. Rather a mix of efficiency, alternates, and sequestration will be necessary to achieve the goal. Sequestration is likely to make the difference in the longer run, because it alone has the potential to address the scale of long-term energy needs.

Paths to a Carbon-Neutral Future

Solutions for a carbon-neutral future must provide terawatts of power and be safe, environmentally acceptable, and stable. These are basic principles on which all may agree, and that biologists may offer as a first point in policy dialog. From these principles, paths to future greenhouse gas stabilization begin to emerge.

Environmentally acceptable solutions are particularly important, because it makes little sense to trade one environmental problem for another. Ocean disposal is one of the earliest proposed options for carbon sequestration, but carries the strong possibility of negative effects on marine biota, as has been discussed in Chapter 18. Ocean fertilization carries negative environmental consequences, and also results in release of greenhouse gases as a by-product—it therefore seems doomed as a viable alternative. Nuclear and hydropower energy sources also probably carry too many environmental problems to be large-scale contributors to a solution. Many renewable energy technologies that are environmentally benign at small scales have major environmental consequences when applied at the scale necessary to displace current fossil fuel consumption. For instance, both solar and wind energy would require huge land areas, which would certainly have an impact on the remaining natural areas of the Earth.

Stability is important because systems that do not offer long-term solutions may trade one problem for another or merely delay impacts. Ocean disposal of CO_2 fails on stability criteria as well as environmental acceptability, since it allows leakage of CO_2 back into the atmosphere (Fig. 24.1). As CO_2 production and sequestration mounted, this would lead to an "echo boom" of leaking CO_2, creating a climate change problem for future generations. Deep injection wells for sequestration also raise leakage questions, although some geologic formations would certainly have low rates of leakage.

Safety is a major concern with power substitution using nuclear energy. Safety may also be a concern with certain injection strategies, since large amounts of CO_2 can be deadly if it escapes near the surface (CO_2 release from African lakes after sudden overturning has been responsible for scores of human deaths).

The largest constraint, and the one likely to require a multi-faceted strategy, is that of scale. Current global annual fossil fuel consumption is about 6 gigatons of carbon (see Fig. 24.1) and will have to

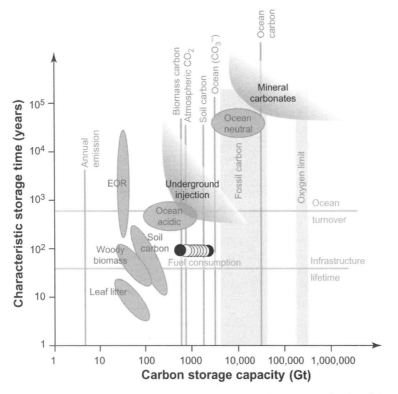

Figure 24.1. Carbon storage capacity versus carbon storage time for different sequestration options. Carbon sequestration requires storage sites that can hold large amounts of carbon (gigatons) for a long period of time. Total leakage rate from all storage sites must be small (fraction of gigaton) for lasting stabilization of atmospheric greenhouse gases. Storage of thousands of gigatons of carbon therefore requires a minimum storage lifetime of thousands of years. The figure gives rough ranges for a number of potential storage schemes (shaded elipses), including enhanced oil recovery (EOR), sequestration in vegetation, ocean disposal, underground injection, and mineral sequestration. For comparison, emissions levels and natural limits on the consumption of fossil fuel are illustrated by the vertical lines. The approximate present annual emissions level of 6 GtC/yr is shown on the left. Future fossil fuel consumption (black circles) is expressed as a possible range of carbon consumption over the course of the 21st century (from 600 Gt, present consumption held constant, to 2400 Gt). Carbon pools in vegetation, soils, and the ocean are shown in solid vertical lines in the center and right of the panel. Gray shaded bars show the limits on fossil fuel consumption. The total amount of available fossil fuels ("fossil carbon") range is bounded on the lower end by typical estimates of coal, tar, shale oil, and gas, and includes at its upper end methane hydrates from the ocean floor. The "oxygen limit" equals the amount of fossil carbon that would use up all oxygen available in the air for its combustion. This limit is represented by a range, because the oxygen consumption per unit of carbon spans a range of roughly 3 mass units per mass unit of carbon in the case of coal and 5 mass units per unit of carbon in the case of natural gas. Note that the high end of available fossil carbon comes close enough to the "oxygen limit" that oxygen may already be the rate limiting resource. On the time axis, we note the turnover time of the ocean and the typical lifetime of energy infrastructure. The latter sets a lower limit on what would be a relevant carbon sequestration technology. *Source:* Lackner (2003a). Copyright American Association for the Advancement of Science. Reprinted with permission.

grow by a factor of 2 or 3 to meet developing country energy growth needs in this century, even assuming large increases in efficiency, changes in lifestyle, and substitution of renewable energy sources such as sun and wind. Without changes in efficiency and increased use of renewables, present energy consumption would need to increase by about a factor of 10 to provide all residents of the planet with the current standard of living enjoyed in the United States and Europe (Lackner 2003b).

Replacing or sequestering this amount of fossil fuel carbon is a huge task. Tackling it with renewables alone would likely mean a transition period of many decades, since it would entail the complete restructuring of the energy sector, including supply, transportation, and end-use infrastructure. A long transition from fossil fuel would mean further buildup of greenhouse gases and unacceptable damage to biodiversity due to climate change. A more rapid response is needed if changes in biodiversity are to be kept within manageable limits.

Two major opportunities exist for rapid change in greenhouse gas levels. The first is implementation of existing efficiency and biological sequestration options on a large scale. The automotive fleet can be converted to hybrid power, for example, with an efficiency savings of 50 percent or more. Reforestation in natural ecosystems and plantations on degraded land (not on natural habitat or land suitable for restoration to natural habitat) is a proven technology that can absorb significant amounts of carbon.

In the long term, however, efficiency and biological sequestration will probably be able to do little better than displace growth in energy consumption. It is physically impossible for efficiency improvements to provide carbon-neutral energy supply, because they are by definition based in fossil fuels (Hoffert et al. 2002). The amount of biomass on the planet is relatively small in relation to future sequestration needs (see Fig. 24.1), and environmentally acceptable biomass options are limited. So while efficiency and biological sequestration are proven and can make a short-term difference, other medium-term and long-term strategies are needed.

The other short-term rapid CO_2 reduction option is geophysical sequestration.

There are proven technologies for sequestration through well injection and prototype mineral sequestration methods that have the potential to lock up large quantities of CO_2 (Lackner 2003a). Injection involves pumping of CO_2 at pressure into abandoned oil wells or saline (nonpotable) aquifers. CO_2 injection is used commercially in oil recovery, and injection in saline aquifers has been proven. Mineral sequestration involves chemical reaction of CO_2 with rocks such as serpentine or peridotite that are abundant (see Fig. 24.1). The processing required for this reaction is energy-consuming and refinements are needed to make it practical, but it offers the advantage of safe, compact storage (Lackner 2003a).

However, CO_2 capture to feed these disposal options is less well tested. Capture would require a new generation of electric power plants for which some prototypes have been tested, and free-air removal which may be possible but hasn't been demonstrated to be practical. The advantage of capture and sequestration is that it allows most of the current energy infrastructure to be maintained, greatly reducing costs, changes in energy production, and the time needed for transition (Lackner 2003a). It is therefore a sound interim step, allowing rapid reduction of atmospheric CO_2.

Sequestration offers the additional benefit of being able to drive down atmospheric CO_2 to pre-industrial levels (Lackner 2003a). Whether this is advisable for biodiversity is a debate for the future. For highly sensitive species, going through a double range shift (once as CO_2 increases, and again in reverse as it decreases) may be less desirable than a single transition, since habitat loss will continue to reduce chances for successful range shifts. Other species, such as some long-lived trees (e.g., Sequoia) or corals, may be able to survive a double transient but would be doomed to extinction if greenhouse gas levels do not return to pre-industrial levels. Since a double transient may also double the risk of destabilizing the climate system, whether to return to pre-industrial CO_2 levels or not is a complex social question. That sequestration provides the option is nonetheless an important attribute.

Sequestration is not a final solution, however, as fossil fuel supplies are finite and the environmental costs of extraction of coal and tar sands are high (Hoffert et al. 2002). Eventually transition to a renewable energy economy is necessary, and earlier transition pays high environmental benefits in avoided damage from coal and tar sand mining.

One path to a carbon-neutral future may then be described that relies on three main strategies or steps:

1. Immediate implementation of efficiency improvements and terrestrial biological sequestration

2. Medium-term transition to renewables while relying on capture and sequestration of fossil fuel CO_2 (deep well injection initially, possibly followed by terrestrial mineral sequestration)

3. Long-term energy supply from fully renewable sources such as solar and wind power

This is a practical course of action that balances environmental costs of fossil fuel production with the damage to biodiversity caused by climate change. It initially relies on biological sequestration options that are not damaging to biodiversity, namely terrestrial biological sequestration on degraded lands, and then on geophysical sequestration with relatively small environmental costs. To be fully effective this strategy for CO_2 reduction must be coupled with programs for reduction of other greenhouse gases. The general outline of the path is flexible, with levels of investment and timetables dependent on the target level for greenhouse gas stabilization. It is therefore a strategy that biologists may advocate as a practical alternative to destruction of biodiversity due to climate change. Its implementation, however, requires agreement on a target or "acceptable" level of atmospheric greenhouse gases.

DEFINING AN "ACCEPTABLE" LEVEL OF GREENHOUSE GAS CONCENTRATIONS

Marine systems may be among the first hit and most vulnerable. Coral reefs seem to be particularly sensitive to climate change, as evidenced by major bleaching events already experienced (Hoegh-Guldberg 1999). Ocean acidification, the direct effect of CO_2 absorbed from the atmosphere by the world's oceans, threatens coral reefs and many other marine organisms (Chapter 18). Rising acidity reduces the available carbonate in seawater, reducing the ability of corals and other organisms to form calcium carbonate shells and skeletons.

Calcium carbonate saturation is temperature dependent, so corals will be less likely to be able to secrete and maintain their external skeletons in more temperate waters. Corals may then be squeezed between bleaching and the deep blue sea—unable to escape bleaching in the warming tropics because of inadequate calcium carbonate saturation poleward (Feely, Sabine et al. 2004). In the long term, this may undermine coral reef structures critical to protecting coastlines from storm surge and tsunami damage, as well as the biological impact of wholesale replacement of corals with algae and other organisms throughout the tropics.

Acidification has serious effects in isolation as well. Many organisms are seriously impacted by changes in pH and resultant reductions in calcium carbonate saturation (Orr et al 2005). From squid that suffer physiological breakdown due to changing pH to phytoplankton that are unable to secrete shells under high CO_2 conditions, acidification threatens all levels of marine food webs, including, perhaps most significantly, their planktonic foundations (Chapter 18). These impacts have serious implications for human welfare and economic activity through effects on fisheries production, tourism, and ecosystem services.

Terrestrial systems are at risk as well. The Succulent Karoo and Cape Floristic Region biodiversity hotspots of South Africa may be among the most sensitive due to lack of poleward land (Hannah, Midgley et al. 2002; Malcolm, Liu et al. 2005). The Tropical Andes may be the most threatened tropical montane hotspot (Malcolm, Liu et al. 2005). For individual species, climate change interactions with chytrid fungal disease may have been the most damaging, accounting for dozens of amphibian extinctions and threatening hundreds more (Pounds 2006). Other interactions with pests such as bark beetles are destroying millions of acres of forests in North America and accelerating tree range shifts (Breshears, Cobb et al. 2005).

This suggests that biological systems are the most sensitive of the systems cited in the UN Framework Convention on Climate Change in its commitments to avoid "dangerous interference" in the climate system. It is likely that as regional climate modeling improves and understanding of the impact of climate change deepens, there will be further examples of critical and negative effects at double pre-industrial CO_2 levels. Prudence might suggest, therefore, that society act as if 450 ppm is probably the acceptable limit and take steps as if that were the case. It may well be that 450 ppm is too high a limit to avoid major disruption to biological systems, but it is probably the lowest limit that is technically and socially feasible. It is consistent with the 2°C target favored in European policy.

In addition, sheer common sense would suggest the vital importance of achieving agreement about "dangerous interference" and hence an acceptable target level. Without such a target, it is difficult to know how hard or urgently to try to reduce emissions. Since the UNFCCC sets "allowing ecosystems to adapt naturally" as one benchmark for avoiding dangerous interference, biology is already on the stage for this debate. Unless this debate is joined, an acceptable target agreed, and action rapidly taken, the sixth great extinction event on Earth will be ensured by increasingly fragmented habitat combined with the biological dynamics resulting from climate change.

REFERENCES

Breshears, D. D., N. S. Cobb et al. 2005. Regional vegetation die-off in response to global-change-type drought. *Proceedings of the National Academy of Sciences of the United States of America* 102(42):15144–48.

Feely, R. A., C. L. Sabine et al. 2004. Impact of anthropogenic CO_2 on the $CaCO_3$ system in the oceans. *Science* 305(5682):362–66.

Hannah, L., G. F. Midgley et al. (2002). Conservation of biodiversity in a changing climate. *Conservation Biology* 16(1):11–15.

Hoegh-Guldberg, O. (1999). "Climate change, coral bleaching and the future of the world's coral reefs." *Marine and Freshwater Research* 50: 839–66.

Hoffert, M. I., K. Caldeira, G. Benford, D. R. Criswell, C. Green, H. Herzog, A. K. Jain, H. S. Kheshgi, K. S. Lackner, J. S. Lewis, H. D. Lightfoot, W. Manheimer, J. C. Mankins, M. E. Mauel, L. J. Perkins, M. E. Schlesinger, T. Volk, and T. M. L. Wigley. 2002. Advanced technology paths to global climate stability: Energy for a greenhouse planet. *Science* 298(5595):981–87.

IPCC. 2001. *Climate Change 2001: The Scientific Basis.* Contribution of Working Group I to the Third Assessment Report of the Intergovernmental Panel on Climate Change. Port Chester, N.H.: Cambridge University Press.

Lackner, K. S. 2003a. A guide to CO_2 sequestration. *Science* 300(13 June):1677–78.

Lackner, K. S. 2003b. Can carbon fuel the 21st century? *International Geology Review* 44:1122–33.

Malcolm, J. R., C. Liu et al. 2005. Global warming and extinctions of endemic species from biodiversity hotspots. *Conservation Biology.* In press.

O'Neill, B. C., and M. Oppenheimer. 2002. Climate

change: Dangerous climate impacts and the Kyoto Protocol. *Science* 296:1971–72.

Orr et. al. (2005) Anthropogenic ocean acidification over the twenty-first century and its impact on calcifying organisms. *Nature* 437:681–86.

Peters, R. L., and T. E. Lovejoy. 1992. *Global Warming and Biological Diversity*. London: Yale University Press.

The Royal Society. 2005. Ocean acidification due to increasing atmospheric carbon dioxide. Policy document of the Royal Society, pg 60.

Sabine et. al. (2004) The oceanic sink for anthropogenic CO_2. *Science* 305: 367–71.

Seibel, B., and V. Fabry. 2003. Marine response to elevated carbon dioxide. In *Climate Change and Biodiversity: Synergistic Impacts*, L. Hannah and T. Lovejoy, eds. Washington, D.C.: Center for Applied Biodiversity Science.

Contributors

J. David Allan
Professor of Conservation Biology and
Ecosystem Management
School of Natural Resources and
Environment
University of Michigan

Carter Atkinson
Microbiologist
Pacific Island Ecosystems Research Center
United States Geological Survey

Patrick J. Bartlein
Professor
Department of Geography
University of Oregon

Tracy L. Benning
Department of Environmental Science
University of San Francisco

Richard A. Betts
Manager, Ecosystems and Climate Imapcts
Hadley Centre for Climate Prediction and
Research
Meteorological Office (UK)

Paulo A. Buckup
Professor Adjunto
Museu Nacional
Universidade Federal do Rio de Janeiro
Brazil

Mark B. Bush
Professor
Department of Biological Sciences
Florida Institute of Technology

Jim Clark
H.L. Blomquist Professor of Biology and

Faculty Director of the Center on Global
Change
Nicholas School of the Environment and
Department of Biology
Duke University

Mark A. Cochrane
Senior Research Scientist
Center for Global Change and Earth
Observations
Michigan State University

Julia Cole
Associate Professor
Department of Geosciences
University of Arizona

Bert G. Drake
Plant Physiologist
Smithsonian Environmental Research
Center
Smithsonian Institution

Victoria J. Fabry
Department of Biological Sciences
California State University, San Marcos

Michael P. L. Fogden
Golden Toad Laboratory for Conservation
Monteverde Cloud Forest Preserve and
Tropical Science Center, Costa Rica

Gustavo A. B. da Fonseca
Executive Vice President for Programs and
Science
Conservation International
Professor of Zoology
Federal University of Minas Gerais, Brazil

Michael Garaci
Faculty of Forestry
University of Toronto, Canada

Filippo Giorgi
Senior Scientist
Physics of Weather and Climate Section
The Abdus Salam International Centre for
Theoretical Physics, Italy

Lee Hannah
Senior Fellow in Climate Change Biology
Center for Applied Biodiversity Science
Conservation International

Lara Hansen
Chief Scientist
Climate Change Program
World Wildlife Fund

Godfrey M. Hewitt
Professor
School of Biological Sciences
University of East Anglia, UK

Ove Hoegh-Guldberg
Professor and Director
Centre for Marine Science
University of Queensland, Australia

Henry Hooghiemstra
Professor of Palynology and Quaternary
Ecology
Institute for Biodiversity and Ecosystem
Dynamics (IBED)
Palynology and Paleo/Actua-ecology
(formerly "Hugo de Vries Laboratory")
Faculty of Science
University of Amsterdam, The
Netherlands

Lesley Hughes
Senior Lecturer
Department of Biological Sciences
Macquarie University
Australia

Mike Hulme
Professor, School of Environmental
Sciences, University of East Anglia
Executive Director, Tyndall Centre for
Climate Change Research, UK

Brian Huntley
Professor of Plant Ecology and
Palaeoecology
School of Biological and Biomedical
Sciences
University of Durham, UK

E. A. Johnson
G8 Legacy Chair in Wildlife Ecology and
Professor of Ecology
University of Calgary, Canada

Thomas R. Karl
Director
National Climatic Data Center
National Oceanic and Atmospheric
Administration

Dennis LaPointe
Ecologist
Pacific Island Ecosystems Research Center
United States Geological Survey

Thomas E. Lovejoy
President
The H. John Heinz III Center for Science,
Economics and the Environment

Jay R. Malcolm
Assistant Professor
Faculty of Forestry
University of Toronto, Canada

Vera Markgraf
Fellow of Institute of Artic and Alpine
Research (INSTAAR) and
Emeritus Research Professor of Geography
University of Colorado at Boulder

Adam Markham
Executive Director
Clean Air-Cool Planet

Enrique Martinez-Meyer
Departamento de Zoología
Instituto de Biología
Universidad Nacional Autónoma de
México

Karen L. Masters
Monteverde Conservation League and
Children's Eternal Rainforest
San Jose, Costa Rica

Matt McGlone
Senior Research Scientist
Landcare Research
New Zealand

Marcelo R. S. Melo
Departamento de Vertebrados
Museu Nacional
Universidade Federal do Rio de Janeiro,
Brazil

Guy F. Midgley
Climate Change Group
Ecology and Conservation
Kirstenbosch Research Center
National Botanical Institute of South
Africa

Dinah Millar
Climate Change Group
Kirstenbosch Research Center
National Botanical Institute of South
Africa

Ronald P. Neilson
Bioclimatologist
Pacific Northwest Research Station
USDA Forest Service

Richard A. Nichols
Professor of Evolutionary Genetics
School of Biological Sciences
Queen Mary, University of London, UK

Judy M. Oglethorpe
Director, Ecoregion Support Unit
World Wildlife Fund

Jonathan Overpeck
Director, Institute for the Study of Planet
Earth
Professor of Geosciences
University of Arizona

Margaret A. Palmer
Professor of Entomology and Biology
Department of Biology
University of Maryland

John M. Pandolfi
Curator
Department of Paleobiology
National Museum of Natural History
Smithsonian Institution

Camille Parmesan
Assistant Professor, Integrative Biology
University of Texas at Austin

Sam H. Pearsall, III
Director of Science and Roanoke River
Project Director
The Nature Conservancy, North Carolina
Chapter

David A. Perry
Professor
Department of Forest Science
Oregon State University

A. Townsend Peterson
Curator in Charge/Associate Professor
Natural History Museum and Biodiversity
Research Center
The University of Kansas

N. Leroy Poff
Associate Professor
Department of Biology and Graduate
Degree Program in Ecology
Colorado State University

J. Alan Pounds
Resident Biologist
Golden Toad Laboratory for Conservation
Monteverde Cloud Forest Preserve and
Tropical Science Center, Costa Rica

Sarah C. B. Raper
Senior Scientist
Alfred Wegener Institute
Foundation for Polar and Marine Research
Germany

Daniel Rasse
Laboratory for Continental Biogeochemistry
Institut National de la Recherche
Agronomique
France

Terry L. Root
Senior Fellow, Center for Environmental
Science and Policy
Stanford Institute for International Studies
Professor, by courtesy, Department of
Biological Sciences
Stanford University

Kaustuv Roy
Associate Professor
Section of Ecology, Behavior and Evolution
Division of Biological Sciences
University of California, San Diego

Rod Salm
Director, Transforming Coral Reef
Conservation
Marine Initiative
The Nature Conservancy

Víctor Sánchez-Cordero
Departamento de Zoología
Instituto de Biología
Universidad Nacional Autónoma de México

Stephen Schneider
Professor, Department of Biological
Sciences
Senior Fellow, Stanford Institute for
International Studies
Professor, by courtesy, Civil and
Environmental Engineering
Co-Director, Center for Environmental
Science and Policy
Co-Director, Interdisciplinary Program in
Environment and Resources
Stanford University

Daniel Scott
Canada Research Chair in Global Change
and Tourism
Faculty of Environmental Studies
University of Waterloo, Canada

Wes Sechrest
Scientific Officer, IUCN Species Survival
Commission
Department of Biology, University of
Virginia

Brad A. Seibel
Assistant Professor of Biological Sciences
Department of Biological Sciences
University of Rhode Island

Henry H. Shugart
W.W. Corcoran Professor
Department of Environmental Sciences
University of Virginia

Jorge Soberón
Executive Secretary
National Commission for the Knowledge
and Use of Biodiversity (CONABIO)
Mexico

Chris D. Thomas
Professor of Conservation Biology
Department of Biology
University of York, UK

Hanqin Tian
Associate Professor
Department of Ecology and Evolutionary
Biology
University of Kansas

Kevin E. Trenberth
Head, Climate Analysis Section
National Center for Atmospheric Research

Diana H. Wall
Director and Professor, Natural Resource
Ecology Laboratory
Colorado State University

Robert T. Watson
Chief Scientist & Senior Advisor,
Environmentally and Socially Sustainable
Development
The World Bank

Index